Mathematics Lecture Series

11

Invariance Theory, The Heat Equation, And the Atiyah-Singer Index Theorem

Peter B. Gilkey

University of Oregon

Publish or Perish, Inc.
Wilmington, Delaware (U.S.A.)

INVARIANCE THEORY, THE HEAT EQUATION, AND THE ATIYAH-SINGER INDEX THEOREM

Manufactured in the United States of America
ISBN **0-914098-20-9**

INTRODUCTION

This book treats the Atiyah-Singer index theorem using heat equation methods. The heat equation gives a local formula for the index of any elliptic complex. We use invariance theory to identify the integrand of the index theorem for the four classical elliptic complexes with the invariants of the heat equation. Since the twisted signature complex provides a sufficiently rich family of examples, this approach yields a proof of the Atiyah-Singer theorem in complete generality. We also use heat equation methods to discuss Lefschetz fixed point formulas, the Gauss-Bonnet theorem for a manifold with smooth boundary, and the twisted eta invariant. We shall not include a discussion of the signature theorem for manifolds with boundary.

The first chapter reviews results from analysis. Sections 1.1 through 1.7 represent standard elliptic material. Sections 1.8 through 1.10 contain the material necessary to discuss Lefschetz fixed point formulas and other topics.

Invariance theory and differential geometry provide the necessary link between the analytic formulation of the index theorem given by heat equation methods and the topological formulation of the index theorem contained in the Atiyah-Singer theorem. Sections 2.1 through 2.3 are a review of characteristic classes from the point of view of differential forms. Section 2.4 gives an invariant-theoretic characterization of the Euler form which is used to give a heat equation proof of the Gauss-Bonnet theorem. Sections 2.5 and 2.6 discuss the Pontrjagin forms of the tangent bundle and the Chern forms of the coefficient bundle using invariance theory.

The third chapter combines the results of the first two chapters to prove the Atiyah-Singer theorem for the four classical elliptic complexes. We first present a heat equation proof of the Hirzebruch signature theorem. The twisted spin complex provides a unified way of discussing the signature, Dolbeault, and de Rham complexes. In sections 3.2–3.4, we discuss the half-spin representations, the spin complex, and derive a formula for the $\hat{A}$ genus. We then discuss the Riemann-Roch formula for an almost complex manifold in section 3.5 using the $SPIN_c$ complex. In sections 3.6–3.7 we give a second derivation of the Riemann-Roch formula for holomorphic Kaehler manifods using a more direct approach. In the final two sections we derive the Atiyah-Singer theorem in its full generality.

The final chapter is devoted to more specialized topics. Sections 4.1–4.2 deal with elliptic boundary value problems and derive the Gauss-Bonnet theorem for manifolds with boundary. In sections 4.3–4.4 we discuss the twisted eta invariant on a manifold without boundary and we derive the Atiyah-Patodi-Singer twisted index formula. Section 4.5 gives a brief discussion of Lefschetz fixed point formulas using heat equation methods. In section 4.6 we use the eta invariant to calculate the K-theory of spherical space forms. In section 4.7, we discuss Singer's conjecture for the Euler form and related questions. In section 4.8, we discuss the local formulas for the invariants of the heat equation which have been derived by several authors, and in section 4.9 we apply these results to questions of spectral geometry.

The bibliography at the end of this book is not intended to be exhaustive but rather to provide the reader with a list of a few of the basic papers which have appeared. We refer the reader to the bibliography of Berger and Berard for a more complete list of works on spectral geometry.

This book is organized into four chapters. Each chapter is divided into a number of sections. Each Lemma or Theorem is indexed according to this subdivision. Thus, for example, Lemma 1.2.3 is the third Lemma of section 2 of Chapter 1.

CONTENTS

Introduction v

Chapter 1. Pseudo-Differential Operators

Introduction 1
1.1. Fourier Transform, Schwartz Class, and Sobolev Spaces 2
1.2. Pseudo-Differential Operators on $\mathbf{R}^m$ 11
1.3. Ellipticity and Pseudo-Differential Operators on Manifolds . . . 23
1.4. Fredholm Operators and the Index of a Fredholm Operator . . . 31
1.5 Elliptic Complexes, The Hodge Decomposition Theorem,
        and Poincaré Duality 37
1.6. The Heat Equation 42
1.7. Local Formula for the Index of an Elliptic Operator 50
1.8. Lefschetz Fixed Point Theorems 62
1.9. Elliptic Boundary Value Problems 70
1.10. Eta and Zeta Functions 78

Chapter 2. Characteristic Classes

Introduction 87
2.1. Characteristic Classes of a Complex Bundle 89
2.2 Characteristic Classes of a Real Vector Bundle.
        Pontrjagin and Euler Classes 98
2.3. Characteristic Classes of Complex Projective Space 104
2.4. The Gauss-Bonnet Theorem 117
2.5 Invariance Theory and the Pontrjagin Classes
        of the Tangent Bundle 123
2.6 Invariance Theory and Mixed Characteristic Classes
        of the Tangent Space and of a Coefficient Bundle 141

Chapter 3. The Index Theorem

Introduction 147
3.1. The Hirzebruch Signature Formula 148
3.2. Spinors and their Representations 159
3.3. Spin Structures on Vector Bundles 165
3.4. The Spin Complex 176
3.5. The Riemann-Roch Theorem for Almost Complex Manifolds . 180
3.6. A Review of Kaehler Geometry 193
3.7 An Axiomatic Characterization of the Characteristic Forms
        for Holomorphic Manifolds with Kaehler Metrics 204
3.8. The Chern Isomorphism and Bott Periodicity 215
3.9. The Atiyah-Singer Index Theorem 224

Chapter 4. Generalized Index Theorems and Special Topics

Introduction 241
4.1. The de Rham Complex for Manifolds with Boundary 243
4.2. The Gauss-Bonnet Theorem for Manifolds with Boundary . . . 250
4.3. The Regularity at $s = 0$ of the Eta Invariant 258
4.4. The Eta Invariant with Coefficients in a Locally Flat Bundle . 270
4.5. Lefschetz Fixed Point Formulas 284
4.6. The Eta Invariant and the K-Theory of Spherical Space Forms 295
4.7. Singer's Conjecture for the Euler Form 307
4.8. Local Formulas for the Invariants of the Heat Equation 314
4.9. Spectral Geometry 331

Bibliography 339

Index 347

Invariance Theory, The Heat Equation, And the Atiyah-Singer Index Theorem

CHAPTER 1

PSEUDO-DIFFERENTIAL OPERATORS

Introduction

In the first chapter, we develop the analysis needed to define the index of an elliptic operator and to compute the index using heat equation methods. Sections 1.1 and 1.2 are brief reviews of Sobolev spaces and pseudo-differential operators on Euclidean spaces. In section 1.3, we transfer these notions to compact Riemannian manifolds using partition of unity arguments. In section 1.4 we review the facts concerning Fredholm operators needed in section 1.5 to prove the Hodge decomposition theorem and to discuss the spectral theory of self-adjoint elliptic operators. In section 1.6 we introduce the heat equation and in section 1.7 we derive the local formula for the index of an elliptic operator using heat equation methods. Section 1.8 generalizes the results of section 1.7 to find a local formula for the Lefschetz number of an elliptic complex. In section 1.9, we discuss the index of an elliptic operator on a manifold with boundary and in section 1.10, we discuss the zeta and eta invariants.

Sections 1.1 and 1.4 review basic facts we need, whereas sections 1.8 through 1.10 treat advanced topics which may be omitted from a first reading. We have attempted to keep this chapter self-contained and to assume nothing beyond a first course in analysis. An exception is the de Rham theorem in section 1.5 which is used as an example.

A number of people have contributed to the mathematical ideas which are contained in the first chapter. We were introduced to the analysis of sections 1.1 through 1.7 by a course taught by L. Nirenberg. Much of the organization in these sections is modeled on his course. The idea of using the heat equation or the zeta function to compute the index of an elliptic operator seems to be due to R. Bott. The functional calculus used in the study of the heat equation contained in section 1.7 is due to R. Seeley as are the analytic facts on the zeta and eta functions of section 1.10.

The approach to Lefschetz fixed point theorems contained in section 1.8 is due to T. Kotake for the case of isolated fixed points and to S. C. Lee and the author in the general case. The analytic facts for boundary value problems discussed in section 1.9 are due to P. Greiner and R. Seeley.

1.1. Fourier Transform, Schwartz Class, And Sobolev Spaces.

The Sobolev spaces and Fourier transform provide the basic tools we shall need in our study of elliptic partial differential operators. Let $x = (x_1, \ldots, x_m) \in \mathbf{R}^m$. If $x, y \in \mathbf{R}^m$, we define:

$$x \cdot y = x_1 y_1 + \cdots + x_m y_m \qquad \text{and} \qquad |x| = (x \cdot x)^{1/2}$$

as the Euclicean dot product and length. Let $\alpha = (\alpha_1, \ldots, \alpha_m)$ be a multi-index. The α_j are non-negative integers. We define:

$$|\alpha| = \alpha_1 + \cdots + \alpha_m, \qquad \alpha! = \alpha_1! \ldots \alpha_m!, \qquad x^\alpha = x_1^{\alpha_1} \ldots x_m^{\alpha_m}.$$

Finally, we define:

$$d_x^\alpha = \left(\frac{\partial}{\partial x_1}\right)^{\alpha_1} \ldots \left(\frac{\partial}{\partial x_m}\right)^{\alpha_m} \qquad \text{and} \qquad D_x^\alpha = (-i)^{|\alpha|} d_x^\alpha$$

as a convenient notation for multiple partial differentiation. The extra factors of $(-i)$ defining D_x^α are present to simplify later formulas. If $f(x)$ is a smooth complex valued function, then Taylor's theorem takes the form:

$$f(x) = \sum_{|\alpha| \leq n} d_x^\alpha f(x_0) \frac{(x - x_0)^\alpha}{\alpha!} + O(|x - x_0|^{n+1}).$$

The Schwartz class S is the set of all smooth complex valued functions f on $\mathbf{R}^m$ such that for all α, β there are constants $C_{\alpha,\beta}$ such that

$$|x^\alpha D_x^\beta f| \leq C_{\alpha,\beta}.$$

This is equivalent to assuming there exist estimates of the form:

$$|D_x^\beta f| \leq C_{n,\beta}(1 + |x|)^{-n}$$

for all (n, β). The functions in S have all their derivatives decreasing faster at ∞ than the inverse of any polynomial.

For the remainder of Chapter 1, we let dx, dy, $d\xi$, etc., denote Lebesgue measure on $\mathbf{R}^m$ with an additional normalizing factor of $(2\pi)^{-m/2}$. With this normalization, the integral of the Gaussian distribution becomes:

$$\int e^{-\frac{1}{2}|x|^2} \, dx = 1.$$

We absorb the normalizing constant into the measure in order to simplify the formulas of the Fourier transform. If $C_0^\infty(\mathbf{R}^m)$ denotes the set of smooth functions of compact support on $\mathbf{R}^m$, then this is a subset of S. Since $C_0(\mathbf{R}^m)$ is dense in $L^2(\mathbf{R}^m)$, S is dense in $L^2(\mathbf{R}^m)$.

We define the convolution product of two elements of S by:

$$(f * g)(x) = \int f(x - y)g(y) \, dy = \int f(y)g(x - y) \, dy.$$

This defines an associative and commutative multiplication. Although there is no identity, there do exist approximate identities:

LEMMA 1.1.1. Let $f \in S$ with $\int f(x)\,dx = 1$. Define $f_u(x) = u^{-m} f(\frac{x}{u})$. Then for any $g \in S$, $f_u * g$ converges uniformly to g as $u \to 0$.

PROOF: Choose C so $\int |f(x)|\,dx \leq C$ and $|g(x)| \leq C$. Because the first derivatives of g are uniformly bounded, g is uniformly continuous. Let $\varepsilon > 0$ and choose $\delta > 0$ so $|x - y| \leq \delta$ implies $|g(x) - g(y)| \leq \varepsilon$. Because $\int f_u(x)\,dx = 1$, we compute:

$$|f_u * g(x) - g(x)| = \left| \int f_u(y)\{g(x - y) - g(x)\}\,dy \right|$$

$$\leq \int |f_u(y)\{g(x - y) - g(x)\}|\,dy.$$

We decompose this integral into two pieces. If $|y| \leq \delta$ we bound it by $C\varepsilon$. The integral for $|y| \geq \delta$ can be bounded by:

$$2C \int_{|y| \geq \delta} |f_u(y)|\,dy = 2C \int_{|y| \geq \delta/u} |f(y)|\,dy.$$

This converges to zero as $u \to 0$ so we can bound this by $C\varepsilon$ if $u < u(\varepsilon)$. This completes the proof.

A similar convolution smoothing can be applied to approximate any element of L^p arbitrarily well in the L^p norm by a smooth function of compact support.

We define the Fourier transform $\hat{f}(\xi)$ by:

$$\hat{f}(\xi) = \int e^{-ix \cdot \xi} f(x)\,dx \qquad \text{for } f \in S.$$

For the moment $\xi \in \mathbf{R}^m$; when we consider operators on manifolds, it will be natural to regard ξ as an element of the fiber of the cotangent space. By integrating by parts and using Lebesgue dominated convergence, we compute:

$$D_\xi^\alpha\{\hat{f}(\xi)\} = (-1)^{|\alpha|}\{\widehat{x^\alpha f}\} \qquad \text{and} \qquad \xi^\alpha \hat{f}(\xi) = \{\widehat{D_x^\alpha f}\}.$$

This implies $\hat{f} \in S$ so Fourier transform defines a map $S \to S$.

We compute the Fourier transform of the Gaussian distribution. Let $f_0(x) = \exp(-\frac{1}{2}|x|^2)$, then $f_0 \in S$ and $\int f_0(x)\,dx = 1$. We compute:

$$\hat{f}_0(\xi) = \int e^{-ix \cdot \xi} e^{-\frac{1}{2}|x|^2}\,dx$$

$$= e^{-\frac{1}{2}|\xi|^2} \int e^{-(x+i\xi) \cdot (x+i\xi)/2}\,dx.$$

We make a change of variables to replace $x + i\xi$ by x and to shift the contour in $\mathbf{C}^m$ back to the original contour $\mathbf{R}^m$. This shows the integral is 1 and $\hat{f}_0(\xi) = \exp\left(-\frac{1}{2}|\xi|^2\right)$ so the function f_0 is its own Fourier transform.

In fact, the Fourier transform is bijective and the Fourier inversion formula gives the inverse expressing f in terms of $\hat{f}$ by:

$$f(x) = \int e^{ix\cdot\xi}\hat{f}(\xi)\,d\xi = \hat{\hat{f}}(-x).$$

We define $T(f) = \hat{\hat{f}}(-x) = \int e^{ix\cdot\xi}\hat{f}(\xi)\,d\xi$ as a linear map from $S \to S$. We must show that $T(f) = f$ to prove the Fourier inversion formula.

Suppose first $f(0) = 0$. We expand:

$$f(x) = \int_0^1 \frac{d}{dt}\{f(tx)\}\,dt = \sum_j x_j \int_0^1 \frac{\partial f}{\partial x_j}(tx)\,dt = \sum_j x_j g_j$$

where the g_j are smooth. Let $\phi \in C_0^\infty(\mathbf{R}^m)$ be identically 1 near $x = 0$. Then we decompose:

$$f(x) = \phi f(x) + (1 - \phi)f(x) = \sum_j x_j \phi g_j + \sum_j x_j \left\{\frac{x_j(1-\phi)f}{|x|^2}\right\}.$$

Since ϕg_j has compact support, it is in S. Since ϕ is identically 1 near $x = 0$, $x_j(1-\phi)f/|x|^2 \in S$. Thus we can decompose $f = \sum x_j h_j$ for $h_j \in S$. We Fourier transform this identity to conclude:

$$\hat{f} = \sum_j \{\widehat{x_j h_j}\} = \sum_j i\frac{\partial \hat{h}_j}{\partial \xi_j}.$$

Since this is in divergence form, $T(f)(0) = \int \hat{f}(\xi)\,d\xi = 0 = f(0)$.

More generally, let $f \in S$ be arbitrary. We decompose $f = f(0)f_0 + (f - f(0)f_0)$ for $f_0 = \exp(-\frac{1}{2}|x|^2)$. Since $\hat{f}_0 = f_0$ is an even function, $T(f_0) = f_0$ so that $T(f)(0) = f(0)f_0(0) + T(f - f(0)f_0) = f(0)f_0(0) = f(0)$ since $(f - f(0)f_0)(0) = 0$. This shows $T(f)(0) = f(0)$ in general.

We use the linear structure on $\mathbf{R}^m$ to complete the proof of the Fourier inversion formula. Let $x_0 \in \mathbf{R}^m$ be fixed. We let $g(x) = f(x + x_0)$ then:

$$f(x_0) = g(0) = T(g)(0) = \int e^{-ix\cdot\xi}f(x + x_0)\,dx\,d\xi$$

$$= \int e^{-ix\cdot\xi}e^{ix_0\cdot\xi}f(x)\,dx\,d\xi$$

$$= T(f)(x_0).$$

This shows the Fourier transform defines a bijective map $S \to S$. If we use the constants $C_{\alpha,\beta} = \sup_{x \in \mathbf{R}^m} |x^\alpha D_x^\beta f|$ to define a Frechet structure on S, then the Fourier transform is a homeomorphism of topological vector spaces. It is not difficult to show $C_0^\infty(\mathbf{R}^m)$ is a dense subset of S in this topology. We can use either pointwise multiplication or convolution to define a multiplication on S and make S into a ring. The Fourier transform interchanges these two ring structures. We compute:

$$\hat{f} \cdot \hat{g} = \int e^{-ix \cdot \xi} f(x) e^{-iy \cdot \xi} g(y) \, dx \, dy$$

$$= \int e^{-i(x-y) \cdot \xi} f(x-y) e^{-iy \cdot \xi} g(y) \, dx \, dy$$

$$= \int e^{-ix \cdot \xi} f(x-y) g(y) \, dx \, dy.$$

The integral is absolutely convergent so we may interchange the order of integration to compute $\hat{f} \cdot \hat{g} = (\widehat{f * g})$. If we replace f by $\hat{f}$ and g by $\hat{g}$ we see $(f \cdot g)(-x) = (\widehat{\hat{f} * \hat{g}})$ using the Fourier inversion formula. We now take the Fourier transform and use the Fourier inversion formula to see $(\widehat{f \cdot g})(-\xi) = (\hat{f} * \hat{g})(-\xi)$ so that $(\widehat{f \cdot g}) = \hat{f} * \hat{g}$.

The final property we shall need of the Fourier transform is related to the L^2 inner product $(f,g) = \int f(x) \overline{g}(x) \, dx$. We compute:

$$(\hat{f}, g) = \int f(x) e^{-ix \cdot \xi} \overline{g}(\xi) \, dx \, d\xi = \int f(x) e^{-ix \cdot \xi} \overline{g}(\xi) \, d\xi \, dx$$

$$= (f, \hat{g}(-x)).$$

If we replace g by $\hat{g}$ then $(\hat{f}, \hat{g}) = (f, \hat{\hat{g}}(-x)) = (f, g)$ so the Fourier transform is an isometry with respect to the L^2 inner product. Since S is dense in L^2, it extends to a unitary map $L^2(\mathbf{R}^m) \to L^2(\mathbf{R}^m)$. We summarize these properties of the Fourier transform as follows:

LEMMA 1.1.2. *The Fourier transform is a homeomorphism* $S \to S$ *such that:*
(a) $f(x) = \int e^{ix \cdot \xi} \hat{f}(\xi) \, d\xi = \int e^{i(x-y) \cdot \xi} f(y) \, dy \, d\xi$ *(Fourier inversion formula);*
(b) $D_x^\alpha f(x) = \int e^{ix \cdot \xi} \xi^\alpha \hat{f}(\xi) \, d\xi$ *and* $\xi^\alpha \hat{f}(\xi) = \int e^{-ix \cdot \xi} D_x^\alpha f(x) \, dx;$
(c) $\hat{f} \cdot \hat{g} = (\widehat{f * g})$ *and* $\hat{f} * \hat{g} = (\widehat{f \cdot g});$
(d) *The Fourier transform extends to a unitary map of* $L^2(\mathbf{R}^m) \to L^2(\mathbf{R}^m)$ *such that* $(f,g) = (\hat{f}, \hat{g})$. *(Plancherel theorem).*

We note that without the normalizing constant of $(2\pi)^{-m/2}$ in the definition of the measures dx and $d\xi$ there would be various normalizing constants appearing in these identities. It is property (b) which will be of the most interest to us since it will enable us to interchange differentiation and multiplication.

We define the Sobolev space $H_s(\mathbf{R}^m)$ to measure L^2 derivatives. If s is a real number and $f \in S$, we define:

$$|f|_s^2 = \int (1 + |\xi|^2)^s |\hat{f}(\xi)|^2 \, d\xi.$$

The Sobolev space $H_s(\mathbf{R}^m)$ is the completion of S with respect to the norm $|_s$. The Plancherel theorem shows $H_0(\mathbf{R}^m)$ is isomorphic to $L^2(\mathbf{R}^m)$. More generally, $H_s(\mathbf{R}^m)$ is isomorphic to L^2 with the measure $(1 + |\xi|^2)^{s/2} \, d\xi$. Replacing $(1 + |\xi|^2)^s$ by $(1 + |\xi|)^{2s}$ in the definition of $|_s$ gives rise to an equivalent norm since there exist positive constants c_i such that:

$$c_1(1 + |\xi|^2)^s \leq (1 + |\xi|)^{2s} \leq c_2(1 + |\xi|^2)^s.$$

In some sense, the subscript "s" counts the number of L^2 derivatives. If $s = n$ is a positive integer, there exist positive constants c_1, c_2 so:

$$c_1(1 + |\xi|^2)^n \leq \sum_{|\alpha| \leq n} |\xi^\alpha|^2 \leq c_2(1 + |\xi|^2)^n.$$

This implies that we could define

$$|f|_n^2 = \sum_{|\alpha| \leq n} \int |\xi^\alpha \hat{f}|^2 \, d\xi = \sum_{|\alpha| \leq n} \int |D_x^\alpha f|^2 \, dx$$

as an equivalent norm for $H_n(\mathbf{R}^m)$. With this interpretation in mind, it is not surprising that when we extend D_x^α to H_s, that $|\alpha|$ L^2 derivatives are lost.

LEMMA 1.1.3. D_x^α extends to define a continuous map $D_x^\alpha \colon H_s \to H_{s-|\alpha|}$.

PROOF: Henceforth we will use C to denote a generic constant. C can depend upon certain auxiliary parameters which will usually be supressed in the interests of notational clarity. In this proof, for example, C depends on (s, α) but not of course upon f. The estimate:

$$|\xi^\alpha|^2 (1 + |\xi|^2)^{s-|\alpha|} \leq C(1 + |\xi|^2)^s$$

implies that:

$$|D_x^\alpha f|_{s-\alpha}^2 = \int |\xi^\alpha \hat{f}(\xi)|^2 (1 + |\xi|^2)^{s-|\alpha|} \, d\xi \leq C|f|_s^2$$

for $f \in S$. Since H_s is the closure of S in the norm $|_s$, this completes the proof.

We can also use the sup norm to measure derivatives. If k is a non-negative integer, we define:

$$|f|_{\infty,k} = \sup_{x \in \mathbf{R}^m} \sum_{|\alpha| \leq k} |D_x^\alpha f| \qquad \text{for} \qquad f \in S.$$

The completion of S with respect to this norm is a subset of $C^k(\mathbf{R}^m)$ (the continuous functions on $\mathbf{R}^m$ with continuous partial derivatives up to order k). The next lemma relates the two norms $|_s$ and $|_{\infty,k}$. It will play an important role in showing the weak solutions we will construct to differential equations are in fact smooth.

LEMMA 1.1.4. Let k be a non-negative integer and let $s > k + \frac{m}{2}$. If $f \in H_s$, then f is C^k and there is an estimate $|f|_{\infty,k} \leq C|f|_s$. (Sobolev Lemma).

PROOF: Suppose first $k = 0$ and $f \in S$. We compute

$$f(x) = \int e^{ix \cdot \xi} \hat{f}(\xi) \, d\xi$$
$$= \int \{e^{ix \cdot \xi} \hat{f}(\xi)(1 + |\xi|^2)^{s/2}\} \cdot \{(1 + |\xi|^2)^{-s/2}\} \, d\xi.$$

We apply the Cauchy-Schwarz inequality to estimate:

$$|f(x)|^2 \leq |f|_s^2 \int (1 + |\xi|^2)^{-s} \, d\xi.$$

Since $2s > m$, $(1 + |\xi|^2)^{-s}$ is integrable so $|f(x)| \leq C|f|_s$. We take the sup over $x \in \mathbf{R}^m$ to conclude $|f|_{\infty,0} \leq C|f|_s$ for $f \in S$. Elements of H_s are the limits in the $|_s$ norm of elements of S. The uniform limit of continuous functions is continuous so the elements of H_s are continuous and the same norm estimate extends to H_s. If $k > 0$, we use the estimate:

$$|D_x^\alpha f|_{\infty,0} \leq C|D_x^\alpha f|_{s-|\alpha|} \leq C|f|_s \qquad \text{for } |\alpha| \leq k \text{ and } s - k > \frac{m}{2}$$

to conclude $|f|_{\infty,k} \leq C|f|_s$ for $f \in S$. A similar argument shows that the elements of H_s must be C^k and that this estimate continues to hold.

If $s > t$, we can estimate $(1 + |\xi|^2)^s \geq (1 + |\xi|^2)^t$. This implies that $|f|_s \geq |f|_t$ so the identity map on S extends to define an injection of $H_s \to H_t$ which is norm non-increasing. The next lemma shows that this injection is compact if we restrict the supports involved.

LEMMA 1.1.5. *Let $\{f_n\} \in S$ be a sequence of functions with support in a fixed compact set K. We suppose there is a constant C so $|f_n|_s \leq C$ for all n. Let $s > t$. There exists a subsequence f_{n_k} which converges in H_t. (Rellich lemma).*

PROOF: Choose $g \in C_0(\mathbf{R}^m)$ which is identically 1 on a neighborhood of K. Then $gf_n = f_n$ so by Lemma 1.1.2(c) $\hat{f}_n = \hat{g} * \hat{f}_n$. We let $\partial_j = \dfrac{\partial}{\partial \xi_j}$ then $\partial_j(\hat{g} * \hat{f}_n) = \partial_j \hat{g} * \hat{f}_n$ so that:

$$|\partial_j \hat{f}_n(\xi)| \leq \int |\{\partial_j \hat{g}(\xi - \varsigma)\} \cdot \hat{f}_n(\varsigma)| \, d\varsigma.$$

We apply the Cauchy-Schwarz inequality to estimate:

$$|\partial_j \hat{f}_n(\xi)| \leq |f_n|_s \cdot \left\{ \int |\partial_j \hat{g}(\xi - \varsigma)|^2 (1 + |\varsigma|^2)^{-s} \, d\varsigma \right\}^{1/2} \leq C \cdot h(\xi)$$

where h is some continuous function of ξ. A similar estimate holds for $|\hat{f}_n(\xi)|$. This implies that the $\{\hat{f}_n\}$ form a uniformly bounded equi-continuous family on compact ξ subsets. We apply the Arzela-Ascoli theorem to extract a subsequence we again label by f_n so that $\hat{f}_n(\xi)$ converges uniformly on compact subsets. We complete the proof by verifying that f_n converges in H_t for $s > t$. We compute:

$$|f_j - f_k|_t^2 = \int |\hat{f}_j - \hat{f}_k|^2 (1 + |\xi|^2)^t \, d\xi.$$

We decompose this integral into two parts, $|\xi| \geq r$ and $|\xi| \leq r$. On $|\xi| \geq r$ we estimate $(1 + |\xi|^2)^t \leq (1 + r^2)^{t-s}(1 + |\xi|^2)^s$ so that:

$$\int_{|\xi| \geq r} |\hat{f}_j - \hat{f}_k|^2 (1 + |\xi|^2)^t \, d\xi \leq (1 + r^2)^{t-s} \int |\hat{f}_j - \hat{f}_k|^2 (1 + |\xi|^2)^s \, d\xi$$

$$\leq 2C(1 + r^2)^{t-s}.$$

If $\varepsilon > 0$ is given, we choose r so that $2C(1 + r^2)^{t-s} < \varepsilon$. The remaining part of the integral is over $|\xi| \leq r$. The $\hat{f}_j$ converge uniformly on compact subsets so this integral can be bounded above by ε if $j, k > j(\varepsilon)$. This completes the proof.

The hypothesis that the supports are uniformly bounded is essential. It is easy to construct a sequence $\{f_n\}$ with $|f_n|_s = 1$ for all n and such that the supports are pair-wise disjoint. In this case we can find $\varepsilon > 0$ so that $|f_j - f_k|_t > \varepsilon$ for all (j, k) so there is no convergent subsequence.

We fix $\phi \in S$ and let $\phi_\varepsilon(x) = \phi(\varepsilon x)$. We suppose $\phi(0) = 1$ and fix $f \in S$. We compute:

$$D_x^\alpha(f - \phi_\varepsilon f) = (1 - \phi_\varepsilon)D_x^\alpha f + \text{terms of the form } \varepsilon^j D_x^\beta \phi(\varepsilon x)D_x^\gamma f.$$

As $\varepsilon \to 0$, these other terms go to zero in L^2. Since $\phi_\varepsilon \to 1$ pointwise, $(1 - \phi_\varepsilon)D_x f$ goes to zero in L^2. This implies $\phi_\varepsilon f \to f$ in H_n for any $n \geq 0$ as $\varepsilon \to 0$ and therefore $\phi_\varepsilon f \to f$ in H_s for any s. If we take $\phi \in C_0^\infty(\mathbf{R}^m)$, this implies $C_0^\infty(\mathbf{R}^m)$ is dense in H_s for any s.

Each H_s space is a Hilbert space so it is isomorphic to its dual. Because there is no preferred norm for H_s, it is useful to obtain an invariant alternative characterization of the dual space H_s^*:

LEMMA 1.1.6. *The L^2 pairing which maps $S \times S \to \mathbf{C}$ extends to a map of $H_s \times H_{-s} \to \mathbf{C}$ which is a perfect pairing and which identifies H_{-s} with H_s^*. That is:*

(a) *$|(f,g)| \leq |f|_s|g|_{-s}$ for $f,g \in S$,*
(b) *given $f \in S$ there exists $g \in S$ so $(f,g) = |f|_s|g|_{-s}$ and we can define*

$$|f|_s = \sup_{g \in S, \, g \neq 0} \frac{|(f,g)|}{|g|_{-s}}.$$

PROOF: This follows from the fact that H_s is L^2 with the weight function $(1+|\xi|^2)^s$ and H_{-s} is L^2 with the weight function $(1+|\xi|^2)^{-s}$. We compute:

$$(f,g) = (\hat{f}, \hat{g}) = \int \hat{f}(\xi)(1 + |\xi|^2)^{s/2}\overline{\hat{g}}(\xi)(1 + |\xi|^2)^{-s/2} \, d\xi$$

and apply the Cauchy-Schwartz inequality to prove (a).

To prove part (b), we note $|f|_s \geq \sup\limits_{g \in S, \, g \neq 0} \dfrac{|(f,g)|}{|g|_{-s}}$. We take g to be defined by:

$$\hat{g} = \hat{f}(1 + |\xi|^2)^s \in S$$

and note that $(f,g) = (\hat{f}, \hat{g}) = |f|_s^2$ and that $|g|_{-s}^2 = |f|_s^2$ to see that equality can occur in (a) which proves (b)

If $s > t > u$ then we can estimate:

$$(1 + |\xi|)^{2t} \leq \varepsilon(1 + |\xi|)^{2s} + C(\varepsilon)(1 + |\xi|)^{2u}$$

for any $\varepsilon > 0$. This leads immediately to the useful estimate:

LEMMA 1.1.7. *Let $s > t > u$ and let $\varepsilon > 0$ be given. Then*

$$|f|_t \leq \varepsilon |f|_s + C(\varepsilon)|f|_u.$$

If V is a finite dimensional vector space, let $C^\infty(V)$ be the space of smooth complex valued maps of $\mathbf{R}^m \to V$. We choose a fixed Hermitian inner product on V and define $S(V)$ and $H_s(V)$ as in the scalar case. If $\dim(V) = k$ and if we choose a fixed orthonormal basis for V, then $S(V)$ and $H_s(V)$ become isomorphic to the direct sum of k copies of S and of H_s. Lemmas 1.1.1 through 1.1.7 extend in the obvious fashion.

We conclude this subsection with an extremely useful if elementary estimate:

LEMMA 1.1.8. (PEETRE'S INEQUALITY). *Let s be real and $x, y \in \mathbf{R}^m$. Then $(1 + |x + y|)^s \leq (1 + |y|)^s (1 + |x|)^{|s|}$.*

PROOF: We suppose first $s > 0$. We raise the triangle inequality:

$$1 + |x + y| < 1 + |x| + |y| \leq (1 + |y|)(1 + |x|)$$

to the sth power to deduce the desired inequality. We now suppose $s < 0$. A similar inequality:

$$(1 + |y|)^{-s} \leq (1 + |x + y|)^{-s}(1 + |x|)^{-s}$$

yields immediately:

$$(1 + |x + y|)^s \leq (1 + |y|)^s (1 + |x|)^{-s}$$

to complete the proof.

1.2. Pseudo-Differential Operators on $\mathbf{R}^m$.

A linear partial differential operator of order d is a polynomial expression $P = p(x, D) = \sum_{|\alpha| \le d} a_\alpha(x) D_x^\alpha$ where the $a_\alpha(x)$ are smooth. The symbol $\sigma P = p$ is defined by:

$$\sigma P = p(x, \xi) = \sum_{|\alpha| \le d} a_\alpha(x) \xi^\alpha$$

and is a polynomial of order d in the dual variable ξ. It is convenient to regard the pair (x, ξ) as defining a point of the cotangent space $T^*(\mathbf{R}^m)$; we will return to this point again when we discuss the effect of coordinate transformations. The leading symbol $\sigma_L P$ is the highest order part:

$$\sigma_L P(x, \xi) = \sum_{|\alpha| = d} a_\alpha(x) \xi^\alpha$$

and is a homogeneous polynomial of order d in ξ.

We can use the Fourier inversion formula to express:

$$P f(x) = \int e^{ix \cdot \xi} p(x, \xi) \hat{f}(\xi) \, d\xi = \int e^{i(x-y) \cdot \xi} p(x, \xi) f(y) \, dy \, d\xi$$

for $f \in S$. We note that since the second integral does not converge absolutely, we cannot interchange the dy and $d\xi$ orders of integration. We use this formalism to define the action of pseudo-differential operators (ΨDO's) for a wider class of symbols $p(x, \xi)$ than polynomials. We make the following

DEFINITION. $p(x, \xi)$ is a symbol of order d and we write $p \in S^d$ if

(a) $p(x, \xi)$ is smooth in $(x, \xi) \in \mathbf{R}^m \times \mathbf{R}^m$ with compact x support,
(b) for all (α, β) there are constants $C_{\alpha,\beta}$ such that

$$|D_x^\alpha D_\xi^\beta p(x, \xi)| \le C_{\alpha,\beta} (1 + |\xi|)^{d - |\beta|}.$$

For such a symbol p, we define the associated operator $P(x, D)$ by:

$$P(x, D)(f)(x) = \int e^{ix \cdot \xi} p(x, \xi) \hat{f}(\xi) \, d\xi = \int e^{(x-y) \cdot \xi} p(x, \xi) f(y) \, dy \, d\xi$$

as a linear operator mapping $S \to S$.

A differential operator has as its order a positive integer. The order of a pseudo-differential operator is not necessarily an integer. For example, if $f \in C_0^\infty(\mathbf{R}^m)$, define:

$$p(x, \xi) = f(x)(1 + |\xi|^2)^{d/2} \in S^d \qquad \text{for any } d \in \mathbf{R}.$$

This will be a symbol of order d. If $p \in S^d$ for all d, then we say that $p \in S^{-\infty}$ is infinitely smoothing. We adopt the notational convention of letting p, q, r denote symbols and P, Q, R denote the corresponding ΨDO's.

Because we shall be interested in problems on compact manifolds, we have assumed the symbols have compact x support to avoid a number of technical complications. The reader should note that there is a well defined theory which does not require compact x support.

When we discuss the heat equation, we shall have to consider a wider class of symbols which depend on a complex parameter. We postpone discussion of this class until later to avoid unnecessarily complicating the discussion at this stage. We shall phrase the theorems and proofs of this section in such a manner that they will generalize easily to the wider class of symbols.

Our first task is to extend the action of P from S to H_s.

LEMMA 1.2.1. Let $p \in S^d$ then $|Pf|_{s-d} \leq C|f|_s$ for $f \in S$. P extends to a continuous map $P\colon H_s \to H_{s-d}$ for all s.

PROOF: We compute $Pf(x) = \int e^{ix\cdot\xi} p(x,\xi) \hat{f}(\xi)\,d\xi$ so that the Fourier transform is given by:

$$\widehat{Pf}(\varsigma) = \int e^{ix\cdot(\xi-\varsigma)} p(x,\xi)\hat{f}(\xi)\,d\xi\,dx.$$

This integral is absolutely convergent since p has compact x support so we may interchange the order of integration. If we define

$$q(\varsigma,\xi) = \int e^{-ix\cdot\varsigma} p(x,\xi)\,dx$$

as the Fourier transform in the x direction, then

$$\widehat{Pf}(\varsigma) = \int q(\varsigma-\xi,\xi)\hat{f}(\xi)\,d\xi.$$

By Lemma 1.1.6, $|Pf|_{s-d} = \sup\limits_{g\in S} \dfrac{|(Pf,g)|}{|g|_{d-s}}$. We compute:

$$(Pf,g) = \int q(\varsigma-\xi,\xi)\hat{f}(\xi)\bar{\hat{g}}(\varsigma)\,d\xi\,d\varsigma.$$

Define:

$$K(\varsigma,\xi) = q(\varsigma-\xi,\xi)(1+|\xi|)^{-s}(1+|\varsigma|)^{s-d}$$

then:

$$(Pf,g) = \int K(\varsigma,\xi)\hat{f}(\xi)(1+|\xi|)^s\bar{\hat{g}}(\varsigma)(1+|\varsigma|)^{d-s}\,d\xi\,d\varsigma.$$

We apply the Cauchy-Schwarz inequality to estimate:

$$|(Pf,g)| \leq \left\{ \int |K(\varsigma,\xi)| \,|\hat{f}(\xi)|^2 (1+|\xi|)^{2s} \, d\xi \, d\varsigma \right\}^{1/2}$$

$$\times \left\{ \int |K(\varsigma,\xi)| \,|\hat{g}(\varsigma)|^2 (1+|\varsigma|)^{2d-2s} \, d\xi \, d\varsigma \right\}^{1/2}.$$

We complete the proof by showing

$$\int |K(\varsigma,\xi)| \, d\xi \leq C \qquad \text{and} \qquad \int |K(\varsigma,\xi)| \, d\varsigma \leq C$$

since then $|(Pf,g)| \leq C|f|_s |g|_{d-s}$.

By hypothesis, p has suppport in a compact set K and we have estimates:

$$|D_x^\alpha p(x,\xi)| \leq C_\alpha (1+|\xi|)^d.$$

Therefore:

$$|\varsigma^\alpha q(\varsigma,\xi)| = \left| \int e^{-ix\cdot\varsigma} D_x^\alpha p(x,\xi) \, dx \right| \leq C_\alpha (1+|\xi|)^d \operatorname{vol}(K).$$

Therefore, for any integer k, $|q(\varsigma,\xi)| \leq C_k (1+|\xi|)^d (1+|\varsigma|)^{-k} \operatorname{vol}(K)$ and:

$$|K(\varsigma,\xi)| \leq C_k (1+|\xi|)^{d-s}(1+|\varsigma|)^{s-d}(1+|\varsigma-\xi|)^{-k} \operatorname{vol}(K).$$

We apply Lemma 1.1.8 with $x+y=\xi$ and $y=\varsigma$ to estimate:

$$|K(\varsigma,\xi)| \leq C_k (1+|\varsigma-\xi|)^{|d-s|-k} \operatorname{vol}(K).$$

If we choose $k > \frac{m}{2} + |d-s|$, then this will be integrable and complete the proof.

Our next task is to show that the class of ΨDO's forms an algebra under the operations of composition and taking adjoint. Before doing that, we study the situation with respect to differential operators to motivate the formulas we shall derive. Let $P = \sum_\alpha p_\alpha(x)D_x^\alpha$ and let $Q = \sum_\alpha q_\alpha(x)D_x^\alpha$ be two differential operators. We assume p and q have compact x support. It is immediate that:

$$P^* = \sum_\alpha D_x^\alpha p_\alpha^* \qquad \text{and} \qquad PQ = \sum_{\alpha,\beta} p_\alpha(x)D_x^\alpha q_\beta(x)D_x^\beta$$

are again differential operators in our class. Furthermore, using Leibnitz's rule

$$D_x^\alpha(fg) = \sum_{\beta+\gamma=\alpha} D_x^\beta(f) \cdot D_x^\gamma(g) \cdot \frac{\alpha!}{\beta!\gamma!},$$

$$d_\xi^\beta(\xi^{\beta+\gamma}) = \xi^\gamma \cdot \frac{(\beta+\gamma)!}{\gamma!},$$

it is an easy combinatorial exercise to compute that:

$$\sigma(P^*) = \sum_\alpha d_\xi^\alpha D_x^\alpha p^* / \alpha! \quad\text{and}\quad \sigma(PQ) = \sum_\alpha d_\xi^\alpha p \cdot D_x^\alpha q / \alpha! \,.$$

The perhaps surprising fact is that these formulas remain true in some sense for ΨDO's, only the sums will become infinite rather than finite.

We introduce an equivalence relation on the class of symbols by defining $p \sim q$ if $p - q \in S^{-\infty}$. We note that if $p \in S^{-\infty}$ then $P: H_s \to H_t$ for all s and t by Lemma 1.2.1. Consequently by Lemma 1.1.4, $P: H_s \to C_0^\infty$ for all s so that P is infinitely smoothing in this case. Thus we mod out by infinitely smoothing operators.

Given symbols $p_j \in S^{d_j}$ where $d_j \to -\infty$, we write

$$p \sim \sum_{j=1}^\infty p_j$$

if for every d there is an integer $k(d)$ such that $k \geq k(d)$ implies that $p - \sum_{j=1}^k p_j \in S^d$. We emphasize that this sum does not in fact need to converge. The relation $p \sim \sum p_j$ simply means that the difference between P and the partial sums of the P_j is as smoothing as we like. It will turn out that this is the appropriate sense in which we will generalize the formulas for $\sigma(P^*)$ and $\sigma(PQ)$ from differential to pseudo-differential operators.

Ultimately, we will be interested in operators which are defined on compact manifolds. Consequently, it poses no difficulties to restrict the domain and the range of our operators. Let U be a open subset of $\mathbf{R}^m$ with compact closure. Let $p(x, \xi) \in S^d$ have x support in U. We restrict the domain of the operator P to $C_0^\infty(U)$ so $P: C_0^\infty(U) \to C_0^\infty(U)$. Let $\Psi_d(U)$ denote the space of all such operators. For $d \leq d'$, then $\Psi_d(U) \subseteq \Psi_{d'}(U)$. We define

$$\Psi(U) = \bigcup_d \Psi_d(U) \quad\text{and}\quad \Psi_{-\infty}(U) = \bigcap_d \Psi_d(U)$$

to be the set of all pseudo-differential operators on U and the set of infinitely smoothing pseudo-differential operators on U.

More generally, let $p(x, \xi)$ be a matrix valued symbol; we suppose the components of p all belong to S^d. The corresponding operator P is given by a matrix of pseudo-differential operators. P is a map from vector valued functions with compact support in U to vector valued functions with compact support in U. We shall not introduce separate notation for the shape of p and shall continue to denote the collection of all such operators by $\Psi_d(U)$. If p and q are matrix valued and of the proper shape, we define $p \cdot q$ and also the operator $P \cdot Q$ by matrix product and by composition. We also define p^* and P^* to be the matrix adjoint and the operator adjoint so that $(P^* f, g) = (f, P^* g)$ where f and g are vector valued and of compact support. Before studying the algebra structure on $\Psi(U)$, we must enlarge the class of symbols which we can admit:

LEMMA 1.2.2. *Let $r(x, \xi, y)$ be a matrix valued symbol which is smooth in (x, ξ, y). We suppose r has compact x support inside U and that there are estimates:*

$$|D_x^\alpha D_\xi^\beta D_y^\gamma r| \le C_{\alpha, \beta, \gamma} (1 + |\xi|)^{d - |\beta|}$$

for all multi-indices (α, β, γ). If f is vector valued with compact support in U, we define:

$$Rf(x) = \int e^{i(x-y)\cdot\xi} r(x, \xi, y) f(y) \, dy \, d\xi.$$

Then this operator is in $\Psi_d(U)$ and the symbol is given by:

$$\sigma R(x, \xi) \sim \left\{ \sum_\alpha d_\xi^\alpha D_y^\alpha r(x, \xi, y) / \alpha! \right\} \Big|_{x=y}.$$

PROOF: We note that any symbol in S^d belongs to this class of operators if we define $r(x, \xi, y) = p(x, \xi)$. We restricted to vector valued functions with compact support in U. By multiplying r by a cut-off function in y with compact support which is 1 over U, we may assume without loss of generality the y support of r is compact as well. Define:

$$q(x, \xi, \varsigma) = \int e^{-iy\cdot\varsigma} r(x, \xi, y) \, dy$$

to be the Fourier transform of r in the y variable. Using Lemma 1.1.2 we see $\widehat{(rf)} = \hat{r} * \hat{f}$. This implies that:

$$\int e^{-iy\cdot\xi} r(x, \xi, y) f(y) \, dy = \int q(x, \xi, \xi - \varsigma) \hat{f}(\varsigma) \, d\varsigma.$$

The argument given in the proof of Lemma 1.2.1 gives estimates of the form:

$$|q(x, \xi, \varsigma)| \le C_k (1 + |\xi|)^d (1 + |\varsigma|)^{-k} \quad \text{and} \quad |\hat{f}(\varsigma)| \le C_k (1 + |\varsigma|)^{-k}$$

for any k. Consequently:

$$|q(x, \xi, \xi - \varsigma) \hat{f}(\varsigma)| \le C_k (1 + |\xi|)^d (1 + |\xi - \varsigma|)^{-k} (1 + |\varsigma|)^{-k}.$$

We apply Lemma 1.1.8 to estimate:

$$|q(x, \xi, \xi - \varsigma) \hat{f}(\varsigma)| \le C_k (1 + |\xi|)^{|d| - k} (1 + |\varsigma|)^{|d| - k}$$

so this is absolutely integrable. We change the order of integration and express:

$$Rf(x) = \int e^{ix\cdot\xi} q(x,\xi,\xi-\varsigma)\hat{f}(\varsigma)\,d\xi\,d\varsigma.$$

We define:

$$p(x,\varsigma) = \int e^{ix(\xi-\varsigma)} q(x,\xi,\xi-\varsigma)\,d\xi$$

and compute:

$$Rf(x) = \int e^{ix\cdot\varsigma} p(x,\varsigma)\hat{f}(\varsigma)\,d\varsigma$$

is a pseudo-differential operator once it is verified that $p(x,\varsigma)$ is a symbol in the correct form.

We change variables to express:

$$p(x,\varsigma) = \int e^{ix\cdot\xi} q(x,\xi+\varsigma,\xi)\,d\xi$$

and estimate:

$$|q(x,\xi+\varsigma,\xi)| \le C_k(1+|\xi+\varsigma|)^d(1+|\xi|)^{-k}$$
$$\le C_k(1+|\varsigma|)^d(1+|\xi|)^{|d|-k}.$$

This is integrable so $|p(x,\varsigma)| \le C'_k(1+|\varsigma|)^d$. Similar estimates on $|D_x^\alpha D_\varsigma^\beta q(x,\xi+\varsigma,\xi)|$ which arise from the given estimates for r show that $p \in S^d$ so that R is a pseudo-differential operator.

We use Taylor's theorem on the middle variable of $q(x,\xi+\varsigma,\xi)$ to expand:

$$q(x,\xi+\varsigma,\xi) = \sum_{|\alpha|\le k} \frac{d_\varsigma^\alpha q(x,\varsigma,\xi)\xi^\alpha}{\alpha!} + q_k(x,\varsigma,\xi).$$

The remainder q_k decays to arbitrarily high order in (ξ,ς) and after integration gives rise to a symbol in S^{d-k} which may therefore be ignored. We integrate to conclude

$$p(x,\varsigma) = \sum_{|\alpha|\le k} \int e^{ix\cdot\xi} \frac{d_\varsigma^\alpha q(x,\varsigma,\xi)\xi^\alpha}{\alpha!}\,d\xi + \text{remainder}$$

$$= \sum_{|\alpha|\le k} \frac{d_\varsigma^\alpha D_y^\alpha r(x,\varsigma,y)}{\alpha!}\Bigg|_{x=y} + \text{a remainder}$$

using Lemma 1.1.2. This completes the proof of the lemma.

We use this technical lemma to show that the pseudo-differential operators form an algebra:

LEMMA 1.2.3. *Let $P \in \Psi_d(U)$ and let $Q \in \Psi_e(U)$. Then:*
(a) If U' is any open set with compact closure containing $\overline{U}$, then $P^* \in \Psi_d(U')$ and $\sigma(P^*) \sim \sum_\alpha d_\xi^\alpha D_x^\alpha p^*/\alpha!$.
(b) Assume that P and Q have the proper shapes so PQ and pq are defined. Then $PQ \in \Psi_{d+e}(U)$ and $\sigma(PQ) \sim \sum_\alpha d_\xi^\alpha p \cdot D_x^\alpha q/\alpha!$.

PROOF: The fact that P^* lies in a larger space is only a slight bit of technical bother; this fact plays an important role in considering boundary value problems of course. Let $(f, g) = f \cdot g$ be the pointwise Hermitian inner product. Fix $\phi \in C_0^\infty(U')$ to be identically 1 on U and compute:

$$(Pf, g) = \int e^{i(x-y) \cdot \xi} p(x, \xi) \phi(y) f(y) \cdot g(x) \, dy \, d\xi \, dx$$
$$= \int f(y) \cdot e^{i(y-x) \cdot \xi} p^*(x, \xi) \phi(y) g(x) \, dy \, d\xi \, dx$$

since the inner product is Hermitian. By approximating $p^*(x, \xi)$ by functions with compact ξ support, we can justify the use of Fubini's theorem to replace $dy \, d\xi \, dx$ by $dx \, d\xi \, dy$ and express:

$$(Pf, g) = \int f(y) \cdot e^{i(y-x) \cdot \xi} p^*(x, \xi) \phi(y) g(x) \, dx \, d\xi \, dy$$
$$= (f, P^* g)$$

where we define:

$$P^* g(y) = \int e^{i(y-x) \cdot \xi} p^*(x, \xi) \phi(y) g(x) \, dx \, d\xi.$$

This is an operator of the form discussed in Lemma 1.2.2 so $P^* \in \Psi_d(U')$ and we compute:

$$\sigma(P^*) \sim \sum_\alpha d_\xi^\alpha D_x^\alpha p^*/\alpha!$$

since $\phi = 1$ on the support of p. This completes the proof of (a). We note that we can delete the factor of ϕ from the expression for $P^* g$ since it was only needed to prove P^* was a ΨDO.

We use (a) to prove (b). Since:

$$Q^* g(y) = \int e^{i(y-x) \cdot \xi} q^*(x, \xi) g(x) \, dx \, d\xi$$

the Fourier inversion formula implies:

$$(\widehat{Q^* g}) = \int e^{-ix \cdot \xi} q^*(x, \xi) g(x) \, dx.$$

If $\tilde{q}$ is the symbol of Q^*, then we interchange the roles of Q and Q^* to see:

$$\widehat{(Qg)} = \int e^{-ix\cdot\xi}\tilde{q}^*(x,\xi)g(x)\,dx.$$

Therefore:

$$PQg(x) = \int e^{ix\cdot\xi}p(x,\xi)\widehat{(Qg)}(\xi)\,d\xi$$

$$= \int e^{i(x-y)\cdot\xi}p(x,\xi)\tilde{q}^*(y,\xi)g(y)\,dy\,d\xi$$

which is an operator of the form discussed in Lemma 1.2.2 if $r(x,\xi,y) = p(x,\xi)\tilde{q}^*(y,\xi)$. This proves PQ is a pseudo-differential operator of the correct order. We compute the symbol of PQ to be:

$$\sim \sum_\alpha d_\xi^\alpha D_y^\alpha(p(x,\xi)\tilde{q}^*(y,\xi))/\alpha! \quad \text{evaluated at } x = y.$$

We use Leibnitz's formula and expand this in the form:

$$\sim \sum_{\beta,\gamma} d_\xi^\beta p(x,\xi)D_y^\beta d_\xi^\gamma D_y^\gamma \tilde{q}^*/\beta!\gamma!\,.$$

The sum over γ yields the symbol of $Q^{**} = Q$ so we conclude finally

$$\sigma(PQ) \sim \sum_\beta d_\xi^\beta p(x,\xi)D_x^\beta q(x,\xi)/\beta!$$

which completes the proof.

Let $K(x,y)$ be a smooth matrix valued function with compact x support in U. If f is vector valued with compact support in U, we define:

$$P(K)(f)(x) = \int K(x,y)f(y)\,dy.$$

LEMMA 1.2.4. Let $K(x,y)$ be smooth with compact x support in U, then $P(K) \in \Psi_{-\infty}(U)$.

PROOF: We let $\phi(\xi) \in C_0^\infty(\mathbf{R}^m)$ with $\int \phi(\xi)\,d\xi = 1$. Define:

$$r(x,\xi,y) = e^{i(y-x)\cdot\xi}\phi(\xi)K(x,y)$$

then this is a symbol in $S^{-\infty}$ of the sort discussed in Lemma 1.2.2. It defines an infinitely smoothing operator. It is immediate that:

$$P(K)(f)(x) = \int e^{i(x-y)\cdot\xi}r(x,\xi,y)f(y)\,dy\,d\xi.$$

Conversely, it can be shown that any infinitely smoothing map has a smooth kernel. In general, of course, it is not possible to represent an arbitrary pseudo-differential operator by a kernel. If P is smoothing enough, however, we can prove:

LEMMA 1.2.5. *Let r satisfy the hypothesis of Lemma 1.2.2 where $d <$ $-m - k$. We define $K(x,y) = \int e^{i(x-y)\cdot\xi} r(x,\xi,y) \, d\xi$. Then K is C^k in (x,y) and $Rf(x) = \int K(x,y)f(y) \, dy$.*

PROOF: If we can show K is well defined, then the representation of R in terms of the kernel K will follow from Fubini's theorem. We estimate:

$$D_x^\alpha D_y^\beta K(x,y) = \sum_{\substack{\alpha = \alpha_1 + \alpha_2 \\ \beta = \beta_1 + \beta_2}} \frac{\alpha! \, \beta!}{\alpha_1! \, \alpha_2! \, \beta_1! \, \beta_2!} (-1)^{|\beta_1|}$$

$$\times \left\{ \int e^{i(x-y)\cdot\xi} \xi^{\alpha_1 + \beta_1} D_x^{\alpha_2} D_y^{\beta_2} r(x,\xi,y) \, d\xi \right\}.$$

Since we can estimate:

$$|\xi^{\alpha_1 + \beta_1} D_x^{\alpha_2} D_y^{\beta_2} r(x,\xi,y)| \le C(1 + |\xi|)^{d + |\alpha| + |\beta|}$$

this will be integrable for $|\alpha| + |\beta| \le k$. Thus K is C^k and the representation of R follows immediately.

In Lemma 1.2.2 we computed the symbol of the pseudo-differential operator defined by $r(x,\xi,y)$ in terms of $d_\xi^\alpha D_y^\alpha r$ when $x = y$. This implies the singular (i.e., the non-smoothing part) of R is concentrated near the diagonal $x = y$. We make this more precise:

LEMMA 1.2.6. *Let $r(x,\xi,y)$ satisfy the hypothesis of Lemma 1.2.2. Suppose the x support of r is disjoint from the y support of r, then R is infinitely smoothing and is represented by a smooth kernel function $K(x,y)$.*

PROOF: We would like to define $K(x,y) = \int e^{(x-y)\cdot\xi} r(x,\xi,y) \, d\xi$. Unfortunately, this integral need not converge in general. By hypothesis, $|x-y| \ge \varepsilon > 0$ on the support of r. We define the Laplacian $\Delta_\xi = \sum_\nu D_{\xi_\nu}^2$. Since $\Delta_\xi e^{i(x-y)\cdot\xi} = |x-y|^2 e^{i(x-y)\cdot\xi}$ we integrate by parts in a formal sense k times to express:

$$Rf(x) = \int e^{i(x-y)\cdot\xi} |x - y|^{-2k} \Delta_\xi^k r(x,\xi,y) f(y) \, dy \, d\xi.$$

This formal process may be justified by first approximating r by a function with compact ξ support. We now define

$$K(x,y) = \int e^{i(x-y)\cdot\xi} |x - y|^{-2k} \Delta_\xi^k r(x,\xi,y) \, d\xi$$

for any k sufficiently large. Since $\Delta_\xi^k r$ decays to arbitrarily high order in ξ, we use the same argument as that given in Lemma 1.2.5 to show that

$K(x, y)$ is arbitrarily smooth in (x, y) and hence is C^∞. This completes the proof.

We note that in general $K(x, y)$ will become singular at $x = y$ owing to the presence of the terms $|x - y|^{-2k}$ if we do not assume the support of x is disjoint from the support of y.

A differential operator P is local in the sense that if $f = 0$ on some open subset of U, then $Pf = 0$ on that same subset since differentiation is a purely local process. ΨDO's are not local in general since they are defined by the Fourier transform which smears out the support. Nevertheless, they do have a somewhat weaker property, they do not smear out the singular support of a distribution f. More precisely, let $f \in H_s$. If $\phi \in C_0^\infty(U)$, we define the map $f \mapsto \phi f$. If we take $r(x, \xi, y) = \phi(x)$ and apply Lemma 1.2.2, then we see that this is a pseudo-differential operator of order 0. Therefore $\phi f \in H_s$ as well. This gives a suitable notion of restriction. We say that f is smooth on an open subset U' of U if and only if $\phi f \in C^\infty$ for every such ϕ. An operator P is said to be pseudo-local if f is smooth on U' implies Pf is smooth on U'.

LEMMA 1.2.7. *Pseudo-differential operators are pseudo-local.*

PROOF: Let $P \in \Psi_d(U)$ and let $f \in H_s$. Fix $x \in U'$ and choose $\phi \in C_0^\infty(U')$ to be identically 1 near x. Choose $\psi \in C_0^\infty(U')$ with support contained in the set where ϕ is identically 1. We must verify that ψPf is smooth. We compute:

$$\psi Pf = \psi P\phi f + \psi P(1 - \phi)f.$$

By hypothesis, ϕf is smooth so $\psi P\phi f$ is smooth. The operator $\psi P(1 - \phi)$ is represented by a kernel of the form $\psi(x)p(x, \xi)(1 - \phi(y))$ which has disjoint x and y support. Lemma 1.2.6 implies $\psi P(1 - \phi)f$ is smooth which completes the proof.

In Lemmas 1.2.2 and 1.2.3 we expressed the symbol of an operator as an infinite asymptotic series. We show that the algebra of symbols is complete in a certain sense:

LEMMA 1.2.8. *Let $p_j \in S^{d_j}(U)$ where $d_j \to -\infty$. Then there exists $p \sim \sum_j p_j$ which is a symbol in our class. p is a unique modulo $S^{-\infty}$.*

PROOF: We may assume without loss of generality that $d_1 > d_2 > \cdots \to -\infty$. We will construct $p \in S^{d_1}$. The uniqueness is clear so we must prove existence. The p_j all have support inside U; we will construct p with support inside U' where U' is any open set containing the closure of U.

Fix a smooth function ϕ such that:

$$0 \le \phi \le 1, \qquad \phi(\xi) = 0 \text{ for } |\xi| \le 1, \qquad \phi(\xi) = 1 \text{ for } |\xi| \ge 2.$$

We use ϕ to cut away the support near $\xi = 0$. Let $t_j \to 0$ and define:

$$p(x, \xi) = \sum_j \phi(t_j \xi) p_j(x, \xi).$$

For any fixed ξ, $\phi(t_j \xi) = 0$ for all but a finite number of j so this sum is well defined and smooth in (x, ξ). For $j > 1$ we have

$$|p_j(x, \xi)| \leq C_j (1 + |\xi|)^{d_j} = C_j (1 + |\xi|)^{d_1} (1 + |\xi|)^{d_j - d_1}.$$

If $|\xi|$ is large enough, $(1 + |\xi|)^{d_j - d_1}$ is as small as we like and therefore by passing to a subsequence of the t_j we can assume

$$|\phi(t_j \xi) p_j(x, \xi)| \leq 2^{-j} (1 - |\xi|)^{d_1} \qquad \text{for } j > 1.$$

This implies that $|p(x, \xi)| \leq (C_1 + 1)(1 + |\xi|)^{d_1}$. We use a similar argument with the derivatives and use a diagonalization argument on the resulting subsequences to conclude $p \in S^d$. The supports of all the p_j are contained compactly in U so the support of p is contained in $\overline{U}$ which is contained in U'.

We now apply exactly the same argument to $p_{d_2} + \cdots$ to assume that $p_{d_2} + \cdots \in S^{d_2}$. We continue in this fashion and use a diagonalization argument on the resulting subsequences to conclude in the end that

$$\sum_{j=j_0}^{\infty} \phi(t_j \xi) p_j(x, \xi) \in S^k \qquad \text{for } k = d_{j_0}.$$

Since $p_j - \phi(t_j \xi) p_j \in S^{-\infty}$, this implies $p - \sum_{j=1}^{j_0} p_j \in S^k$ and completes the proof.

If $K(x, y)$ is smooth with compact x, y support in U, then $P(K) \in \Psi_{-\infty}(U)$ defines a continuous operator from $H_s \to H_t$ for any s, t. Let $|P|_{s,t}$ denote the operator norm so $|Pf|_t \leq |P|_{s,t} |f|_s$ for any $f \in S$. It will be convenient to be able to estimate $|K|_{\infty,k}$ in terms of these norms:

LEMMA 1.2.9. Let $K(x, y)$ be a smooth kernel with compact x, y support in U. Let $P = P(K)$ be the operator defined by K. If k is a non-negative integer, then $|K|_{\infty,k} \leq C(k) |P|_{-k,k}$

PROOF: By arguing separately on each entry in the matrix K, we may reduce ourselves to the scalar case. Suppose first $k = 0$. Choose $\phi \in C_0^\infty(\mathbf{R}^m)$ positive with $\int \phi(x) \, dx = 1$. Fix points $(x_0, y_0) \in U \times U$ and define:

$$f_n(x) = n^m \phi(n(x - x_0)) \qquad \text{and} \qquad g_n(y) = n^m \phi(n(y - y_0)).$$

Then if n is large, f_n and g_n have compact support in U. Then:

$$K(x_0, y_0) = \lim_{n\to\infty} \int f_n(x) K(x,y) g_n(y)\, dy\, dx = \lim_{n\to\infty} (f_n, Pg_n)$$

by Lemma 1.1.1. We estimate

$$|(f_n, Pg_n)| \leq |P|_{0,0} |f_n|_0 |g_n|_0 = |P|_{0,0} |\phi|_0^2$$

to complete the proof in this case.

If $|\alpha| \leq k$, $|\beta| \leq k$ then:

$$\begin{aligned}
D_x^\alpha D_y^\beta K(x,y) &= \lim_{n\to\infty} \int f_n(x) \{ D_x^\alpha D_y^\beta K(x,y) \} g_n(y)\, dy\, dx \\
&= \lim_{n\to\infty} \int (D_x^\alpha f_n) K(x,y)(D_y^\beta g_n(y))\, dy\, dx \\
&= \lim_{n\to\infty} (D_x^\alpha f_n, P D_y^\beta g_n).
\end{aligned}$$

We use Lemma 1.1.6 to estimate this by

$$\begin{aligned}
|D_x^\alpha f_n|_{-k} |P D_y^\beta g_n|_k &\leq |f_n|_0 |P|_{-k,k} |D_y^\beta g_n|_{-k} \\
&\leq |f_n|_0 |g_n|_0 |P|_{-k,k} = |P|_{-k,k} |\phi|_0^2
\end{aligned}$$

to complete the proof.

1.3. Ellipticity and Pseudo-Differential Operators on Manifolds.

The norms we have given to define the spaces H_s depend upon the Fourier transform. In order to get a more invariant definition which can be used to extend these notions to manifolds, we must consider elliptic pseudo-differential operators.

Let $p \in S^d(U)$ be a square matrix and let U_1 be an open set with $\overline{U}_1 \subset U$. We say that p is elliptic on U_1 if there exists an open subset U_2 with $\overline{U}_1 \subset U_2 \subset \overline{U}_2 \subset U$ and if there exists $q \in S^{-d}$ such that $pq - I \in S^{-\infty}$ and $qp - I \in S^{-\infty}$ over U_2. (To say that $r \in S^{-\infty}$ over U_2 simply means the estimates of section 1.2 hold over U_2. Equivalently, we assume $\phi r \in S^{-\infty}$ for every $\phi \in C_0^\infty(U_2)$). This constant technical fuss over domains will be eliminated very shortly when we pass to considering compact manifolds; the role of U_2 is to ensure uniform estimates over U_1.

It is clear that p is elliptic over U_1 if and only if there exists constants C_0 and C_1 such that $p(x, \xi)$ is invertible for $|\xi| \geq C_0$ and

$$|p(x, \xi)^{-1}| \leq C_1(1 + |\xi|)^{-d} \qquad \text{for } |\xi| \geq C_0, \ x \in U_2.$$

We define $q = \phi(\xi)p^{-1}(x, \xi)$ where $\phi(\xi)$ is a cut-off function identically 0 near $\xi = 0$ and identically 1 near $\xi = \infty$. We used similar cutoff functions in the proof of Lemma 1.2.8. Furthermore, if $p_0 \in S^{d-1}$, then p is elliptic if and only if $p + p_0$ is elliptic; adding lower order terms does not alter the ellipticity. If p is a polynomial and P is a differential operator, then p is elliptic if and only if the leading symbol $\sigma_L(p) = \sum_{|\alpha|=d} p_\alpha(x)\xi^\alpha$ is invertible for $\xi \neq 0$.

There exist elliptic operators of all orders. Let $\phi(x) \in C_0^\infty$ and define the symbol $p(x, \xi) = \phi(x)(1 + |\xi|^2)^{d/2}I$, then this is an elliptic symbol of order d whenever $\phi(x) \neq 0$.

LEMMA 1.3.1. Let $P \in \Psi^d(U)$ be elliptic over U_1 then:
(a) There exists $Q \in \Psi^{-d}(U)$ such that $PQ - I \sim 0$ and $QP - I \sim 0$ over U_1 (i.e., $\phi(PQ - I)$ and $\phi(QP - I)$ are infinitely smoothing for any $\phi \in C_0^\infty(U_2)$).
(b) P is hypo-elliptic over U_1, i.e., if $f \in H_s$ and if Pf is smooth over U_1 then f is smooth over U_1.
(c) There exists a constant C such that $|f|_d \leq C(|f|_0 + |Pf|_0)$ for $f \in C_0^\infty(U_1)$. (Gärding's inequality).

PROOF: We will define Q to have symbol $q_0 + q_1 + \cdots$ where $q_j \in S^{-d-j}$. We try to solve the equation

$$\sigma(PQ - I) \sim \sum_{\alpha,j} d_\xi^\alpha p \cdot D_x^\alpha q_j/\alpha! - I \sim 0.$$

When we decompose this sum into elements of S^{-k}, we conclude we must solve

$$\sum_{|\alpha|+j=k} d_\xi^\alpha p \cdot D_x^\alpha q_j / \alpha! = \begin{cases} I & \text{if } k = 0 \\ 0 & \text{if } k \neq 0. \end{cases}$$

We define $q_0 = q$ and then solve the equation inductively to define:

$$q_k = -q \cdot \sum_{\substack{|\alpha|+j=k \\ j<k}} d_\xi^\alpha p \cdot D_x^\alpha q_j / \alpha!.$$

This defines Q so $\sigma(PQ-I) \sim 0$ over U_2. Similarly we could solve $\sigma(Q_1 P - I) \sim 0$ over U_2. We now compute $\sigma(Q - Q_1) = \sigma(Q - Q_1 PQ) + \sigma(Q_1 PQ - Q_1) = \sigma((I - Q_1 P)Q) + \sigma(Q_1(QP - I)) \sim 0$ over U_2 so that in fact Q and Q_1 agree modulo infinitely smoothing operators. This proves (a).

Let $f \in H_s$ with Pf smooth over U_1, and choose $\phi \in C_0(U_1)$. We compute:

$$\phi f = \phi(I - QP)f + \phi QPf.$$

As $\phi(I - QP) \sim 0$, $\phi(I - QP)f$ is smooth. Since Pf is smooth over U_1, ϕQPf is smooth since Q is pseudo-local. Thus ϕf is smooth which proves (b).

Finally, we choose $\phi \in C_0^\infty(U_2)$ to be identically 1 on U_1. Then if $f \in C_0^\infty(U_1)$,

$$|f|_d = |\phi f|_d = |\phi(I - QP)f + \phi QPf|_d \leq |\phi(I - QP)f|_d + |\phi QPf|_d.$$

We estimate the first norm by $C|f|_0$ since $\phi(I - QP)$ is an infinitely smoothing operator. We estimate the second norm by $C|Pf|_0$ since ϕQ is a bounded map from L^2 to H^d. This completes the proof.

We note (c) is immediate if $d < 0$ since $|_d \leq |_0$. If $d > 0$, $|f|_0 + |Pf|_0 \leq C(|f|_d)$ so this gives a equivalent norm on H^d.

We now consider the effect of changes of coordinates on our class of pseudo-differential operators. Let $h: U \to \tilde{U}$ be a diffeomorphism. We define $h^*: C^\infty(\tilde{U}) \to C^\infty(U)$ by $h^* f(x) = f(h(x))$. If P is a linear operator on $C^\infty(U)$, we define $h_* P$ acting on $C^\infty(\tilde{U})$ by $(h_* P)f = (h^{-1})^* P(h^* f)$. The fundamental lemma we shall need is the following:

LEMMA 1.3.2. Let $h: U \to \tilde{U}$ be a diffeomorphism. Then:
(a) If $P \in \Psi^d(U)$ then $h_* P \in \Psi^d(\tilde{U})$. Let $p = \sigma(P)$ and define $h(x) = x_1$ and $dh(x)^t \xi_1 = \xi$. Let $p_1(x_1, \xi_1) = p(x, \xi)$ then $\sigma(h_* P) - p_1 \in S^{d-1}$.
(b) Let U_1 be an open subset with $\overline{U}_1 \subset U$. There exists a constant C such that $|h^* f|_d \leq C|f|_d$ for all $f \in C_0^\infty(h(U_1))$. In other words, the Sobolev spaces are invariant.

PROOF: The first step is to localize the problem. Let $\{\phi_i\}$ be a partition of unity and let $P_{ij} = \phi_i P \phi_j$ so $P = \sum_{i,j} P_{ij}$. If the support of ϕ_i is disjoint from the support of ϕ_j, then P_{ij} is an infinitely smoothing operator with a smooth kernel $K_{ij}(x,y)$ by Lemma 1.2.6. Therefore h_*P_{ij} is also given by a smooth kernel and is a pseudo-differential operator by Lemma 1.2.4. Consequently, we may restrict attention to pairs (i,j) such that the supports of ϕ_i and ϕ_j intersect. We assume henceforth P is defined by a symbol $p(x,\xi,y)$ where p has arbitrarily small support in (x,y).

We first suppose h is linear to motivate the constructions of the general case. Let $h(x) = hx$ where h is a constant matrix. We equate:

$$hx = x_1, \qquad hy = y_1, \qquad h^t\xi_1 = \xi$$

and define

$$p_1(x_1,\xi_1,y_1) = p(x,\xi,y).$$

(In the above, h^t denotes the matrix transpose of h). If $f \in C_0^\infty(\widetilde{U})$, we compute:

$$(h_*P)f(x_1) = \int e^{i(x-y)\cdot\xi}p(x,\xi,y)f(hy)\,dy\,d\xi$$

$$= \int e^{ih^{-1}(x_1-y_1)\cdot\xi}p(h^{-1}x_1,\xi,h^{-1}y_1)f(y_1)$$

$$\times |\det(h)|^{-1}\,dy_1\,d\xi.$$

We now use the identities $h^{-1}(x_1-y_1)\cdot\xi = (x_1-y_1)\cdot\xi_1$ and $|\det(h)|\,d\xi_1 = d\xi$ to write:

$$(h_*P)f(x_1) = \int e^{i(x_1-y_1)\cdot\xi_1}p(h^{-1}x_1,h^t\xi_1,h^{-1}y_1)f(y_1)\,dy_1\,d\xi_1$$

$$= \int e^{i(x_1-y_1)\cdot\xi_1}p_1(x_1,\xi_1,y_1)f(y_1)\,dy_1\,d\xi_1.$$

This proves that (h_*P) is a pseudo-differential operator on $\widetilde{U}$. Since we don't need to localize in this case, we compute directly that

$$\sigma(h_*P)(x_1,\xi_1) = p(h^{-1}x_1,h^t\xi_1).$$

We regard (x,ξ) as giving coordinates for T^*M when we expand any co-vector in the form $\sum \xi_i\,dx^i$. This is exactly the transformation for the cotangent space so we may regard σP as being invariantly defined on $T^*\mathbf{R}^m$.

If h is not linear, the situation is somewhat more complicated. Let

$$x - y = h^{-1}(x_1) - h^{-1}(y_1) = \int_0^1 \frac{d}{dt}\{h^{-1}(tx_1 + (1-t)y_1)\}\,dt$$

$$= \int_0^1 d(h^{-1})(tx_1 + (1-t)y_1)\cdot(x_1 - y_1)\,dt = T(x_1,y_1)(x_1 - y_1)$$

where $T(x_1, y_1)$ is a square matrix. If $x_1 = y_1$, then $T(x_1, y_1) = d(h^{-1})$ is invertible since h is a diffeomorphism. We localize using a partition of unity to suppose henceforth the supports are small enough so $T(x_1, y_1)$ is invertible for all points of interest.

We set $\xi_1 = T(x_1, y_1)^t \xi$ and compute:

$$
\begin{aligned}
(h_* P)(f)(x_1) &= \int e^{i(x-y)\cdot\xi} p(x, \xi, y) f(hy)\, dy\, d\xi \\
&= \int e^{iT(x_1,y_1)(x_1-y_1)\cdot\xi} p(h^{-1}x_1, \xi, h^{-1}y_1) f(y_1) J\, dy_1\, d\xi \\
&= \int e^{i(x_1-y_1)\cdot\xi_1} p_1(x_1, \xi_1, y_1) f(y_1) \\
&\qquad\qquad \times J |\det T(x_1, y_1)|^{-1}\, dy_1\, d\xi_1
\end{aligned}
$$

where $J = |\det(dh^{-1})| = |\det T(y_1, y_1)|$. By Lemma 1.2.2, this defines a pseudo-differential operator of order d such that $\sigma(h_* P) = p_1$ modulo S^{d-1} which completes the proof of (a). Since $|dh|$ is uniformly bounded on U_1, $|f|_0 \leq C|h^* f|_0$ and $|h^* f|_0 \leq C|f|_0$. If P is elliptic, then $h_* P$ is elliptic of the same order. For $d > 0$, choose P elliptic of order d and compute:

$$
|h^* f|_d \leq C(|h^* f|_0 + |Ph^* f|_0) \leq C(|f|_0 + |(h_* P)f|_0) \leq C|f|_d
$$

which completes the proof of (b) if $d \geq 0$. The result for $d \leq 0$ follows by duality using Lemma 1.1.6(b).

We introduce the spaces S^d/S^{d-1} and define $\sigma_L(P)$ to be the element defined by $\sigma(P)$ in this quotient. Let P and Q be pseudo-differential operators of order d_1 and d_2. Then PQ is a pseudo-differential operator of order $d_1 + d_2$ and $\sigma_L(PQ) = \sigma_L(P)\sigma_L(Q)$ since the remaining terms in the asymptotic series are of lower order. Similarly $\sigma_L(P^*) = \sigma_L(P)^*$. If we define (x, ξ) as coordinates for $T^*(\mathbf{R}^m)$ by representing a cotangent vector at a point x in the form $\sum \xi_i dx^i$, then Lemma 1.3.2 implies $\sigma_L(P)$ is invariantly defined on $T^*(\mathbf{R}^m)$. If $P = \sum_\alpha p_\alpha D_x^\alpha$ is a differential operator there is a natural identification of $\sum_{|\alpha|=d} p_\alpha \xi^\alpha$ with the image of p in S^d/S^{d-1} so this definition of the leading symbol agrees with that given earlier.

We now extend the results of section 1.2 to manifolds. Let M be a smooth compact Riemannian manifold without boundary. Let m be the dimension of M and let dvol or sometimes simply dx denote the Riemanian measure on M. In Chapter 2, we will use the notation $|\text{dvol}|$ to denote this measure in order to distinguish between measures and m-forms, but we shall not bother with this degree of formalism here. We restrict to scalars first. Let $C^\infty(M)$ be the space of smooth functions on M and let $P: C^\infty(M) \to C^\infty(M)$ be a linear operator. We say that P is a pseudo-differential operator of order d and write $P \in \Psi_d(M)$ if for every open chart

U on M and for every $\phi, \psi \in C_0^\infty(U)$, the localized operator $\phi P \psi \in \Psi_d(U)$. We say that P is elliptic if $\phi P \psi$ is elliptic where $\phi\psi(x) \neq 0$. If $Q \in \Psi_d(U)$, we let $P = \phi Q \psi$ for $\phi, \psi \in C_0^\infty(U)$. Lemma 1.3.2 implies P is a pseudo-differential operator on M so there exists operators of all orders on M. We define:

$$\Psi(M) = \bigcup_d \Psi_d(M) \qquad \text{and} \qquad \Psi_{-\infty}(M) = \bigcap_d \Psi_d(M)$$

to be the set of all pseudo-differential operators on M and the set of infinitely smoothing operators on M.

In any coordinate system, we define $\sigma(P)$ to the symbol of the operator $\phi P \phi$ where $\phi = 1$ near the point in question; this is unique modulo $S^{-\infty}$. The leading symbol is invariantly defined on T^*M, but the total symbol changes under the same complicated transformation that the total symbol of a differential operator does under coordinate transformations. Since we shall not need this transformaton law, we omit the statement; it is implicit in the computations performed in Lemma 1.3.2.

We define $L^2(M)$ using the L^2 inner product

$$(f, g) = \int_M f(x)\overline{g}(x)\,dx, \qquad |f|_0^2 = (f, f).$$

We let $L^2(M)$ be the completion of $C^\infty(M)$ in this norm. Let $P: C^\infty(M) \to C^\infty(M)$. We let P^* be defined by $(Pf, g) = (f, P^*g)$ if such a P^* exists. Lemmas 1.3.2 and 1.2.3 imply that:

LEMMA 1.3.3.
(a) If $P \in \Psi_d(M)$, then $P^* \in \Psi_d(M)$ and $\sigma_L(P^*) = \sigma_L(P)^*$. In any coordinate chart, $\sigma(P^*)$ has a asymptotic expansion given by Lemma 1.2.3(a).
(b) If $P \in \Psi_d(M)$ and $Q \in \Psi_e(M)$, then $PQ \in \Psi_{d+e}(M)$ and $\sigma_L(PQ) = \sigma_L(P)\sigma_L(Q)$. In any coordinate chart $\sigma(PQ)$ has an asymptotic expansion given in Lemma 1.2.3(b).

We use a partition of unity to define the Sobolev spaces $H_s(M)$. Cover M by a finite number of coordinate charts U_i with diffeomorphisms $h_i: O_i \to U_i$ where the O_i are open subsets of $\mathbf{R}^m$ with compact closure. If $f \in C_0^\infty(U_i)$, we define:

$$|f|_s^{(i)} = |h_i^* f|_s$$

where we shall use the superscript $^{(i)}$ to denote the localized norm. Let $\{\phi_i\}$ be a partition of unity subordinate to this cover and define:

$$|f|_s = \sum |\phi_i f|_s^{(i)}.$$

If $\psi \in C^\infty(M)$, we note that $|\phi_i \psi f|_s^{(i)} \le C|\phi_i f|_s^{(i)}$ since multiplication by ψ defines a ΨDO of order 0. Suppose $\{U_j', O_j', h_j', \phi_j'\}$ is another possible choice to define $|_s'$. We estimate:

$$|\phi_j' f|_s'^{(j)} \le \sum_i |\phi_j' \phi_i f|_s'^{(j)}.$$

Since $\phi_j' \phi_i f \in C_0^\infty(U_i \cap U_j')$, we can use Lemma 1.3.2 (b) to estimate $|_s'^{(j)}$ by $|_s^{(i)}$ so

$$|\phi_j' f|_s'^{(j)} \le \sum_i C|\phi_j' \phi_i f|_s^{(i)} \le C|\phi_i f|_s^{(i)} \le C|f|_s$$

so that $|f|_s' \le C|f|_s$. Similarly $|f|_s \le C|f|_s'$. This shows these two norms are equivalent so $H_s(M)$ is defined independent of the choices made.

We note that:

$$\sum_i \{|\phi_i f|_s^{(i)}\}^2 \le |f|_s^2 \le C \sum_i \{|\phi_i f|_s^{(i)}\}^2$$

so by using this equivalent norm we conclude the $H_s(M)$ are topologically Hilbert spaces. If $\{\psi_i\}$ are given subordinate to the cover U_i with $\psi_i \ge 0$ and $\psi = \sum_i \psi_i > 0$, we let $\phi_i = \psi_i/\psi$ and compute:

$$\sum_i |\psi_i f|_s^{(i)} = \sum_i |\psi \phi_i f|_s^{(i)} \le C \sum_i |\phi_i f|_s^{(i)} = C|f|_s$$

$$\sum_i |\phi_i f|_s^{(i)} = \sum_i |\psi^{-1} \psi_i f|_s^{(i)} \le C \sum_i |\psi_i f|_s^{(i)}$$

to see the norm defined by $\sum_i |\psi_i f|_s^{(i)}$ is equivalent to the norm $|f|_s$ as well.

Lemma 1.1.5 implies that $|f|_t \le |f|_s$ if $t < s$ and that the inclusion of $H_s(M) \to H_t(M)$ is compact. Lemma 1.1.7 implies given $s > t > u$ and $\varepsilon > 0$ we can estimate:

$$|f|_t \le \varepsilon|f|_s + C(\varepsilon)|f|_u.$$

We assume the coordinate charts U_i are chosen so the union $U_i \cup U_j$ is also contained in a larger coordinate chart for all (i, j). We decompose $p \in \Psi_d(M)$ as $P = \sum_{i,j} P_{i,j}$ for $P_{i,j} = \phi_i P \phi_j$. By Lemma 1.2.1 we can estimate:

$$|\phi_i P \phi_j f|_s^{(i)} \le C|\phi_j f|_{s+d}^{(j)}$$

so $|Pf|_s \le C|f|_{s+d}$ and P extends to a continuous map from $H_{s+d}(M) \to H_s(M)$ for all s.

We define:

$$|f|_{\infty,k} = \sum_i |\phi_i f|^{(i)}_{\infty,k}$$

as a measure of the sup norm of the kth derivatives of f. This is independent of particular choices made. Lemma 1.1.4 generalizes to:

$$|f|_{\infty,k} \leq C|f|_s \qquad \text{for } s > \frac{m}{2} + k.$$

Thus $H_s(M)$ is a subset of $C^k(M)$ in this situation.

We choose $\psi_i \in C^\infty(U_i)$ with $\sum_i \psi_i^2 = 1$ then:

$$|(f,g)| = |\sum_i (\psi_i f, \psi_i g)| \leq \sum_i |(\psi_i f, \psi_i g)|$$

$$\leq C \sum_i |\psi_i f|^{(i)}_s |\psi_i g|^{(i)}_{-s} \leq C|f|_s |g|_{-s}.$$

Thus the L^2 inner product gives a continuous map $H_s(M) \times H_{-s}(M) \to \mathbf{C}$.

LEMMA 1.3.4.

(a) The natural inclusion $H_s \to H_t$ is compact for $s > t$. Furthermore, if $s > t > u$ and if $\varepsilon > 0$, then $|f|_t \leq \varepsilon|f|_s + C(\varepsilon)|f|_u$.

(b) If $s > k + \frac{m}{2}$ then $H_s(M)$ is contained in $C^k(M)$ and we can estimate $|f|_{\infty,k} \leq C|f|_s$.

(c) If $P \in \Psi_d(M)$ then $P: H_{s+d}(M) \to H_s(M)$ is continuous for all s.

(d) The pairing $H_s(M) \times H_{-s}(M) \to \mathbf{C}$ given by the L^2 inner product is a perfect pairing.

PROOF: We have proved every assertion except the fact (d) that the pairing is a perfect pairing. We postpone this proof briefly until after we have discussed elliptic ΨDO's.

The sum of two elliptic operators need not be elliptic. However, the sum of two elliptic operators with positive symbols is elliptic. Let P_i have symbol $(1+|\xi_i|^2)^{d/2}$ on U_i and let ϕ_i be a partition of unity. $P = \sum_i \phi_i P_i \phi_i$; this is an elliptic ΨDO of order d for any d, so elliptic operators exist. We let P be an elliptic ΨDO of order d and let ψ_i be identically 1 on the support of ϕ_i. We use these functions to construct $Q \in \Psi_{-d}(M)$ so $PQ - I \in \Psi_{-\infty}(M)$ and $QP - I \in \Psi_{-\infty}(M)$. In each coordinate chart, let $P_i = \psi_i P \psi_i$ then $P_i - P_j \in \Psi_{-\infty}$ on the support of $\phi_i \phi_j$. We construct Q_i as the formal inverse to P_i on the support of ϕ_i, then $Q_i - Q_j \in \Psi_{-\infty}$ on the support of $\phi_i \phi_j$ since the formal inverse is unique. Modulo $\Psi_{-\infty}$ we have $P = \sum_i \phi_i P \sim \sum_i \phi_i P_i$. We define $Q = \sum_j Q_j \phi_j$ and note Q has the desired properties.

It is worth noting we could also construct the formal inverse using a Neumann series. By hypothesis, there exists q so $qp - I \in S^{-1}$ and $pq - I \in$

S^{-1}, where $p = \sigma_L(P)$. We construct Q_1 using a partition of unity so $\sigma_L(Q_1) = q$. We let Q^r and Q^l be defined by the formal series:

$$Q^l = Q_1 \left\{ \sum_k (-1)^k (PQ_1 - I)^k \right\}$$

$$Q^r = \left\{ \sum_k (-1)^k (Q_1 P - I)^k \right\} Q_1$$

to construct formal left and right inverses so $Q = Q^r = Q^l$ modulo $\Psi_{-\infty}$.

Let P be elliptic of order $d > 0$. We estimate:

$$|f|_d \leq |(QP - I)f|_d + |QPf|_d \leq C|f|_0 + C|Pf|_0 \leq C|f|_d$$

so we could define H_d using the norm $|f|_d = |f|_0 + |Pf|_0$. We specialize to the following case. Let Q be elliptic of order $d/2$ and let $P = Q^*Q + 1$. Then Q^*Q is self-adjoint and non-negative so we can estimate $|f|_0 \leq |Pf|_0$ and we can define $|f|_d = |Pf|_0$ in this case. Consequently:

$$|f|_d^2 = (Pf, Pf) = (f, P^*Pf) = (f, g),$$

for $g = P^*Pf$. Since $|P^*Pf|_{-d} \leq C|f|_d$, we conclude:

$$|f|_d = (Pf, Pf)/|f|_d \leq C(f, g)/|g|_{-d} \leq C \sup_h |(f, h)|/|h|_{-d}.$$

Since the pairing of H_d with H_{-d} is continuous, this proves $H_d = H^*_{-d}$. Topologically these are Hilbert spaces so we see dually that $H_d^* = H_{-d}$. This completes the proof of Lemma 1.3.4. We also note that we have proved:

LEMMA 1.3.5. *Let $P \in \Psi_d$ be elliptic. Then there exists $Q \in \Psi_{-d}$ so $PQ - I \in \Psi_{-\infty}$ and $QP - I \in \Psi_{-\infty}$. P is hypoelliptic. If $d > 0$, we can define h_d by using the norm $|f|_0 + |Pf|_0$ and define H_{-d} by duality.*

If V is a vector bundle, we cover M by coordinate charts U_i over which V is trivial. We use this cover to define $H_s(V)$ using a partition of unity. We shall always assume V has a given fiber metric so $L^2(V)$ is invariantly defined. $P: C^\infty(V) \to C^\infty(W)$ is a ΨDO of order d if $\phi P \psi$ is given by a matrix of dth order ΨDO's for $\phi, \psi \in C_0^\infty(U)$ for any coordinate chart U over which V and W are trivial. Lemmas 1.3.4 and 1.3.5 generalize immediately to this situation.

1.4. Fredholm Operators and the Index of a Fredholm Operator.

Elliptic ΨDO's are invertible modulo $\Psi_{-\infty}$. Lemma 1.3.4 will imply that elliptic ΨDO's are invertible modulo compact operators and that such operators are Fredholm. We briefly review the facts we shall need concerning Fredholm and compact operators.

Let H be a Hilbert space and let $\text{END}(H)$ denote the space of all bounded linear maps $T\colon H \to H$. There is a natural norm on $\text{END}(H)$ defined by:

$$|T| = \sup_{x \in H} \frac{|Tx|}{|x|}$$

where the sup ranges over $x \neq 0$. $\text{END}(H)$ becomes a Banach space under this norm. The operations of addition, composition, and taking adjoint are continuous. We let $\text{GL}(H)$ be the subset of $\text{END}(H)$ consisting of maps T which are 1-1 and onto. The inverse boundedness theorem shows that if $T \in \text{END}(H)$ is 1-1 and onto, then there exists $\varepsilon > 0$ such that $|Tx| \geq \varepsilon|x|$ so T^{-1} is bounded as well. The Neuman series:

$$(1 - z)^{-1} = \sum_{k=0}^{\infty} z^k$$

converges for $|z| < 1$. If $|I - T| < 1$, we may express $T = I - (I - T)$. If we define

$$S = \sum_{k=0}^{\infty} (I - T)^k$$

then this converges in $\text{END}(H)$ to define an element $S \in \text{END}(H)$ so $ST = TS = I$. Furthermore, this shows $|T|^{-1} \leq (1 - |I - T|)^{-1}$ so $\text{GL}(H)$ contains an open neighborhood of I and the map $T \to T^{-1}$ is continuous there. Using the group operation on $\text{GL}(H)$, we see that $\text{GL}(H)$ is a open subset of $\text{END}(H)$ and is a topological group.

We say that $T \in \text{END}(H)$ is compact if T maps bounded sets to precompact sets—i.e., if $|x_n| \leq C$ is a bounded sequence, then there exists a subsequence x_{n_k} so $Tx_{n_k} \to y$ for some $y \in H$. We let $\text{COM}(H)$ denote the set of all compact maps.

LEMMA 1.4.1. $\text{COM}(H)$ is a closed 2-sided $*$-ideal of $\text{END}(H)$.

PROOF: It is clear the sum of two compact operators is compact. Let $T \in \text{END}(H)$ and let $C \in \text{COM}(H)$. Let $\{x_n\}$ be a bounded sequence in H then $\{Tx_n\}$ is also a bounded sequence. By passing to a subsequence, we may assume $Cx_n \to y$ and $CTx_n \to z$. Since $TCx_n \to Ty$, this implies CT and TC are compact so $\text{COM}(H)$ is a ideal. Next let $C_n \to C$ in $\text{END}(H)$, and let x_n be a bounded sequence in H. Choose a subsequence

x_n^1 so $C_1 x_n^1 \to y^1$. We choose a subsequence of the x_n^1 so $C_2 x_n^2 \to y^2$. By continuing in this way and then using the diagonal subsequence, we can find a subsequence we denote by x_n^n so $C_k(x_n^n) \to y^k$ for all k. We note $|C x_n^n - C_k x_n^n| \le |C - C_k| c$. Since $|C - C_k| \to 0$ this shows the sequence $C x_n^n$ is Cauchy so C is compact and $\mathrm{COM}(H)$ is closed. Finally let $C \in \mathrm{COM}(H)$ and suppose $C^* \notin \mathrm{COM}(H)$. We choose $|x_n| \le 1$ so $|C^* x_n - C^* x_m| \ge \varepsilon > 0$ for all n, m. We let $y_n = C^* x_n$ be a bounded sequence, then $(C y_n - C y_m, x_n - x_m) = |C^* x_n - C^* x_m|^2 \ge \varepsilon^2$. Therefore $\varepsilon^2 \le |C y_n - C y_m||x_n - x_m| \le 2|C y_n - C y_m|$ so $C y_n$ has no convergent subsequence. This contradicts the assumption $C \in \mathrm{COM}(H)$ and proves $C^* \in \mathrm{COM}(H)$.

We shall assume henceforth that H is a separable infinite dimensional space. Although any two such Hilbert spaces are isomorphic, it is convenient to separate the domain and range. If E and F are Hilbert spaces, we define $\mathrm{HOM}(E, F)$ to be the Banach space of bounded linear maps from E to F with the operator norm. We let $\mathrm{ISO}(E, F)$ be the set of invertible maps in $\mathrm{HOM}(E, F)$ and let $\mathrm{COM}(E, F)$ be the closed subspace of $\mathrm{HOM}(E, F)$ of compact maps. If we choose a fixed isomorphism of E with F, we may identify $\mathrm{HOM}(E, F) = \mathrm{END}(E)$, $\mathrm{ISO}(E, F) = \mathrm{GL}(E)$, and $\mathrm{COM}(E, F) = \mathrm{COM}(E)$. $\mathrm{ISO}(E, F)$ is a open subset of $\mathrm{HOM}(E, F)$ and the operation of taking the inverse is a continuous map from $\mathrm{ISO}(E, F)$ to $\mathrm{ISO}(F, E)$. If $T \in \mathrm{HOM}(E, F)$, we define:

$$N(T) = \{\, e \in E : T(E) = 0 \,\} \qquad \text{(the null space)}$$
$$R(T) = \{\, f \in F : f = T(e) \text{ for some } e \in E \,\} \qquad \text{(the range)}.$$

$N(T)$ is always closed, but $R(T)$ need not be. If $\perp$ denotes the operation of taking orthogonal complement, then $R(T)^{\perp} = N(T^*)$. We let $\mathrm{FRED}(E, F)$ be the subset of $\mathrm{HOM}(E, F)$ consisting of operators invertible modulo compact operators:

$$\mathrm{FRED}(E, F) = \{\, T \in \mathrm{HOM}(E, F) : \exists S_1, S_2 \in \mathrm{HOM}(F, E) \text{ so}$$
$$S_1 T - I \in \mathrm{COM}(E) \text{ and } T S_2 - I \in \mathrm{COM}(F) \,\}.$$

We note this condition implies $S_1 - S_2 \in \mathrm{COM}(F, E)$ so we can assume $S_1 = S_2$ if we like. The following lemma provides another useful characterization of $\mathrm{FRED}(E, F)$:

LEMMA 1.4.2. *The following are equivalent:*
(a) $T \in \mathrm{FRED}(E, F)$;
(b) $T \in \mathrm{END}(E, F)$ *has* $\dim N(T) < \infty$, $\dim N(T^*) < \infty$, $R(T)$ *is closed, and* $R(T^*)$ *is closed.*

PROOF: Let $T \in \mathrm{FRED}(E, F)$ and let $x_n \in N(T)$ with $|x_n| = 1$. Then $x_n = (I - S_1 T) x_n = C x_n$. Since C is compact, there exists a convergent subsequence. This implies the unit sphere in $N(T)$ is compact so $N(T)$ is finite dimensional. Next let $y_n = Tx_n$ and $y_n \to y$. We may assume without loss of generality that $x_n \in N(T)^\perp$. Suppose there exists a constant C so $|x_n| \le C$. We have $x_n = S_1 y_n + (I - S_1 T)x_n$. Since $S_1 y_n \to S_1 y$ and since $(I - S_1 T)$ is compact, we can find a convergent subsequence so $x_n \to x$ and hence $y = \lim_n y_n = \lim_n Tx_n = Tx$ is in the range of T so $R(T)$ will be closed. Suppose instead $|x_n| \to \infty$. If $x_n' = x_n/|x_n|$ we have $Tx_n' = y_n/|x_n| \to 0$. We apply the same argument to find a subsequence $x_n' \to x$ with $Tx = 0$, $|x| = 1$, and $x \in N(T)^\perp$. Since this is impossible, we conclude $R(T)$ is closed (by passing to a subsequence, one of these two possibilities must hold). Since $T^* S_1^* - I = C_1^*$ and $S_2^* T^* - I = C_2^*$ we conclude $T^* \in \mathrm{FRED}(F, E)$ so $N(T^*)$ is finite dimensional and $R(T^*)$ is closed. This proves (a) implies (b).

Conversely, suppose $N(T)$ and $N(T^*)$ are finite dimensional and that $R(T)$ is closed. We decompose:

$$E = N(T) \oplus N(T)^\perp \qquad F = N(T^*) \oplus R(T)$$

where $T \colon N(T)^\perp \to R(T)$ is 1-1 and onto. Consequently, we can find a bounded linear operator S so $ST = I$ on $N(T)^\perp$ and $TS = I$ on $R(T)$. We extend S to be zero on $N(T^*)$ and compute

$$ST - I = \pi_{N(T)} \qquad \text{and} \qquad TS - I = \pi_{N(T^*)}$$

where π denotes orthogonal projection on the indicated subspace. Since these two projections have finite dimensional range, they are compact which proves $T \in \mathrm{FRED}(E, F)$.

If $T \in \mathrm{FRED}(E, F)$ we shall say that T is Fredholm. There is a natural law of composition:

LEMMA 1.4.3.
(a) If $T \in \mathrm{FRED}(E, F)$ then $T^* \in \mathrm{FRED}(F, E)$.
(b) If $T_1 \in \mathrm{FRED}(E, F)$ and $T_2 \in \mathrm{FRED}(F, G)$ then $T_2 T_1 \in \mathrm{FRED}(E, G)$.

PROOF: (a) follows from Lemma 1.4.2. If $S_1 T_1 - I \in C(E)$ and $S_2 T_2 - I \in C(F)$ then $S_1 S_2 T_2 T_1 - I = S_1(S_2 T_2 - I)T_1 + (S_1 T_1 - I) \in C(E)$. Similarly $T_2 T_1 S_1 S_2 - I \in C(E)$.

If $T \in \mathrm{FRED}(E, F)$, then we define:

$$\mathrm{index}(T) = \dim N(T) - \dim N(T^*).$$

We note that $\mathrm{ISO}(E, F)$ is contained in $\mathrm{FRED}(E, F)$ and that $\mathrm{index}(T) = 0$ if $T \in \mathrm{ISO}(E, F)$.

Lemma 1.4.4.
(a) $\operatorname{index}(T) = -\operatorname{index}(T^*)$.
(b) If $T \in \operatorname{FRED}(E,F)$ and $S \in \operatorname{FRED}(F,G)$ then

$$\operatorname{index}(ST) = \operatorname{index}(T) + \operatorname{index}(S).$$

(c) $\operatorname{FRED}(E,F)$ is a open subset of $\operatorname{HOM}(E,F)$.
(d) $\operatorname{index}\colon \operatorname{FRED}(E,F) \to \mathbf{Z}$ is continuous and locally constant.

Proof: (a) is immediate from the definition. We compute

$$N(ST) = N(T) \oplus T^{-1}(R(T) \cap N(S))$$
$$N(T^*S^*) = N(S^*) \oplus (S^*)^{-1}(R(S^*) \cap N(T^*))$$
$$= N(S^*) \oplus (S^*)^{-1}(R(T)^\perp \cap N(S)^\perp)$$

so that

$$\begin{aligned}
\operatorname{index}(ST) &= \dim N(T) + \dim(R(T) \cap N(S)) - \dim N(S^*) \\
&\quad - \dim(R(T)^\perp \cap N(S)^\perp) \\
&= \dim N(T) + \dim(R(T) \cap N(S)) + \dim(R(T)^\perp \cap N(S)) \\
&\quad - \dim N(S^*) - \dim(R(T)^\perp \cap N(S)^\perp) \\
&\quad - \dim(R(T)^\perp \cap N(S)) \\
&= \dim N(T) + \dim N(S) - \dim N(S^*) - \dim(R(T)^\perp) \\
&= \dim N(T) + \dim N(S) - \dim N(S^*) - \dim N(T^*) \\
&= \operatorname{index}(S) + \operatorname{index}(T)
\end{aligned}$$

We prove (c) and (d) as follows. Fix $T \in \operatorname{FRED}(E,F)$. We decompose:

$$E = N(T) \oplus N(T)^\perp \qquad \text{and} \qquad F = N(T^*) \oplus R(T)$$

where $T\colon N(T)^\perp \to R(T)$ is 1-1 onto. We let $\pi_1\colon E \to N(T)$ be orthogonal projection. We define $E_1 = N(T^*) \oplus E$ and $F_1 = N(T) \oplus F$ to be Hilbert spaces by requiring the decompositon to be orthogonal. Let $S \in \operatorname{HOM}(E,F)$, we define $S_1 \in \operatorname{HOM}(E_1, F_1)$ by:

$$S_1(f_0 \oplus e) = \pi_1(e) \oplus (f_0 + S_1(e)).$$

It is clear $|S_1 - S_1'| = |S - S'|$ so the map $S \to S_1$ defines a continuous map from $\operatorname{HOM}(E,F) \to \operatorname{HOM}(E_1, F_1)$. Let $i_1\colon E \to E_1$ be the natural inclusion and let $\pi_2\colon F_1 \to F$ be the natural projection. Since $N(T)$ and $N(T^*)$ are finite dimensional, these are Fredholm maps. It is immediate from the definition that if $S \in \operatorname{HOM}(E,F)$ then

$$S = \pi_2 S_1 i_1.$$

If we let $T = S$ and decompose $e = e_0 \oplus e_1$ and $f = f_0 \oplus f_1$ for $e_0 \in N(T)$ and $f_0 \in N(T^*)$, then:

$$T_1(f_0 \oplus e_0 \oplus e_1) = e_0 \oplus f_0 \oplus Te_1$$

so that $T_1 \in \mathrm{ISO}(E_1, F_1)$. Since $\mathrm{ISO}(E_1, F_1)$ is an open subset, there exists $\varepsilon > 0$ so $|T - S| < \varepsilon$ implies $S_1 \in \mathrm{ISO}(E_1, F_1)$. This implies S_1 is Fredholm so $S = \pi_2 S_1 i_1$ is Fredholm and $\mathrm{FRED}(E, F)$ is open. Furthermore, we can compute $\mathrm{index}(S) = \mathrm{index}(\pi_2) + \mathrm{index}(S_1) + \mathrm{index}(i_1) = \mathrm{index}(\pi_2) + \mathrm{index}(i_1) = \dim N(T) - \dim N(T^*) = \mathrm{index}(T)$ which proves index is locally constant and hence continuous. This completes the proof of the lemma.

We present the following example of an operator with index 1. Let ϕ_n be orthonormal basis for L^2 as $n \in \mathbf{Z}$. Define the one sided shift

$$T\phi_n = \begin{cases} \phi_{n-1} & \text{if } n > 0 \\ 0 & \text{if } n = 0 \\ \phi_n & \text{if } n < 0 \end{cases}$$

then T is surjective so $N(T^*) = \{0\}$. Since $N(T)$ is one dimensional, $\mathrm{index}(T) = 1$. Therefore $\mathrm{index}(T^n) = n$ and $\mathrm{index}((T^*)^n) = -n$. This proves $\mathrm{index} \colon \mathrm{FRED}(E, F) \to \mathbf{Z}$ is surjective. In the next chapter, we will give several examples of differential operators which have non-zero index.

If we specialize to the case $E = F$ then $\mathrm{COM}(E)$ is a closed two-sided ideal of $\mathrm{END}(E)$ so we can pass to the quotient algebra $\mathrm{END}(E)/\mathrm{COM}(E)$. If $\mathrm{GL}(\mathrm{END}(E)/\mathrm{COM}(E))$ denotes the group of invertible elements and if $\pi \colon \mathrm{END}(E) \to \mathrm{END}(E)/\mathrm{COM}(E)$ is the natural projection, then $\mathrm{FRED}(E)$ is π^{-1} of the invertible elements. If C is compact and T Fredholm, $T + tC$ is Fredholm for any t. This implies $\mathrm{index}(T) = \mathrm{index}(T + tC)$ so we can extend $\mathrm{index} \colon \mathrm{GL}(\mathrm{END}(E)/\mathrm{COM}(E)) \to \mathbf{Z}$ as a surjective group homomorphism.

Let $P \colon C^\infty(V) \to C^\infty(W)$ be a elliptic ΨDO of order d. We construct an elliptic ΨDO S of order $-d$ with $S \colon C^\infty(W) \to C^\infty(V)$ so that $SP - I$ and $PS - I$ are infinitely smoothing operators. Then

$$P \colon H_s(V) \to H_{d-s}(W) \qquad \text{and} \qquad S \colon H_{d-s}(W) \to H_s(V)$$

are continuous. Since $SP - I \colon H_s(V) \to H_t(V)$ is continuous for any t, it is compact. Similarly $PS - I$ is a compact operator so both P and S are Fredholm. If $f \in N(P)$, then f is smooth by Lemma 1.3.5. Consequently, $N(P)$ and $N(P^*)$ are independent of the choice of s and $\mathrm{index}(P)$ is invariantly defined. Furthermore, if P_τ is a smooth 1-parameter family of such operators, then $\mathrm{index}(P_\tau)$ is independent of the parameter τ by Lemma 1.4.4. In particular, $\mathrm{index}(P)$ only depends on the homotopy type of the leading symbol of P. In Chapter 3, we will give a topological formula for $\mathrm{index}(P)$ in terms of characteristic classes.

We summarize our conclusions about $\mathrm{index}(P)$ in the following

LEMMA 1.4.5. *Let* $P: C^\infty(V) \to C^\infty(W)$ *be an elliptic* ΨDO *of order* d *over a compact manifold without boundary. Then:*

(a) $N(P)$ *is a finite dimensional subset of* $C^\infty(V)$.

(b) $P: H_s(V) \to H_{s-d}(W)$ *is Fredholm.* P *has closed range.* $\mathrm{index}(P)$ *does not depend on the particular* s *chosen.*

(c) $\mathrm{index}(P)$ *only depends on the homotopy type of the leading symbol of* P *within the class of elliptic* ΨDO's *of order* d.

1.5. Elliptic Complexes,
The Hodge Decomposition Theorem,
And Poincaré Duality.

Let V be a graded vector bundle. V is a collection of vector bundles $\{V_j\}_{j \in \mathbf{Z}}$ such that $V_j \neq \{0\}$ for only a finite number of indices j. We let P be a graded ΨDO of order d. P is a collection of dth order pseudo-differential operators $P_j: C^\infty(V_j) \to C^\infty(V_{j+1})$. We say that (P, V) is a complex if $P_{j+1} P_j = 0$ and $\sigma_L P_{j+1} \sigma_L P_j = 0$ (the condition on the symbol follows from $P^2 = 0$ for differential operators). We say that (P, V) is elliptic if:

$$N(\sigma_L P_j)(x, \xi) = R(\sigma_L P_{j-1})(x, \xi) \qquad \text{for } \xi \neq 0$$

or equivalently if the complex is exact on the symbol level.

We define the cohomology of this complex by:

$$H^j(V, P) = N(P_j) / R(P_{j-1}).$$

We shall show later in this section that $H^j(V, P)$ is finite dimensional if (P, V) is an elliptic complex. We then define

$$\text{index}(P) = \sum_j (-1)^j \dim H^j(V, P)$$

as the Euler characteristic of this elliptic complex.

Choose a fixed Hermitian inner product on the fibers of V. We use that inner product together with the Riemannian metric on M to define $L^2(V)$. We define adjoints with respect to this structure. If (P, V) is an elliptic complex, we construct the associated self-adjoint Laplacian:

$$\Delta_j = (P^* P)_j = P_j^* P_j + P_{j-1} P_{j-1}^* : C^\infty(V_j) \to C^\infty(V_j).$$

If $p_j = \sigma_L(P_j)$, then $\sigma_L(\Delta_j) = p_j^* p_j + p_{j-1} p_{j-1}^*$. We can also express the condition of ellipticity in terms of Δ_j:

LEMMA 1.5.1. *Let (P, V) be a dth order partial differential complex. Then (P, V) is elliptic if and only if Δ_j is an elliptic operator of order $2d$ for all j.*

PROOF: We suppose that (P, V) is elliptic; we must check $\sigma_L(\Delta_j)$ is non-singular for $\xi \neq 0$. Suppose $(p_j^* p_j + p_{j-1} p_{j-1}^*)(x, \xi) v = 0$. If we dot this equation with v, we see $p_j v \cdot p_j v + p_{j-1}^* v \cdot p_{j-1}^* v = 0$ so that $p_j v = p_{j-1}^* v = 0$. Thus $v \in N(p_j)$ so $v \in R(p_{j-1})$ so we can write $v = p_{j-1} w$. Since $p_{j-1}^* p_{j-1} w = 0$, we dot this equation with w to see $p_{j-1} w \cdot p_{j-1} w = 0$ so $v = p_{j-1} w = 0$ which proves Δ_j is elliptic. Conversely, let $\sigma_L(\Delta_j)$ be non-singular for $\xi \neq 0$. Since (P, V) is a complex, $R(p_{j-1})$ is a subset of $N(p_j)$.

Conversely, let $v \in N(p_j)$. Since $\sigma_L(\Delta_j)$ is non-singular, we can express $v = (p_j^* p_j + p_{j-1} p_{j-1}^*) w$. We apply p_j to conclude $p_j p_j^* p_j w = 0$. We dot this equation with $p_j w$ to conclude $p_j p_j^* p_j w \cdot p_j w = p_j^* p_j w \cdot p_j^* p_j w = 0$ so $p_j^* p_j w = 0$. This implies $v = p_{j-1} p_{j-1}^* w \in R(p_{j-1})$ which completes the proof.

We can now prove the following:

THEOREM 1.5.2 (HODGE DECOMPOSITION THEOREM). *Let (P,V) be a dth order ΨDO elliptic complex. Then*
(a) We can decompose $L^2(V_j) = N(\Delta_j) \oplus R(P_{j-1}) \oplus R(P_j^*)$ as an orthogonal direct sum.
(b) $N(\Delta_j)$ is a finite dimensional vector space and there is a natural isomorphism of $H^j(P,V) \simeq N(\Delta_j)$. The elements of $N(\Delta_j)$ are smooth sections to V_j.

PROOF: We regard $\Delta_j: H_{2d}(V_j) \to L^2(V_j)$. Since this is elliptic, it is Fredholm. This proves $N(\Delta_j)$ is finite dimensional. Since Δ_j is hypoelliptic, $N(\Delta_j)$ consists of smooth sections to V. Since Δ_j is self adjoint and Fredholm, $R(\Delta_j)$ is closed so we may decompose $L^2(V_j) = N(\Delta_j) \oplus R(\Delta_j)$. It is clear $R(\Delta_j)$ is contained in the span of $R(P_{j-1})$ and $R(P_j^*)$. We compute the L^2 inner product:

$$(P_j^* f, P_{j-1} g) = (f, P_j P_{j-1} g) = 0$$

since $P_j P_{j-1} = 0$. This implies $R(P_{j-1})$ and $R(P_j^*)$ are orthogonal. Let $f \in N(\Delta_j)$, we take the L^2 inner product with f to conclude

$$0 = (\Delta_j f, f) = (P_j f, P_j f) + (P_{j-1}^* f, P_{j-1}^* f)$$

so $N(\Delta_j) = N(P_j) \cap N(P_{j-1}^*)$. This implies $R(\Delta_j)$ contains the span of $R(P_{j-1})$ and $R(P_j^*)$. Since these two subspaces are orthogonal and $R(\Delta_j)$ is closed, $R(P_{j-1})$ and $R(P_j^*)$ are both closed and we have an orthogonal direct sum:
$$L^2(V_j) = N(\Delta_j) \oplus R(P_{j-1}) \oplus R(P_j^*).$$

This proves (a).

The natural inclusion of $N(\Delta_j)$ into $N(P_j)$ defines a natural map of $N(\Delta_j) \to H^j(P,V) = N(P_j)/R(P_{j-1})$. Since $R(P_{j-1})$ is orthogonal to $N(\Delta_j)$, this map is injective. If $f \in C^\infty(V_j)$ and $P_j f = 0$, we can decompose $f = f_0 \oplus \Delta_j f_1$ for $f_0 \in N(\Delta_j)$. Since f and f_0 are smooth, $\Delta_j f_1$ is smooth so $f_1 \in C^\infty(V_j)$. $P_j f = P_j f_0 + P_j \Delta_j f_1 = 0$ implies $P_j \Delta_j f_1 = 0$. We dot this equation with $P_j f_1$ to conclude:

$$0 = (P_j f_1, P_j P_j^* P_j f_1 + P_j P_{j-1} P_{j-1}^* f_1) = (P_j^* P_j f_1, P_j^* P_j f_1)$$

so $P_j^* P_j f_1 = 0$ so $\Delta_j f_1 = P_{j-1} P_{j-1}^* f_1 \in R(P_{j-1})$. This implies f and f_0 represent the same element of $H^j(P,V)$ so the map $N(\Delta_j) \to H^j(P,V)$ is surjective. This completes the proof.

To illustrate these concepts, we discuss briefly the de Rham complex. Let T^*M be the cotangent space of M. The exterior algebra $\Lambda(T^*M)$ is the universal algebra generated by T^*M subject to the relation $\xi \wedge \xi = 0$ for $\xi \in T^*M$. If $\{e_1, \ldots, e_m\}$ is a basis for T^*M and if $I = \{1 \le i_1 < i_2 < \cdots < i_p \le m\}$, we define $e_I = e_{i_1} \wedge \cdots \wedge e_{i_p}$. The $\{e_I\}$ form a basis for $\Lambda(T^*M)$ which has dimension 2^m. If we define $|I| = p$, then $\Lambda^p(T^*M)$ is the span of the $\{e_I\}_{|I|=p}$; this is the bundle of p-forms. A section of $C^\infty(\Lambda^p T^*M)$ is said to be a smooth p-form over M.

Let $x = (x_1, \ldots, x_m)$ be local coordinates on M and let $\{dx_1, \ldots, dx_m\}$ be the corresponding frame for T^*M. If $f \in C^\infty(M) = C^\infty(\Lambda^0(T^*M))$, we define:

$$df = \sum_k \frac{\partial f}{\partial x_k} dx_k.$$

If $y = (y_1, \ldots, y_m)$ is another system of local coordinates on M, the identity:

$$dy_j = \sum_k \frac{\partial y_j}{\partial x_k} dx_k$$

means that d is well defined and is independent of the coordinate system chosen. More generally, we define $d(f dx_I) = df \wedge dx_I$ so that, for example,

$$d\left(\sum_j f_j dx^j\right) = \sum_{j<k} \left\{\frac{\partial f_k}{\partial x_j} - \frac{\partial f_j}{\partial x_k}\right\} dx_j \wedge dx_k.$$

Again this is well defined and independent of the coordinate system. Since mixed partial derivatives commute, $d^2 = 0$ so

$$d: C^\infty(\Lambda^p(T^*M)) \to C^\infty(\Lambda^{p+1}(T^*M))$$

forms a complex.

Let $\xi \in T^*M$ and let $\mathrm{ext}(\xi): \Lambda^p(T^*M) \to \Lambda^{p+1}(T^*M)$ be defined by exterior multiplication, i.e., $\mathrm{ext}(\xi)\omega = \xi \wedge \omega$. If we decompose $\xi = \sum \xi_j dx_j$ relative to a local coordinate frame, then $df = \sum_j \partial f/\partial x_j dx_j$ implies $\sigma(d) = i \, \mathrm{ext}(\xi)$; the symbol of exterior differentiation is exterior multiplication up to a factor of i. Fix $\xi \ne 0$ and choose a basis $\{e_1, \ldots, e_m\}$ for T^*M such that $\xi = e_1$. Then

$$\mathrm{ext}(\xi)e_I = \begin{cases} 0 & \text{if } i_1 = 1 \\ e_J \text{ for } J = \{1, i_1, \ldots, i_p\} & \text{if } i_1 > 1. \end{cases}$$

From this it is clear that $N(\mathrm{ext}(\xi)) = R(\mathrm{ext}(\xi))$ so the de Rham complex is an elliptic complex.

A Riemannian metric on M defines fiber metrics on $\Lambda^p(T^*M)$. If $\{e_i\}$ is an orthonormal local frame for TM we take the dual frame $\{e_i^*\}$ for T^*M. For notational simplicity, we will simply denote this frame again by $\{e_i\}$.

The corresponding $\{e_I\}$ define an orthonormal local frame for $\Lambda(T^*M)$. We define interior multiplication $\text{int}(\xi) \colon \Lambda^p(T^*M) \to \Lambda^{p-1}(T^*M)$ to be the dual of exterior multiplication. Then:

$$\text{int}(e_1)e_I = \begin{cases} e_j & \text{for } J = \{i_2, \ldots, i_p\} & \text{if } i_1 = 1 \\ 0 & & \text{if } i_1 > 1 \end{cases}$$

so

$$\text{int}(e_1)\text{ext}(e_1) + \text{ext}(e_1)\text{int}(e_1) = I.$$

If $|\xi|^2$ denotes the length of the covector ξ, then more generally:

$$(i\,\text{ext}(\xi) - i\,\text{int}(\xi))^2 = |\xi|^2 I.$$

If $\delta \colon C^\infty(\Lambda^p(T^*M)) \to C^\infty(\Lambda^{p-1}(T^*M))$ is the adjoint of d, then $\sigma_L\delta = i\,\text{int}(\xi)$. We let $\delta d + d\delta = (d+\delta)^2 = \Delta$. $\sigma_L\Delta = |\xi|^2$ is elliptic. The de Rham theorem gives a natural isomorphism from the cohomology of M to $H^*(\Lambda, d)$:

$$H^p(M; \mathbf{C}) = N(d_p)/R(d_{p-1})$$

where we take closed modulo exact forms. The Hodge decomposition theorem implies these groups are naturally isomorphic to the harmonic p-forms $N(\Delta_p)$ which are finite dimensional. The Euler-Poincaré characteristic $\chi(M)$ is given by:

$$\chi(M) = \sum(-1)^p \dim H^p(M; \mathbf{C}) = \sum(-1)^p \dim N(\Delta_p) = \text{index}(d)$$

is therefore the index of an elliptic complex.

If M is oriented, we let dvol be the oriented volume element. The Hodge $*$ operator $* \colon \Lambda^p(T^*M) \to \Lambda^{m-p}(T^*M)$ is defined by the identity:

$$\omega \wedge *\omega = (\omega \cdot \omega)\,\text{dvol}$$

where "$\cdot$" denotes the inner product defined by the metric. If $\{e_i\}$ is an oriented local frame for T^*M, then $\text{dvol} = e_1 \wedge \cdots \wedge e_m$ and

$$*(e_1 \wedge \cdots \wedge e_p) = e_{p+1} \wedge \cdots \wedge e_m.$$

The following identities are immediate consequences of Stoke's theorem:

$$** = (-1)^{p(m-p)} \quad \text{and} \quad \delta = (-1)^{mp+m+1} * d *.$$

Since $\Delta = (d+\delta)^2 = d\delta + \delta d$ we compute $*\Delta = \Delta*$ so $*: N(\Delta_p) \to N(\Delta_{m-p})$ is an isomorphism. We may regard $*$ as an isomorphism $*: H^p(M; \mathbf{C}) \to H^{m-p}(M; \mathbf{C})$; in this description it is Poincaré duality.

The exterior algebra is *not* very suited to computations owing to the large number of signs which enter in the discussion of $*$. When we discuss the signature and spin complexes in the third chapter, we will introduce Clifford algebras which make the discussion of Poincaré duality much easier.

It is possible to "roll up" the de Rham complex and define:

$$(d + \delta)_e: C^\infty(\Lambda^e(T^*M)) \to C^\infty(\Lambda^o(T^*M))$$

where

$$\Lambda^e(T^*M) = \bigoplus_{2k} \Lambda^{2k}(T^*M) \qquad \text{and} \qquad \Lambda^o(T^*M) = \bigoplus_{2k+1} \Lambda^{2k+1}(T^*M)$$

denote the differential forms of even and odd degrees. $(d + \delta)$ is an elliptic operator since $(d + \delta)_e^*(d + \delta)_e = \Delta$ is elliptic since $\dim \Lambda^e = \dim \Lambda^o$. (In this representation $(d + \delta)_e$ is not self-adjoint since the range and domain are distinct, $(d + \delta)_e^* = (d + \delta)_o$). It is clear $\mathrm{index}(d + \delta)_e = \dim N(\Delta^e) = \dim N(\Delta^o) = \chi(M)$. We can always "roll up" any elliptic complex to form an elliptic complex of the same index with two terms. Of course, the original elliptic complex does not depend upon the choice of a fiber metric to define adjoints so there is some advantage in working with the full complex occasionally as we shall see later.

We note finally that if m is even, we can always find a manifold with $\chi(M)$ arbitrary so there exist lots of elliptic operators with non-zero index. We shall see that $\mathrm{index}(P) = 0$ if m is odd (and one must consider pseudo-differential operators to get a non-zero index in that case).

We summarize these computations for the de Rham complex in the following:

LEMMA 1.5.3. *Let $\Lambda(M) = \Lambda(T^*M)$ be the complete exterior algebra.*
(a) $d: C^\infty(\Lambda^p(T^*M)) \to C^\infty(\Lambda^{p+1}(T^*M))$ is an elliptic complex. The symbol is $\sigma_L(d)(x, \xi) = i \, \mathrm{ext}(\xi)$.
(b) If δ is the adjoint, then $\sigma_L(\delta)(x, \xi) = -i \, \mathrm{int}(\xi)$.
(c) If $\Delta_p = (d\delta + \delta d)_p$ is the associated Laplacian, then $N(\Delta_p)$ is finite dimensional and there are natural identifications:

$$N(\Delta_p) \simeq N(d_p) / R(d_{p-1}) \simeq H^p(M; \mathbf{C}).$$

(d) $\mathrm{index}(d) = \chi(M)$ is the Euler-Poincaré characteristic of M.
(e) If M is oriented, we let $*$ be the Hodge operator. Then

$$** = (-1)^{p(m-p)} \qquad \text{and} \qquad \delta = (-1)^{mp+m+1} * d * .$$

Furthermore, $*: N(\Delta_p) \simeq N(\Delta_{m-p})$ gives Poincaré duality.

In the next chapter, we will discuss the Gauss-Bonnet theorem which gives a formula for $\mathrm{index}(d) = \chi(M)$ in terms of curvature.

1.6. The Heat Equation.

Before proceeding with our discussion of the index of an elliptic operator, we must discuss spectral theory. We restrict ourselves to the context of a compact self-adjoint operator to avoid unnecessary technical details. Let $T \in \text{COM}(H)$ be a self-adjoint compact operator on the Hilbert space H. Let

$$\text{spec}(T) = \{\, \lambda \in \mathbf{C} : (T - \lambda) \notin \text{GL}(H) \,\}.$$

Since $\text{GL}(H)$ is open, $\text{spec}(T)$ is a closed subset of C. If $|\lambda| > |T|$, the series

$$g(\lambda) = \sum_{n=0}^{\infty} T^n / \lambda^{n+1}$$

converges to define an element of $\text{END}(H)$. As $(T-\lambda)g(\lambda) = g(\lambda)(T-\lambda) = -I$, $\lambda \notin \text{spec}(T)$. This shows $\text{spec}(T)$ is bounded.

Since T is self-adjoint, $\text{N}(T - \bar{\lambda}) = \{0\}$ if $\lambda \notin \mathbf{R}$. This implies $\text{R}(T - \lambda)$ is dense in H. Since $Tx \cdot x$ is always real, $|(T - \lambda)x \cdot x| \geq \text{im}(\lambda)\,|x|^2$ so that $|(T - \lambda)x| \geq \text{im}(\lambda)|\,x|$. This implies $\text{R}(T - \lambda)$ is closed so $T - \lambda$ is surjective. Since $T - \lambda$ is injective, $\lambda \notin \text{spec}(T)$ if $\lambda \in \mathbf{C} - \mathbf{R}$ so $\text{spec}(T)$ is a subset of $\mathbf{R}$. If $\lambda \in [-|T|, |T|]$, we define $E(\lambda) = \{\, x \in H : Tx = \lambda x \,\}$.

LEMMA 1.6.1. *Let $T \in \text{COM}(H)$ be self-adjoint. Then*

$$\dim\{E(-|T|) \oplus E(|T|)\} > 0.$$

PROOF: If $|T| = 0$ then $T = 0$ and the result is clear. Otherwise choose $|x_n| = 1$ so that $|Tx_n| \to |T|$. We choose a subsequence so $Tx_n \to y$. Let $\lambda = |T|$, we compute:

$$|T^2 x_n - \lambda^2 x_n|^2 = |T^2 x_n|^2 + |\lambda^2 x_n|^2 - 2\lambda^2 T^2 x_n \cdot x_n$$
$$\leq 2\lambda^4 - 2\lambda^2 |Tx_n|^2 \to 0.$$

Since $Tx_n \to y$, $T^2 x_n \to Ty$. Thus $\lambda^2 x_n \to Ty$. Since $\lambda^2 \neq 0$, this implies $x_n \to x$ for $x = Ty/\lambda^2$. Furthermore $|T^2 x - \lambda^2 x| = 0$. Since $(T^2 - \lambda^2) = (T - \lambda)(T + \lambda)$, either $(T + \lambda)x = 0$ so $x \in E(-\lambda) \neq \{0\}$ or $(T - \lambda)(y_1) = 0$ for $y_1 = (T + \lambda)x \neq 0$ so $E(\lambda) \neq \{0\}$. This completes the proof.

If $\lambda \neq 0$, the equation $Tx = \lambda x$ implies the unit disk in $E(\lambda)$ is compact and hence $E(\lambda)$ is finite dimensional. $E(\lambda)$ is T-invariant. Since T is self-adjoint, the orthogonal complement $E(\lambda)$ is also T-invariant. We take an orthogonal decomposition $H = E(|T|) \oplus E(-|T|) \oplus H_1$. T respects this decomposition; we let T_1 be the restriction of T to H_1. Clearly $|T_1| \leq |T|$. If we have equality, then Lemma 1.6.1 implies there exists $x \neq 0$ in H_1

so $T_1 x = \pm |T| x$, which would be false. Thus $|T_1| < |T|$. We proceed inductively to decompose:

$$H = E(\lambda_1) \oplus E(-\lambda_1) \oplus \cdots \oplus E(\lambda_n) \oplus E(-\lambda_n) \oplus H_n$$

where $\lambda_1 > \lambda_2 > \cdots > \lambda_n$ and where $|T_n| < \lambda_n$. We also assume $E(\lambda_n) \oplus E(-\lambda_n) \neq \{0\}$. We suppose that the λ_n do not converge to zero but converge to ε positive. For each n, we choose x_n so $|x_n| = 1$ and $T x_n = \pm \lambda_n x_n$. Since T is self-adjoint, this is an orthogonal decomposition; $|x_j - x_k| = \sqrt{2}$. Since T is compact, we can choose a convergent subsequence $\lambda_n x_n \to y$. Since $\lambda_n \to \varepsilon$ positive, this implies $x_n \to x$. This is impossible so therefore the $\lambda_n \to 0$. We define $H_0 = \bigcap_n H_n$ as a closed subset of H. Since $|T| < \lambda_n$ for all n on H_0, $|T| = 0$ so $T = 0$ and $H_0 = E(0)$. This defines a direct sum decomposition of the form:

$$H = \bigoplus_k E(\mu_k) \oplus E(0)$$

where the $\mu_k \in \mathbf{R}$ are the non-zero subspace of the $E(\lambda_n)$ and $E(-\lambda_n)$. We construct a complete orthonormal system $\{\phi_n\}_{n=1}^{\infty}$ for H with either $\phi_n \in E(0)$ or $\phi_n \in E(\mu_k)$ for some k. Then $T\phi_n = \lambda_n \phi_n$ so ϕ_n is an eigenvector of T. This proves the spectral decomposition for self-adjoint compact operators:

LEMMA 1.6.2. Let $T \in \mathrm{COM}(H)$ be self-adjoint. We can find a complete orthonormal system for H consisting of eigenvectors of T.

We remark that this need not be true if T is self-adjoint but not compact or if T is compact but not self-adjoint.

We can use this lemma to prove the following:

LEMMA 1.6.3. Let $P: C^{\infty}(V) \to C^{\infty}(V)$ be an elliptic self-adjoint ΨDO of order $d > 0$.
(a) We can find a complete orthonormal basis $\{\phi_n\}_{n=1}^{\infty}$ for $L^2(V)$ of eigenvectors of P. $P\phi_n = \lambda_n \phi_n$.
(b) The eigenvectors ϕ_n are smooth and $\lim_{n \to \infty} |\lambda_n| = \infty$.
(c) If we order the eigenvalues $|\lambda_1| \leq |\lambda_2| \leq \cdots$ then there exists a constant $C > 0$ and an exponent $\delta > 0$ such that $|\lambda_n| \geq C n^{\delta}$ if $n > n_0$ is large.
Remark: The estimate (c) can be improved to show $|\lambda_n| \sim n^{d/m}$ but the weaker estimate will suffice and is easier to prove.

PROOF: $P: H_d(V) \to L^2(V)$ is Fredholm. $P: \mathrm{N}(P)^{\perp} \cap H_d(M) \to \mathrm{N}(P)^{\perp} \cap L^2(V)$ is 1-1 and onto; by Gårding's inequality $P\phi \in L^2$ implies $\phi \in H_d$. We let S denote the inverse of this map and extend S to be zero on the finite dimensional space $\mathrm{N}(P)$. Since the inclusion of $H_d(M)$ into $L^2(V)$ is compact, S is a compact self-adjoint operator. S is often referred to as the Greens operator. We find a complete orthonormal system $\{\phi_n\}$ of eigenvectors of S. If $S\phi_n = 0$ then $P\phi_n = 0$ since $N(S) = N(P)$. If $S\phi_n = \mu_n\phi_n \neq 0$ then $P\phi_n = \lambda_n\phi_n$ for $\lambda_n = \mu_n^{-1}$. Since the $\mu_n \to 0$, the $|\lambda_n| \to \infty$. If k is an integer such that $dk > 1$, then $P^k - \lambda_n^k$ is elliptic. Since $(P^k - \lambda_n^k)\phi_n = 0$, this implies $\phi_n \in C^\infty(V)$. This completes the proof of (a) and (b).

By replacing P with P^k we replace λ_n by λ_n^k. Since $kd > \frac{m}{2}$ if k is large, we may assume without loss of generality that $d > \frac{m}{2}$ in the proof of (c). We define:

$$|f|_{\infty,0} = \sup_{x \in M} |f(x)| \qquad \text{for } f \in C^\infty(V).$$

We estimate:

$$|f|_{\infty,0} \leq C|f|_d \leq C(|Pf|_0 + |f|_0).$$

Let $F(a)$ be the space spanned by the ϕ_j where $|\lambda_j| \leq a$ and let $n = \dim F(a)$. We estimate $n = n(a)$ as follows. We have

$$|f|_{\infty,0} \leq C(1+a)|f|_0 \qquad \text{on } F(a).$$

Suppose first $V = M \times \mathbf{C}$ is the trivial line bundle. Let c_j be complex constants, then this estimate shows:

$$\left|\sum_{j=1}^{n} c_j\phi_j(x)\right| \leq C(1+a)\left\{\sum_{j=1}^{n} |c_j|^2\right\}^{1/2}.$$

If we take $c_j = \bar{\phi}_j(x)$ then this yields the estimate:

$$\sum_{j=1}^{n} \phi_j(x)\bar{\phi}_j(x) \leq C(1+a)\left\{\sum_{j=1}^{n} \phi_j(x)\bar{\phi}_j(x)\right\}^{1/2},$$

i.e.,

$$\sum_{j=1}^{n} \phi_j(x)\bar{\phi}_j(x) \leq C^2(1+a)^2.$$

We integrate this estimate over M to conclude

$$n \leq C^2(1+a)^2 \operatorname{vol}(M)$$

or equivalently

$$C_1(n - C_2)^{1/2} \leq a = |\lambda_n|$$

from which the desired estimate follows.

If $\dim V = k$, we choose a local orthonormal frame for V to decompose ϕ_j into components ϕ_j^u for $1 \leq u \leq k$. We estimate:

$$\left| \sum_{j=1}^{n} c_j^u \phi_j^u(x) \right| \leq C(1+a) \left\{ \sum_{j=1}^{n} |c_j^u|^2 \right\}^{1/2} \qquad \text{for } u = 1, \ldots, k.$$

If we let $c_j^u = \bar{\phi}_j^u(x)$, then summing over u yields:

$$\left| \sum_{j=1}^{n} \phi_j \bar{\phi}_j(x) \right| \leq kC(1+a) \left| \sum_{j=1}^{n} \phi_j \bar{\phi}_j(x) \right|^{1/2}$$

since each term on the left hand side of the previous inequality can be estimated seperately by the right hand side of this inequality. This means that the constants which arise in the estimate of (c) depend on k, but this causes no difficulty. This completes the proof of (c).

The argument given above for (c) was shown to us by Prof. B. Allard (Duke University) and it is a clever argument to avoid the use of the Schwarz kernel theorem.

Let $P : C^\infty(V) \to C^\infty(V)$ be an elliptic ΨDO of order $d > 0$ which is self-adjoint. We say P has positive definite leading symbol if there exists $p(x, \xi) : T^*M \to \text{END}(V)$ such that $p(x, \xi)$ is a positive definite Hermitian matrix for $\xi \neq 0$ and such that $\sigma P - p \in S^{d-1}$ in any coordinate system. The spectrum of such a P is not necessarily non-negative, but it is bounded from below as we shall show. We construct Q_0 with leading symbol $\sqrt{p}$ and let $Q = Q_0^* Q_0$. Then by hypothesis $P - Q \in S^{d-1}$. We compute:

$$(Pf, f) = ((P - Q)f, f) + (Qf, f)$$
$$|((P - Q)f, f)| \leq C|f|_{d/2} |(P - Q)f|_{d/2} \leq C|f|_{d/2} |f|_{d/2-1}$$
$$(Qf, f) = (Q_0 f, Q_0 f) \qquad \text{and} \qquad |f|_{d/2}^2 < C|f|_0^2 + C|Q_0 f|_0^2.$$

We use this to estimate for any $\varepsilon > 0$ that:

$$|((P - Q)f, f)| \leq C|f|_{d/2}(|f|_{d/2-1}) \leq \varepsilon|f|_{d/2}^2 + C(\varepsilon)|f|_{d/2}|f|_0$$
$$\leq 2\varepsilon|f|_{d/2}^2 + C(\varepsilon)|f|_0^2$$
$$\leq 2C\varepsilon|Q_0 f|_0^2 + C(\varepsilon)|f|_0^2.$$

Choose ε so $2C\varepsilon \leq 1$ and estimate:

$$(Pf, f) \geq (Q_0 f, Q_0 f) - |((P - Q)f, f)| \geq -C(\varepsilon)|f|_0^2.$$

This implies:

LEMMA 1.6.4. *Let* $P: C^\infty(V) \to C^\infty(V)$ *be an elliptic ΨDO of order* $d > 0$ *which is self-adjoint with positive definite leading symbol. Then* $\mathrm{spec}(P)$ *is contained in* $[-C, \infty)$ *for some constant* C.

We fix such a P henceforth. The heat equation is the partial differential equation:

$$\left(\frac{d}{dt} + P\right) f(x, t) = 0 \qquad \text{for } t \geq 0 \text{ with } f(x, 0) = f(x).$$

At least formally, this has the solution $f(x, t) = e^{-tP} f(x)$. We decompose $f(x) = \sum c_n \phi_n(x)$ for $c_n = (f, \phi_n)$ in a generalized Fourier series. The solution of the heat equation is given by:

$$f(x, t) = \sum_n e^{-t\lambda_n} c_n \phi_n(x).$$

Proceeding formally, we define:

$$K(t, x, y) = \sum_n e^{-t\lambda_n} \phi_n(x) \otimes \bar{\phi}_n(y) : V_y \to V_x$$

so that:

$$e^{-tP} f(x) = \int_M K(t, x, y) f(y) \ \mathrm{dvol}(y)$$

$$= \sum_n e^{-t\lambda_n} \phi_n(x) \int_M f(y) \cdot \phi_n(y) \ \mathrm{dvol}(y).$$

We regard $K(t, x, y)$ as an endomorphism from the fiber of V over y to the fiber of V over x.

We justify this purely formal procedure using Lemma 1.3.4. We estimate that:

$$|\phi_n|_{\infty,k} \leq C(|\phi_n|_0 + |P^j \phi|_0) = C(1 + |\lambda_n|^j) \qquad \text{where } jd > k + \tfrac{m}{2}.$$

Only a finite number of eigenvalues of P are negative by Lemma 1.6.4. These will not affect convergence questions, so we may assume $\lambda > 0$. Estimate:

$$e^{-t\lambda} \lambda^j \leq t^{-j} C(j) e^{-t\lambda/2}$$

to compute:

$$|K(t, x, y)|_{\infty,k} \leq t^{-j(k)} C(k) \sum_n e^{-t\lambda_n/2}.$$

Since $|\lambda| \geq Cn^\delta$ for $\delta > 0$ and n large, the series can be bounded by

$$\sum_{n>0} e^{-tn^{\delta/2}}$$

which converges. This shows $K(t, x, y)$ is an infinitely smooth function of (t, x, y) for $t > 0$ and justifies all the formal procedures involved. We compute

$$\mathrm{Tr}_{L^2} e^{-tP} = \sum e^{-t\lambda_n} = \int_M \mathrm{Tr}_{V_x} K(t, x, x) \, \mathrm{dvol}(x).$$

We can use this formula to compute the index of Q:

LEMMA 1.6.5. Let $Q: C^\infty(V) \to C^\infty(W)$ be an elliptic ΨDO of order $d > 0$. Then for $t > 0$, e^{-tQ^*Q} and e^{-tQQ^*} are in $\Psi_{-\infty}$ with smooth kernel functions and

$$\mathrm{index}(Q) = \mathrm{Tr}\, e^{-tQ^*Q} - \mathrm{Tr}\, e^{-tQQ^*} \qquad \text{for any } t > 0.$$

PROOF: Since e^{-tQ^*Q} and e^{-tQQ^*} have smooth kernel functions, they are in $\Psi_{-\infty}$ so we must only prove the identity on $\mathrm{index}(Q)$. We define $E_0(\lambda) = \{\phi \in L^2(V) : Q^*Q\phi = \lambda\phi\}$ and $E_1(\lambda) = \{\phi \in L^2(W) : QQ^*\phi = \lambda\phi\}$. These are finite dimensional subspaces of smooth sections to V and W. Because $Q(Q^*Q) = (QQ^*)Q$ and $Q^*(QQ^*) = (Q^*Q)Q^*$, Q and Q^* define maps:

$$Q: E_0(\lambda) \to E_1(\lambda) \qquad \text{and} \qquad Q^*: E_1(\lambda) \to E_0(\lambda).$$

If $\lambda \neq 0$, $\lambda = Q^*Q: E_0(\lambda) \to E_1(\lambda) \to E_0(\lambda)$ is an isomorphism so $\dim E_0(\lambda) = \dim E_1(\lambda)$. We compute:

$$\mathrm{Tr}\, e^{-tQ^*Q} - \mathrm{Tr}\, e^{-tQQ^*} = \sum_\lambda e^{-t\lambda}\{\dim E_0(\lambda) - \dim E_1(\lambda)\}$$
$$= e^{-t0}\{\dim E_0(0) - \dim E_1(0)\}$$
$$= \mathrm{index}(Q)$$

which completes the proof.

If (Q, V) is a elliptic complex, we use the same reasoning to conclude $e^{-t\Delta_i}$ is in $\Psi_{-\infty}$ with a smooth kernel function. We define $E_i(\lambda) = \{\phi \in L^2(V_i) : \Delta_i\phi = \lambda\phi\}$. Then $Q_i: E_i(\lambda) \to E_{i+1}(\lambda)$ defines an acyclic complex (i.e., $\mathrm{N}(Q_i) = \mathrm{R}(Q_{i-1})$) if $\lambda \neq 0$ so that $\sum(-1)^i \dim E_i(\lambda) = 0$ for $\lambda \neq 0$. Therefore

$$\mathrm{index}(Q) = \sum_i (-1)^i \mathrm{Tr}(e^{-t\Delta_i})$$

and Lemma 1.6.5 generalizes to the case of elliptic complexes which are not just two term complexes.

Let P be an elliptic ΨDO of order $d > 0$ which is self-adjoint with positive definite leading symbol. Then $\mathrm{spec}(P)$ is contained in $[-C, \infty)$ for some constant C. For $\lambda \notin \mathrm{spec}(P)$, $(P - \lambda)^{-1} \in \mathrm{END}(L^2(V))$ satisfies:

$$|(P - \lambda)^{-1}| = \mathrm{dist}(\lambda, \mathrm{spec}(P))^{-1} = \inf_n |\lambda - \lambda_n|^{-1}.$$

The function $|(P-\lambda)^{-1}|$ is a continuous function of $\lambda \in \mathbf{C} - \mathrm{spec}(P)$. We use the Riemann integral with values in the Banach space $\mathrm{END}(L^2(V))$ to define:

$$e^{-tP} = \frac{1}{2\pi i} \int_\gamma e^{-t\lambda}(P-\lambda)^{-1}\, d\lambda$$

where γ is the path about $[-C, \infty)$ given by the union of the appropriate pieces of the straight lines $\mathrm{Re}(\lambda+C+1) = \pm\,\mathrm{Im}(\lambda)$ pictured below. We let $\mathcal{R}$ be the closed region of $\mathbf{C}$ consisting of γ together with that component of $\mathbf{C} - \gamma$ which does not contain $[-C, \infty)$.

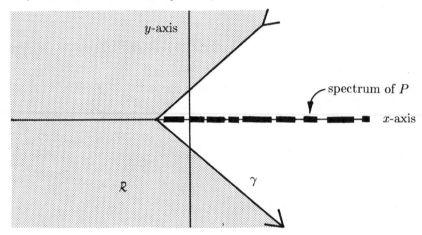

We wish to extend $(P-\lambda)^{-1}$ to H_s. We note that

$$|\lambda - \mu|^{-1} \le C \qquad \text{for } \lambda \in \mathcal{R},\ \mu \in \mathrm{spec}(P)$$

so that $|(P-\lambda)^{-1}f|_0 < C|f|_0$. We use Lemma 1.3.5 to estimate:

$$
\begin{aligned}
|(P-\lambda)^{-1}f|_{kd} &\le C\{|P^k(P-\lambda)^{-1}f|_0 + |(P-\lambda)^{-1}f|_0\} \\
&\le C\{|P^{k-1}f|_0 + |\lambda P^{k-1}(P-\lambda)^{-1}f|_0 + |f|_0\} \\
&\le C\{|f|_{kd-d} + |\lambda|\,|(P-\lambda)^{-1}f|_{kd-d}\}.
\end{aligned}
$$

If $k = 1$, this implies $|(P-\lambda)^{-1}f|_d \le C(1+|\lambda|)|f|_0$. We now argue by induction to estimate:

$$|(P-\lambda)^{-1}f|_{kd} \le C(1+|\lambda|)^{k-1}|f|_{kd-d}.$$

We now interpolate. If $s > 0$, choose k so $kd \ge s > kd - d$ and estimate:

$$
\begin{aligned}
|(P-\lambda)^{-1}f|_s &\le C|(P-\lambda)^{-1}f|_{kd} \le C(1+|\lambda|)^{k-1}|f|_{kd-d} \\
&\le C(1+|\lambda|)^{k-1}|f|_s.
\end{aligned}
$$

Similarly, if $s < 0$, we use duality to estimate:

$$|((P-\lambda)^{-1}f, g)| = |(f, (P-\bar{\lambda})^{-1}g)| \le C(1+|\lambda|)^{k-1}|f|_s|g|_{-s}$$

which by Lemma 1.3.4 shows

$$|(P-\lambda)^{-1}f|_s < C(1+|\lambda|)^{k-1}|f|_s$$

in this case as well.

LEMMA 1.6.6. *Let P be a elliptic ΨDO of order $d > 0$ which is self-adjoint. We suppose the leading symbol of P is positive definite. Then:*
(a) Given s there exists $k = k(s)$ and $C = C(s)$ so that

$$|(P-\lambda)^{-1}f|_s \le C(1+|\lambda|)^k|f|_s$$

for all $\lambda \in \mathcal{R}$.
(b) Given j there exists $k = k(j, d)$ so $(P^k - \lambda)^{-1}$ represents a smoothing operator with C^j kernel function which is of trace class for any $\lambda \in \mathcal{R}$.

PROOF: (a) follows from the estimates previously. To prove (b) we let

$$K_k(x, y, \lambda) = \sum_n \frac{1}{\lambda_n^k - \lambda}\phi_n(x) \otimes \phi_n(y)$$

be the kernel of $(P^k - \lambda)^{-1}$. The region $\mathcal{R}$ was chosen so $|\tilde{\lambda} - \lambda| \ge \varepsilon|\tilde{\lambda}|$ for some $\varepsilon > 0$ and $\tilde{\lambda} \in \mathbf{R}^+$. Therefore $|(\lambda_n^k - \lambda)|^{-1} \le \varepsilon|\lambda_n^k|^{-1} \le \varepsilon^{-1}n^{-k\delta}$ by Lemma 1.6.3. The convergence of the sum defining K_k then follows using the same arguments as given in the proof that e^{-tP} is a smoothing operator.

This technical lemma will be used in the next subsection to estimate various error terms which occur in the construction of a parametrix.

1.7. Local Formula for
The Index of an Elliptic Operator.

In this subsection, let $P: C^\infty(V) \to C^\infty(V)$ be a self-adjoint elliptic partial differential operator of order $d > 0$. Decompose the symbol $\sigma(P) = p_d + \cdots + p_0$ into homogeneous polymonials p_j of order j in $\xi \in T^*M$. We assume p_d is a positive definite Hermitian matrix for $\xi \neq 0$. Let the curve γ and the region $\mathcal{R}$ be as defined previously.

The operator $(P - \lambda)^{-1}$ for $\lambda \in \mathcal{R}$ is not a pseudo-differential operator. We will approximate $(P - \lambda)^{-1}$ by a pseudo-differential operator $R(\lambda)$ and then use that approximation to obtain properties of $\exp(-tP)$. Let U be an open subset of $\mathbf{R}^m$ with compact closure. Fix $d \in \mathbf{Z}$. We make the following definition to generalize that given in section 1.2:

DEFINITION. $q(x, \xi, \lambda) \in S^k(\lambda)(U)$ is a symbol of order k depending on the complex parameter $\lambda \in \mathcal{R}$ if

(a) $q(x, \xi, \lambda)$ is smooth in $(x, \xi, \lambda) \in \mathbf{R}^m \times \mathbf{R}^m \times \mathcal{R}$, has compact x-support in U and is holomorphic in λ.

(b) For all (α, β, γ) there exist constants $C_{\alpha,\beta,\gamma}$ such that:

$$|D_x^\alpha D_\xi^\beta D_\lambda^\gamma q(x, \xi, \gamma)| \leq C_{\alpha,\beta,\gamma}(1 + |\xi| + |\lambda|^{1/d})^{k-|\beta|-d|\gamma|}.$$

We say that $q(x, \xi, \gamma)$ is homogeneous of order k in (ξ, λ) if

$$q(x, t\xi, t^d\lambda) = t^k q(x, \xi, \gamma) \qquad \text{for } t \geq 1.$$

We think of the parameter λ as being of order d. It is clear if q is homogeneous in (ξ, λ), then it satisfies the decay conditions (b). Since the p_j are polynomials in ξ, they are regular at $\xi = 0$ and define elements of $S^j(\lambda)$. If the p_j were only pseudo-differential, we would have to smooth out the singularity at $\xi = 0$ which would destroy the homogeneity in (ξ, λ) and they would not belong to $S^j(\lambda)$; equivalently, $d_\xi^\alpha p_j$ will not exhibit decay in λ for $|\alpha| > j$ if p_j is not a polynomial. Thus the restriction to differential operators is an essential one, although it is possible to discuss some results in the pseudo-differential case by taking more care with the estimates involved.

We also note that $(p_d - \lambda)^{-1} \in S^{-d}(\lambda)$ and that the spaces $S^*(\lambda)$ form a symbol class closed under differentiation and multiplication. They are suitable generalizations of ordinary pseudo-differential operators discussed earlier.

We let $\Psi_k(\lambda)(U)$ be the set of all operators $Q(\lambda): C_0^\infty(U) \to C_0^\infty(U)$ with symbols $q(x, \xi, \gamma)$ in $S^k(\lambda)$ having x-support in U; we let $q = \sigma Q$. For any fixed λ, $Q(\lambda) \in \Psi_k(U)$ is an ordinary pseudo-differential operator of order k. The new features arise from the dependence on the parameter λ. We extend the definition of $\sim$ given earlier to this wider class by:

$$q \sim \sum_j q_j$$

if for every $k > 0$ there exists $n(k)$ so $n \geq n(k)$ implies $q - \sum_{j \leq n} q_j \in$ $S^{-k}(\lambda)$. Lemmas 1.2.1, 1.2.3(b), and 1.3.2 generalize easily to yield the following Lemma. We omit the proofs in the interests of brevity as they are identical to those previously given with only minor technical modifications.

LEMMA 1.7.1.

(a) Let $Q_i \in \Psi_{k_i}(\lambda)(U)$ with symbol q_i. Then $Q_1 Q_2 \in \Psi_{k_1 + k_2}(\lambda)(U)$ has symbol q where:

$$q \sim \sum_\alpha d_\xi^\alpha q_1 D_x^\alpha q_2 / \alpha! \,.$$

(b) Given $n > 0$ there exists $k(n) > 0$ such that $Q \in \Psi_{-k(n)}(\lambda)(U)$ implies $Q(\lambda)$ defines a continuous map from $H_{-n} \to H_n$ and we can estimate the operator norm by:

$$|Q(\lambda)|_{-n,n} \leq C(1 + |\lambda|)^{-n}.$$

(c) If $h: U \to \widetilde{U}$ is a diffeomorphism, then $h_*: \Psi_k(\lambda)(U) \to \Psi_k(\lambda)(\widetilde{U})$ and

$$\sigma(h_* P) - p(h^{-1}x_1, (dh^{-1}(x_1))^t \xi_1, \lambda) \in S^{k-1}(\lambda)(\widetilde{U}).$$

As before, we let $\Psi(\lambda)(U) = \bigcup_n \Psi_n(\lambda)(U)$ be the set of all pseudo-differential operators depending on a complex parameter $\lambda \in \mathcal{R}$ defined over U. This class depends on the order d chosen and also on the region $\mathcal{R}$, but we supress this dependence in the interests of notational clarity. There is no analogue of the completeness of Lemma 1.2.8 for such symbols since we require analyticity in λ. Thus in constructing an approximation to the parametrix, we will always restrict to a finite sum rather than an infinite sum.

Using (c), we extend the class $\Psi(\lambda)$ to compact manifolds using a partition of unity argument. (a) and (b) generalize suitably. We now turn to the question of ellipticity. We wish to solve the equation:

$$\sigma(R(\lambda)(P - \lambda)) - I \sim 0.$$

Inductively we define $R(\lambda)$ with symbol $r_0 + r_1 + \cdots$ where $r_j \in S^{-d-j}(\lambda)$. We define $p_j'(x, \xi, \lambda) = p_j(x, \xi)$ for $j < d$ and $p_d'(x, \xi, \lambda) = p_d(x, \xi) - \lambda$. Then $\sigma(P - \lambda) = \sum_{j=0}^d p_j'$. We note that $p_j' \in S^j(\lambda)$ and that $p_d'^{-1} \in$ $S^{-d}(\lambda)$ so that $(P - \lambda)$ is elliptic in a suitable sense. The essential feature of this construction is that the parameter λ is absorbed in the leading symbol and not treated as a lower order perturbation.

The equation $\sigma(R(\lambda)(P - \lambda)) \sim I$ yields:

$$\sum_{\alpha, j, k} d_\xi^\alpha r_j \cdot D_x^\alpha p_k' / \alpha! \sim I.$$

We decompose this series into terms homogeneous of order $-n$ to write:

$$\sum_{n} \sum_{|\alpha|+j+d-k=n} d_\xi^\alpha r_j \cdot D_x^\alpha p_k' / \alpha! \sim I$$

where $j, k \geq 0$ and $k \leq d$. There are no terms with $n < 0$, i.e., positive order. If we investigate the term with $n = 0$, we arrive at the condition $r_0 p_d' = I$ so $r_0 = (p_d - \lambda)^{-1}$ and inductively:

$$r_n = -r_0 \sum_{\substack{|\alpha|+j+d-k=n \\ j<n}} d_\xi^\alpha r_j D_x^\alpha p_k' / \alpha!.$$

If $k = d$ then $|\alpha| > 0$ so $D_x^\alpha p_k' = D_x^\alpha p_k$ in this sum. Therefore we may replace p_k' by p_k and write

$$r_n = -r_0 \sum_{\substack{|\alpha|+j+d-k=n \\ j<n}} d_\xi^\alpha r_j D_x^\alpha p_k / \alpha!.$$

In a similar fashion, we can define $\widetilde{R}(\lambda)$ so $\sigma((P - \lambda)\widetilde{R}(\lambda) - I) \sim 0$. This implies $\sigma(R(\lambda) - \widetilde{R}(\lambda)) \sim 0$ so $R(\lambda)$ provides a formal left and right inverse.

Since such an inverse is unique modulo lower order terms, $R(\lambda)$ is well defined and unique modulo lower order terms in any coordinate system. We define $R(\lambda)$ globally on M using a partition of unity argument. To avoid convergence questions, we shall let $R(\lambda)$ have symbol $r_0 + \cdots + r_{n_0}$ where n_0 is chosen to be very large. $R(\lambda)$ is unique modulo $\Psi_{-n_0-d}(\lambda)$. For notational convenience, we supress the dependence of $R(\lambda)$ upon n_0.

$R(\lambda)$ gives a good approximation to $(P - \lambda)^{-1}$ in the following sense:

LEMMA 1.7.2. *Let $k > 0$ be given. We can choose $n_0 = n_0(k)$ so that*

$$|\{(P - \lambda)^{-1} - R(\lambda)\}f|_k \leq C_k(1 + |\lambda|)^{-k}|f|_{-k} \qquad \text{for } \lambda \in \mathcal{R}, \ f \in C^\infty(V).$$

Thus $(P - \lambda)^{-1}$ is approximated arbitrarily well by the parametrix $R(\lambda)$ in the operator norms as $\lambda \to \infty$.

PROOF: We compute:

$$|\{(P - \lambda)^{-1} - R(\lambda)\}f|_k = |(P - \lambda)^{-1}\{I - (P - \lambda)R(\lambda)\}f|_k$$
$$\leq C_k(1 + |\lambda|^\nu)|(I - (P - \lambda)R(\lambda))f|_k$$

by Lemma 1.6.6. Since $I - (P - \lambda)R(\lambda) \in S^{-n_0}$, we use Lemma 1.7.1 to complete the proof.

We define $E(t) = \frac{1}{2\pi i} \int_\gamma e^{-t\lambda} R(\lambda) \, d\lambda$. We will show shortly that this has a smooth kernel $K'(t, x, y)$. Let $K(t, x, y)$ be the smooth kernel of e^{-tP}. We will use Lemma 1.2.9 to estimate the difference between these two kernels. We compute:

$$E(t) - e^{-tP} = \frac{1}{2\pi i} \int_\gamma e^{-t\lambda} (R(\lambda) - (P - \lambda)^{-1}) \, d\lambda.$$

We assume $0 < t < 1$ and make a change of variables to replace $t\lambda$ by λ. We use Cauchy's theorem to shift the resulting path $t\gamma$ inside $\mathcal{R}$ back to the original path γ where we have uniform estimates. This expresses:

$$E(t) - e^{-tP} = \frac{1}{2\pi i} \int_\gamma e^{-\lambda} (R(t^{-1}\lambda) - (P - t^{-1}\lambda)) t^{-1} \, d\lambda$$

We estimate therefore:

$$|E(t) - e^{-tP}|_{-k,k} \le C_k \int_\gamma |e^{-\lambda} (1 + t^{-1}|\lambda|)^{-k} \, d\lambda|$$
$$\le C_k t^k$$

provided n_0 is large enough. Lemma 1.2.9 implies

LEMMA 1.7.3. *Let k be given. If n_0 is large enough, we can estimate:*

$$|K(t, x, y) - K'(t, x, y)|_{\infty, k} \le C_k t^k \qquad \text{for } 0 < t < 1.$$

This implies that K' approximates K to arbitrarily high jets as $t \to 0$.

We now study the operator $E(t)$. We define

$$e_n(t, x, \xi) = \frac{1}{2\pi i} \int_\gamma e^{-t\lambda} r_n(x, \xi, \lambda) \, d\lambda,$$

then $E(t)$ is a ΨDO with symbol $e_0 + \cdots + e_{n_0}$. If we integrate by parts in λ we see that we can express

$$e_n(t, x, \xi) = \frac{1}{2\pi i} \frac{1}{t^k} \int_\gamma e^{-t\lambda} \frac{d^k}{d\lambda^k} r_n(x, \xi, \lambda) \, d\lambda.$$

Since $\dfrac{d^k}{d\lambda^k} r_n$ is homogeneous of degree $-d - n - kd$ in (ξ, λ), we see that $e_n(t, x, \xi) \in S^{-\infty}$ for any $t > 0$. Therefore we can apply Lemma 1.2.5 to conclude $E_n(t)$ which has symbol $e_n(t)$ is represented by a kernel function defined by:

$$K_n(t, x, y) = \int e^{i(x-y)\cdot\xi} e_n(t, x, \xi) \, d\xi.$$

We compute:

$$K_n(t, x, x) = \frac{1}{2\pi i} \iint_\gamma e^{-t\lambda} r_n(x, \xi, \lambda) \, d\lambda \, d\xi.$$

We make a change of variables to replace λ by $t^{-1}\lambda$ and ξ by $t^{-1/d}\xi$ to compute:

$$K_n(t, x, x) = t^{-\frac{m}{d}-1} \frac{1}{2\pi i} \iint_\gamma e^{-\lambda} r_n(x, t^{\frac{-1}{d}}\xi, t^{-1}\lambda) \, d\lambda \, d\xi$$

$$= t^{-\frac{m}{d}-1+\frac{n+d}{d}} \frac{1}{2\pi i} \iint_\gamma e^{-\lambda} r_n(x, \xi, \lambda) \, d\lambda \, d\xi$$

$$= t^{\frac{n-m}{d}} e_n(x)$$

where we let this integral define $e_n(x)$.

Since $p_d(x, \xi)$ is assumed to be positive definite, d must be even since p_d is a polynomial in ξ. Inductively we express r_n as a sum of terms of the form:

$$r_0^{j_1} q_1 r_0^{j_2} q_2 \ldots r_0^{j_k}$$

where the q_k are polynomials in (x, ξ). The sum of the degrees of the q_k is odd if n is odd and therefore $r_n(x, -\xi, \lambda) = -r_n(x, \xi, \lambda)$. If we replace ξ by $-\xi$ in the integral defining $e_n(x)$, we conclude $e_n(x) = 0$ if n is odd.

LEMMA 1.7.4. *Let P be a self-adjoint elliptic partial differential operator of order $d > 0$ such that the leading symbol of P is positive definite for $\xi \neq 0$. Then:*
(a) If we choose a coordinate system for M near a point $x \in M$ and choose a local frame for V, we can define $e_n(x)$ using the complicated combinatorial recipe given above. $e_n(x)$ depends functorially on a finite number of jets of the symbol $p(x, \xi)$.
(b) If $K(t, x, y)$ is the kernel of e^{-tP} then

$$K(t, x, x) \sim \sum_{n=0}^\infty t^{\frac{n-m}{d}} e_n(x) \qquad \text{as } t \to 0^+$$

i.e., given any integer k there exists $n(k)$ such that:

$$\left| K(t, x, x) - \sum_{n \leq n(k)} t^{\frac{n-m}{d}} e_n(x) \right|_{\infty, k} < C_k t^k \qquad \text{for } 0 < t < 1.$$

(c) $e_n(x) \in \text{END}(V, V)$ is invariantly defined independent of the coordinate system and local frame for V.

(d) $e_n(x) = 0$ if n is odd.

PROOF: (a) is immediate. We computed that

$$K'(t, x, x) = \sum_{n=0}^{n_0} t^{\frac{n-m}{d}} e_n(x)$$

so (b) follows from Lemma 1.7.3. Since $K(t, x, x)$ does not depend on the coordinate system chosen, (c) follows from (b). We computed (d) earlier to complete the proof.

We remark that this asymptotic representation of $K(t, x, x)$ exists for a much wider class of operators P. We refer to the literature for further details. We shall give explicit formulas for e_0, e_2, and e_4 in section 4.8 for certain examples arising in geometry.

The invariants $e_n(x) = e_n(x, P)$ are sections to the bundle of endomorphisms, $\text{END}(V)$. They have a number of functorial properties. We sumarize some of these below.

LEMMA 1.7.5.
(a) Let $P_i: C^\infty(V_i) \to C^\infty(V_i)$ be elliptic self-adjoint partial differential operators of order $d > 0$ with positive definite leading symbol. We form $P = P_1 \oplus P_2: C^\infty(V_1 \oplus V_2) \to C^\infty(V_1 \oplus V_2)$. Then P is an elliptic self-adjoint partial differential operator of order $d > 0$ with positive definite leading symbol and $e_n(x, P_1 \oplus P_2) = e_n(x, P_1) \oplus e_n(x, P_2)$.
(b) Let $P_i: C^\infty(V_i) \to C^\infty(V_i)$ be elliptic self-adjoint partial differential operators of order $d > 0$ with positive definite leading symbol defined over different manifolds M_i. We let

$$P = P_1 \otimes 1 + 1 \otimes P_2: C^\infty(V_1 \otimes V_2) \to C^\infty(V_1 \otimes V_2)$$

over $M = M_1 \times M_2$. Then P is an elliptic self-adjoint partial differential operator of order $d > 0$ with positive definite leading symbol over M and

$$e_n(x, P) = \sum_{p+q=n} e_p(x_1, P_1) \otimes e_q(x_2, P_2).$$

(c) Let $P: C^\infty(V) \to C^\infty(V)$ be an elliptic self-adjoint partial differential operator of order $d > 0$ with positive definite leading symbol. We decompose the total symbol of P in the form:

$$p(x, \xi) = \sum_{|\alpha| \le d} p_\alpha(x) \xi^\alpha.$$

Fix a local frame for V and a system of local coordinates on M. We let indices a, b index the local frame and let $p_\alpha = p_{ab\alpha}$ give the components of the matrix p_α. We introduce formal variables $p_{ab\alpha/\beta} = D_x^\beta p_{ab\alpha}$ for the jets of the symbol of P. Define $\text{ord}(p_{ab\alpha/\beta}) = |\beta| + d - |\alpha|$. Then $e_n(x, P)$ can be expressed as a sum of monomials in the $\{p_{ab\alpha/\beta}\}$ variables which are homogeneous of order n in the jets of the symbol as discussed above and with coefficients which depend smoothly on the leading symbol $\{p_{ab\alpha}\}_{|\alpha|=d}$.

(d) If the leading symbol of P is scalar, then the invariance theory can be simplified. We do not need to introduce the components of the symbol explicitly and can compute $e_n(x, P)$ as a non-commutative polynomial which is homogeneous of order n in the $\{p_{\alpha/\beta}\}$ variables with coefficients which depend smoothly upon the leading symbol.

Remark: The statement of this lemma is somewhat technical. It will be convenient, however, to have these functorial properties precisely stated for later reference. This result suffices to study the index theorem. We shall give one more result which generalizes this one at the end of this section useful in studying the eta invariant. The objects we are studying have a bigrading; one grading comes from counting the number of derivatives in the jets of the symbol, while the other measures the degree of homogeneity in the (ξ, λ) variables.

PROOF: (a) and (b) follow from the identities:

$$e^{-t(P_1 \oplus P_2)} = e^{-tP_1} \oplus e^{-tP_2}$$

$$e^{-t(P_1 \otimes 1 + 1 \otimes P_2)} = e^{-tP_1} \otimes e^{-tP_2}$$

so the kernels satisfy the identities:

$$K(t, x, x, P_1 \oplus P_2) = K(t, x, x, P_1) \oplus K(t, x, x, P_2)$$

$$K(t, x, x, P_1 \otimes 1 + 1 \otimes P_2) = K(t, x_1, x_1, P_1) \otimes K(t, x_2, x_2, P_2).$$

We equate equal powers of t in the asymptotic series:

$$\sum t^{\frac{n-m}{d}} e_n(x, P_1 \oplus P_2)$$
$$\sim \sum t^{\frac{n-m}{d}} e_n(x, P_1) \oplus \sum t^{\frac{n-m}{d}} e_n(x, P_2)$$
$$\sum t^{\frac{n-m}{d}} e_n(x, P_1 \otimes 1 + 1 \otimes P_2)$$
$$\sim \left\{ \sum t^{\frac{p-m_1}{d}} e_p(x_1, P_1) \right\} \otimes \left\{ \sum t^{\frac{q-m_2}{d}} e_q(x_2, P_2) \right\}$$

to complete the proof of (a) and (b). We note that the multiplicative property (b) is a direct consequence of the identity $e^{-t(a+b)} = e^{-ta} e^{-tb}$; it was for this reason we worked with the heat equation. Had we worked instead with the zeta function to study $\text{Tr}(P^{-s})$ the corresponding multiplicative property would have been much more difficult to derive.

We prove (c) as follows: expand $p(x, \xi) = \sum_j p_j(x, \xi)$ into homogeneous polynomials where $p_j(x, \xi) = \sum_{|\alpha|=j} p_\alpha(x)\xi^\alpha$. Suppose for the moment that $p_d(x, \xi) = p_d(x, \xi)I_V$ is scalar so it commutes with every matrix. The approximate resolvent is given by:

$$r_0 = (p_d(x, \xi) - \lambda)^{-1}$$

$$r_n = -r_0 \sum_{\substack{|\alpha|+j+d-k=n \\ j<n}} d_\xi^\alpha r_j D_x^\alpha p_k / \alpha!.$$

Since p_d is scalar, r_0 is scalar so it and all its derivatives commute with any matrix. If we assume inductively r_j is of order j in the jets of the symbol, then $d_\xi^\alpha r_j D_x^\alpha p_k / \alpha!$ is homogeneous of order $j + |\alpha| + d - k = n$ in the jets of the symbol. We observe inductively we can decompose r_n in the form:

$$r_n = \sum_{\substack{n=dj+d-|\alpha| \\ |\alpha| \leq n}} r_0^j r_{n,j,\alpha}(x)\xi^\alpha$$

where the $r_{n,j,\alpha}(x)$ are certain non-commutative polynomials in the jets of the total symbol of P which are homogeneous of order n in the sense we have defined.

The next step in the proof of Lemma 1.7.4 was to define:

$$e_n(t, x, \xi) = \frac{1}{2\pi i} \int_\gamma e^{-t\lambda} r_n(x, \xi, \lambda) \, d\lambda$$

$$= \frac{1}{2\pi i} \sum_{j,\alpha} r_{n,j,\alpha}(x)\xi^\alpha \int_\gamma e^{-t\lambda} r_0^j \, d\lambda$$

$$e_n(x) = \frac{1}{2\pi i} \sum_{j,\alpha} r_{n,j,\alpha}(x) \int \left(\int_\gamma e^{-\lambda} r_0^j \, d\lambda \right) \xi^\alpha \, d\xi.$$

We note again that $e_n = 0$ if n is odd since the resulting function of ξ would be odd. The remaining coefficients of $r_{n,j,\alpha}(x)$ depend smoothly on the leading symbol p_d. This completes the proof of the lemma.

If the leading symbol is not scalar, then the situation is more complicated. Choose a local frame to represent the $p_\alpha = p_{ab\alpha}$ as matrices. Let $h(x, \xi, \lambda) = \det(p_d(x, \xi) - \lambda)^{-1}$. By Cramer's rule, we can express $r_0 = r_{0ab} = \{(p_d - \lambda)^{-1}\}_{ab}$ as polynomials in the $\{h, \lambda, \xi, p_{ab\alpha}(x)\}$ variables. We noted previously that r_n was a sum of terms of the form $r_0 q_0 r_0 q_1 \ldots r_0 q_k r_0$. The matrix components of such a product can in turn be decomposed as a sum of terms $h^v \tilde{q}_v$ where $\tilde{q}_v$ is a polynomial in the $(\lambda, \xi, p_{ab\alpha/\beta})$ variables. The same induction argument which was used in the scalar case shows the $\tilde{q}_v$ will be homogeneous of order n in the jets of the symbol. The remainder of the argument is the same; performing the $d\lambda\, d\xi$ integral yields a smooth function of the leading symbol as a coefficient of such a term, but this function is not in general rational of course. The scalar case is much simpler as we don't need to introduce the components of the matrices explicitly (although the frame dependence is still there of course) since r_0 can be commuted. This is a technical point, but one often useful in making specific calculations.

We define the scalar invariants

$$a_n(x, P) = \operatorname{Tr} e_n(x, P)$$

where the trace is the fiber trace in V over the point x. These scalar invariants $a_n(x, P)$ inherit suitable functorial properties from the functorial properties of the invariants $e_n(x, P)$. It is immediate that:

$$\operatorname{Tr}_{L^2} e^{-tP} = \int_M \operatorname{Tr}_{V_x} K(t, x, x)\, \mathrm{dvol}(x)$$

$$\sim \sum_{n=0}^{\infty} t^{\frac{m-n}{d}} \int_M a_n(x, P)\, \mathrm{dvol}(x).$$

$$\sim \sum_{n=0}^{\infty} t^{\frac{m-n}{d}} a_n(P)$$

where $a_n(P) = \int_M a_n(x, P)\, \mathrm{dvol}(x)$ is the integrated invariant. This is a spectral invariant of P which can be computed from local information about the symbol of P.

Let (P, V) be an elliptic complex of differential operators and let Δ_i be the associated Laplacians. We define:

$$a_n(x, P) = \sum_i (-1)^i \operatorname{Tr} e_n(x, \Delta_i)$$

then Lemma 1.6.5 and the remark which follows this lemma imply

$$\operatorname{index}(P) = \sum_i (-1)^i \operatorname{Tr} e^{-t\Delta_i} \sim \sum_{n=0}^{\infty} t^{\frac{n-m}{d}} \int_M a_n(x, P)\, \mathrm{dvol}(x).$$

Since the left hand side does not depend on the parameter t, we conclude:

THEOREM 1.7.6. Let (P, V) be a elliptic complex of differential operators.

(a) $a_n(x, P)$ can be computed in any coordinate system and relative to any local frames as a complicated combinatorial expression in the jets of P and of P^* up to some finite order. $a_n = 0$ if n is odd.

(b)

$$\int_M a_n(x, P)\, \mathrm{dvol}(x) = \begin{cases} \mathrm{index}(P) & \text{if } n = m \\ 0 & \text{if } n \neq m. \end{cases}$$

We note as an immediate consequence that $\mathrm{index}(P) = 0$ if m is odd. The local nature of the invariants a_n will play a very important role in our discussion of the index theorem. We will develop some of their functorial properties at that point. We give one simple example to illustrate this fact.

Let $\pi: M_1 \to M_2$ be a finite covering projection with fiber F. We can choose the metric on M_1 to be the pull-back of a metric on M_2 so that π is an isometry. Let d be the operator of the de Rham complex. $a_n(x, d) = a_n(\pi x, d)$ since a_n is locally defined. This implies:

$$\chi(M_1) = \int_{M_1} a_m(x, d)\, \mathrm{dvol}(x) = |F| \int_{M_2} a_m(x, d)\, \mathrm{dvol}(x) = |F|\chi(M_2)$$

so the Euler-Poincare characteristic is multiplicative under finite coverings. A similar argument shows the signature is multiplicative under orientation preserving coverings and that the arithmetic genus is multiplicative under holomorphic coverings. (We will discuss the signature and arithmetic genus in more detail in Chapter 3).

We conclude this section with a minor generalization of Lemma 1.7.5 which will be useful in discussing the eta invariant in section 1.10:

LEMMA 1.7.7. *Let* $P: C^\infty(V) \to C^\infty(V)$ *be an elliptic self-adjoint partial differential operator of order* $d > 0$ *with positive definite leading symbol and let* Q *be an auxilary partial differential operator on* $C^\infty(V)$ *of order* $a \geq 0$. Qe^{-tP} *is an infinitely smoothing operator with kernel* $QK(t, x, y)$. *There is an asymptotic expansion on the diagonal:*

$$\{QK(t, x, y)\}|_{x=y} \sim \sum_{n=0}^{\infty} t^{(n-m-a)/d} e_n(x, Q, P).$$

The e_n *are smooth local invariants of the jets of the symbols of* P *and* Q *and* $e_n = 0$ *if* $n + a$ *is odd. If we let* $Q = \sum q_\alpha D_x^\alpha$, *we define* $\mathrm{ord}(q_\alpha) = a - |\alpha|$. *If the leading symbol of* P *is scalar, we can compute* $e_n(x, Q, P)$ *as a non-commutative polynomial in the variables* $\{q_\alpha, p_{\alpha/\beta}\}$ *which is homogeneous of order* n *in the jets with coefficients which depend smoothly on the* $\{p_\alpha\}_{|\alpha|=d}$ *variables. This expression is* <u>linear</u> *in the* $\{q_\alpha\}$ *variables and does not involve the higher jets of these variables. If the leading symbol of* P *is not scalar, there is a similar expression for the matrix components of* e_n *in the matrix components of these variables.* $e_n(x, Q, P)$ *is additive and multiplicative in the sense of Lemma 1.7.5(a) and (b) with respect to direct sums of operators and tensor products over product manifolds.*

Remark: In fact it is not necessary to assume P is self-adjoint to define e^{-tP} and to define the asymptotic series. It is easy to generalize the techniques we have developed to prove this lemma continues to hold true if we only asume that $\det(p_d(x, \xi) - \lambda) \neq 0$ for $\xi \neq 0$ and $\mathrm{Im}(\lambda) \leq 0$. This implies that the spectrum of P is pure point and contained in a cone about the positive real axis. We omit the details.

PROOF: Qe^{-tP} has smooth kernel $QK(t, x, y)$ where Q acts as a differential operator on the x variables. One representative piece is:

$$q_\alpha(x)D_x^\alpha K_n(t, x, y) = q_\alpha(x)D_x^\alpha \int e^{i(x-y)\cdot\xi}e_n(t, x, \xi)\,d\xi$$

$$= q_\alpha(x) \sum_{\beta+\gamma=\alpha} \frac{\alpha!}{\beta!\,\gamma!} \int e^{i(x-y)\cdot\xi}\xi^\beta D_x^\gamma e_n(t, x, \xi)\,d\xi.$$

We define $q_{\beta,\gamma}(x) = q_\alpha(x)\dfrac{\alpha!}{\beta!\,\gamma!}$ where $\beta + \gamma = \alpha$, then a representative term of this kernel has the form:

$$q_{\beta,\gamma}(x) \int e^{i(x-y)\cdot\xi}\xi^\beta D_x^\gamma e_n(t, x, \xi)\,d\xi$$

where $q_{\beta,\gamma}(x)$ is homogeneous of order $a - |\beta| - |\gamma|$ in the jets of the symbol of Q.

We suppose the leading symbol is scalar to simplify the computations; the general case is handled similarly using Cramer's rule. We express

$$e_n = \frac{1}{2\pi i} \int e^{-t\lambda}r_n(x, \xi, \lambda)\,d\lambda$$

where r_n is a sum of terms of the form $r_{n,j,\alpha}(x)\xi^\alpha r_0^j$. When we differentiate such a term with respect to the x variables, we change the order in the jets of the symbol, but do not change the (ξ, λ) degree of homogeneity. Therefore $QK(t, x, y)$ is a sum of terms of the form:

$$q_{\beta,\gamma}(x) \int \left\{ \int e^{-t\lambda}r_0^j(x, \xi, \lambda)\,d\lambda \right\} e^{i(x-y)\cdot\xi}\xi^\beta \xi^\delta \tilde{r}_{n,j,\delta}(x)\,d\xi$$

where:

$r_{n,j,\delta}$ is of order $n + |\gamma|$ in the jets of the symbol,

$-d - n = -jd + |\delta|$.

We evaluate on the diagonal to set $x = y$. This expression is homogeneous of order $-d - n + |\beta|$ in (ξ, λ), so when we make the appropriate change of variables we calculate this term becomes:

$$t^{(n-m-|\beta|)/d}q_{\beta,\gamma}(x)\tilde{r}_{n,j,\delta}(x) \cdot \left\{ \int \left\{ e^{-\lambda}r_0^j(x, \xi, \lambda)\,d\lambda \right\} \xi^\beta \xi^\delta\,d\xi \right\}.$$

This is of order $n+|\gamma|+a-|\beta|-|\gamma| = n+a-|\beta| = \nu$ in the jets of the symbol. The exponent of t is $(n-m-|\beta|)/d = (\nu-a+|\beta|-m-|\beta|)/d = (\nu-a-m)/d$. The asymptotic formula has the proper form if we index by the order in the jets of the symbol ν. It is linear in q and vanishes if $|\beta| + |\delta|$ is odd. As d is even, we compute:

$$\nu + a = n + a - |\beta| + a = jd - d - |\delta| + 2a - |\beta| \equiv |\delta| + |\beta| \mod 2$$

so that this term vanishes if $\nu + a$ is odd. This completes the proof.

1.8. Lefschetz Fixed Point Theorems.

Let $T: M \to M$ be a continuous map. T^* defines a map on the cohomology of M and the Lefschetz number of T is defined by:

$$L(T) = \sum (-1)^p \operatorname{Tr}(T^* \text{ on } H^p(M; \mathbf{C})).$$

This is always an integer. We present the following example to illustrate the concepts involved. Let $M = S^1 \times S^1$ be the two dimensional torus which we realize as $\mathbf{R}^2$ modulo the integer lattice $\mathbf{Z}^2$. We define $T(x, y) = (n_1 x + n_2 y, n_3 x + n_4 y)$ where the n_i are integers. Since this preserves the integer lattice, this defines a map from M to itself. We compute:

$$
\begin{aligned}
T^*(1) &= 1, & T^*(dx \wedge dy) &= (n_1 n_4 - n_2 n_3) dx \wedge dy \\
T^*(dx) &= n_1 \, dx + n_2 \, dy, & T^*(dy) &= n_3 \, dx + n_4 \, dy \\
L(T) &= 1 + (n_1 n_4 - n_2 n_3) - (n_1 + n_4).
\end{aligned}
$$

Of course, there are many other interesting examples.

We computed $L(T)$ in the above example using the de Rham isomorphism. We let $T^* = \Lambda^p(dT): \Lambda^p(T^*M) \to \Lambda^p(T^*M)$ to be the pull-back operation. It is a map from the fiber over $T(x)$ to the fiber over x. Since $dT^* = T^*d$, T^* induces a map on $H^p(M, \mathbf{C}) = \ker(d_p)/\operatorname{image}(d_{p-1})$. If T is the identity map, then $L(T) = \chi(M)$. The perhaps somewhat surprising fact is that Lemma 1.6.5 can be generalized to compute $L(T)$ in terms of the heat equation.

Let $\Delta_p = (d\delta + \delta d)_p: C^\infty(\Lambda^p T^*M) \to C^\infty(\Lambda^p T^*M)$ be the associated Laplacian. We decompose $L^2(\Lambda^p T^*M) = \bigoplus_\lambda E_p(\lambda)$ into the eigenspaces of Δ_p. We let $\pi(p, \lambda)$ denote orthogonal projection on these subspaces, and we define $T^*(p, \lambda) = \pi(p, \lambda)T^*: E_p(\lambda) \to E_p(\lambda)$. It is immediate that:

$$\operatorname{Tr}(T^* e^{-t\Delta_p}) = \sum_\lambda e^{-t\lambda} \operatorname{Tr}(T^*(p, \lambda)).$$

Since $dT^* = T^*d$ and $d\pi = \pi d$, for $\lambda \neq 0$ we get a chain map between the exact sequences:

$$
\begin{array}{ccccccc}
\cdots & E_{p-1}(\lambda) & \xrightarrow{d} & E_p(\lambda) & \xrightarrow{d} & E_{p+1}(\lambda) & \cdots \\
& \downarrow{\scriptstyle T^*} & & \downarrow{\scriptstyle T^*} & & \downarrow{\scriptstyle T^*} & \\
\cdots & L^2(\Lambda^{p-1}) & \xrightarrow{d} & L^2(\Lambda^p) & \xrightarrow{d} & L^2(\Lambda^{p+1}) & \cdots \\
& \downarrow{\scriptstyle \pi} & & \downarrow{\scriptstyle \pi} & & \downarrow{\scriptstyle \pi} & \\
\cdots & E_{p-1}(\lambda) & \xrightarrow{d} & E_p(\lambda) & \xrightarrow{d} & E_{p+1}(\lambda) & \cdots
\end{array}
$$

Since this diagram commutes and since the two rows are long exact sequences of finite dimensional vector spaces, a standard result in homological algebra implies:

$$\sum_p (-1)^p \operatorname{Tr}(T^*(p, \lambda)) = 0 \qquad \text{for } \lambda \neq 0.$$

It is easy to see the corresponding sum for $\lambda = 0$ yields $L(T)$, so Lemma 1.6.5 generalizes to give a heat equation formula for $L(T)$.

This computation was purely formal and did not depend on the fact that we were dealing with the de Rham complex.

LEMMA 1.8.1. *Let (P, V) be an elliptic complex over M and let $T: M \to M$ be smooth. We assume given linear maps $V_i(T): V_i(Tx) \to V_i(x)$ so that $P_i(V_i(T)) = V_i(T)P_i$. Then T induces a map on $H^p(P, V)$. We define*

$$L(T)_P = \sum_p (-1)^p \operatorname{Tr}(T \text{ on } H^p(P, V)).$$

Then we can compute $L(T)_P = \sum_p \operatorname{Tr}(V_i(T)e^{-t\Delta_p})$ where $\Delta_p = (P^*P + PP^*)_p$ is the associated Laplacian.

The $V_i(T)$ are a smooth linear action of T on the bundles V_i. They will be given by the representations involved for the de Rham, signature, spin, and Dolbeault complexes as we shall discuss in Chapter 4. We usually denote $V_i(T)$ by T^* unless it is necessary to specify the action involved.

This Lemma implies Lemma 1.6.5 if we take $T = I$ and $V_i(T) = I$. To generalize Lemma 1.7.4 and thereby get a local formula for $L(T)_P$, we must place some restrictions on the map T. We assume the fixed point set of T consists of the finite disjoint union of smooth submanifolds N_i. Let $\nu(N_i) = T(M)/T(N_i)$ be the normal bundle over the submanifold N_i. Since dT preserves $T(N_i)$, it induces a map dT_ν on the bundle $\nu(N_i)$. We suppose $\det(I - dT_\nu) \neq 0$ as a non-degeneracy condition; there are no normal directions left fixed infinitesimally by T.

If T is an isometry, this condition is automatic. We can construct a non-example by defining $T(z) = z/(z+1): S^2 \to S^2$. The only fixed point is at $z = 0$ and $dT(0) = I$, so this fixed point is degenerate.

If K is the kernel of e^{-tP}, we pull back the kernel to define $T^*(K)(t, x, y) = T^*(x)K(t, Tx, y)$. It is immediate T^*K is the kernel of T^*e^{-tP}.

LEMMA 1.8.2. *Let P be an elliptic partial differential operator of order $d > 0$ which is self-adjoint and which has a positive definite leading symbol for $\xi \neq 0$. Let $T: M \to M$ be a smooth non-degenerate map and let $T^*: V_{T(x)} \to V_x$ be a smooth linear action. If K is the kernel of e^{-tP} then $\operatorname{Tr}(T^*e^{-tP}) = \int_M \operatorname{Tr}(T^*K)(t, x, x) \, \mathrm{dvol}(x)$. Furthermore:*
(a) If T has no fixed points, $|\operatorname{Tr}(T^*e^{-tP})| \leq C_n t^{-n}$ as $t \to 0^+$ for any n.

(b) If the fixed point set of T consists of submanifolds N_i of dimension m_i, we will construct scalar invariants $a_n(x)$ which depend functorially upon a finite number of jets of the symbol and of T. The $a_n(x)$ are defined over N_i and

$$\text{Tr}(T^* e^{-tP}) \sim \sum_i \sum_{n=0}^{\infty} t^{\frac{n-m_i}{d}} \int_{N_i} a_n(x)\, \text{dvol}_i(x).$$

$\text{dvol}_i(x)$ denotes the Riemannian measure on the submanifold. It follows that if T has no fixed points, then $L(T)_P = 0$.

PROOF: Let $\{e_n, r_n, K_n\}$ be as defined in section 1.7. The estimates of Lemma 1.7.3 show that $|T^* K - \sum_{n \le n_0} T^* K_n|_{\infty,k} \le C(k) t^{-k}$ as $t \to 0^+$ for any k if $n_0 = n_0(k)$. We may therefore replace K by K_n in proving (a) and (b). We recall that:

$$e_n(t, x, \xi) = \frac{1}{2\pi i} \int_\gamma e^{-t\lambda} r_n(x, \xi, \lambda)\, d\lambda.$$

We use the homogeneity of r_n to express:

$$e_n(t, x, \xi) = \frac{1}{2\pi i} \int_\gamma e^{-\lambda} r_n(x, \xi, t^{-1}\lambda) t^{-1}\, d\lambda$$

$$= t^{\frac{n}{d}} \frac{1}{2\pi i} \int_\gamma e^{-\lambda} r_n(x, t^{\frac{1}{d}}\xi, \lambda)\, d\lambda$$

$$= t^{\frac{n}{d}} e_n(x, t^{\frac{1}{d}}\xi)$$

where we define $e_n(x, \xi) = e_n(1, x, \xi) \in S^{-\infty}$. Then:

$$K_n(t, x, y) = \int e^{i(x-y)\cdot\xi} e_n(t, x, \xi)\, d\xi$$

$$= t^{\frac{n-m}{d}} \int e^{i(x-y)\cdot t^{-\frac{1}{d}}\xi} e_n(x, \xi)\, d\xi.$$

This shows that:

$$T^* K_n(t, x, x) = t^{\frac{n-m}{d}} \int T^*(x) e^{i(Tx-x)\cdot t^{-1/d}\xi} e_n(Tx, \xi)\, d\xi.$$

We must study terms which have the form:

$$\int e^{i(Tx-x)\cdot t^{-1/d}\xi} \text{Tr}(T^*(x) e_n(Tx, \xi))\, d\xi\, dx$$

where the integral is over the cotangent space $T^*(M)$. We use the method of stationary phase on this highly oscillatory integral. We first bound

$|Tx - x| \geq \varepsilon > 0$. Using the argument developed in Lemma 1.2.6, we integrate by parts to bound this integral by $C(n, k, \varepsilon)t^k$ as $t \to 0$ for any k. If T has no fixed points, this proves (a). There is a slight amount of notational sloppiness here since we really should introduce partitions of unity and coordinate charts to define $Tx - x$, but we supress these details in the interests of clarity.

We can localize the integral to an arbitrarily small neighborhood of the fixed point set in proving (b). We shall assume for notational simplicity that the fixed point set of T consists of a single submanifold N of dimension m_1. The map dT_ν on the normal bundle has no eigenvalue 1. We identify ν with the span of the generalized eigenvectors of dT on $T(M)|_N$ which correspond to eigenvalues not equal to 1. This gives a direct sum decomposition over N:

$$T(M) = T(N) \oplus \nu \qquad \text{and} \qquad dT = I \oplus dT_\nu.$$

We choose a Riemannian metric for M so this splitting is orthogonal. We emphasize that these are bundles over N and not over the whole manifold M.

We describe the geometry near the fixed manifold N using the normal bundle ν. Let $y = (y_1, \ldots, y_{m_1})$ be local coordinates on N and let $\{\vec{s}_1, \ldots, \vec{s}_{m-m_1}\}$ be a local orthonormal frame for ν. We use this local orthonormal frame to introduce fiber coordinates $z = (z_1, \ldots, z_{m-m_1})$ for ν by decomposing any $\vec{s} \in \nu$ in the form:

$$\vec{s} = \sum_j z_j \vec{s}_j(y).$$

We let $x = (y, z)$ be local coordinates for ν. The geodesic flow identifies a neighborhood of the zero section of the bundle ν with a neighborhood of N in M so we can also regard $x = (y, z)$ as local coordinates on M.

We decompose

$$T(x) = (T_1(x), T_2(x))$$

into tangential and fiber coordinates. Because the Jacobian matrix has the form:

$$dT(y, 0) = \begin{pmatrix} I & 0 \\ 0 & dT_\nu \end{pmatrix}$$

we conclude that $T_1(x) - y$ must vanish to second order in z along N.

We integrate $\operatorname{Tr}(T^* K_n)(t, x, x)$ along a small neighborhood of the zero section of ν. We shall integrate along the fibers first to reduce this to an integral along N. We decompose $\xi = (\xi_1, \xi_2)$ corresponding to the decomposition of $x = (y, z)$. Let

$$D(\nu) = \{ (y, z) : |z| \leq 1 \}$$

be the unit disk bundle of the normal bundle. Let $U = T^*(D(\nu))$ be the cotangent bundle of the unit disk bundle of the normal bundle. We assume the metric chosen so the geodesic flow embeds $D(\nu)$ in M. We parametrize U by $\{(y, z, \xi_1, \xi_2) : |z| \leq 1\}$. Let $s = t^{\frac{1}{d}}$. Modulo terms which vanish to infinite order in s, we compute:

$$I \stackrel{\text{def}}{=} \int_M \text{Tr}((T^*K_n)(t, x, x))\, dx$$

$$= s^{n-m} \int_U e^{i(T_1(y,z)-y)\cdot\xi_1 s^{-1}} e^{i(T_2(y,z)-z)\cdot\xi_2 s^{-1}}$$
$$\times \text{Tr}(T^*(x)e_n(Tx, \xi_1, \xi_2))\, d\xi_1\, d\xi_2\, dz\, dy.$$

The non-degeneracy assumption on T means the phase function $\overline{w} = T_2(y, z) - z$ defines a non-degenerate change of variables if we replace (y, z) by $(y, \overline{w})$. This transforms the integral into the form:

$$I = s^{n-m} \int_{\overline{U}} e^{i(T_1(y,\overline{w})-y)\cdot\xi_1 s^{-1}} e^{i\overline{w}\cdot\xi_2 s^{-1}}$$
$$\times |\det(I - dT_2)|^{-1} \text{Tr}(T^*(y,\overline{w})e_n(T(y,\overline{w}), \xi_1, \xi_2)\, d\xi_1\, d\xi_2\, d\overline{w}\, dy,$$

where $\overline{U}$ is the image of U under this change. We now make another change of coordinates to let $w = s^{-1}\overline{w}$. As $s \to 0$, $s^{-1}\overline{U}$ will converge to $T^*(v)$. Since $d\overline{w} = s^{m-m_1} dw$ this transforms the integral I to the form:

$$I = s^{n-m_1} \int_{s^{-1}\overline{U}} e^{i(T_1(y,sw)-y)\cdot\xi_1 s^{-1}} e^{iw\cdot\xi_2} |\det(I - dT_2)^{-1}|(y, sw)$$
$$\times \text{Tr}(T^*(y, sw)e_n(T(y, sw), \xi_1, \xi_2))\, d\xi_1\, d\xi_2\, dw\, dy.$$

Since the phase function $T_1(y, \overline{w}) - y$ vanishes to second order at $\overline{w} = 0$, the function $(T_1(y, sw) - y)s^{-1}$ is regular at $s = 0$. Define:

$$e'_n(y, w, \xi_1, \xi_2, s) = e^{i(T_1(y,sw)-y)\cdot\xi_1 s^{-1}} |\det(I - dT_2)(y, sw)|$$
$$\times \text{Tr}(T^*(y, sw)e_n(T(y, sw), \xi_1, \xi_2)).$$

This vanishes to infinite order in ξ_1, ξ_2 at ∞ and is regular at $s = 0$. To complete the proof of Lemma 1.8.2, we must evaluate:

$$s^{(n-m_1)} \int_{s^{-1}\overline{U}} e'_n(y, w, \xi_1, \xi_2, s)e^{iw\cdot\xi_2}\, d\xi_1\, d\xi_2\, dw\, dy.$$

We expand e'_n in a Taylor series in s centered at $s = 0$. If we differentiate e'_n with respect to s a total of k times and evaluate at $s = 0$, then the exponential term disappears and we are left with an expression which is polynomial in w and of degree at worst $2k$. It still vanishes to infinite order in (ξ_1, ξ_2). We decompose:

$$e'_n = \sum_{j \leq j_0} s^j \sum_{|\alpha| \leq 2j} c_{j,\alpha}(y, \xi_1, \xi_2) w^\alpha + s^{j_0} \varepsilon(y, w, \xi_1, \xi_2, s)$$

where ε is the remainder term.

We first study a term of the form:

$$s^{n+j-m_1} \int_{|w| < s^{-1}} c_{j,\alpha}(y, \xi_1, \xi_2) w^\alpha e^{i\xi_2 \cdot w} \, d\xi_1 \, d\xi_2 \, dw \, dy.$$

Since $c_{j,\alpha}$ vanishes to infinite order in ξ, the $d\xi$ integral creates a function which vanishes to infinite order in w. We can let $s \to 0$ to replace the domain of integration by the normal fiber. The error vanishes to infinite order in s and gives a smooth function of y. The dw integral just yields the inverse Fourier transform with appropriate terms and gives rise to asymptotics of the proper form. The error term in the Taylor series grows at worst polynomially in w and can be bounded similarly. This completes the proof. If there is more than one component of the fixed point set, we sum over the components since each component makes a separate contibution to the asymptotic series.

The case of isolated fixed points is of particular interest. Let $d = 2$ and $\sigma_L(P) = |\xi|^2 I_V$. We compute the first term at a fixed point:

$$r_0(x, \xi, \lambda) = (|\xi|^2 - \lambda)^{-1} \qquad e_0(x, \xi, t) = e^{-t|\xi|^2}$$

$$\int \operatorname{Tr}(T^* K_0)(t, x, x) \, \mathrm{dvol}(x) = \int \operatorname{Tr}(T^*(x)) e^{i(Tx-x)\cdot\xi} e^{-t|\xi|^2} \, d\xi \, dx.$$

We assume $T(0) = 0$ is an isolated non-degenerate fixed point. We let $y = Tx - x$ be a change of variables and compute the first term:

$$\int \operatorname{Tr}(T^*(y)) e^{iy\cdot\xi} |\det(I - dT(y))|^{-1} e^{-t|\xi|^2} \, d\xi \, dy.$$

We make a change of variables $\xi \to \xi t^{-1/2}$ and $y \to y t^{1/2}$ to express the first term:

$$\int \operatorname{Tr}(T^*(yt^{1/2})) e^{iy\cdot\xi} |\det(I - dT(yt^{1/2}))| e^{-|\xi|^2} \, d\xi \, dy.$$

The $d\xi$ integral just gives $e^{-|y|^2}$ so this becomes

$$\int \operatorname{Tr}(T^*(yt^{1/2})) |\det(I - dT(yt^{1/2}))|^{-1} e^{-|y|^2} \, dy.$$

We expand this in a Taylor series at $t = 0$ and evaluate to get

$$\operatorname{Tr} T^*(0) \, |\det(I - dT(O))|^{-1}.$$

This proves:

Lemma 1.8.3. *Let P be a second order elliptic partial differential operator with leading symbol $|\xi|^2 I$. Let $T: M \to M$ be smooth with non-degenerate isolated fixed points. Then:*

$$\operatorname{Tr} T^* e^{-tP} = \sum_i \operatorname{Tr}(T^*) \, |\det(I - dT)|^{-1}(x_i)$$

summed over the fixed point set.

We combine Lemmas 1.8.1 and 1.8.2 to constuct a local formula for $L(T)_P$ to generalize the local formula for $\operatorname{index}(P)$ given by Theorem 1.7.6; we will discuss this further in the fourth chapter.

We can use Lemma 1.8.3 to prove the classical Lefschetz fixed point formula for the de Rham complex. Let $T: V \to V$ be a linear map, then it is easily computed that:

$$\sum_p (-1)^p \operatorname{Tr} \Lambda^p(T) = \det(I - T).$$

We compute:

$$\begin{aligned}
L(T) &= \sum_p (-1)^p \operatorname{Tr}(T^* e^{-t\Delta_p}) \\
&= \sum_{p,i} (-1)^p \operatorname{Tr}(\Lambda^p dT) \, |\det(I - dT)|^{-1}(x_i) \\
&= \sum_i \det(I - dT) \, |\det(I - dT)|^{-1}(x_i) \\
&= \sum_i \operatorname{sign} \det(I - dT)(x_i)
\end{aligned}$$

summed over the fixed point set of T. This proves:

Theorem 1.8.4 (Classical Lefschetz Fixed Point Formula). *Let $T: M \to M$ be smooth with isolated non-degenerate fixed points. Then:*

$$\begin{aligned}
L(T) &= \sum_p (-1)^p \operatorname{Tr}(T^* \text{ on } H^p(M; \mathbf{C})) \\
&= \sum_i \operatorname{sign} \det(I - dT)(x_i)
\end{aligned}$$

summed over the fixed point set.

Remark: We can generalize Lemma 1.8.2 to study $\operatorname{Tr}(T^* Q e^{-tP})$ where Q is an auxilary differential operator of order a. Just as in lemma 1.7.7 we may obtain an asymptotic series:

$$\operatorname{Tr}(T^* Q e^{-tP}) \sim \sum_i \sum_{n=0}^{\infty} t^{(n-m_i-a)/d} a_n(x, Q, P) \, \mathrm{dvol}_i(x).$$

We shall omit the details as the additional terms created by the operator Q are exactly the same as those given in the proof of Lemma 1.7.7. Each term a_n is homogeneous of order n in the jets of the symbols of (Q, P) and of the map T in a suitable sense.

1.9. Elliptic Boundary Value Problems.

In section 1.7 we derived a local formula for the index of an elliptic partial-differential complex using heat equation methods. This formula will lead to a heat equation proof of the Atiyah-Singer index theorem which we shall discuss later. Unfortunately, it is not known at present how to give a heat equation proof of the Atiyah-Bott index theorem for manifolds with boundary in full generality. We must adopt a much stronger notion of ellipticity to deal with the analytic problems involved. This will yield a heat equation proof of the Gauss-Bonnet theorem for manifolds with boundary which we shall discuss in the fourth chapter.

Let M be a smooth compact manifold with smooth boundary dM and let $P: C^\infty(V) \to C^\infty(V)$ be a partial differential operator of order $d > 0$. We let $p = \sigma_L(P)$ be the leading symbol of P. We assume henceforth that p is self-adjoint and elliptic—i.e., $\det p(x, \xi) \neq 0$ for $\xi \neq 0$. Let $\mathbf{R}_\pm$ denote the non-zero positive/negative real numbers. It is immediate that

$$\det\{p(x,\xi) - \lambda\} \neq 0 \qquad \text{for } (\xi, \lambda) \neq (0,0) \in T^*(M) \times \{\mathbf{C} - \mathbf{R}_+ - \mathbf{R}_-\},$$

since p is self-adjoint and elliptic.

We fix a fiber metric on V and a volume element on M to define the global inner product $(\cdot, \cdot)$ on $L^2(V)$. We assume that P is formally self-adjoint:

$$(Pf, g) = (f, Pg)$$

for f and g smooth section with supports disjoint from the boundary dM. We must impose boundary conditions to make P self-adjoint. For example, if $P = -\partial^2/\partial x^2$ on the line segment $[0, A]$, then P is formally self-adjoint and elliptic, but we must impose Neumann or Dirichlet boundary conditions to ensure that P is self-adjoint with discrete spectrum.

Near dM we let $x = (y, r)$ where $y = (y_1, \ldots, y_{m-1})$ is a system of local coordinates on dM and where r is the normal distance to the boundary. We assume $dM = \{x : r(x) = 0\}$ and that $\partial/\partial r$ is the inward unit normal. We further normalize the choice of coordinate by requiring the curves $x(r) = (y_0, r)$ for $r \in [0, \delta)$ are unit speed geodesics for any $y_0 \in dM$. The inward geodesic flow identifies a neighborhood of dM in M with the collar $dM \times [0, \delta)$ for some $\delta > 0$. The collaring gives a splitting of $T(M) = T(dM) \oplus T(\mathbf{R})$ and a dual splitting $T^*(M) = T^*(dM) \oplus T^*(\mathbf{R})$. We let $\xi = (\varsigma, z)$ for $\varsigma \in T^*(dM)$ and $z \in T^*(\mathbf{R})$ reflect this splitting.

It is convenient to discuss boundary conditions in the context of graded vector bundles. A graded bundle U over M is a vector bundle U together with a fixed decomposition

$$U = U_0 \oplus \cdots \oplus U_{d-1}$$

into sub-bundles U_j where $U_j = \{0\}$ is permitted in this decomposition. We let $W = V \otimes 1^d = V \oplus \cdots \oplus V$ restricted to dM be the bundle of

Cauchy data. If $W_j = V|_{dM}$ is the $(j+1)^{\text{st}}$ factor in this decomposition, then this defines a natural grading on W; we identify W_j with the bundle of jth normal derivatives. The restriction map:

$$\underline{\gamma} \colon C^\infty(V) \to C^\infty(W)$$

defined by:

$$\underline{\gamma}(f) = (f_0, \ldots, f_{d-1}) \qquad \text{where } f_j = D_r^j f|_{dM} = (-i)^j \left. \frac{\partial^j f}{\partial r^j} \right|_{dM}$$

assigns to any smooth section its Cauchy data.

Let W' be an auxiliary graded vector bundle over dM. We assume that $\dim W = d \cdot \dim V$ is even and that $2 \dim W' = \dim W$. Let $B \colon C^\infty(W) \to C^\infty(W')$ be a tangential differential operator over dM. Decompose $B = B_{ij}$ for

$$B_{ij} \colon C^\infty(W_i) \to C^\infty(W'_j)$$

and assume that

$$\operatorname{ord}(B_{ij}) \le j - i.$$

It is natural to regard a section to $C^\infty(W_i)$ as being of order i and to define the graded leading symbol of B by:

$$\sigma_g(B)_{ij}(y, \varsigma) = \begin{cases} \sigma_L(B_{ij})(y, \varsigma) & \text{if } \operatorname{ord}(B_{ij}) = j - i \\ 0 & \text{if } \operatorname{ord}(B_{ij}) < j - i. \end{cases}$$

We then regard $\sigma_g(B)$ as being of graded homogeneity 0.

We let P_B be the operator P restricted to those $f \in C^\infty(V)$ such that $B\underline{\gamma}f = 0$. For example, let $P = -\partial^2/\partial x^2$ on the interval $[0, A]$. To define Dirichlet boundary conditions at $x = 0$, we would set:

$$W' = \mathbf{C} \oplus 0 \qquad \text{and} \quad B_{0,0} = 1,\, B_{1,1} = B_{0,1} = B_{1,0} = 0$$

while to define Neumann boundary conditions at $x = 0$, we would set:

$$W' = 0 \oplus \mathbf{C} \qquad \text{and} \quad B_{1,1} = 1,\, B_{0,0} = B_{0,1} = B_{1,0} = 0.$$

To define the notion of ellipticity we shall need, we consider the ordinary differential equation:

$$p(y, 0, \varsigma, D_r) f(r) = \lambda f(r) \qquad \text{with } \lim_{r \to \infty} f(r) = 0$$

where

$$(\varsigma, \lambda) \ne (0, 0) \in T^*(dM) \times \mathbf{C} - \mathbf{R}_+.$$

We say that (P, B) is elliptic with respect to $\mathbf{C} - \mathbf{R}_+$ if $\det(p(x, \xi) - \lambda) \neq 0$ on the interior for all $(\varsigma, \lambda) \neq (0, 0) \in T^*(M) \times \mathbf{C} - \mathbf{R}_+$ and if on the boundary there always exists a unique solution to this ordinary differential equation such that $\sigma_g(B)(y, \varsigma)\gamma f = f'$ for any prescribed $f' \in W'$. In a similar fashion, we define ellipticity with respect to $\mathbf{C} - \mathbf{R}_+ - \mathbf{R}_-$ if these conditions hold for $\lambda \in \mathbf{C} - \mathbf{R}_+ - \mathbf{R}_-$.

Again, we illustrate these notions for the operator $P = -\partial^2/\partial x^2$ and a boundary condition at $x = 0$. Since $\dim M = 1$, there is no dependence on ς and we must simply study the ordinary differential equation:

$$-f'' = \lambda f \qquad \text{with } \lim_{r \to \infty} f(r) = 0 \quad (\lambda \neq 0).$$

If $\lambda \in \mathbf{C} - \mathbf{R}_+$, then we can express $\lambda = \mu^2$ for $\text{Im}(\mu) > 0$. Solutions to the equation $-f'' = \lambda f$ are of the form $f(r) = ae^{i\mu r} + be^{-i\mu r}$. The decay at ∞ implies $b = 0$ so $f(r) = ae^{i\mu r}$. Such a function is uniquely determined by either its Dirichlet or Neumann data at $r = 0$ and hence P is elliptic with respect to either Neumann or Dirichlet boundary conditions.

We assume henceforth that P_B is self-adjoint and that (P, B) is elliptic with respect to either the cone $\mathbf{C} - \mathbf{R}_+$ or the cone $\mathbf{C} - \mathbf{R}_+ - \mathbf{R}_-$. It is beyond the scope of this book to develop the analysis required to discuss elliptic boundary value problems; we shall simply quote the required results and refer to the appropriate papers of Seeley and Greiner for further details. Lemmas 1.6.3 and 1.6.4 generalize to yield:

LEMMA 1.9.1. *Let $P: C^\infty(V) \to C^\infty(V)$ be an elliptic partial differential operator of order $d > 0$. Let B be a boundary condition. We assume (P, B) is self-adjoint and elliptic with respect to $\mathbf{C} - \mathbf{R}_+ - \mathbf{R}_-$.*
(a) We can find a complete orthonormal system $\{\phi_n\}_{n=1}^\infty$ for $L^2(V)$ with $P\phi_n = \lambda_n \phi_n$.
(b) $\phi_n \in C^\infty(V)$ and satisfy the boundary condition $B\gamma\phi_n = 0$.
(c) $\lambda_n \in \mathbf{R}$ and $\lim_{n \to \infty} |\lambda_n| = \infty$. If we order the λ_n so $|\lambda_1| \leq |\lambda_2| \cdots$ then there exists n_0 and $\delta > 0$ so that $|\lambda_n| > n^\delta$ for $n > n_0$.
(d) If (P, B) is elliptic with respect to the cone $\mathbf{C} - \mathbf{R}_+$, then the λ_n are bounded from below and $\text{spec}(P_B)$ is contained in $[-C, \infty)$ for some C.

For example, if $P = -\dfrac{\partial^2}{\partial x^2}$ on $[0, \pi]$ with Dirichlet boundary conditions, then the spectral resolution becomes $\left\{ \sqrt{\frac{2}{\pi}} \sin(nx) \right\}_{n=1}^\infty$ and the corresponding eigenvalues are n^2.

If (P, B) is elliptic with respect to the cone $\mathbf{C} - \mathbf{R}_+$, then e^{-tP_B} is a smoothing operator with smooth kernel $K(t, x, x')$ defined by:

$$K(t, x, x') = \sum_n e^{-t\lambda_n} \phi_n(x) \otimes \bar{\phi}_n(x').$$

Lemma 1.7.4 generalizes to yield:

LEMMA 1.9.2. *Let (P, B) be elliptic with respect to the cone $\mathbf{C} - \mathbf{R}_+$, and be of order $d > 0$.*
(a) e^{-tP_B} is an infinitely smoothing operator with smooth kernel $K(t, x, x')$.
(b) On the interior we define $a_n(x, P) = \operatorname{Tr} e_n(x, P)$ be as in Lemma 1.7.4. On the boundary we define $a_n(y, P, B)$ using a complicated combinatorial recipe which depends functorially on a finite number of jets of the symbols of P and of B.
(c) As $t \to 0^+$ we have an asymptotic expansion:

$$\operatorname{Tr} e^{-tP_B} = \sum_n e^{-t\lambda_n} = \int_M \operatorname{Tr} K(t, x, x)\, \mathrm{dvol}(x)$$

$$\sim \sum_{n=0}^{\infty} t^{\frac{n-m}{d}} \int_M a_n(x, P)\, \mathrm{dvol}(x)$$

$$+ \sum_{n=0}^{\infty} t^{\frac{n-m+1}{d}} \int_{dM} a_n(y, P, B)\, \mathrm{dvol}(y).$$

(d) $a_n(x, P)$ and $a_n(y, P, B)$ are invariantly defined scalar valued functions which do not depend on the coordinate system chosen nor on the local frame chosen. $a_n(x, P) = 0$ if n is odd, but $a_n(y, P, B)$ is in general non-zero for all values of n.

The interior term $a_n(x, P)$ arises from the calculus described previously. We first construct a parametrix on the interior. Since this parametrix will not have the proper boundary values, it is necessary to add a boundary correction term which gives rise to the additional boundary integrands $a_n(y, P, B)$.

To illustrate this asymptotic series, we let $P = -\dfrac{\partial^2}{\partial x^2} - e(x)$ on the interval $[0, A]$ where $e(x)$ is a real potential. Let B be the modified Neumann boundary conditions: $f'(0) + s(0)f(0) = f'(A) + s(A)f(A) = 0$. This is elliptic and self-adjoint with respect to the cone $\mathbf{C} - \mathbf{R}_+$. It can be computed that:

$$\operatorname{Tr} e^{-tP_B}$$

$$\sim (4\pi t)^{-1/2} \int \{1 + t \cdot e(x) + t^2 \cdot (e''(x) + 3e(x)^2)/6 + \cdots\}\, dx$$

$$+ \frac{1}{2} + \left(\frac{t}{\pi}\right)^{1/2} (s(0) - s(A))$$

$$+ \frac{t}{4} \cdot (e(0) + e(A) + 2s^2(0) + 2s^2(A)) + \cdots$$

The terms arising from the interior increase by integer powers of t while the terms from the boundary increase by powers of $t^{1/2}$. In this integral, dx is ordinary unnormalized Lebesgue measure.

In practice, we shall be interested in first order operators which are elliptic with respect to the cone $\mathbf{C} - \mathbf{R}_+ - \mathbf{R}_-$ and which have both positive and negative spectrum. An interesting measure of the spectral asymmetry of such an operator is obtained by studying $\text{Tr}(P_B e^{-tP_B^2})$ which will be discussed in more detail in the next section. To study the index of an elliptic operator, however, it suffices to study $e^{-tP_B^2}$; it is necessary to work with P_B^2 of course to ensure that the spectrum is positive so this converges to define a smoothing operator. There are two approaches which are available. The first is to work with the function $e^{-t\lambda^2}$ and integrate over a path of the form:

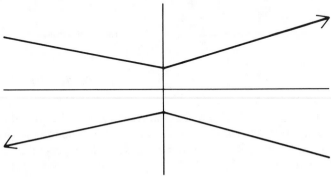

to define $e^{-tP_B^2}$ directly using the functional calculus. The second approach is to define $e^{-tP_B^2}$ using the operator P^2 with boundary conditions $B\gamma f = B\underline{\gamma}Pf = 0$. These two approaches both yield the same operator and give an appropriate asymptotic expansion which generalizes Lemma 1.9.2.

We use the heat equation to construct a local formula for the index of certain elliptic complexes. Let $Q: C^\infty(V_1) \to C^\infty(V_2)$ be an elliptic differential operator of order $d > 0$. Let $Q^*: C^\infty(V_2) \to C^\infty(V_1)$ be the formal adjoint. Let $B_1: C^\infty(W_1) \to C^\infty(W_1')$ be a boundary condition for the operator Q. We assume there exists a boundary condition B_2 for Q^* of the same form so that

$$(Q_{B_1})^* = (Q^*)_{B_2}.$$

We form:

$$P = Q \oplus Q^* \qquad \text{and} \qquad B = B_1 \oplus B_2$$

so $P: C^\infty(V_1 \oplus V_2) \to C^\infty(V_1 \oplus V_2)$. Then P_B will be self-adjoint. We assume that (P, B) is elliptic with respect to the cone $\mathbf{C} - \mathbf{R}_+ - \mathbf{R}_-$.

This is a very strong assmption which rules out many interesting cases, but is necessary to treat the index theorem using heat equation methods. We emphasize that the Atiyah-Bott theorem in its full generality does not follow from these methods.

Since $P^2 = Q^*Q + QQ^*$, it is clear that P_B^2 decomposes as the sum of two operators which preserve $C^\infty(V_1)$ and $C^\infty(V_2)$. We let S be the endomorphism $+1$ on V_1 and -1 on V_2 to take care of the signs; it is clear $P_B^2 S = S P_B^2$. The same cancellation lemma we have used previously yields:

$$\text{index}(Q_1, B_1) = \dim \text{N}(Q_{B_1}) - \dim \text{N}(Q_{B_2}^*) = \text{Tr}(S e^{-t P_B^2}).$$

(The fact that this definition of the index agrees with the definition given in the Atiyah-Bott paper follows from the fact that $B: C^\infty(W) \to C^\infty(W_1)$ is surjective). Consequently, the application of Lemma 1.9.2 yields:

THEOREM 1.9.3. Let $Q: C^\infty(V_1) \to C^\infty(V_2)$ be an elliptic differential operator of order $d > 0$. Let $Q^*: C^\infty(V_2) \to C^\infty(V_1)$ be the formal adjoint and define $P = Q \oplus Q^*: C^\infty(V_1 \oplus V_2) \to C^\infty(V_1 \oplus V_2)$. Let $B = B_1 \oplus B_2$ be a boundary condition such that (P, B) is elliptic with respect to the cone $\mathbf{C} - \mathbf{R}_+ - \mathbf{R}_-$ and so that P_B is self-adjoint. Define $S = +1$ on V_1 and -1 on V_2 then:
(a) $\text{index}(Q_{B_1}) = \text{Tr}(S e^{-t P_B^2})$ for all t.
(b) There exist local invariants $a_n(x, Q)$ and $a_n(y, Q, B_1)$ such that:

$$\text{Tr}(S e^{-t P_B^2}) \sim \sum_{n=0}^{\infty} t^{\frac{n-m}{2d}} \int_M a_n(x, Q)\, \text{dvol}(x)$$

$$+ \sum_{n=0}^{\infty} t^{\frac{n-m+1}{2d}} \int_{dM} a_n(y, Q, B_1)\, \text{dvol}(y).$$

(c)

$$\int_M a_n(x, Q)\, \text{dvol}(x) + \int_{dM} a_{n-1}(y, Q, B_1)\, \text{dvol}(y)$$

$$= \begin{cases} \text{index}(Q_{B_1}) & \text{if } n = m \\ 0 & \text{if } n \neq m. \end{cases}$$

This gives a local formula for the index; there is an analogue of Lemma 1.7.5 giving various functorial properties of these invariants we will discuss in the fourth chapter where we shall discuss the de Rham complex and the Gauss-Bonnet theorem for manifolds with boundary.

We now specialize henceforth to first order operators. We decompose

$$p(x, \xi) = \sum_j e_j(x) \xi_j$$

and near the boundary express:

$$p(y, 0, \varsigma, z) = e_0(y) \cdot z + \sum_{j=1}^{m-1} e_j(y) \varsigma_j = e_0(y) \cdot z + p(y, 0, \varsigma, 0).$$

We study the ordinary differential equation:

$$\{ip_0\partial/\partial r + p(y,0,\varsigma,0)\}f(r) = \lambda f(r)$$

or equivalently:

$$-ip_0\{\partial/\partial r + ip_0^{-1}p(y,0,\varsigma,0) - ip_0^{-1}\lambda)\}f(r) = 0.$$

With this equation in mind, we define:

$$\tau(y,\varsigma,\lambda) = ip_0^{-1}(p(y,0,\varsigma,0) - \lambda).$$

LEMMA 1.9.4. *Let* $p(x,\xi)$ *be self-adjoint and elliptic. Then* $\tau(y,\varsigma,\lambda)$ *has no purely imaginary eigenvalues for* $(\varsigma,\lambda) \neq (0,0) \in T^*(dM) \times (\mathbf{C} - \mathbf{R}_+ - \mathbf{R}_-)$.

PROOF: We suppose the contrary and set $\tau(y,\varsigma,\lambda)v = izv$ where z is real and $v \neq 0$. This implies that:

$$(p(y,0,\varsigma,0) - \lambda)v = p_0zv$$

or equivalently that:

$$p(y,0,\varsigma,-z)v = \lambda v.$$

Since p is self-adjoint, this implies $\lambda = 0$ so $\varsigma \neq 0$ which contradicts the ellipticity of p.

We define bundles $\Pi_\pm(\tau)$ over $T^*(dM) \times \{\mathbf{C} - \mathbf{R}_+ - \mathbf{R}_-\} - (0,0)$ to be the span of the generalized eigenvectors of τ which correspond to eigenvalues with positive/negative real part; $\Pi_+ \oplus \Pi_- = V$. The differential equation has the form:

$$\{\partial/\partial r + \tau\}f = 0$$

so the condition $\lim_{r\to\infty} f(r) = 0$ implies that $f(0) \in \Pi_+(\tau)$. Thus $\Pi_+(\tau)$ is the bundle of Cauchy data corresponding to solutions to this ODE. Since $d = 1$, the boundary condition B is just an endomorphism:

$$B: V_{|dM} \to W'$$

and we conclude:

LEMMA 1.9.5. *Let* P *be a first order formally self-adjoint elliptic differential operator. Let* B *be a* 0th *order boundary condition. Define*

$$\tau(y,\varsigma,\lambda) = ip_0^{-1}(p(y,0,\varsigma,0) - \lambda).$$

Then τ *has no purely imaginary eigenvalues and we define* $\Pi_\pm(\tau)$ *to be bundles of generalized eigenvectors corresponding to eigenvalues with positive real and negative real parts.* (P, B) *is elliptic with respect to the cone*

$\mathbf{C} - \mathbf{R}_+ - \mathbf{R}_-$ if and only if $B\colon \Pi_+(\tau) \to W'$ is an isomorphism for all $(\varsigma, \lambda) \neq (0,0) \in T^*(dM) \times \{\mathbf{C} - \mathbf{R}_+ - \mathbf{R}_-\}$. P_B is self-adjoint if and only if $p_0\, N(B)$ is perpendicular to $N(B)$.

PROOF: We have checked everything except the condition that P_B be self-adjoint. Since P is formally self-adjoint, it is immediate that:

$$(Pf, g) - (f, Pg) = \int_{dM} (-ip_0 f, g).$$

We know that on the boundary, both f and g have values in $N(B)$ since they satisfy the boundary condition. Thus this vanishes identically if and only if $(p_0 f, g) = 0$ for all $f, g \in N(B)$ which completes the proof.

We emphasize that these boundary conditions are much stronger than those required in the Atiyah-Bott theorem to define the index. They are much more rigid and avoid some of the pathologies which can occur otherwise. Such boundary conditions do not necessarily exist in general as we shall see later.

We shall discuss the general case in more detail in Chapter 4. We complete this section by discussing the one-dimensional case case to illustrate the ideas involved. We consider $V = [0,1] \times \mathbf{C}^2$ and let P be the operator:

$$P = -i\frac{\partial}{\partial r} \begin{pmatrix} 0 & 1 \\ 1 & 0 \end{pmatrix}.$$

At the point 0, we have:

$$\tau(\lambda) = -i\lambda \begin{pmatrix} 0 & 1 \\ 1 & 0 \end{pmatrix}.$$

Thus if $\mathrm{Im}(\lambda) > 0$ we have $\Pi_+(\tau) = \left\{ \begin{pmatrix} a \\ a \end{pmatrix} \right\}$ while if $\mathrm{Im}(\lambda) < 0$ we have $\Pi_+(\tau) = \left\{ \begin{pmatrix} a \\ -a \end{pmatrix} \right\}$. We let B be the boundary projection which projects on the first factor—$N(B) = \left\{ \begin{pmatrix} 0 \\ a \end{pmatrix} \right\}$; we take the same boundary condition at $x = 1$ to define an elliptic self-adjoint operator P_B. Let $P^2 = -\dfrac{\partial^2}{\partial x^2}$ with boundary conditions: Dirichlet boundary conditions on the first factor and Neumann boundary conditions on the second factor. The index of the problem is -1.

1.10. Eta and Zeta Functions.

We have chosen to work with the heat equation for various technical reasons. However, much of the development of the subject has centered on the zeta function so we shall briefly indicate the relationship between these two in this section. We shall also define the eta invariant which plays an important role in the Atiyah-Singer index theorem for manifolds with boundary.

Recall that Γ is defined by:

$$\Gamma(s) = \int_0^\infty t^{s-1} e^{-t} dt$$

for $\text{Re}(s) > 0$. We use the functional equation $s\Gamma(s) = \Gamma(s+1)$ to extend Γ to a meromorphic function on C with isolated simple poles at $s = 0, -1, -2, \ldots$. Let $P: C^\infty(V) \to C^\infty(V)$ be an elliptic self-adjoint partial differential operator of order $d > 0$ with positive definite leading symbol. We showed in section 1.6 that this implies $\text{spec}(P)$ is contained in $[-C, \infty)$ for some constant C. We now assume that P itself is positive definite—i.e., $\text{spec}(P)$ is contained in $[\varepsilon, \infty)$ for some $\varepsilon > 0$.

Proceeding formally, we define P^s by:

$$P^s = \frac{1}{2\pi i} \int_\alpha \lambda^s (P - \lambda)^{-1} d\lambda$$

where α is a suitable path in the half-plane $\text{Re}(\lambda) > 0$. The estimates of section 1.6 imply this is smoothing operator if $\text{Re}(s) \ll 0$ with a kernel function given by:

$$L(s, x, y) = \sum_n \lambda_n^s \phi_n(x) \otimes \bar{\phi}_n(y);$$

this converges to define a C^k kernel if $\text{Re}(s) < s_0(k)$.

We use the Mellin transform to relate the zeta and heat kernels.

$$\int_0^\infty t^{s-1} e^{-\lambda t} dt = \lambda^{-s} \int_0^\infty (\lambda t)^{s-1} e^{-\lambda t} d(\lambda t) = \lambda^{-s} \Gamma(s).$$

This implies that:

$$\Gamma(s) \operatorname{Tr}(P^{-s}) = \int_0^\infty t^{s-1} \operatorname{Tr} e^{-tP} dt.$$

We define $\varsigma(s, P) = \operatorname{Tr}(P^{-s})$; this is holomorphic for $\text{Re}(s) \gg 0$. We decompose this integral into $\int_0^1 + \int_1^\infty$. We have bounded the eigenvalues of

P away from zero. Using the growth estimates of section 1.6 it is immediate that:

$$\int_1^\infty t^{s-1} \operatorname{Tr} e^{-tP} \, dt = r(s, P)$$

defines an entire function of s. If $0 < t < 1$ we use the results of section 1.7 to expand

$$\operatorname{Tr} e^{-tP} = \sum_{n \le n_0} t^{\frac{n-m}{d}} a_n(P) + O\left(t^{\frac{n_0-m}{d}}\right)$$

$$a_n(P) = \int_M a_n(x, P) \, \mathrm{dvol}(x)$$

where $a_n(x, P)$ is a local scalar invariant of the jets of the total symbol of P given by Lemma 1.7.4.

If we integrate the error term which is $O\left(t^{\frac{n_0-m}{d}}\right)$ from 0 to 1, we define a holomorphic function of s for $\operatorname{Re}(s) + \frac{n_0-m}{d} > 0$. We integrate $t^{s-1} t^{\frac{(n-m)}{d}}$ between 0 and 1 to conclude:

$$\Gamma(s) \operatorname{Tr}(P^{-s}) = \Gamma(s)\varsigma(s, P) = \sum_{n < n_0} d(sd + n - m)^{-1} a_n(P) + r_{n_0}$$

where r_{n_0} is holomorphic for $\operatorname{Re}(s) > -\frac{(n_0-m)}{d}$. This proves $\Gamma(s)\varsigma(s, P)$ extends to a meromorphic function on C with isolated simple poles. Furthermore, the residue at these poles is given by a local formula. Since Γ has isolated simple poles at $s = 0, -1, -2, \ldots$ we conclude that:

LEMMA 1.10.1. *Let P be a self-adjoint, positive, elliptic partial differential operator of order $d > 0$ with positive definite symbol. We define:*

$$\varsigma(s, P) = \operatorname{Tr}(P^{-s}) = \Gamma(s)^{-1} \int_0^\infty t^{s-1} \operatorname{Tr} e^{-tP} \, dt.$$

This is well defined and holomorphic for $\operatorname{Re}(s) \gg 0$. It has a meromorphic extension to C with isolated simple poles at $s = (m - n)/d$ for $n = 0, 1, 2, \ldots$. The residue of ς at these local poles is $a_n(P)\Gamma((m - n)/d)^{-1}$. If $s = 0, -1, \ldots$ is a non-positive integer, then $\varsigma(s, P)$ is regular at this value of s and its value is given by $a_n(P) \operatorname{Res}_{s=(m-n)/d} \Gamma(s)$. $a_n(P)$ is the invariant given in the asymptotic expansion of the heat equation. $a_n(P) = \int_M a_n(x, P) \, \mathrm{dvol}(x)$ where $a_n(x, P)$ is a local invariant of the jets of the total symbol of P. a_n vanishes if n is odd.

Remark: If A is an auxiliary differential operator of order a, we can define

$$\varsigma(s, A, P) = \operatorname{Tr}(AP^{-s}) = \Gamma(s)^{-1} \int_0^\infty t^{s-1} \operatorname{Tr}(Ae^{-tP}) \, dt.$$

We apply Lemma 1.7.7 to see this has a meromorphic extension to $\mathbf{C}$ with isolated simple poles at $s = (m + a - n)/d$. The residue at these poles is given by the generalized invariants of the heat equation

$$\frac{a_n(A, P)}{\Gamma\{(m + a - n)/d\}}.$$

There is a similar theorem if $dM \neq \emptyset$ and if B is an elliptic boundary condition as discussed in section 1.9.

If P is only positive semi-definite so it has a finite dimensional space corresponding to the eigenvalue 0, we define:

$$\varsigma(s, P) = \sum_{\lambda_n > 0} \lambda_n^{-s}$$

and Lemma 1.10 extends to this case as well with suitable modifications. For example, if $M = S^1$ is the unit circle and if $P = -\partial^2/\partial\theta^2$ is the Laplacian, then:

$$\varsigma(s, P) = 2 \sum_{n > 0} n^{-2s}$$

is essentially just the Riemann zeta function. This has a simple isolated pole at $s = 1/2$.

If $Q: C^\infty(V_1) \rightarrow C^\infty(V_2)$, then the cancellation lemma of section 1.6 implies that:

$$\varepsilon^{-s} \operatorname{index}(Q) = \varsigma(s, Q^*Q + \varepsilon) - \varsigma(s, QQ^* + \varepsilon)$$

for any $\varepsilon > 0$ and for any s. ς is regular at $s = 0$ and its value is given by a local formula. This gives a local formula for $\operatorname{index}(Q)$. Using Lemma 1.10.1 and some functorial properties of the heat equation, it is not difficult to show this local formula is equivalent to the local formula given by the heat equation so no new information has resulted. The asymptotics of the heat equation are related to the values and residues of the zeta function by Lemma 1.10.1.

If we do not assume that P is positive definite, it is possible to define a more subtle invariant which measures the difference between positive and negative spectrum. Let P be a self-adjoint elliptic partial differential operator of order $d > 0$ which is not necessarily positive definite. We define the eta invariant by:

$$\eta(s, P) = \sum_{\lambda_n > 0} \lambda_n^{-s} - \sum_{\lambda_n < 0} (-\lambda_n)^{-s} = \sum_{\lambda_n \neq 0} \operatorname{sign}(\lambda_n)|\lambda_n|^{-s}$$
$$= \operatorname{Tr}(P \cdot (P^2)^{-(s+1)/2}).$$

Again, this is absolutely convergent and defines a holomorphic function if $\operatorname{Re}(s) \gg 0$.

We can also discuss the eta invariant using the heat equation. The identity:

$$\int_0^\infty t^{(s-1)/2} \lambda e^{-\lambda^2 t}\, dt = \Gamma((s+1)/2)\operatorname{sign}(\lambda)|\lambda|^{-s}$$

implies that:

$$\eta(s,P) = \Gamma((s+1)/2)^{-1} \int_0^\infty t^{(s-1)/2} \operatorname{Tr}\{Pe^{-tP^2}\}\, dt.$$

Again, the asymptotic behavior at $t = 0$ is all that counts in producing poles since this decays exponentially at ∞ assuming P has no zero eigenvector; if $\dim N(P) > 0$ a seperate argument must be made to take care of this eigenspace. This can be done by replacing P by $P + \varepsilon$ and letting $\varepsilon \to 0$.

Lemma 1.7.7 shows that there is an asymptotic series of the form:

$$\operatorname{Tr}(Pe^{-tP^2}) \sim \sum_{n=0}^\infty t^{\frac{n-m-d}{2d}} a_n(P,P^2)$$

for

$$a_n(P,P^2) = \int_M a_n(x,P,P^2)\, \mathrm{dvol}(x).$$

This is a local invariant of the jets of the total symbol of P.

We substitute this asymptotic expansion into the expression for η to see:

$$\eta(s,P)\Gamma((s+1)/2)^{-1} = \sum_{n \leq n_0} \frac{2d}{ds + n - m} a_n(P,P^2) + r_{n_0}$$

where r_{n_0} is holomorphic on a suitable half-plane. This proves η has a suitable meromorphic extension to $\mathbf{C}$ with locally computable residues.

Unfortunately, while it was clear from the analysis that ς was regular at $s = 0$, it is not immediate that η is regular at $s = 0$ since Γ does not have a pole at $\frac{1}{2}$ to cancel the pole which may be introduced when $n = m$. in fact, if one works with the local invariants involved, $a_n(x,P,P^2) \neq 0$ in general so the local poles are in fact present at $s = 0$. However, it is a fact which we shall discuss later that η is regular at $s = 0$ and we will define

$$\tilde{\eta}(p) = \frac{1}{2}\{\eta(0,p) + \dim N(p)\} \quad \mod \mathbf{Z}$$

(we reduce modulo $\mathbf{Z}$ since η has jumps in $2\mathbf{Z}$ as eigenvalues cross the origin).

We compute a specific example to illustrate the role of η in measuring spectral asymmetry. Let $P = -i\partial/\partial\theta$ on $C^\infty(S^1)$, then the eigenvalues of p are the integers so $\eta(s, p) = 0$ since the spectrum is symmetric about the origin. Let $a \in \mathbf{R}$ and define:

$$P_a = P - a, \qquad \eta(s, P_a) = \sum_{n \in \mathbf{Z}} \text{sign}\{n - a\}|n - a|^{-s}.$$

We differentiate this with respect to the parameter a to conclude:

$$\frac{d}{da}\eta(s, P(a)) = \sum_{n \in \mathbf{Z}} s((n - a)^2)^{-(s+1)/2}.$$

If we compare this with the Riemann zeta function, then

$$\sum_{n \in \mathbf{Z}} ((n - a)^2)^{-(s+1)/2}$$

has a simple pole at $s = 0$ with residue 2 and therefore:

$$\frac{d}{da}\eta(s, P(a))\bigg|_{s=0} = 2.$$

Since η vanishes when $a = 0$, we integrate this with respect to a to conclude:

$$\eta(0, P(a)) = 2a + 1 \mod 2\mathbf{Z} \qquad \text{and} \qquad \tilde{\eta}(P(a)) = a + \frac{1}{2}$$

is non-trivial. (We must work modulo $\mathbf{Z}$ since $\text{spec } P(a)$ is periodic with period 1 in a).

We used the identity:

$$\frac{d}{da}\eta(s, P_a) = -s\,\text{Tr}(\dot{P}_a(P_a^2)^{-(s+1)/2})$$

in the previous computation; it in fact holds true in general:

LEMMA 1.10.2. *Let $P(a)$ be a smooth 1-parameter family of elliptic self-adjoint partial differential operators of order $d > 0$. Assume $P(a)$ has no zero spectrum for a in the parameter range. Then if "·" denotes differentiation with respect to the parameter a,*

$$\dot{\eta}(s, P(a)) = -s\,\text{Tr}(\dot{P}(a)(P(a)^2)^{-(s+1)/2}).$$

If $P(a)$ has zero spectrum, we regard $\eta(s, P(a)) \in \mathbf{C}/\mathbf{Z}$.

PROOF: If we replace P by P^k for an odd positive integer k, then $\eta(s, P^k) = \eta(ks, P)$. This shows that it suffices to prove Lemma 1.10.2 for $d \gg 0$.

By Lemma 1.6.6, $(P - \lambda)^{-1}$ is smoothing and hence of trace class for d large. We can take trace inside the integral to compute:

$$\eta(s, P(a)) = \frac{1}{2\pi i} \left\{ \int_{\gamma_1} \lambda^{-s} \operatorname{Tr}((P(a) - \lambda)^{-1}) \, d\lambda \right.$$
$$\left. - \int_{\gamma_2} (-\lambda)^{-s} \operatorname{Tr}((P(a) - \lambda)^{-1} \, d\lambda \right\}$$

where γ_1 and γ_2 are paths about the positive/negative real axis which are oriented suitably. We differentiate with respect to the parameter a to express:

$$\frac{d}{da}((P(a) - \lambda)^{-1}) = -(P(a) - \lambda)^{-1} \dot{P}(a)(P(a) - \lambda)^{-1}.$$

Since the operators involved are of Trace class,

$$\operatorname{Tr}\left(\frac{d}{da}(P(a) - \lambda)^{-1}\right) = -\operatorname{Tr}(\dot{P}(a)(P(a) - \lambda)^{-2}).$$

We use this identity and bring trace back outside the integral to compute:

$$\eta(s, P(a)) = \frac{1}{2\pi i} \operatorname{Tr}(\dot{P}(a)) \left\{ \int_{\gamma_1} -\lambda^{-s}(P(a) - \lambda)^{-2} \, d\lambda \right.$$
$$\left. + \int_{\gamma_2} (-\lambda)^{-s}(P(a) - \lambda)^{-2} \, d\lambda \right\}.$$

We now use the identity:

$$\frac{d}{d\lambda}((P(a) - \lambda)^{-1}) = (P(a) - \lambda)^{-2}$$

to integrate by parts in λ in this expression. This leads immediately to the desired formula.

We could also calculate using the heat equation. We proceed formally:

$$\frac{d\eta}{da}(s, P(a)) = \Gamma\{(s+1)/2\}^{-1} \int_0^\infty t^{(s-1)/2} \operatorname{Tr}\left(\frac{d}{da}(P(a)e^{-tP(a)^2})\right) dt$$
$$= \Gamma\{(s+1)/2\}^{-1} \int_0^\infty t^{(s-1)/2} \operatorname{Tr}\left(\frac{dP}{da}(1 - 2tP^2)e^{-tP^2}\right) dt$$
$$= \Gamma\{(s+1)/2\}^{-1} \int_0^\infty t^{(s-1)/2} \left(1 + 2t\frac{d}{dt}\right) \operatorname{Tr}\left(\frac{dP}{da}e^{-tP^2}\right) dt.$$

We now integrate by parts to compute:

$$= \Gamma\{(s+1)/2\}^{-1} \int_0^\infty -st^{(s-1)/2} \operatorname{Tr}\left(\frac{dP}{da}e^{-tP^2}\right) dt$$

$$\sim -s\Gamma\{(s+1)/2\}^{-1} \sum_n \int_0^1 t^{(s-1)/2} t^{(n-m-d)/2d} a_n\left(\frac{dP}{da},P^2\right) dt$$

$$\sim -s\Gamma\{(s+1)/2\}^{-1} \sum_n 2/(s+(n-m)/d) a_n\left(\frac{dP}{da},P^2\right).$$

In particular, this shows that $\dfrac{d\tilde{\eta}}{da}$ is regular at $s=0$ and the value

$$-\Gamma\left(\tfrac{1}{2}\right)^{-1} a_m\left(\frac{dP}{da},P^2\right)$$

is given by a local formula.

The interchange of order involved in using global trace is an essential part of this argument. $\operatorname{Tr}(\dot{P}(a)(P^2)^{-(s+1)/2})$ has a meromorphic extension to $\mathbf{C}$ with isolated simple poles. The pole at $s=-1/2$ can be present, but is cancelled off by the factor of $-s$ which multiplies the expression. Thus $\dfrac{d}{da}\operatorname{Res}_{s=0}\eta(s,P(a))=0$. This shows the global residue is a homotopy invariant; this fact will be used in Chapter 4 to show that in fact η is regular at the origin. This argument does not go through locally, and in fact it is possible to construct operators in any dimension $m \geq 2$ so that the local eta function is not regular at $s=0$.

We have assumed that $P(a)$ has no zero spectrum. If we supress a finite number of eigenvalues which may cross the origin, this makes no contribution to the residue at $s=0$. Furthermore, the value at $s=0$ changes by jumps of an even integer as eigenvalues cross the origin. This shows $\tilde{\eta}$ is in fact smooth in the parameter a and proves:

LEMMA 1.10.3.

(a) *Let P be a self-adjoint elliptic partial differential operator of positive order d. Define:*

$$\eta(s,P) = \operatorname{Tr}(P(P^2)^{-(s+1)/2}) = \Gamma\{(s+1)/2\}^{-1}\int_0^\infty t^{(s-1)/2}\operatorname{Tr}(Pe^{-tP^2})\,dt.$$

This admits a meromorphic extension to $\mathbf{C}$ with isolated simple poles at $s=(m-n)/d$. The residue of η at such a pole is computed by:

$$\operatorname{Res}_{s=(m-n)/d} = 2\Gamma\{(m+d-n)/2d\}^{-1}\int_M a_n(x,P,P^2)\,\mathrm{dvol}(x)$$

where $a_n(x,P,P^2)$ is defined in Lemma 1.7.7; it is a local invariant in the jets of the symbol of P.

(b) Let $P(a)$ be a smooth 1-parameter family of such operators. If we assume eigenvalues do not cross the origin, then:

$$\frac{d}{da}\eta(s, P(a)) = -s\operatorname{Tr}\left(\frac{dP}{da} \cdot (P(a)^2)^{-(s+1)/2}\right)$$

$$= -2s\Gamma\{(s+1)/2\}^{-1}\int_0^\infty t^{(s-1)/2}\operatorname{Tr}\left(\frac{dP}{da} \cdot e^{-tP^2}\right)dt.$$

Regardless of whether or not eigenvalues cross the origin, $\tilde\eta(P(a))$ is smooth in $\mathbf{R}$ mod $\mathbf{Z}$ in the parameter a and

$$\frac{d}{da}\tilde\eta(P(a)) = -\Gamma(\tfrac{1}{2})^{-1}a_m\left(\frac{dP}{da}, P^2\right)$$

$$= -\Gamma(\tfrac{1}{2})^{-1}\int_M a_m\left(x, \frac{dP}{da}, P^2\right)\operatorname{dvol}(x)$$

is the local invariant in the jets of the symbols of $\left(\dfrac{dP}{da}, P^2\right)$ given by Lemma 1.7.7.

In the example on the circle, the operator $P(a)$ is locally isomorphic to P. Thus the value of η at the origin is not given by a local formula. This is a global invariant of the operator; only the derivative is locally given.

It is not necessary to assume that P is a differential operator to define the eta invariant. If P is an elliptic self-adjoint pseudo-differential operator of order $d > 0$, then the sum defining $\eta(s, P)$ converges absolutely for $s \gg 0$ to define a holomorphic function. This admits a meromorphic extension to the half-plane $\operatorname{Re}(s) > -\delta$ for some $\delta > 0$ and the results of Lemma 1.10.3 continue to apply. This requires much more delicate estimates than we have developed and we shall omit details. The reader is referred to the papers of Seeley for the proofs.

We also remark that it is not necessary to assume that P is self-adjoint; it suffices to assume that $\det(p(x, \xi) - it) \neq 0$ for $(\xi, t) \neq (0, 0) \in T^*M \times \mathbf{R}$. Under this ellipticity hypothesis, the spectrum of P is discrete and only a finite number of eigenvalues lie on or near the imaginary axis. We define:

$$\eta(s, P) = \sum_{\operatorname{Re}(\lambda_i)>0}(\lambda_i)^{-s} - \sum_{\operatorname{Re}(\lambda_i)<0}(-\lambda_i)^{-s}$$

$$\tilde\eta(P) = \frac{1}{2}\left\{\eta(s, P) - \frac{1}{s}\operatorname{Res}_{s=0}\eta(s, P) + \dim\operatorname{N}(i\mathbf{R})\right\}_{s=0} \quad \mod \mathbf{Z}$$

where $\dim\operatorname{N}(i\mathbf{R})$ is the dimension of the finite dimensional vector space of generalized eigenvectors of P corresponding to purely imaginary eigenvalues.

In section 4.3, we will discuss the eta invariant in further detail and use it to define an index with coefficients in a locally flat bundle using secondary characteristic classes.

If the leading symbol of P is positive definite, the asymptotics of $\operatorname{Tr}(e^{-tP})$ as $t \to 0^+$ are given by local formulas integrated over the manifold. Let $A(x) = a_0 x^a + \cdots + a_a$ and $B(x) = b_0 x^b + \cdots + b_b$ be polynomials where $b_0 > 0$. The operator $A(P)e^{-tB(P)}$ is infinitely smoothing. The asymptotics of $\operatorname{Tr}(A(P)e^{-tB(P)})$ are linear combinations of the invariants $a_N(P)$ giving the asymptotics of $\operatorname{Tr}(e^{-tP})$. Thus there is no new information which results by considering more general operators of heat equation type. If the leading symbol of P is indefinite, one must consider both the zeta and eta function or equivalently:

$$\operatorname{Tr}(e^{-tP^2}) \quad \text{and} \quad \operatorname{Tr}(Pe^{-tP^2}).$$

Then if $b_0 > 0$ is even, we obtain enough invariants to calculate the asymptotics of $\operatorname{Tr}(A(P)e^{-tB(P)})$. We refer to Fegan-Gilkey for further details.

CHAPTER 2

CHARACTERISTIC CLASSES

Introduction

In the second chapter, we develop the theory of characteristic classes. In section 2.1, we discuss the Chern classes of a complex vector bundle and in section 2.2 we discuss the Pontrjagin and Euler classes of a real vector bundle. We shall define the Todd class, the Hirzebruch L-polynomial, and the A-roof genus which will play a central role in our discussion of the index theorem. We also discuss the total Chern and Pontrjagin classes as well as the Chern character.

In section 2.3, we apply these ideas to the tangent space of a real manifold and to the holomorphic tangent space of a holomorphic manifold. We compute several examples defined by Clifford matrices and compute the Chern classes of complex projective space. We show that suitable products of projective spaces form a dual basis to the space of characteristic classes. Such products will be used in chapter three to find the normalizing constants which appear in the formula for the index theorem.

In section 2.4 and in the first part of section 2.5, we give a heat equation proof of the Gauss-Bonnet theorem. This is based on first giving an abstract characterization of the Euler form in terms of invariance theory. This permis us to identify the integrand of the heat equation with the Euler integrand. This gives a more direct proof of Theorem 2.4.8 which was first proved by Patodi using a complicated cancellation lemma. (The theorem for dimension $m = 2$ is due to McKean and Singer).

In the remainder of section 2.5, we develop a similar characterization of the Pontrjagin forms of the real tangent space. We shall wait until the third chapter to apply these results to obtain the Hirzebruch signature theorem. There are two different approaches to this result. We have presented both our original approach and also one modeled on an approach due to Atiyah, Patodi and Bott. This approach uses H. Weyl's theorem on the invariants of the orthogonal group and is not self-contained as it also uses facts about geodesic normal coordinates we shall not develop. The other approach is more combinatorial in flavor but is self-contained. It also generalizes to deal with the holomorphic case for a Kaehler metric.

The signature complex with coefficients in a bundle V gives rise to invariants which depend upon both the metric on the tangent space of M and on the connection 1-form of V. In section 2.6, we extend the results of section 2.5 to cover more general invariants of this type. We also construct dual bases for these more general invariants similar to those constructed in section 2.3 using products of projective spaces and suitable bundles over these spaces.

The material of sections 2.1 through 2.3 is standard and reviews the theory of characteristic classes in the context we shall need. Sections 2.4 through 2.6 deal with less standard material. The chapter is entirely self-contained with the exception of some material in section 2.5 as noted above. We have postponed a discussion of similar material for the holomorphic Kaehler case until sections 3.6 and 3.7 of chapter three since this material is not needed to discuss the signature and spin complexes.

2.1. Characteristic Classes
Of a Complex Vector Bundle.

The characteristic classes are topological invariants of a vector bundle which are represented by differential forms. They are defined in terms of the curvature of a connection.

Let M be a smooth compact manifold and let V be a smooth complex vector bundle of dimension k over M. A connection ∇ on V is a first order partial differential operator $\nabla: C^\infty(V) \to C^\infty(T^*M \otimes V)$ such that:

$$\nabla(fs) = df \otimes s + f\nabla s \qquad \text{for } f \in C^\infty(M) \text{ and } s \in C^\infty(V).$$

Let $(s_1, \ldots, s_k)$ be a local frame for V. We can decompose any section $s \in C^\infty(V)$ locally in the form $s(x) = f_i(x)s_i(x)$ for $f_i \in C^\infty(M)$. We adopt the convention of summing over repeated indices unless otherwise specified in this subsection. We compute:

$$\nabla s = df_i \otimes s_i + f_i \nabla s_i = df_i \otimes s_i + f_i \omega_{ij} \otimes s_j \qquad \text{where } \nabla s_i = \omega_{ij} \otimes s_j.$$

The connection 1-form ω defined by

$$\omega = \omega_{ij}$$

is a matrix of 1-forms. The connection ∇ is uniquely determined by the derivation property and by the connection 1-form. If we specify ω arbitrarily locally, we can define ∇ locally. The convex combination of connections is again a connection, so using a partition of unity we can always construct connections on a bundle V.

If we choose another frame for V, we can express $s'_i = h_{ij}s_j$. If $h_{ij}^{-1}s'_j = s_i$ represents the inverse matrix, then we compute:

$$\nabla s'_i = \omega'_{ij} \otimes s'_j = \nabla(h_{ik}s_k) = dh_{ik} \otimes s_k + h_{ik}\omega_{kl} \otimes s_l$$
$$= (dh_{ik}h_{kj}^{-1} + h_{ik}\omega_{kl}h_{lj}^{-1}) \otimes s'_j.$$

This shows ω' obeys the transformation law:

$$\omega'_{ij} = dh_{ik}h_{kj}^{-1} + h_{ik}\omega_{kl}h_{lj}^{-1} \qquad \text{i.e., } \omega' = dh \cdot h^{-1} + h\omega h^{-1}$$

in matrix notation. This is, of course, the manner in which the 0th order symbol of a first order operator transforms.

We extend ∇ to be a derivation mapping

$$C^\infty(\Lambda^p T^*M \otimes V) \to C^\infty(\Lambda^{p+1}T^*M \otimes V)$$

so that:

$$\nabla(\theta_p \otimes s) = d\theta_p \otimes s + (-1)^p \theta_p \wedge \nabla s.$$

We compute that:

$$\nabla^2(fs) = \nabla(df \otimes s + f \nabla s) = d^2 f \otimes s - df \wedge \nabla s + df \wedge \nabla s + f \nabla^2 s = f \nabla^2 s$$

so instead of being a second order operator, ∇^2 is a 0th order operator. We may therefore express

$$\nabla^2(s)(x_0) = \Omega(x_0)s(x_0)$$

where the curvature Ω is a section to the bundle $\Lambda^2(T^*M) \otimes \text{END}(V)$ is a 2-form valued endomorphism of V.

In local coordinates, we compute:

$$\Omega(s_i) = \Omega_{ij} \otimes s_j = \nabla(\omega_{ij} \otimes s_j) = d\omega_{ij} \otimes s_j - \omega_{ij} \wedge \omega_{jk} \otimes s_k$$

so that:

$$\Omega_{ij} = d\omega_{ij} - \omega_{ik} \wedge \omega_{kj} \qquad \text{i.e.,} \quad \Omega = d\omega - \omega \wedge \omega.$$

Since ∇^2 is a 0th order operator, Ω must transform like a tensor:

$$\Omega'_{ij} = h_{ik} \Omega_{kl} h_{lj}^{-1} \qquad \text{i.e.,} \quad \Omega' = h\Omega h^{-1}.$$

This can also be verified directly from the transition law for ω'. The reader should note that in some references, the curvature is defined by $\Omega = d\omega + \omega \wedge \omega$. This sign convention results from writing $V \otimes T^*M$ instead of $T^*M \otimes V$ and corresponds to studying left invariant rather than right invariant vector fields on $\text{GL}(k, C)$.

It is often convenient to normalize the choice of frame.

LEMMA 2.1.1. *Let ∇ be a connection on a vector bundle V. We can always choose a frame s so that at a given point x_0 we have*

$$\omega(x_0) = 0 \qquad \text{and} \qquad d\Omega(x_0) = 0.$$

PROOF: We find a matrix $h(x)$ defined near x_0 so that $h(x_0) = I$ and $dh(x_0) = -\omega(x_0)$. If $s'_i = h_{ij}s_j$, then it is immediate that $\omega'(x_0) = 0$. Similarly we compute $d\Omega'(x_0) = d(d\omega' - \omega' \wedge \omega')(x_0) = \omega'(x_0) \wedge d\omega'(x_0) - d\omega'(x_0) \wedge \omega'(x_0) = 0$.

We note that as the curvature is invariantly defined, we cannot in general find a parallel frame s so ω vanishes in a neighborhood of x_0 since this would imply $\nabla^2 = 0$ near x_0 which need not be true.

Let A_{ij} denote the components of a matrix in $\text{END}(C^k)$ and let $P(A) = P(A_{ij})$ be a polynomial mapping $\text{END}(C^k) \to C$. We assume that P is invariant—i.e., $P(hAh^{-1}) = P(A)$ for any $h \in \text{GL}(k, C)$. We define $P(\Omega)$ as an even differential form on M by substitution. Since P is invariant and since the curvature transforms tensorially, $P(\Omega) \stackrel{\text{def}}{=} P(\nabla)$ is defined independently of the particular local frame which is chosen.

There are two examples which are of particular interest and which will play an important role in the statement of the Atiyah-Singer index theorem. We define:

$$c(A) = \det \left(I + \frac{i}{2\pi} A \right)$$
$$= 1 + c_1(A) + \cdots + c_k(A) \qquad \text{(the total Chern form)}$$
$$ch(A) = \operatorname{Tr} e^{iA/2\pi} = \sum_j \operatorname{Tr} \left(\frac{iA}{2\pi} \right)^j \Big/ j! \qquad \text{(the total Chern character)}$$

The $c_j(A)$ represent the portion of $c(A)$ which is homogeneous of order j. $c_j(\nabla) \in \Lambda^{2j}(M)$. In a similar fashion, we decompose $ch(A) = \sum_j ch_j(A)$ for $ch_j(A) = \operatorname{Tr}(iA/2\pi)^j/j!$

Strictly speaking, $ch(A)$ is not a polynomial. However, when we substitute the components of the curvature tensor, this becomes a finite sum since $\operatorname{Tr}(\Omega^j) = 0$ if $2j > \dim M$. More generally, we can define $P(\Omega)$ if $P(A)$ is an invariant formal power series by truncating the power series appropriately.

As a differential form, $P(\nabla)$ depends on the connection chosen. We show $P(\nabla)$ is always closed. As an element of de Rham cohomology, $P(\nabla)$ is independent of the connection and defines a cohomology class we shall denote by $P(V)$.

LEMMA 2.1.2. Let P be an invariant polynomial.
(a) $dP(\nabla) = 0$ so $P(\nabla)$ is a closed differential form.
(b) Given two connections ∇_0 and ∇_1, we can define a differential form $TP(\nabla_0, \nabla_1)$ so that $P(\nabla_1) - P(\nabla_0) = d\{TP(\nabla_1, \nabla_0)\}$.

PROOF: By decomposing P as a sum of homogeneous polynomials, we may assume without loss of generality that P is homogeneous of order k. Let $P(A_1, \ldots, A_k)$ denote the complete polarization of P. We expand $P(t_1 A_1 + \cdots + t_k A_k)$ as a polynomial in the variables $\{t_j\}$. $1/k!$ times the coefficient of $t_1 \ldots t_k$ is the polarization of P. P is a multi-linear symmetric function of its arguments. For example, if $P(A) = \operatorname{Tr}(A^3)$, then the polarization is given by $\frac{1}{2} \operatorname{Tr}(A_1 A_2 A_3 + A_2 A_1 A_3)$ and $P(A) = P(A, A, A)$.

Fix a point x_0 of M and choose a frame so $\omega(x_0) = d\Omega(x_0) = 0$. Then

$$dP(\Omega)(x_0) = dP(\Omega, \ldots, \Omega)(x_0) = kP(d\Omega, \Omega, \ldots, \Omega)(x_0) = 0.$$

Since x_0 is arbitrary and since $dP(\Omega)$ is independent of the frame chosen this proves $dP(\Omega) = 0$ which proves (a).

Let $\nabla_t = t\nabla_1 + (1-t)\nabla_0$ have connection 1-form $\omega_t = \omega_0 + t\theta$ where $\theta = \omega_1 - \omega_0$. The transformation law for ω implies relative to a new frame:

$$\theta' = \omega_1' - \omega_0' = (dh \cdot h^{-1} + h\omega_1 h^{-1}) - (dh \cdot h^{-1} + h\omega_0 h^{-1})$$
$$= h(\omega_1 - \omega_0)h^{-1} = h\theta h^{-1}$$

so θ transforms like a tensor. This is of course nothing but the fact that the difference between two first order operators with the same leading symbol is a 0th order operator.

Let Ω_t be the curvature of the connection ∇_t. This is a matrix valued 2-form. Since θ is a 1-form, it commutes with 2-forms and we can define

$$P(\theta, \Omega_t, \ldots, \Omega_t) \in \Lambda^{2k-1}(T^*M)$$

by substitution. Since P is invariant, its complete polarization is also invariant so $P(h^{-1}\theta h, h^{-1}\Omega_t h, \ldots, h^{-1}\Omega_t h) = P(\theta, \Omega_t, \ldots, \Omega_t)$ is invariantly defined independent of the choice of frame.

We compute that:

$$P(\nabla_1) - P(\nabla_0) = \int_0^1 \frac{d}{dt} P(\Omega_t, \ldots, \Omega_t)\, dt = k \int_0^1 P(\Omega_t', \Omega_t, \ldots, \Omega_t)\, dt.$$

We define:

$$TP(\nabla_1, \nabla_0) = k \int_0^1 P(\theta, \Omega_t, \ldots, \Omega_t)\, dt.$$

To complete the proof of the Lemma, it suffices to check

$$dP(\theta, \Omega_t, \ldots, \Omega_t) = P(\Omega_t', \Omega_t, \ldots, \Omega_t).$$

Since both sides of the equation are invariantly defined, we can choose a suitable local frame to simplify the computation. Let $x_0 \in M$ and fix t_0. We choose a frame so $\omega_t(x_0, t_0) = 0$ and $d\Omega_t(x_0, t_0) = 0$. We compute:

$$\Omega_t' = \{d\omega_0 + t d\theta - \omega_t \wedge \omega_t\}'$$
$$= d\theta - \omega_t' \wedge \omega_t - \omega_t \wedge \omega_t'$$
$$\Omega_t'(x_0, t_0) = d\theta$$

and

$$dP(\theta, \Omega_t, \ldots, \Omega_t)(x_0, t_0) = P(d\theta, \Omega_t, \ldots, \Omega_t)(x_0, t_0)$$
$$= P(\Omega_t', \Omega_t, \ldots, \Omega_t)(x_0, t_0)$$

which completes the proof.

TP is called the transgression of P and will play an important role in our discussion of secondary characteristic classes in Chapter 4 when we discuss the eta invariant with coefficients in a locally flat bundle.

We suppose that the matrix $A = \text{diag}(\lambda_1, \ldots, \lambda_k)$ is diagonal. Then modulo suitable normalizing constants $\left(\frac{i}{2\pi}\right)^j$, it is immediate that $c_j(A)$ is the j th elementary symmetric function of the λ_i's since

$$\det(I + A) = \prod(1 + \lambda_j) = 1 + s_1(\lambda) + \cdots + s_k(\lambda).$$

If $P(\cdot)$ is any invariant polynomial, then $P(A)$ is a symmetric function of the λ_i's. The elementary symmetric functions form an algebra basis for the symmetric polynomials so there is a unique polynomial Q so that

$$P(A) = Q(c_1, \ldots, c_k)(A)$$

i.e., we can decompose P as a polynomial in the c_i's for diagonal matrices. Since P is invariant, this is true for diagonalizable matrices. Since the diagonalizable matrices are dense and P is continuous, this is true for all A. This proves:

LEMMA 2.1.3. *Let $P(A)$ be a invariant polynomial. There exists a unique polynomial Q so that $P = Q(c_1(A), \ldots, c_k(A))$.*

It is clear that any polynomial in the c_k's is invariant, so this completely characterizes the ring of invariant polynomials; it is a free algebra in k variables $\{c_1, \ldots, c_k\}$.

If $P(A)$ is homogeneous of degree k and is defined on $k \times k$ matrices, we shall see that if $P(A) \neq 0$ as a polynomial, then there exists a holomorphic manifold M so that if $T_c(M)$ is the holomorphic tangent space, then $\int_M P(T_c(M)) \neq 0$ where $\dim M = 2k$. This fact can be used to show that in general if $P(A) \neq 0$ as a polynomial, then there exists a manifold M and a bundle V over M so $P(V) \neq 0$ in cohomology.

We can apply functorial constructions on connections. Define:

$$\begin{array}{ccc} \nabla_1 \oplus \nabla_2 & \text{on} & C^\infty(V_1 \oplus V_2) \\ \nabla_1 \otimes \nabla_2 & \text{on} & C^\infty(V_1 \otimes V_2) \\ \nabla_1^* & \text{on} & C^\infty(V_1^*) \end{array}$$

by:

$$(\nabla_1 \oplus \nabla_2)(s_1 \oplus s_2) = (\nabla_1 s_1) \oplus (\nabla_2 s_2)$$
$$\text{with } \omega = \omega_1 \oplus \omega_2 \text{ and } \Omega = \Omega_1 \oplus \Omega_2$$
$$(\nabla_1 \otimes \nabla_2)(s_1 \otimes s_2) = (\nabla_1 s_1) \otimes (\nabla_2 s_2)$$
$$\text{with } \omega = \omega_1 \otimes 1 + 1 \otimes \omega_2 \text{ and } \Omega = \Omega_1 \otimes 1 + 1 \otimes \Omega_2$$
$$(\nabla_1 s_1, s_1^*) + (s_1, \nabla_1^* s_1^*) = d(s_1, s_1^*)$$
$$\text{relative to the dual frame } \omega = -\omega_1^t \text{ and } \Omega = -\Omega_1^t.$$

In a similar fashion we can define an induced connection on $\Lambda^p(V)$ (the bundle of p-forms) and $S^p(V)$ (the bundle of symmetric p-tensors). If V has a given Hermitian fiber metric, the connection ∇ is said to be unitary or Riemannian if $(\nabla s_1, s_2) + (s_1, \nabla s_2) = d(s_1, s_2)$. If we identify V with V^* using the metric, this simply means $\nabla = \overline{\nabla}^*$. This is equivalent to the condition that ω is a skew adjoint matrix of 1-forms relative to a local orthonormal frame. We can always construct Riemannian connections locally by taking $\omega = 0$ relative to a local orthonormal frame and then using a partition of unity to construct a global Riemannian connection.

If we restrict to Riemannian connections, then it is natural to consider polynomials $P(A)$ which are defined for skew-Hermitian matrices $A + A^* = 0$ and which are invariant under the action of the unitary group. Exactly the same argument as that given in the proof of Lemma 2.1.3 shows that we can express such a P in the form $P(A) = Q(c_1(A), \ldots, c_k(A))$ so that the $\mathrm{GL}(\cdot, \mathbf{C})$ and the $\mathrm{U}(\cdot)$ characteristic classes coincide. This is not true in the real category as we shall see; the Euler form is a characteristic form of the special orthogonal group which can not be defined as a characteristic form using the general linear group $\mathrm{GL}(\cdot, \mathbf{R})$.

The Chern form and the Chern character satisfy certain identities with respect to functorial constructions.

LEMMA 2.1.4.
(a)
$$c(V_1 \oplus V_2) = c(V_1)c(V_2)$$
$$c(V^*) = 1 - c_1(V) + c_2(V) - \cdots + (-1)^k c_k(V).$$

(b)
$$ch(V_1 \oplus V_2) = ch(V_1) + ch(V_2)$$
$$ch(V_1) \oplus V_2) = ch(V_1)ch(V_2).$$

PROOF: All the identities except the one involving $c(V^*)$ are immediate from the definition if we use the direct sum and product connections. If we fix a Hermitian structure on V, the identification of V with V^* is conjugate linear. The curvature of ∇^* is $-\Omega^t$ so we compute:

$$c(V^*) = \det\left(I - \frac{i}{2\pi}\Omega^t\right) = \det\left(I - \frac{i}{2\pi}\Omega\right)$$

which gives the desired identity.

If we choose a Riemannian connection on V, then $\Omega + \overline{\Omega}^t = 0$. This immediately yields the identities:

$$\overline{ch(V)} = ch(V) \qquad \text{and} \qquad \overline{c(V)} = c(V)$$

so these are real cohomology classes. In fact the normalizing constants were chosen so $c_k(V)$ is an integral cohomology class—i.e., if N_{2k} is any oriented submanifold of dimension $2k$, then $\int_{N_{2k}} c_k(V) \in \mathbf{Z}$. The $ch_k(V)$ are not integral for $k > 1$, but they are rational cohomology classes. As we shall not need this fact, we omit the proof.

The characteristic classes give cohomological invariants of a vector bundle. We illustrate this by constructing certain examples over even dimensional spheres; these examples will play an important role in our later development of the Atiyah-Singer index theorem.

DEFINITION. *A set of matrices $\{e_0, \ldots, e_m\}$ are Clifford matrices if the e_i are self-adjoint and satisfy the commutation relations $e_i e_j + e_j e_i = 2\delta_{ij}$ where δ_{ij} is the Kronecker symbol.*

For example, if $m = 2$ we can take:

$$e_0 = \begin{pmatrix} 1 & 0 \\ 0 & -1 \end{pmatrix}, \qquad e_1 = \begin{pmatrix} 0 & 1 \\ 1 & 0 \end{pmatrix}, \qquad e_2 = \begin{pmatrix} 0 & i \\ -i & 0 \end{pmatrix}$$

to be the Dirac matrices. More generally, we can take the e_i's to be given by the spin representations. If $x \in \mathbf{R}^{m+1}$, we define:

$$e(x) = \sum_i x_i e_i \qquad \text{so} \quad e(x)^2 = |x|^2 I.$$

Conversely let $e(x)$ be a linear map from $\mathbf{R}^m$ to the set of self-adjoint matrices with $e(x) = |x|^2 I$. If $\{v_0, \ldots, v_m\}$ is any orthonormal basis for $\mathbf{R}^m$, $\{e(v_0), \ldots, e(v_m)\}$ forms a set of Clifford matrices.

If $x \in S^m$, we let $\Pi_{\pm}(x)$ be the range of $\frac{1}{2}(1 + e(x)) = \pi_{\pm}(x)$. This is the span of the ± 1 eigenvectors of $e(x)$. If $e(x)$ is a $2k \times 2k$ matrix, then $\dim \Pi_{\pm}(x) = k$. We have a decomposition $S^m \times \mathbf{C}^{2k} = \Pi_+ \oplus \Pi_-$. We project the flat connection on $S^m \times \mathbf{C}^{2k}$ to the two sub-bundles to define connections $\nabla_{\pm}$ on $\Pi_{\pm}$. If $e_{\pm}^0$ is a local frame for $\Pi_{\pm}(x_0)$, we define $e_{\pm}(x) = \pi_{\pm} e_{\pm}^0$ as a frame in a neighborhood of x_0. We compute

$$\nabla_{\pm} e_{\pm} = \pi_{\pm} \, d\pi_{\pm} e_{\pm}^0, \qquad \Omega_{\pm} e_{\pm} = \pi_{\pm} \, d\pi_{\pm} \, d\pi_{\pm} e_{\pm}^0.$$

Since $e_{\pm}^0 = e_{\pm}(x_0)$, this yields the identity:

$$\Omega_{\pm}(x_0) = \pi_{\pm} \, d\pi_{\pm} \, d\pi_{\pm}(x_0).$$

Since Ω is tensorial, this holds for all x.

Let $m = 2j$ be even. We wish to compute ch_j. Suppose first $x_0 = (1, 0, \ldots, 0)$ is the north pole of the sphere. Then:

$$\pi_+(x_0) = \frac{1}{2}(1 + e_0)$$

$$d\pi_+(x_0) = \frac{1}{2}\sum_{i \geq 1} dx_i e_i$$

$$\Omega_+(x_0) = \frac{1}{2}(1 + e_0)\left(\frac{1}{2}\sum_{i \geq 1} dx_i e_i\right)^2$$

$$\Omega_+(x_0)^j = \frac{1}{2}(1 + e_0)\left(\frac{1}{2}\sum_{i \geq 1} dx_i e_i\right)^{2j}$$

$$= 2^{-m-1}m!(1 + e_0)(e_1 \ldots e_m)(dx_1 \wedge \cdots \wedge dx_m).$$

The volume form at x_0 is $dx_1 \wedge \cdots \wedge dx_m$. Since e_1 anti-commutes with the matrix $e_1 \ldots e_m$, this matrix has trace 0 so we compute:

$$ch_j(\Omega_+)(x_0) = \left(\frac{i}{2\pi}\right)^j 2^{-m-1}m! \operatorname{Tr}(e_0 \ldots e_m) \operatorname{dvol}(x_0)/j!.$$

A similar computation shows this is true at any point x_0 of S^m so that:

$$\int_{S^m} ch_j(\Pi_+) = \left(\frac{i}{2\pi}\right)^j 2^{-m-1}m! \operatorname{Tr}(e_0 \ldots e_m) \operatorname{vol}(s^m)/j!.$$

Since the volume of S^m is $j! 2^{m+1}\pi^j/m!$ we conclude:

LEMMA 2.1.5. *Let $e(x)$ be a linear map from $\mathbf{R}^{m+1}$ to the set of self-adjoint matrices. We suppose $e(x)^2 = |x|^2 I$ and define bundles $\Pi_\pm(x)$ over S^m corresponding to the ± 1 eigenvalues of e. Let $m = 2j$ be even, then:*

$$\int_{S^m} ch_j(\Pi_+) = i^j 2^{-j} \operatorname{Tr}(e_0 \ldots e_m).$$

In particular, this is always an integer as we shall see later when we consider the spin complex. If

$$e_0 = \begin{pmatrix} 1 & 0 \\ 0 & -1 \end{pmatrix}, \qquad e_1 = \begin{pmatrix} 0 & 1 \\ 1 & 0 \end{pmatrix}, \qquad e_2 = \begin{pmatrix} 0 & i \\ -i & 0 \end{pmatrix}$$

then $\operatorname{Tr}(e_0 e_1 e_2) = -2i$ so $\int_{S^2} ch_1(\Pi_+) = 1$ which shows in particular Π_+ is a non-trivial line bundle over S^2.

There are other characteristic classes arising in the index theorem. These are most conveniently discussed using generating functions. Let

$$x_j = \frac{i}{2\pi}\lambda_j$$

be the normalized eigenvalues of the matrix A. We define:

$$c(A) = \prod_{j=1}^{k}(1 + x_j) \quad \text{and} \quad ch(A) = \sum_{j=1}^{k} e^{x_j}.$$

If $P(x)$ is a symmetric polynomial, we express $P(x) = Q(c_1(x), \ldots, c_k(x))$ to show $P(A)$ is a polynomial in the components of A. More generally if P is analytic, we first express P in a formal power series and then collect homogeneous terms to define $P(A)$. We define:

$$\text{Todd class: } Td(A) = \prod_{j=1}^{k} \frac{x_j}{1 - e^{-x_j}}$$

$$= 1 + \frac{1}{2}c_1(A) + \frac{1}{12}(c_1^2 + c_2)(A) + \frac{1}{24}c_1(A)c_2(A) + \cdots$$

The Todd class appears in the Riemann-Roch theorem. It is clear that it is multiplicative with respect to direct sum—$Td(V_1 \oplus V_2) = Td(V_1)\,Td(V_2)$. The Hirzebruch L-polynomial and the $\hat{A}$ polynomial will be discussed in the next subsection. These are real characteristic classes which also are defined by generating functions.

If V is a bundle of dimension k, then $V \oplus 1$ will be a bundle of dimension $k+1$. It is clear $c(V \oplus 1) = c(V)$ and $Td(V \oplus 1) = Td(V)$ so these are stable characteristic classes. $ch(V)$ on the other hand is *not* a stable characteristic class since $ch_0(V) = \dim V$ depends explicitly on the dimension of V and changes if we alter the dimension of V by adding a trivial bundle.

2.2. Characteristic Classes of a Real Vector Bundle.
Pontrjagin and Euler Classes.

Let V be a real vector bundle of dimension k and let $V_c = V \otimes \mathbf{C}$ denote the complexification of V. We place a real fiber metric on V to reduce the structure group from $\mathrm{GL}(k, \mathbf{R})$ to the orthogonal group $O(k)$. We restrict henceforth to Riemannian connections on V, and to local orthonormal frames. Under these assumptions, the curvature is a skew-symmetric matrix of 2-forms:

$$\Omega + \Omega^t = 0.$$

Since V_c arises from a real vector bundle, the natural isomorphism of V with V^* defined by the metric induces a complex linear isomorphism of V_c with V_c^* so $c_j(V_c) = 0$ for j odd by Lemma 2.1.4. Expressed locally,

$$\det\left(I + \frac{i}{2\pi}A\right) = \det\left(I + \frac{i}{2\pi}A^t\right) = \det\left(I - \frac{i}{2\pi}A\right)$$

if $A + A^t = 0$ so $c(A)$ is an even polynomial in A. To avoid factors of i we define the Pontrjagin form:

$$p(A) = \det\left(I + \frac{1}{2\pi}A\right) = 1 + p_1(A) + p_2(A) + \cdots$$

where $p_j(A)$ is homogeneous of order $2j$ in the components of A; the corresponding characteristic class $p_j(V) \in H^{4j}(M; \mathbf{R})$. It is immediate that:

$$p_j(V) = (-1)^j c_{2j}(V_c)$$

where the factor of $(-1)^j$ comes form the missing factors of i.

The set of skew-symmetric matrices is the Lie algebra of $O(k)$. Let $P(A)$ be an invariant polynomial under the action of $O(k)$. We define $P(\Omega)$ for ∇ a Riemannian connection exactly as in the previous subsections. Then the analogue of Lemma 2.1.3 is:

LEMMA 2.2.1. *Let $P(A)$ be a polynomial in the components of a matrix A. Suppose $P(A) = P(gAg^{-1})$ for every skew-symmetric A and for every $g \in O(k)$. Then there exists a polynomial $Q(p_1, \ldots)$ so $P(A) = Q(p_1(A), \ldots)$ for every skew-symmetric A.*

PROOF: It is important to note that we are not asserting that we have $P(A) = Q(p_1(A), \ldots)$ for every matrix A, but only for skew-symmetric A. For example, $P(A) = \mathrm{Tr}(A)$ vanishes for skew symmetric A but does not vanish in general.

It is not possible in general to diagonalize a skew-symmetric real matrix. We can, however, put it in block diagonal form:

$$A = \begin{pmatrix} 0 & -\lambda_1 & 0 & 0 & \cdots\cdots \\ \lambda_1 & 0 & 0 & 0 & \cdots\cdots \\ 0 & 0 & 0 & -\lambda_2 & \cdots\cdots \\ 0 & 0 & \lambda_2 & 0 & \cdots\cdots \\ & \cdots\cdots\cdots\cdots\cdots\cdots \\ & \cdots\cdots\cdots\cdots\cdots\cdots \end{pmatrix}$$

If k is odd, then the last block will be a 1×1 block with a zero in it. We let $x_j = -\lambda_j/2\pi$; the sign convention is chosen to make the Euler form have the right sign. Then:

$$p(A) = \prod_j (1 + x_j^2)$$

where the product ranges from 1 through $\left[\frac{k}{2}\right]$.

If $P(A)$ is any invariant polynomial, then P is a symmetric function in the $\{x_j\}$. By conjugating A by an element of $O(k)$, we can replace any x_j by $-x_j$ so P is a symmetric function of the $\{x_j^2\}$. The remainder of the proof of Lemma 2.2.1 is the same; we simply express $P(A)$ in terms of the elementary symmetric functions of the $\{x_j^2\}$.

Just as in the complex case, it is convenient to define additional characteristic classes using generating functions. The functions:

$$z/\tanh z \qquad \text{and} \qquad z\{2\sinh(z/2)\}$$

are both even functions of the parameter z. We define:

The Hirzebruch L-polynomial:

$$L(x) = \prod_j \frac{x_j}{\tanh x_j}$$

$$= 1 + \frac{1}{3}p_1 + \frac{1}{45}(7p_2 - p_1^2)\cdots$$

The $\hat{A}$ (A-roof) genus:

$$A(x) = \prod_j \frac{x_j}{2\sinh(x_j/2)}$$

$$= 1 - \frac{1}{24}p_1 + \frac{1}{5760}(7p_1^2 - 4p_2) + \cdots .$$

These characteristic classes appear in the formula for the signature and spin complexes.

For both the real and complex case, the characteristic ring is a pure polynomial algebra without relations. Increasing the dimension k just adds generators to the ring. In the complex case, the generators are the Chern classes $\{1, c_1, \ldots, c_k\}$. In the real case, the generators are the Pontrjagin classes $\{1, p_1, \ldots, p_{[k/2]}\}$ where $[k/2]$ is the greatest integer in $k/2$. There is one final structure group which will be of interest to us.

Let V be a vector bundle of real dimension k. V is orientable if we can choose a fiber orientation consistently. This is equivalent to assuming that the real line bundle $\Lambda^k(V)$ is trivial. We choose a fiber metric for V and an orientation for V and restrict attention to oriented orthonormal frames. This restricts the structure group from $O(k)$ to the special orthogonal group $SO(k)$.

If k is odd, no new characteristic classes result from the restriction of the fiber group. We use the final 1×1 block of 0 in the representation of A to replace any λ_i by $-\lambda_i$ by conjugation by an element of $SO(k)$. If n is even, however, we cannot do this as we would be conjugating by a orientation reversing matrix.

Let $P(A)$ be a polymonial in the components of A. Suppose $P(A) = P(gAg^{-1})$ for every skew-symmetric A and every $g \in SO(k)$. Let $k = 2\bar{k}$ be even and let $P(A) = P(x_1, \ldots, x_{\bar{k}})$. Fix $g_0 \in O(k) - SO(k)$ and define:

$$P_0(A) = \frac{1}{2}(P(A) + P(g_0 A g_0^{-1})) \quad \text{and} \quad P_1(A) = \frac{1}{2}(P(A) - P(g_0 A g_0^{-1})).$$

Both P_0 and P_1 are $SO(k)$ invariant. P_0 is $O(k)$ invariant while P_1 changes sign under the action of an element of $O(k) - SO(k)$.

We can replace x_1 by $-x_1$ by conjugating by a suitable orientation reversing map. This shows:

$$P_1(x_1, x_2, \ldots) = -P(-x_1, x_2, \ldots)$$

so that x_1 must divide every monomial of P_0. By symmetry, x_j divides every monomial of P_1 for $1 \leq j \leq \bar{k}$ so we can express:

$$P_1(A) = x_1 \ldots x_{\bar{k}} P_1'(A)$$

where $P_1'(A)$ is now invariant under the action of $O(k)$. Since P_1' is polynomial in the $\{x_j^2\}$, we conclude that both P_0 and P_1' can be represented as polynomials in the Pontrjagin classes. We define:

$$e(A) = x_1 \ldots x_{\bar{k}} \quad \text{so} \quad e(A)^2 = \det(A) = p_{\bar{k}}(A) = \prod_j x_j^2$$

and decompose:

$$P = P_0 + e(A)P_1'.$$

$e(A)$ is a square root of the determinant of A.

It is not, of course, immediate that $e(A)$ is a polynomial in the components of A. Let $\{v_i\}$ be an oriented orthonormal basis for $\mathbf{R}^k$. We let $Av_i = \sum_j A_{ij}v_j$ and define

$$\omega(A) = \frac{1}{2\pi}\sum_{i<j} A_{ij}v_i \wedge v_j \in \Lambda^2(\mathbf{R}^k).$$

We let $v_1 \wedge \cdots \wedge v_k$ be the orientation class of $\mathbf{R}^k$ and define:

$$e(A) = (\omega(A)^{\bar{k}}, v_1 \wedge \cdots \wedge v_k)/k!$$

where $(\ ,\ )$ denotes the natural inner product on $\Lambda^k \mathbf{R}^k \simeq \mathbf{R}$. It is clear from the definition that $e(A)$ is invariant under the action of $SO(k)$ since $\omega(A)$ is invariantly defined for skew-symmetric matrices A. It is clear that $e(A)$ is polynomial in the components of A. If we choose a block basis so that:

$$Av_1 = \lambda_1 v_2, \quad Av_2 = -\lambda_1 v_1, \quad Av_3 = \lambda_2 v_4, \quad Av_4 = -\lambda_2 v_3, \ldots$$

then we compute:

$$\omega(A) = \{-\lambda_1 v_1 \wedge v_2 - \lambda_2 v_3 \wedge v_4 \ldots\}/2\pi = x_1 v_1 \wedge v_2 + x_2 v_3 \wedge v_4 \cdots$$

$$e(A) = x_1 x_2 \ldots$$

This new characteristic class is called the Euler class. While the Pontrjagin classes can be computed from the curvature of an arbitrary connection, the Euler class can only be computed from the curvature of a Riemannian connection. If Ω_{ij} are the matrix components of the curvature of V relative to some oriented orthomormal basis, then:

$$e(\Omega) = (-4\pi)^{-\bar{k}}/k! \sum_{\rho} \text{sign}(\rho)\Omega_{\rho(1)\rho(2)} \cdots \Omega_{\rho(k-1)\rho(k)} \in \Lambda^k(M)$$

for $2\bar{k} = k = \dim(V)$. The sum ranges over all possible permutations ρ.

We define $e(V) = 0$ if $\dim(V)$ is odd. It is immediate that $e(V_1 \oplus V_2) = e(V_1)e(V_2)$ if we give the natural orientation and fiber metric to the direct sum and if we use the direct sum connection.

We illustrate this formula (and check that we have the correct normalizing constants) by studying the following simple example. Let $m = 2$ and let $M = S^2$ be the unit sphere. We calculate $e(TS^2)$. Parametrize M using spherical coordinates $f(u, v) = (\cos u \sin v, \sin u \sin v, \cos v)$ for $0 \leq u \leq 2\pi$ and $0 \leq v \leq \pi$. Define a local orthonormal frame for $T(\mathbf{R}^3)$ over S^2 by:

$$e_1 = (\sin v)^{-1}\partial/\partial u = (-\sin u, \cos u, 0)$$

$$e_2 = \qquad \partial/\partial v = (\cos u \cos v, \sin u \cos v, -\sin v)$$

$$e_3 = \qquad N = (\cos u \sin v, \sin u \sin v, \cos v).$$

The Euclidean connection $\widetilde{\nabla}$ is easily computed to be:

$$\widetilde{\nabla}_{e_1} e_1 = (\sin v)^{-1}(-\cos u, -\sin u, 0) = -(\cos v/\sin v)e_2 - e_3$$
$$\widetilde{\nabla}_{e_1} e_2 = (\cos v/\sin v)(-\sin u, \cos u, 0) = (\cos v/\sin v)e_1$$
$$\widetilde{\nabla}_{e_2} e_1 = 0$$
$$\widetilde{\nabla}_{e_2} e_2 = (-\cos u \sin v, -\sin u \sin v, -\cos v) = -e_3.$$

Covariant differentiation ∇ on the sphere is given by projecting back to $T(S^2)$ so that:

$$\nabla_{e_1} e_1 = (-\cos v/\sin v)e_2 \qquad \nabla_{e_1} e_2 = (\cos v/\sin v)e_1$$
$$\nabla_{e_2} e_1 = 0 \qquad\qquad \nabla_{e_2} e_2 = 0$$

and the connection 1-form is given by:

$$\nabla e_1 = (-\cos v/\sin v)e^1 \otimes e_2 \quad \text{so } \omega_{11} = 0 \text{ and } \omega_{12} = -(\cos v/\sin v)e^1$$
$$\nabla e_2 = (\cos v/\sin v)e^1 \otimes e_1 \qquad \text{so } \omega_{22} = 0 \text{ and } \omega_{21} = (\cos v/\sin v)e^1.$$

As $du = e^1/\sin v$ we compute $\omega_{12} = -\cos v \, du$ so $\Omega_{12} = \sin v \, dv \wedge du = -e^1 \wedge e^2$ and $\Omega_{21} = e^1 \wedge e^2$. From this we calculate that:

$$e(\Omega) = -\frac{1}{4\pi}(\Omega_{12} - \Omega_{21}) = \frac{e^1 \wedge e^2}{2\pi}$$

and consequently $\int_{S^2} E_2 = \text{vol}(S^2)/2\pi = 2 = \chi(S^2)$.

There is a natural relation between the Euler form and $c_{\bar{k}}$. Let V be a complex vector space of dimension $\bar{k}$ and let V_r be the underlying real vector space of dimention $2\bar{k} = k$. If V has a Hermitian inner product, then V_r inherits a natural real inner product. V_r also inherits a natural orientation from the complex structure on V. If $\{e_j\}$ is a unitary basis for V, then $\{e_1, ie_1, e_2, ie_2, \dots\}$ is an oriented orthonormal basis for V_r. Let A be a skew-Hermitian matrix on V. The restriction of A to V_r defines a skew-symmetric matrix A_r on V_r. We choose a basis $\{e_j\}$ for V so $Ae_j = i\lambda_j e_j$. Then:

$$x_j = -\lambda_j/2\pi \qquad \text{and} \qquad c_{\bar{k}}(A) = x_1 \dots x_{\bar{k}}$$

defines the $\bar{k}^{\text{th}}$ Chern class. If $v_1 = e_1$, $v_2 = ie_1$, $v_3 = e_2$, $v_4 = ie_2, \dots$ then:

$$A_r v_1 = \lambda_1 v_2, \quad A_r v_2 = -\lambda_1 v_1, \quad A_r v_3 = \lambda_2 v_4, \quad A_r v_4 = -\lambda_2 v_3, \dots$$

so that

$$e(A_r) = c_{\bar{k}}(A).$$

We summarize these results as follows:

LEMMA 2.2.2. *Let $P(A)$ be a polynomial with $P(A) = P(gAg^{-1})$ for every skew symmetric A and every $g \in SO(k)$.*
(a) If k is odd, then $P(A)$ is invariant under $O(k)$ and is expressible in terms of Pontrjagin classes.
(b) If $k = 2\bar{k}$ is even, then we can decompose $P(A) = P_0(A) + e(A)P_1(A)$ where $P_i(A)$ are $O(k)$ invariant and are expressible in terms of Pontrjagin classes.
(c) $e(A)$ is defined by:

$$e(A) = (-4\pi)^{-\bar{k}} \sum_{\rho} \text{sign}(\rho) A_{\rho(1)\rho(2)} \cdots A_{\rho(k-1)\rho(k)} / k!$$

This satisfies the identity $e(A)^2 = p_{\bar{k}}(A)$. (We define $e(A) = 0$ if k is odd).

This completely describes the characteristic ring. We emphasize the conclusions are only applicable to skew-symmetric real matrices A. We have proved that $e(A)$ has the following functorial properties:

LEMMA 2.2.3.
(a) If we take the direct sum connection and metric, then $e(V_1 \oplus V_2) = e(V_1)e(V_2)$.
(b) If we take the metric and connection of V_r obtained by forgetting the complex structure on a complex bundle V then $e(V_r) = c_{\bar{k}}(V)$.

This lemma establishes that the top dimensional Chern class $c_{\bar{k}}$ does not really depend on having a complex structure but only depends on having an orientation on the underlying real vector bundle. The choice of the sign in computing $\det(A)^{1/2}$ is, of course, motivated by this normalization.

2.3. Characteristic Classes of Complex Projective Space.

So far, we have only discussed covariant differentiation from the point of view of the total covariant derivative. At this stage, it is convenient to introduce covariant differentiation along a direction. Let $X \in T(M)$ be a tangent vector and let $s \in C^\infty(V)$ be a smooth section. We define $\nabla_X s \in C^\infty(V)$ by:

$$\nabla_X s = X \cdot \nabla s$$

where "·" denotes the natural pairing from $T(M) \otimes T^*(M) \otimes V \to V$. Let $[X, Y] = XY - YX$ denote the Lie-bracket of vector fields. We define:

$$\Omega(X, Y) = \nabla_X \nabla_Y - \nabla_Y \nabla_X - \nabla_{[X,Y]}$$

and compute that:

$$\Omega(fX, Y)s = \Omega(X, fY)s = \Omega(X, Y)fs = f\Omega(X, Y)s$$
$$\text{for} \quad f \in C^\infty(M), \ s \in C^\infty(V), \text{ and } X, Y \in C^\infty(TM).$$

If $\{e_i\}$ is a local frame for $T(M)$, let $\{e^i\}$ be the dual frame for $T^*(M)$. It is immediate that:

$$\nabla s = \sum_i e^i \otimes \nabla_{e_i}(s)$$

so we can recover the total covariant derivative from the directional derivatives. We can also recover the curvature:

$$\Omega s = \sum_{i<j} e^i \wedge e^j \otimes \Omega(e_i, e_j)s.$$

If $V = T(M)$, there is a special connection called the Levi-Cività connection on $T(M)$. It is the unique Riemannian connection which is torsion free—i.e.,

$$(\nabla_X Y, Z) + (Y, \nabla_X Z) = X(Y, Z) \qquad \text{(Riemannian)}$$
$$\nabla_X Y - \nabla_Y X - [X, Y] = 0 \qquad \text{(Torsion free)}.$$

Let $x = (x_1, \ldots, x_m)$ be a system of local coordinates on M and let $\{\partial/\partial x_i\}$ be the coordinate frame for $T(M)$. The metric tensor is given by:

$$ds^2 = \sum_{i,j} g_{ij} \, dx^i \otimes dx^j \qquad \text{for} \ g_{ij} = (\partial/\partial x_i, \partial/\partial x_j).$$

We introduce the Christoffel symbols Γ_{ij}^k and Γ_{ijk} by:

$$\nabla_{\partial/\partial x_i} \partial/\partial x_j = \sum_k \Gamma_{ij}^k \partial/\partial x_k$$
$$(\nabla_{\partial/\partial x_i} \partial/\partial x_j, \partial/\partial x_k) = \Gamma_{ijk}.$$

They are related by the formula:

$$\Gamma_{ijk} = \sum_l \Gamma_{ij}^l g_{lk}, \qquad \Gamma_{ij}^k = \sum_l \Gamma_{ijl} g^{lk}$$

where $g^{lk} = (dx_l, dx_k)$ is the inverse of the matrix g_{ij}. It is not difficult to compute that:

$$\Gamma_{ijk} = \tfrac{1}{2}\{g_{jk/i} + g_{ik/j} - g_{ij/k}\}$$

where we use the notation "/" to denote (multiple) partial differentiation.

The complete curvature tensor is defined by:

$$\Omega(\partial/\partial x_i, \partial/\partial x_j)\partial/\partial x_k = \sum_l R_{ijk}{}^l \partial/\partial x_l$$

or equivalently if we lower indices:

$$(\Omega(\partial/\partial x_i, \partial/\partial x_j)\partial/\partial x_k, \partial/\partial x_l) = R_{ijkl}.$$

The expression of the curvature tensor in terms of the derivatives of the metric is very complicated in general. By making a linear change of coordinates we can always normalize the metric so $g_{ij}(x_0) = \delta_{ij}$ is the Kronecker symbol. Similarly, by making a quadratic change of coordinates, we can further normalize the metric so $g_{ij/k}(x_0) = 0$. Relative to such a choice of coordinates:

$$\Gamma_{ij}^k(x_0) = \Gamma_{ijk}(x_0) = 0$$

and

$$R_{ijkl}(x_0) = R_{ijk}{}^l(x_0) = \tfrac{1}{2}\{g_{jl/ik} + g_{ik/lj} - g_{jk/il} - g_{il/jk}\}(x_0).$$

At any other point, of course, the curvature tensor is not as simply expressed. In general it is linear in the second derivatives of the metric, quadratic in the first derivatives of the metric with coefficients which depend smoothly on the g_{ij}'s.

We choose a local orthonormal frame $\{e_i\}$ for $T^*(M)$. Let $m = 2\bar{m}$ be even and let orn $= * \cdot 1 =$ dvol be the oriented volume. Let

$$e = E_m(g)\,\text{orn}$$

be the Euler form. If we change the choice of the local orientation, then e and orn both change signs so $E_m(g)$ is scalar invariant of the metric. In terms of the curvature, if $2\bar{m} = m$, then:

$$c(m) = (-1)^{\bar{m}}(8\pi)^{-\bar{m}}\frac{1}{(\bar{m})!}$$

$$E_m = c(m) \sum_{\rho,\tau} \text{sign}(\rho)\,\text{sign}(\tau) R_{\rho(1)\rho(2)\tau(1)\tau(2)} \cdots R_{\rho(m-1)\rho(m)\tau(m-1)\tau(m)}$$

where the sum ranges over all permutations ρ, τ of the integers 1 through m. For example:

$$E_2 = -(2\pi)^{-1} R_{1212}$$

$$E_4 = (2\pi)^{-2} \sum_{i,j,k,l} \{R_{ijij} R_{klkl} + R_{ijkl} R_{ijkl} - 4R_{ijik} R_{ljlk}\}/8.$$

Let dvol be the Riemannian measure on M. If M is oriented, then

$$\int_M E_m \, \mathrm{dvol} = \int_M e$$

is independent of the orientation of M and of the metric. If M is not orientable, we pass to the double cover to see $\int_M E_m \, \mathrm{dvol}$ is a topological invariant of the manifold M. We shall prove later this integral is the Euler characteristic $\chi(M)$ but for the moment simply note it is not dependent upon a choice of orientation of M and is in fact defined even if M is not orientable.

It is worth computing an example. We let S^2 be the unit sphere in $\mathbf{R}^3$. Since this is homogeneous, E_2 is constant on S^2. We compute E_2 at the north pole $(0,0,1)$ and parametrize S^2 by $(u, v, (1 - u^2 - v^2)^{1/2})$. Then

$$\partial/\partial u = (1, 0, -u/(1 - u^2 - v^2)^{1/2}), \quad \partial/\partial v = (0, 1, -v/(1 - u^2 - v^2)^{1/2})$$

$$g_{11} = 1 + u^2/(1 - u^2 - v^2)$$

$$g_{22} = 1 + v^2/(1 - u^2 - v^2)$$

$$g_{12} = uv/(1 - u^2 - v^2).$$

It is clear $g_{ij}(0) = \delta_{ij}$ and $g_{ij/k}(0) = 0$. Therefore at $u = v = 0$,

$$E_2 = -(2\pi)^{-1} R_{1212} = -(2\pi)^{-1}\{g_{11/22} + g_{22/11} - 2g_{12/12}\}/2$$
$$= -(2\pi)^{-1}\{0 + 0 - 2\}/2 = (2\pi)^{-1}$$

so that

$$\int_{S^2} E_2 \, \mathrm{dvol} = (4\pi)(2\pi)^{-1} = 2 = \chi(S^2).$$

More generally, we can let $M = S^2 \times \cdots \times S^2$ where we have $\bar{m}$ factors and $2\bar{m} = m$. Since $e(V_1 \oplus V_2) = e(V_1)e(V_2)$, when we take the product metric we have $E_m = (E_2)^{\bar{m}} = (2\pi)^{-\bar{m}}$ for this example. Therefore:

$$\int_{S^2 \times \cdots \times S^2} E_m \, \mathrm{dvol} = 2^{\bar{m}} = \chi(S^2 \times \cdots \times S^2).$$

We will use this example later in this chapter to prove $\int_M E_m = \chi(M)$ in general. The natural examples for studying the Euler class are products of two dimensional spheres. Unfortunately, the Pontrjagin classes vanish identically on products of spheres so we must find other examples.

It is convenient at this point to discuss the holomorphic category. A manifold M of real dimension $m = 2\bar{m}$ is said to be holomorphic if we have local coordinate charts $z = (z_1, \ldots, z_{\bar{m}}): M \to \mathbf{C}^{\bar{m}}$ such that two charts are related by holomorphic changes of coordinates. We expand $z_j = x_j + iy_j$ and define:

$$\partial/\partial z_j = \tfrac{1}{2}(\partial/\partial x_j - i\partial/\partial y_j), \qquad \partial/\partial \bar{z}_j = \tfrac{1}{2}(\partial/\partial x_j + i\partial/\partial y_j)$$
$$dz_j = dx_j + i\,dy_j, \qquad\qquad d\bar{z}_j = dx_j + i\,dy_j.$$

We complexify $T(M)$ and $T^*(M)$ and define:

$$T_c(M) = \operatorname{span}\{\partial/\partial z_j\},$$
$$\Lambda^{1,0}(M) = \operatorname{span}\{dz_j\},$$
$$\Lambda^{0,1}(M) = \operatorname{span}\{d\bar{z}_j\}.$$

Then

$$\Lambda^1(M) = \Lambda^{1,0}(M) \oplus \Lambda^{0,1}(M)$$
$$T_c(M)^* = \Lambda^{1,0}(M).$$

As complex bundles, $T_c(M) \simeq \Lambda^{0,1}(M)$. Define:

$$\partial(f) = \sum_j \partial f/\partial z_j\, dz^j : C^\infty(M) \to C^\infty(\Lambda^{1,0}(M))$$
$$\bar{\partial}(f) = \sum_j \partial f/\partial \bar{z}_j\, d\bar{z}^j : C^\infty(M) \to C^\infty(\Lambda^{0,1}(M))$$

then the Cauchy-Riemann equations show f is holomorphic if and only if $\bar{\partial}f = 0$.

This decomposes $d = \partial + \bar{\partial}$ on functions. More generally, we define:

$$\Lambda^{p,q} = \operatorname{span}\{dz_{i_1} \wedge \cdots \wedge dz_{i_p} \wedge d\bar{z}_{j_1} \wedge \cdots \wedge d\bar{z}_{j_q}\}$$
$$\Lambda^n = \bigoplus_{p+q=n} \Lambda^{p,q}.$$

This spanning set of $\Lambda^{p,q}$ is closed. We decompose $d = \partial + \bar{\partial}$ where

$$\partial: C^\infty(\Lambda^{p,q}) \to C^\infty(\Lambda^{p+1,q}) \qquad \text{and} \qquad \bar{\partial}: C^\infty(\Lambda^{p,q}) \to C^\infty(\Lambda^{p,q+1})$$

so that $\partial\partial = \bar{\partial}\bar{\partial} = \bar{\partial}\partial + \partial\bar{\partial} = 0$. These operators and bundles are all invariantly defined independent of the particular holomorphic coordinate system chosen.

A bundle V over M is said to be holomorphic if we can cover M by holomorphic charts U_α and find frames s_α over each U_α so that on $U_\alpha \cap U_\beta$ the transition functions $s_\alpha = f_{\alpha\beta} s_\beta$ define holomorphic maps to $GL(n, \mathbf{C})$. For such a bundle, we say a local section s is holomorphic if $s = \sum_\nu a_\alpha^\nu s_\alpha^\nu$ for holomorphic functions a_α^ν. For example, $T_c(M)$ and $\Lambda^{p,0}(M)$ are holomorphic bundles over M; $\Lambda^{0,1}$ is anti-holomorphic.

If V is holomorphic, we use a partition of unity to construct a Hermitian fiber metric h on V. Let $h_\alpha = (s_\alpha, s_\alpha)$ define the metric locally; this is a positive definite symmetric matrix which satisfies the transition rule $h_\alpha = f_{\alpha\beta} h_\beta \bar{f}_{\alpha\beta}$. We define a connection 1-form locally by:

$$\omega_\alpha = \partial h_\alpha h_\alpha^{-1}$$

and compute the transition rule:

$$\omega_\alpha = \partial\{f_{\alpha\beta} h_\beta \bar{f}_{\alpha\beta}\} \bar{f}_{\alpha\beta}^{-1} h_\beta^{-1} f_{\alpha\beta}^{-1}.$$

Since $f_{\alpha\beta}$ is holomorphic, $\bar{\partial} f_{\alpha\beta} = 0$. This implies $df_{\alpha\beta} = \partial f_{\alpha\beta}$ and $\partial \bar{f}_{\alpha\beta} = 0$ so that:

$$\omega_\alpha = \partial f_{\alpha\beta} f_{\alpha\beta}^{-1} + f_{\alpha\beta} \partial h_\beta h_\beta^{-1} f_{\alpha\beta}^{-1} = df_{\alpha\beta} f_{\alpha\beta}^{-1} + f_{\alpha\beta} \omega_\beta f_{\alpha\beta}^{-1}.$$

Since this is the transition rule for a connection, the $\{\omega_\alpha\}$ patch together to define a connection ∇_h.

It is immediate from the definition that:

$$(\nabla_h s_\alpha, s_\alpha) + (s_\alpha, \nabla_h s_\alpha) = \omega_\alpha h_\alpha + h_\alpha \omega_\alpha^* = \partial h_\alpha + \bar{\partial} h_\alpha = dh_\alpha$$

so ∇_h is a unitary connection on V. Since $\nabla_h s_\alpha \in C^\infty(\Lambda^{1,0} \otimes V)$ we conclude ∇_h vanishes on holomorphic sections when differentiated in anti-holomorphic directions (i.e., ∇_h is a holomorphic connection). It is easily verified that these two properties determine ∇_h.

We shall be particularly interested in holomorphic line bundles. If L is a line bundle, then h_α is a positive function on U_α with $h_\alpha = |f_{\alpha\beta}|^2 h_\beta$. The curvature in this case is a 2-form defined by:

$$\Omega_\alpha = d\omega_\alpha - \omega_\alpha \wedge \omega_\alpha = d(\partial h_\alpha h_\alpha^{-1}) = \bar{\partial}\partial \log(h_\alpha) = -\partial\bar{\partial} \log(h_\alpha).$$

Therefore:

$$c_1(L) = \frac{1}{2\pi i} \partial\bar{\partial} \log(h)$$

is independent of the holomorphic frame chosen for evaluation. $c_1(L) \in C^\infty(\Lambda^{1,1}(M))$ and $dc_1 = \partial c_1 = \bar{\partial} c_1 = 0$ so $c_1(L)$ is closed in all possible senses.

Let CP_n be the set of all lines through 0 in $\mathbf{C}^{n+1}$. Let $\mathbf{C}^* = \mathbf{C} - 0$ act on $\mathbf{C}^{n+1} - 0$ by complex multiplication, then $CP_n = (\mathbf{C}^{n+1} - 0)/\mathbf{C}^*$ and we give CP_n the quotient topology. Let L be the tautological line bundle over CP_n:

$$L = \{ (x, z) \in CP_n \times \mathbf{C}^{n+1} : z \in x \}.$$

Let L^* be the dual bundle; this is called the hyperplane bundle.

Let $(z_0, \ldots, z_n)$ be the usual coordinates on $\mathbf{C}^{n+1}$. Let $U_j = \{ z : z_j \neq 0 \}$. Since U_j is $\mathbf{C}^*$ invariant, it projects to define an open set on CP_n we again shall denote by U_j. We define

$$z_k^j = z_k / z_j$$

over U_j. Since these functions are invariant under the action of $\mathbf{C}^*$, they extend to continuous functions on U_j. We let $z^j = (z_0^j, \ldots, \widehat{z_j^j}, \ldots, z_n^j)$ where we have deleted $z_j^j = 1$. This gives local coordinates on U_j in CP_n. The transition relations are:

$$z_\nu^j = (z_j^k)^{-1} z_\nu^k$$

which are holomorphic so CP_n is a holomorphic manifold.

We let $s^j = (z_0^j, \ldots, z_n^j)$ be a section to L over U_j. Then $s^j = (z_j^k)^{-1} s^k$ transform holomorphically so L is a holomorphic line bundle over CP_n. The coordinates z_j on $\mathbf{C}^{n+1}$ give linear functions on L and represent global holomorphic sections to the dual bundle L^*. There is a natural inner product on the trivial bundle $CP_n \times \mathbf{C}^{n+1}$ which defines a fiber metric on L. We define:

$$x = -c_1(L) = \frac{i}{2\pi} \partial \bar{\partial} \log(1 + |z^j|^2)$$

then this is a closed 2-form over CP_n.

$U(n + 1)$ acts on $\mathbf{C}^{n+1}$; this action induces a natural action on CP_n and on L. Since the metric on L arises from the invariant metric on $\mathbf{C}^{n+1}$, the 2-form x is invariant under the action of $U(n + 1)$.

LEMMA 2.3.1. Let $x = -c_1(L)$ over CP_n. Then:
(a) $\int_{CP_n} x^n = 1$.
(b) $H^*(CP_n; \mathbf{C})$ is a polynomial ring with additive generators $\{1, x, \ldots, x^n\}$.
(c) If $i: CP_{n-1} \to CP_n$ is the natural inclusion map, then $i^*(x) = x$.

PROOF: (c) is immediate from the naturality of the constructions involved. Standard methods of algebraic topology give the additive structure of $H^*(CP_n; \mathbf{C}) = \mathbf{C} \oplus 0 \oplus \mathbf{C} \oplus 0 \cdots \oplus \mathbf{C}$. Since $x \in H^2(CP_n; \mathbf{C})$ satisfies $x^n \neq 0$, (a) will complete the proof of (b). We fix a coordinate chart U_n. Since $CP_n - U_n = CP_{n-1}$, it has measure zero. It suffices to check that:

$$\int_{U_n} x^n = 1.$$

We identify $z \in U_n = \mathbf{C}^n$ with $(z,1) \in \mathbf{C}^{n+1}$. We imbed $U(n)$ in $U(n+1)$ as the isotropy group of the vector $(0,\ldots,0,1)$. Then $U(n)$ acts on $(z,1)$ in $\mathbf{C}^{n+1}$ exactly the same way as $U(n)$ acts on z in $\mathbf{C}^n$. Let $\mathrm{dvol} = dx_1 \wedge dy_1 \wedge \cdots \wedge dx_n \wedge dy_n$ be ordinary Lebesgue measure on $\mathbf{C}^n$ without any normalizing constants of 2π. We parametrize $\mathbf{C}^n$ in the form (r,θ) for $0 \le r < \theta$ and $\theta \in S^{2n-1}$. We express $x^n = f(r,\theta)\,\mathrm{dvol}$. Since x is invariant under the action of $U(n)$, $f(r,\theta) = f(r)$ is spherically symmetric and does not depend on the parameter θ.

We compute that:

$$x = -\frac{1}{2\pi i}\partial\bar{\partial}\log\left(1 + \sum z_j\bar{z}_j\right) = -\frac{1}{2\pi i}\partial\left(\sum_j z_j\,d\bar{z}_j/(1+|z|^2)\right)$$

$$= -\frac{1}{2\pi i}\left\{\sum_j dz_j \wedge d\bar{z}_j/(1+r^2) - \sum_{j,k} z_j\bar{z}_k\,dz_k \wedge d\bar{z}_j/(1+r^2)^2\right\}.$$

We evaluate at the point $z = (r,0,\ldots,0)$ to compute:

$$x = -\frac{1}{2\pi i}\left\{\sum_j dz_j\,d\bar{z}_j/(1+r^2) - r^2\,dz_1 \wedge d\bar{z}_1/(1+r^2)^2\right\}$$

$$= -\frac{1}{2\pi i}\left\{dz_1 \wedge d\bar{z}_1/(1+r^2)^2 + \sum_{j>1} dz_j \wedge d\bar{z}_j/(1+r^2)\right\}.$$

Consequently at this point,

$$x^n = \left(-\frac{1}{2\pi i}\right)^n n!(1+r^2)^{-n-1}\,dz_1 \wedge d\bar{z}_1 \wedge \cdots \wedge dz_n \wedge d\bar{z}_n$$

$$= \pi^{-n}n!(1+r^2)^{-n-1}\,dx_1 \wedge dy_1 \wedge \cdots \wedge dx_n \wedge dy_n$$

$$= \pi^{-n}n!(1+r^2)^{-n-1}\,\mathrm{dvol}.$$

We use the spherical symmetry to conclude this identitiy holds for all z.

We integrate over $U_n = \mathbf{C}^n$ and use spherical coordinates:

$$\int x^n = n!\,\pi^{-n}\int (1+r^2)^{-n-1}\,\mathrm{dvol} = n!\,\pi^{-n}\int (1+r^2)^{-n-1}r^{2n-1}\,dr\,d\theta$$

$$= n!\,\pi^{-n}\mathrm{vol}(S^{2n-1})\int_0^\infty (1+r^2)^{-n-1}r^{2n-1}\,dr$$

$$= \frac{1}{2}n!\,\pi^{-n}\mathrm{vol}(S^{2n-1})\int_0^\infty (1+t)^{-n-1}t^{n-1}\,dt.$$

We compute the volume of S^{2n-1} using the identity $\sqrt{\pi} = \int_{-\infty}^\infty e^{-t^2}\,dt$. Thus:

$$\pi^n = \int e^{-r^2}\,\mathrm{dvol} = \mathrm{vol}(S^{2n-1})\int_0^\infty r^{2n-1}e^{-r^2}\,dr$$

$$= \frac{1}{2}\mathrm{vol}(S^{2n-1})\int_0^\infty t^{n-1}e^{-t}\,dt = \frac{(n-1)!}{2}\mathrm{vol}(S^{2n-1}).$$

We solve this for $\text{vol}(S^{2n-1})$ and substitute to compute:

$$\int x^n = n \int_0^\infty (1+t)^{-n-1} t^{n-1} \, dt = (n-1) \int_0^\infty (1+t)^{-n} t^{n-2} \, dt = \cdots$$
$$= \int_0^\infty (1+t)^{-2} \, dt = 1$$

which completes the proof.

x can also be used to define a $U(n+1)$ invariant metric on $\mathbf{C}P_n$ called the Fubini-Study metric. We shall discuss this in more detail in Chapter 3 as this gives a Kaehler metric for $\mathbf{C}P_n$.

There is a relation between L and $\Lambda^{1,0}(\mathbf{C}P_n)$ which it will be convenient to exploit:

LEMMA 2.3.2. *Let $M = \mathbf{C}P_n$.*
(a) There is a short exact sequence of holomorphic bundles:

$$0 \to \Lambda^{1,0}(M) \to L \otimes 1^{n+1} \to 1 \to 0.$$

(b) There is a natural isomorphism of complex bundles:

$$T_c(M) \oplus 1 \simeq L^* \otimes 1^{n+1} = \underbrace{L^* \oplus \cdots \oplus L^*}_{n+1 \text{ times}}.$$

PROOF: Rather than attempting to give a geometric proof of this fact, we give a combinatorial argument. Over U_k we have functions z_i^k which give coordinates $0 \leq i \leq n$ for $i \neq k$. Furthermore, we have a section s^k to L. We let $\{s_i^k\}_{i=0}^n$ give a frame for $L \otimes 1^{n+1} = L \oplus \cdots \oplus L$. We define $F : \Lambda^{1,0}(M) \to L \otimes 1^{n+1}$ on U_k by:

$$F(dz_i^k) = s_i^k - z_i^k s_k^k.$$

We note that $z_k^k = 1$ and $F(dz_k^k) = 0$ so this is well defined on U_k. On the overlap, we have the relations $z_i^k = (z_k^j)^{-1} z_i^j$ and:

$$s_i^k = (z_k^j)^{-1} s_i^j \qquad \text{and} \qquad dz_i^k = (z_k^j)^{-1} dz_i^j - (z_k^j)^{-2} z_i^j \, dz_k^j.$$

Thus if we compute in the coordinate system U_j we have:

$$F(dz_i^k) = (z_k^j)^{-1}(s_i^j - z_i^j s_j^j) - (z_k^j)^{-2} z_i^j (s_k^j - z_k^j s_j^j)$$
$$= s_i^k - z_i^j s_j^k - z_i^k s_k^k + z_i^j s_j^k = s_i^k - z_i^k s_k^k$$

which agrees with the definition of F on the coordinate system U_k. Thus F is invariantly defined. It is clear F is holomorphic and injective. We let

$\nu = L \otimes 1^{n+1}/\text{image}(F)$, and let π be the natural projection. It is clear that s_k^k is never in the image of F so $\pi s_k^k \neq 0$. Since

$$F(dz_j^k) = s_j^k - z_j^k s_k^k = z_j^k(s_j^j - s_k^k)$$

we conclude that $\pi s_j^j = \pi s_k^k$ so $s = \pi s_k^k$ is a globally defined non-zero section to ν. This completes the proof of (a). We dualize to get a short exact sequence:

$$0 \to 1 \to L^* \otimes 1^{n+1} \to T_c(M) \to 0.$$

These three bundles have natural fiber metrics. Any short exact sequence of vector bundles splits (although the splitting is not holomorphic) and this proves (b).

From assertion (b) it follows that the Chern class of $T_c(CP_n)$ is given by:

$$c(T_c) = c(T_c \oplus 1) = c(L^* \oplus \cdots \oplus L^*) = c(L^*)^{n+1} = (1+x)^{n+1}.$$

For example, if $m = 1$, then $c(T_c) = (1+x)^2 = 1 + 2x$. In this case, $c_1(T_c) = e(T(S^2))$ and we computed $\int_{S^2} e(T(S)) = 2$ so $\int_{S^2} x = 1$ which checks with Lemma 2.3.1.

If we forget the complex structure on $T_c(M)$ when M is holomorphic, then we obtain the real tangent space $T(M)$. Consequently:

$$T(M) \otimes \mathbf{C} = T_c \oplus T_c^*$$

and

$$\begin{aligned}
c(T(CP_n) \otimes \mathbf{C}) &= c(T_c(CP_n) \oplus T_c^*(CP_n)) \\
&= (1+x)^{n+1}(1-x)^{n+1} = (1-x^2)^{n+1}.
\end{aligned}$$

When we take into account the sign changes involved in defining the total Pontrjagin form, we conclude:

LEMMA 2.3.3.
(a) If $M = S^2 \times \cdots \times S^2$ has dimension $2n$, then $\int_M e(T(M)) = \chi(M) = 2^n$.
(b) If $M = CP_n$ has dimension $2n$, then

$$c(T_c(M)) = (1+x)^{n+1} \qquad \text{and} \qquad p(T(M)) = (1+x^2)^{n+1}.$$

$x \in H^2(CP_n; \mathbf{C})$ is the generator given by $x = c_1(L^*) = -c_1(L)$, L is the tautological line bundle over CP_n, and L^* is the dual, the hyperplane bundle.

The projective spaces form a dual basis to both the real and complex characteristic classes. Let ρ be a partition of the positive integer k in the form $k = i_1 + \cdots + i_j$ where we choose the notation so $i_1 \geq i_2 \geq \cdots$. We let $\pi(k)$ denote the number of such partitions. For example, if $k = 4$ then $\pi(4) = 5$ and the possible partitions are:

$$4 = 4, \quad 4 = 3 + 1, \quad 4 = 2 + 2, \quad 4 = 2 + 1 + 1, \quad 4 = 1 + 1 + 1 + 1.$$

We define classifying manifolds:

$$M_\rho^c = \mathbf{CP}_{i_1} \times \cdots \times \mathbf{CP}_{i_j} \qquad \text{and} \qquad M_\rho^r = \mathbf{CP}_{2i_1} \times \cdots \times \mathbf{CP}_{2i_j}$$

to be real manifolds of dimension $2k$ and $4k$.

LEMMA 2.3.4. *Let k be a positive integer. Then:*
(a) Let constants $c(\rho)$ be given. There exists a unique polynomial $P(A)$ of degree k in the components of a $k \times k$ complex matrix which is $\mathrm{GL}(k, \mathbf{C})$ invariant such that the characteristic class defined by P satisfies:

$$\int_{M_\rho^c} P(T_c(M_\rho^c)) = c(\rho)$$

for every such partition ρ of k.
(b) Let constants $c(\rho)$ be given. There exists a polynomial $P(A)$ of degree $2k$ in the components of a $2k \times 2k$ real matrix which is $\mathrm{GL}(2k, \mathbf{R})$ invariant such that the characteristic class defined by P satisfies:

$$\int_{M_\rho^r} P(T(M_\rho^r)) = c(\rho)$$

for every such partition ρ of k. If P' is another such polynomial, then $P(A) = P'(A)$ for every skew-symmetric matrix A.

In other words, the real and complex characteristic classes are completely determined by their values on the appropriate classifying manifolds.

PROOF: We prove (a) first. Let P_k denote the set of all such polynomials $P(A)$. We define $c_\rho = c_{i_1} \ldots c_{i_j} \in P_k$, then by Lemma 2.1.3 the $\{c_\rho\}$ form a basis for P_k so $\dim(P_k) = \pi(k)$. The $\{c_\rho\}$ are not a very convenient basis to work with. We will define instead:

$$H_\rho = ch_{i_1} \ldots ch_{i_j} \in P_k$$

and show that the matrix:

$$a(\rho, \tau) = \int_{M_\tau^c} H_\rho(T_c(M_\tau^c))$$

is a non-singular matrix. This will prove the H_ρ also form a basis for P_k and that the M_τ^c are a dual basis. This will complete the proof.

The advantage of working with the Chern character rather than with the Chern class is that:

$$ch_i(T_c(M_1 \times M_2)) = ch_i(T_cM_1 \oplus T_cM_2) = ch_i(T_cM_1) + ch_i(T_cM_2).$$

Furthermore, $ch_i(T_cM) = 0$ if $2i > \dim(M)$. We define the length $\ell(\rho) = j$ to be the number of elements in the partition ρ. Then the above remarks imply:

$$a(\rho, \tau) = 0 \qquad \text{if } \ell(\tau) > \ell(\rho).$$

Furthermore, if $\ell(\tau) = \ell(\rho)$, then $a(\rho, \tau) = 0$ unless $\tau = \rho$. We define the partial order $\tau > \rho$ if $\ell(\tau) > \ell(\rho)$ and extend this to a total order. Then $a(\rho, \tau)$ is a triangular matrix. To show it is invertible, it suffices to show the diagonal elements are non-zero.

We first consider the case in which $\rho = \tau = k$. Using the identity $T_c(\mathbf{C}P_k) \oplus 1 = (L^* \oplus \cdots \oplus L^*)$ $(k+1$ times), it is clear that $ch_k(T_c\mathbf{C}P_k) = (k+1)ch_k(L^*)$. For a line bundle, $ch_k(L^*) = c_1(L^*)^k/k!$. If $x = c_1(L^*)$ is the generator of $H^2(\mathbf{C}P_k; \mathbf{C})$, then $ch_k(T_c\mathbf{C}P_k) = (k+1)x^k/k!$ which does not integrate to zero. If $\rho = \tau = \{i_1, \ldots, i_j\}$ then:

$$a(\rho, \rho) = c \prod_{\nu=1}^{\ell(\rho)} a(i_\nu, i_\nu)$$

where c is a positive constant related to the multiplicity with which the i_ν appear. This completes the proof of (a).

To prove (b) we replace M_ρ^c by $M_{2\rho}^c = M_\rho^r$ and H_ρ by $H_{2\rho}$. We compute

$$ch_{2i}(T(M)) \overset{\text{def}}{=} ch_{2i}(T(M) \otimes \mathbf{C}) = ch_{2i}(T_cM \oplus T_c^*M)$$
$$= ch_{2i}(T_cM) + ch_{2i}(T_c^*M)$$
$$= 2ch_{2i}(T_cM).$$

Using this fact, the remainder of the proof of (b) is immediate from the calculations performed in (a) and this completes the proof.

The Todd class and the Hirzebruch L-polynomial were defined using generating functions. The generating functions were chosen so that they would be particularly simple on the classifying examples:

LEMMA 2.3.5.
(a) Let $x_j = -\lambda_j/2\pi i$ be the normalized eigenvalues of a complex matrix A. We define $Td(A) = Td(x) = \prod_j x_j/(1 - e^{-x_j})$ as the Todd class. Then:

$$\int_{M_\rho^c} Td(T_c(M_\rho^c)) = 1 \qquad \text{for all } \rho.$$

(b) Let $x_j = \lambda_j/2\pi$ where the eigenvalues of the skew-symmetric real matrix A are $\{\pm\lambda_1,\ldots\}$. We define $L(A) = L(x) = \prod_j x_j/\tanh x_j$ as the Hirzebruch L-polynomial. Then $\int_{M_\rho^r} L(T(M^r)) = 1$ for all ρ.

We will use this calculation to prove the integral of the Todd class gives the arithmetic genus of a complex manifold and that the integral of the Hirzebruch L-polynomial gives the signature of an oriented real manifold. In each case, we only integrate the part of the total class which is of the same degree as the dimension of the manifold.

PROOF: Td is a multiplicative class:

$$Td(T_c(M_1 \times M_2)) = Td(T_c(M_1) \oplus T_c(M_2)) = Td(T_c(M_1))\, Td(T_c(M_2)).$$

Similarly the Hirzebruch polynomial is multiplicative. This shows it suffices to prove Lemma 2.3.5 in the case $\rho = k$ so $M = \mathbf{CP}_k$ or $\mathbf{CP}_{2k}$.

We use the decomposition $T_c(\mathbf{CP}_k) \oplus 1 = L^* \oplus \cdots \oplus L^*$ $(k + 1$ times$)$ to compute $Td(T_c(\mathbf{CP}_k)) = [Td(x)]^{k+1}$ where $x = c_1(L^*)$ is the generator of $H^2(\mathbf{CP}_k; \mathbf{R})$. Since x^k integrates to 1, it suffices to show the coefficient of x^k in $Td(x)^{k+1}$ is 1 or equivalently to show:

$$\text{Res}_{x=0}\, x^{-k-1}[Td(x)^{k+1}] = \text{Res}_{x=0}(1 - e^{-x})^{-k-1} = 1.$$

If $k = 0$, then:

$$(1 - e^{-x})^{-1} = \left(x - \frac{1}{2}x^2 + \cdots\right)^{-1} = x^{-1}\left(1 + \frac{1}{2}x + \cdots\right)$$

and the result follows. Similarly, if $k = 1$

$$(1 - e^{-x})^{-2} = x^{-2}(1 + x + \cdots)$$

and the result follows. For larger values of k, proving this directly would be a combinatorial nightmare so we use instead a standard trick from complex variables. If $g(x)$ is any meromorphic function, then $\text{Res}_{x=0}\, g'(x) = 0$. We apply this to the function $g(x) = (1 - e^{-x})^{-k}$ for $k \geq 1$ to conclude:

$$\text{Res}_{x=0}(1 - e^{-x})^{-k-1}e^{-x} = 0.$$

This implies immediately that:

$$\text{Res}_{x=0}(1 - e^{-x})^{-k-1} = \text{Res}_{x=0}(1 - e^{-x})^{-k-1}(1 - e^{-x})$$
$$= \text{Res}_{x=0}(1 - e^{-x})^{-k} = 1$$

by induction which completes the proof of assertion (a).

We now assume k is even and study $\mathbf{CP}_k$. Again, using the decomposition $T_c(\mathbf{CP}_k) \oplus 1 = L^* \oplus \cdots \oplus L^*$ it follows that

$$L(T(\mathbf{CP}_k)) = [L(x)]^{k+1} = \frac{x^{k+1}}{\tanh^{k+1} x}.$$

Since we are interested in the coefficient of x^k, we must show:

$$\text{Res}_{x=0} \tanh^{-k-1} x = 1 \qquad \text{if } k \text{ is even}$$

or equivalently

$$\text{Res}_{x=0} \tanh^{-k} x = 1 \qquad \text{if } k \text{ is odd}.$$

We recall that

$$\tanh x = \frac{e^x - e^{-x}}{e^x + e^{-x}}$$

$$\tanh^{-1} x = \frac{2 + x^2 + \cdots}{2x + \cdots}$$

so the result is clear if $k = 1$. We now proceed by induction. We differentiate $\tanh^{-k} x$ to compute:

$$(\tanh^{-k} x)' = -k(\tanh^{-k-1} x)(1 - \tanh^2 x).$$

This implies

$$\text{Res}_{x=0} \tanh^{-k+1} x = \text{Res}_{x=0} \tanh^{-k-1} x$$

for any integer k. Consequently $\text{Res}_{x=0} \tanh^{-k} x = 1$ for any odd integer k since these residues are periodic modulo 2. (The residue at k even is, of course, zero). This completes the proof.

2.4. The Gauss-Bonnet Theorem.

Let P be a self-adjoint elliptic partial differential operator of order $d > 0$. If the leading symbol of P is positive definite, we derived an asymptotic expansion for $\mathrm{Tr}\{e^{-tP}\}$ in section 1.7. This is too general a setting in which to work so we shall restrict attention henceforth to operators with leading symbol given by the metric tensor, as this is the natural category in which to work.

Let $P\colon C^\infty(V) \to C^\infty(V)$ be a second order operator. We choose a local frame for V. Let $x = (x_1, \ldots, x_m)$ be a system of local coordinates. Let $ds^2 = g_{ij}\, dx^i\, dx^j$ be the metric tensor and let g^{ij} denote the inverse matrix; $dx^i \cdot dx^j = g^{ij}$ is the metric on the dual space $T^*(M)$. We assume P has the form:

$$P = -g^{ij}\frac{\partial^2}{\partial x_i \partial x_j}I + a_j \frac{\partial}{\partial x_j} + b$$

where we sum over repeated indices. The a_j and b are sections to $\mathrm{END}(V)$.

We introduce formal variables for the derivatives of the symbol of P. Let

$$g_{ij/\alpha} = d_x^\alpha g_{ij}, \qquad a_{j/\alpha} = d_x^\alpha a_j, \qquad b_{/\alpha} = d_x^\alpha b.$$

We will also use the notation $g_{ij/kl\ldots}$, $a_{j/kl\ldots}$, and $b_{/kl\ldots}$ for multiple partial derivatives. We emphasize that these variables are not tensorial but depend upon the choice of a coordinate system (and local frame for V).

There is a natural grading on these variables. We define:

$$\mathrm{ord}(g_{ij/\alpha}) = |\alpha|, \qquad \mathrm{ord}(a_{j/\alpha}) = 1 + |\alpha|, \qquad \mathrm{ord}(b_{/\alpha}) = 2 + |\alpha|.$$

Let $\mathcal{P}$ be the non-commutative polynomial algebra in the variables of positive order. We always normalize the coordinate system so x_0 corresponds to the point $(0, \ldots, 0)$ and so that:

$$g_{ij}(x_0) = \delta_{ij} \qquad \text{and} \qquad g_{ij/k}(x_0) = 0.$$

Let $K(t, x, x)$ be the kernel of e^{-tP}. Expand

$$K(t, x, x) \sim \sum_{n \geq 0} t^{\frac{n-m}{2}} e_n(x, P)$$

where $e_n(x, P)$ is given by Lemma 1.7.4. Lemma 1.7.5(c) implies:

LEMMA 2.4.1. $e_n(x, P) \in \mathcal{P}$ *is a non-commutative polynomial in the jets of the symbol of P which is homogeneous of order n. If $a_n(x, P) = \mathrm{Tr}\, e_n(x, P)$ is the fiber trace, then $a_n(x, P)$ is homogeneous of order n in* *the components (relative to some local frame) of the jets of the total symbol of P.*

We could also have proved Lemma 2.4.1 using dimensional analysis and the local nature of the invariants $e_n(x, P)$ instead of using the combinatorial argument given in Chapter 1.

We specialize to the case where $P = \Delta_p$ is the Laplacian on p-forms. If $p = 0$, then $\Delta_0 = -g^{-1}\partial/\partial x_i\{gg^{ij}\partial/\partial x_j\}$; $\quad g = \det(g_{ij})^{1/2}$ defines the Riemannian volume dvol $= g\,dx$. The leading symbol is given by the metric tensor; the first order symbol is linear in the 1-jets of the metric with coefficients which depend smoothly on the metric tensor. The 0th order symbol is zero in this case. More generally:

LEMMA 2.4.2. *Let $x = (x_1, \ldots, x_m)$ be a system of local coordinates on M. We use the dx^I to provide a local frame for $\Lambda(T^*M)$. Relative to this frame, we expand $\sigma(\Delta_p) = p_2 + p_1 + p_0$. $p_2 = |\xi|^2 I$. p_1 is linear in the 1-jets of the metric with coefficients which depend smoothly on the g_{ij}'s. p_0 is the sum of a term which is linear in the 2-jets of the metric and a term which is quadratic in the 1-jets of the metric with coefficients which depend smoothly on the g_{ij}'s.*

PROOF: We computed the leading symbol in the first chapter. The remainder of the lemma follows from the decomposition $\Delta = d\delta + \delta d = \pm d * d * \pm * d * d$. In flat space, $\Delta = -\sum_i \partial^2/\partial x_i^2$ as computed earlier. If the metric is curved, we must also differentiate the matrix representing the Hodge $*$ operator. Each derivative applied to "$*$" reduces the order of differentiation by one and increases the order in the jets of the metric by one.

Let $a_n(x, P) = \operatorname{Tr} e_n(x, P)$. Then:

$$\operatorname{Tr} e^{-tP} \sim \sum_{n \geq 0} t^{\frac{n-m}{2}} \int_M a_n(x, P)\,\mathrm{dvol}(x).$$

For purposes of illustration, we give without proof the first few terms in the asymptotic expansion of the Laplacian. We shall discuss such formulas in more detail in the fourth chapter.

LEMMA 2.4.3.
(a) $a_0(x, \Delta_p) = (4\pi)^{-1}\dim(\Lambda^p) = (4\pi)^{-1}\binom{m}{p}$.
(b) $a_2(x, \Delta_0) = (4\pi)^{-1}(-R_{ijij})/6$.
(c)

$$a_4(x, \Delta_0) = \frac{1}{4\pi} \cdot \frac{-12R_{ijij;kk} + 5R_{ijij}R_{klkl} - 2R_{ijik}R_{ljlk} + 2R_{ijkl}R_{ijkl}}{360}.$$

In this lemma, "$;$" denotes multiple covariant differentiation. a_0, a_2, a_4, and a_6 have been computed for Δ_p for all p, but as the formulas are extremely complicated, we shall not reproduce them here.

We consider the algebra generated by the variables $\{g_{ij/\alpha}\}_{|\alpha|\geq 2}$. If X is a coordinate system and G a metric and if P is a polynomial in these variables, we define $P(X,G)(x_0)$ by evaluation. We always normalize the choice of X so $g_{ij}(X,G)(x_0) = \delta_{ij}$ and $g_{ij/k}(X,G)(x_0) = 0$, and we omit these variables from consideration. We say that P is invariant if $P(X,G)(x_0) = P(Y,G)(x_0)$ for any two normalized coordinate systems X and Y. We denote the common value by $P(G)(x_0)$. For example, the scalar curvature $K = -\frac{1}{2}R_{ijij}$ is invariant as are the $a_n(x,\Delta_p)$.

We let $\mathcal{P}_m$ denote the ring of all invariant polynomials in the derivatives of the metric for a manifold of dimension m. We defined $\mathrm{ord}(g_{ij/\alpha}) = |\alpha|$; let $\mathcal{P}_{m,n}$ be the subspace of invariant polynomials which are homogeneous of order n. This is an algebraic characterization; it is also useful to have the following coordinate free characterization:

LEMMA 2.4.4. *Let* $P \in \mathcal{P}_m$, *then* $P \in \mathcal{P}_{m,n}$ *if and only if* $P(c^2 G)(x_0) = c^{-n}P(G)(x_0)$ *for every* $c \neq 0$.

PROOF: Fix $c = 0$ and let X be a normalized coordinate system for the metric G at the point x_0. We assume $x_0 = (0,\ldots,0)$ is the center of the coordinate system X. Let $Y = cX$ be a new coordinate system, then:

$$\partial/\partial y_i = c^{-1}\partial/\partial x_i \qquad c^2 G(\partial/\partial y_i, \partial/\partial y_j) = G(\partial/\partial x_i, \partial/\partial x_j)$$
$$d_y^\alpha = c^{-|\alpha|}d_x^\alpha \qquad g_{ij/\alpha}(Y,c^2 G) = c^{-|\alpha|}g_{ij/\alpha}(X,G).$$

This implies that if A is any monomial of P that:

$$A(Y,c^2 G)(x_0) = c^{-\,\mathrm{ord}(A)}A(X,G)(x_0).$$

Since Y is normalized coordinate system for the metric $c^2 G$, $P(c^2 G)(x_0) = P(Y,c^2 G)(x_0)$ and $P(G)(x_0) = P(X,G)(x_0)$. This proves the Lemma.

If $P \in \mathcal{P}_m$ we can always decompose $P = P_0 + \cdots + P_n$ into homogeneous polynomials. Lemma 2.4.4 implies the P_j are all invariant separately. Therefore $\mathcal{P}_m$ has a direct sum decomposition $\mathcal{P}_m = \mathcal{P}_{m,0} \oplus \mathcal{P}_{m,1} \oplus \cdots \oplus \mathcal{P}_{m,n} \oplus \cdots$ and has the structure of a graded algebra. Using Taylor's theorem, we can always find a metric with the $g_{ij/\alpha}(X,G)(x_0) = c_{ij,\alpha}$ arbitrary constants for $|\alpha| \geq 2$ and so that $g_{ij}(X,G)(x_0) = \delta_{ij}$, $g_{ij/k}(X,G)(x_0) = 0$. Consequently, if $P \in \mathcal{P}_m$ is non-zero as a polynomial, then we can always find G so $P(G)(x_0) \neq 0$ so P is non-zero as a formula. It is for this reason we work with the algebra of jets. This is a pure polynomial algebra. If we work instead with the algebra of covariant derivatives of the curvature tensor, we must introduce additional relations which correspond to the Bianchi identities as this algebra is not a pure polynomial algebra.

We note finally that $\mathcal{P}_{m,n}$ is zero if n is odd since we may take $c = -1$. Later in this chapter, we will let $\mathcal{P}_{m,n,p}$ be the space of p-form valued invariants which are homogeneous of order n. A similar argument will show $\mathcal{P}_{m,n,p}$ is zero if $n - p$ is odd.

Lemmas 2.4.1 and 2.4.2 imply:

LEMMA 2.4.5. $a_n(x, \Delta_p)$ defines an element of $P_{m,n}$.

This is such an important fact that we give another proof based on Lemma 2.4.4 to illustrate the power of dimensional analysis embodied in this lemma. Fix $c > 0$ and let $\Delta_p(c^2 G) = c^{-2}\Delta_p(G)$ be the Laplacian corresponding to the new metric. Since $\mathrm{dvol}(c^2 G) = c^m \, \mathrm{dvol}(G)$, we conclude:

$$e^{-t\Delta_p(c^2 G)} = e^{-tc^{-2}\Delta_p(G)}$$

$$K(t, x, x, \Delta_p(c^2 G)) \, \mathrm{dvol}(c^2 G) = K(c^{-2}t, x, x, \Delta_p(G)) \, \mathrm{dvol}(G)$$

$$K(t, x, x, \Delta_p(c^2 G)) = c^{-m} K(c^{-2}t, x, x, \Delta_p(G))$$

$$\sum_n t^{\frac{n-m}{2}} a_n(x, \Delta_p(c^2 G)) \sim \sum_n c^{-m} c^m c^{-n} t^{\frac{n-m}{2}} a_n(x, \Delta_p(G))$$

$$a_n(x, \Delta_p(c^2 G)) = c^{-n} a_n(x, \Delta_p(G)).$$

We expand $a_n(x, \Delta_p(G)) = \sum_\nu a_{n,\nu,p} + r$ in a finite Taylor series about $g_{ij} = \delta_{ij}$ in the $g_{ij/\alpha}$ variables. Then if $a_{n,\nu,p}$ is the portion which is homogeneous of order ν, we use this identity to show $a_{n,\nu,p} = 0$ for $n \neq \nu$ and to show the remainder in the Tayor series is zero. This shows a_n is a homogeneous polynomial of order n and completes the proof.

Since $a_n(x, \Delta_p) = 0$ if n is odd, in many references the authors replace the asymptotic series by $t^{-\frac{m}{2}} \sum_n t^n a_n(x, \Delta_p)$, They renumber this sequence $a_0, a_1, \ldots$ rather than $a_0, 0, a_2, 0, a_4, \ldots$. We shall not adopt this notational convention as it makes dealing with boundary problems more cumbersome.

H. Weyl's theorem on the invariants of the orthogonal group gives a spanning set for the spaces $P_{m,n}$:

LEMMA 2.4.6. We introduce formal variables $R_{i_1 i_2 i_3 i_4; i_5 \ldots i_k}$ for the multiple covariant derivatives of the curvature tensor. The order of such a variable is $k + 2$. We consider the polynomial algebra in these variables and contract on pairs of indices. Then all possible such expressions generate P_m. In particular:

$$\{1\} \text{ spans } P_{m,0}, \qquad \{R_{ijij}\} \text{ spans } P_{m,2}$$
$$\{R_{ijij;kk}, R_{ijij}R_{klkl}, R_{ijik}R_{ljlk}, R_{ijkl}R_{ijkl}\} \text{ spans } P_{m,4}.$$

This particular spanning set for $P_{m,4}$ is linearly independent and forms a basis if $m \geq 4$. If $m = 3$, $\dim(P_{3,4}) = 3$ while if $m = 2$, $\dim(P_{2,4}) = 2$ so there are relations imposed if the dimension is low. The study of these additional relations is closely related to the Gauss-Bonnet theorem.

There is a natural restriction map

$$r: P_{m,n} \to P_{m-1,n}$$

which is defined algebraically as follows. We let

$$\deg_k(g_{ij/\alpha}) = \delta_{i,k} + \delta_{j,k} + \alpha(k)$$

be the number of times an index k appears in the variable $g_{ij/\alpha}$. Let

$$r(g_{ij/\alpha}) = \begin{cases} g_{ij/\alpha} \in P_{m-1} & \text{if } \deg_m(g_{ij/\alpha}) = 0 \\ 0 & \text{if } \deg_m(g_{ij/\alpha}) > 0. \end{cases}$$

We extend $r: P_m \to P_{m-1}$ to be an algebra morphism; $r(P)$ is a polynomial in the derivatives of a metric on a manifold of dimension $m - 1$. It is clear that r preserves the degree of homogeneity.

r is the dual of a natural extension map. Let G' be a metric on a manifold M' of dimension $m - 1$. We define $i(G') = G + d\theta^2$ on the manifold $M = M' \times S^1$ where S^1 is the unit circle with natural parameter θ. If X' is a local coordinate system on M', then $i(X') = (x, \theta)$ is a local coordinate system on M. It is clear that:

$$(rP)(X', G')(x_0') = P(i(X'), i(G'))(x_0', \theta_0)$$

for any $\theta_0 \in S^1$; what we have done by restricting to product manifolds $M' \times S^1$ with product metrics $G' + d\theta^2$ is to introduce the relation which says the metric is flat in the last coordinate. Restiction is simply the dual of this natural extension; rP is invariant if P is invariant. Therefore $r: P_{m,n} \to P_{m,n-1}$.

There is one final description of the restriction map which will be useful. In discussing a H. Weyl spanning set, the indices range from 1 through m. We define the restriction by letting the indices range from 1 through $m - 1$. Thus $R_{ijij} \in P_{m,2}$ is its own restriction in a formal sense; of course $r(R_{ijij}) = 0$ if $m = 2$ since there are no non-trivial local invariants over a circle.

THEOREM 2.4.7.

(a) $r: P_{m,n} \to P_{m-1,n}$ is always surjective.

(b) $r: P_{m,n} \to P_{m-1,n}$ is bijective if $n < m$.

(c) $r: P_{m,m} \to P_{m-1,m}$ has 1-dimensional kernel spanned by the Euler class E_m if m is even. If m is odd, $P_{m,m} = P_{m-1,m} = 0$.

(c) is an axiomatic characterization of the Euler form. It is an expression of the fact that the Euler form is a unstable characteristic class as opposed to the Pontrjagin forms which are stable characteristic classes.

PROOF: (a) is consequence of H. Weyl's theorem. If we choose a H. Weyl spanning set, we let the indices range from 1 to m instead of from 1 to $m - 1$ to construct an element in the inverse image of r. The proof of (b) and of (c) is more complicated and will be postponed until the next subsection. Theorem 2.4.7 is properly a theorem in invariance theory, but we have stated it at this time to illustrate how invariance theory can be used to prove index theorems using heat equation methods:

THEOREM 2.4.8. *Let* $a_n(x, d + \delta) = \sum_p (-1)^p a_n(x, \Delta_p) \in \mathcal{P}_{n,m}$ *be the invariant of the de Rham complex. We showed in Lemma 1.7.6 that:*

$$\int_M a_n(x, d + \delta) \, \mathrm{dvol}(x) = \begin{cases} \chi(M) & \text{if } n = m \\ 0 & \text{if } n \neq m. \end{cases}$$

Then:
(a) $a_n(x, d + \delta) = 0$ *if either* m *is odd or if* $n < m$.
(b) $a_m(x, d + \delta) = E_m$ *is* m *is even so* $\chi(M) = \int_M E_m \, \mathrm{dvol}(x)$ *(Gauss-Bonnet theorem).*

PROOF: We suppose first that m is odd. Locally we can always choose an orientation for $T(M)$. We let $*$ be the Hodge operator then $*\Delta_p = \pm \Delta_{m-p} *$ locally. Since these operators are locally isomorphic, their local invariants are equal so $a_n(x, \Delta_p) = a_n(x, \Delta_{m-p})$. If m is odd, these terms cancel in the alternating sum to give $a_n(x, d + \delta) = 0$. Next we suppose m is even. Let $M = M' \times S^1$ with the product metric. We decompose any $\omega \in \Lambda(T^* M)$ uniquely in the form

$$\omega = \omega_1 + \omega_2 \wedge d\theta \qquad \text{for } \omega_i \in \Lambda(T^* M').$$

We define:

$$F(\omega) = \omega_1 \wedge d\theta + \omega_2$$

and compute easily that $F\Delta = \Delta F$ since the metric is flat in the S^1 direction. If we decompose $\Lambda(M) = \Lambda^e(M) \oplus \Lambda^o(M)$ into the forms of even and odd degree, then F interchanges these two factors. Therefore, $a_n(x, \Delta_e) = a_n(x, \Delta_0)$ so $a_n(x, d + \delta) = 0$ for such a product metric. This implies $r(a_n) = 0$. Therefore $a_n = 0$ for $n < m$ by Theorem 2.4.7. Furthermore:

$$a_m = c_m E_m$$

for some universal constant c_m. We show $c_m = 1$ by integrating over the classifying manifold $M = S^2 \times \cdots \times S^2$. Let $2\bar{m} = m$, then

$$2^{\bar{m}} = \chi(M) = \int_M a_m(x, d + \delta) \, \mathrm{dvol}(x) = \int_M E_m \, \mathrm{dvol}(x)$$

by Lemma 2.3.4. This completes the proof of the Gauss-Bonnet theorem.

2.5. Invariance Theory and the Pontrjagin Classes of the Tangent Bundle.

In the previous subsection, we gave in Theorem 2.4.7 an axiomatic characterization of the Euler class in terms of functorial properties. In this subsection we will complete the proof of Theorem 2.4.7. We will also give a similar axiomatic characterization of the Pontrjagin classes which we will use in our discussion of the signature complex in Chapter 3.

Let $T: \mathbf{R}^m \to \mathbf{R}^m$ be the germ of a diffeomorphism. We assume that:

$$T(0) = 0 \quad \text{and} \quad dT(x) = dT(0) + O(x^2) \quad \text{for } dT(0) \in O(k).$$

If X is any normalized coordinate system for a metric G, then TX is another normalized coordinate system for G. We define an action of the group of germs of diffeomorphisms on the polynomial algebra in the $\{g_{ij/\alpha}\}$, $|\alpha| \geq 2$ variables by defining the evaluation:

$$(T^*P)(X,G)(x_0) = P(TX,G)(x_0).$$

Clearly P is invariant if and only if $T^*P = P$ for every such diffeomorphism T.

Let P be invariant and let A be a monomial. We let $c(A,P)$ be the coefficient of A in P; $c(A,P)$ defines a linear functional on P for any monomial A. We say A is a monomial of P if $c(A,P) \neq 0$. Let T_j be the linear transformation:

$$T_j(x_k) = \begin{cases} -x_j & \text{if } k = j \\ x_k & \text{if } k \neq j. \end{cases}$$

This is reflection in the hyperplane defined by $x_j = 0$. Then

$$T_j^*(A) = (-1)^{\deg_j(A)} A$$

for any monomial A. Since

$$T_j^*P = \sum (-1)^{\deg_j(A)} c(A,P)A = P = \sum c(A,P)A,$$

we conclude $\deg_j(A)$ must be even for any monomial A of P. If A has the form:

$$A = g_{i_1 j_1/\alpha_1} \cdots g_{i_r j_r/\alpha_r}$$

we define the length of A to be:

$$\ell(A) = r.$$

It is clear $2\ell(A) + \text{ord}(A) = \sum_j \deg_j(A)$ so $\text{ord}(A)$ is necessarily even if A is a monomial of P. This provides another proof $P_{m,n} = 0$ if n is odd.

In addition to the hyperplane reflections, it is convenient to consider coordinate permutations. If ρ is a permutation, then $T_\rho^*(A) = A^\rho$ is defined by replacing each index i by $\rho(i)$ in the variables $g_{ij/k...}$. Since $T_\rho^*(P) = P$, we conclude the form of P is invariant under coordinate permutations or equivalently $c(A, P) = c(A^\rho, P)$ for every monomial A of P.

We can use these two remarks to begin the proof of Theorem 2.4.7. Fix $P \neq 0$ with $r(P) = 0$ and $P \in \mathcal{P}_{m,n}$. Let A be a monomial of P. Then $r(P) = 0$ implies $\deg_m(A) > 0$ is even. Since P is invariant under coordinate permutations, $\deg_k(A) \geq 2$ for $1 \leq k \leq m$. We construct the chain of inequalities:

$$2m \leq \sum_k \deg_k(A) = 2\ell(A) + \text{ord}(A)$$

$$= \sum_\nu (2 + |\alpha_\nu|) \leq \sum_\nu 2|\alpha_\nu| = 2\,\text{ord}(A) = 2n.$$

We have used the fact $|\alpha_\nu| \geq 2$ for $1 \leq \nu \leq \ell(A)$. This implies $m \leq n$ so in particular $P = 0$ if $n < m$ which proves assertion (b) of Theorem 2.4.7. If $n = m$, then all of the inequalities in this string must be equalities. This implies $2\ell(A) = m$, $\deg_k(A) = 2$ for all k, and $|\alpha_\nu| = 2$ for all ν. Thus the monomial A must have the form:

$$A = g_{i_1 j_1/k_1 l_1} \cdots g_{i_r j_r/k_r l_r} \qquad \text{for } 2r = m$$

and in particular, P only involves the second derivatives of the metric.

In order to complete the proof of Theorem 2.4.7, we must use some results involving invariance under circle actions:

LEMMA 2.5.1. *Parametrize the circle* $SO(2)$ *by* $z = (a, b)$ *for* $a^2 + b^2 = 1$. *Let* T_z *be the coordinate transformation:*

$$y_1 = ax_1 + bx_2, \qquad y_2 = -bx_1 + ax_2, \qquad y_k = x_k \qquad \text{for } k > 2.$$

Let P *be a polynomial and assume* $T_z^* P = P$ *for all* $z \in S^1$. *Then:*
(a) *If* $g_{12/\alpha}$ *divides a monomial* A *of* P *for some* α, *then* $g_{11/\beta}$ *divides a monomial* B *of* P *for some* β.
(b) *If* $g_{ij/\alpha}$ *divides a monomial of* A *of* P *for some* (i, j), *then* $g_{kl/\beta}$ *divides a monomial* B *of* P *for some* β *and some* (k, l) *where* $\beta(1) = \alpha(1) + \alpha(2)$ *and* $\beta(2) = 0$.
Of course, the use of the indices 1 and 2 is for convenience only. This lemma holds true for any pair of indices under the appropriate invariance assumption.

We postpone the proof of this lemma for the moment and use it to complete the proof of Theorem 2.4.7. Let $P \neq 0 \in \mathcal{P}_{m,m}$ with $r(P) = 0$.

Let A be a monomial of P. We noted $\deg_k(A) = 2$ for $1 \leq k \leq m$ and A is a polynomial in the 2-jets of the metric. We decompose

$$A = g_{i_1 j_1/k_1 l_1} \cdots g_{i_r j_r/k_r l_r} \qquad \text{for } 2r = m.$$

By making a coordinate permutation we can choose A so $i_1 = 1$. If $j_1 \neq 1$, we can make a coordinate permutation to assume $j_1 = 2$ and then apply Lemma 2.5.1(a) to assume $i_1 = j_1 = 1$. We let $P_1 = \sum c(A,P)A$ where the sum ranges over A of the form:

$$A = g_{11/k_1 l_1} \cdots g_{i_r j_r/k_r l_r}$$

which are monomials of P. Then $P_1 \neq 0$ and P_1 is invariant under coordinate transformations which fix the first coordinate. Since $\deg_1(A) = 2$, the index 1 appears nowhere else in A. Thus k_1 is not 1 so we may make a coordinate permutation to choose A so $k_1 = 2$. If $l_2 \neq 2$, then $l_1 \geq 3$ so we may make a coordinate permutation to assume $l_1 = 3$. We then apply Lemma 2.5.1(b) to choose A a monomial of P_1 of the form:

$$A = g_{11/22}A'.$$

We have $\deg_1(A) = \deg_2(A) = 2$ so $\deg_1(A') = \deg_2(A') = 0$ so these indices do not appear in A'. We define $P_2 = \sum c(A,P)A$ where the sum ranges over those monomial A of P which are divisible by $g_{11/22}$, then $P_2 \neq 0$.

We proceed inductively in this fashion to show finally that

$$A_0 = g_{11/22}g_{33/44} \cdots g_{m-1,m-1/mm}$$

is a monomial of P so $c(A_0,P) \neq 0$. The function $c(A_0,P)$ is a separating linear functional on the kernel of r in $\mathcal{P}_{m,m}$ and therefore

$$\dim(\{\, P \in \mathcal{P}_{m,m} : r(P) = 0 \,\}) \leq 1.$$

We complete the proof of Theorem 2.4.7(c) by showing that $r(E_m) = 0$. If E_m is the Euler form and $M = M' \times S^1$, then:

$$E_m(M) = E_m(M' \times S^1) = E_{m-1}(M')E_1(S^1) = 0$$

which completes the proof; $\dim N(r) = 1$ and $E_m \in N(r)$ spans.

Before we begin the proof of Lemma 2.5.1, it is helpful to consider a few examples. If we take $A = g_{11/11}$ then it is immediate:

$$\partial/\partial y_1 = a\partial/\partial x_1 + b\partial/\partial x_2, \qquad \partial/\partial y_2 = -b\partial/\partial x_1 + a\partial/\partial x_2$$
$$\partial/\partial y_k = \partial/\partial x_k \qquad \text{for } k > 2.$$

We compute $T_z^*(A)$ by formally replacing each index 1 by $a1 + b2$ and each index 2 by $-b1 + a2$ and expanding out the resulting expression. Thus, for example:

$$
\begin{aligned}
T_z^*(g_{11/11}) &= g_{a1+b2,a1+b2/a1+b2,a1+b2} \\
&= a^4 g_{11/11} + 2a^3 b g_{11/12} + 2a^3 b g_{12/11} \\
&\quad + a^2 b^2 g_{11/22} + a^2 b^2 g_{22/11} + 4a^2 b^2 g_{12/12} \\
&\quad + 2ab^3 g_{12/22} + 2ab^3 g_{22/12} + b^4 g_{22/22}.
\end{aligned}
$$

We note that those terms involving $a^3 b$ arose from changing exactly one index to another index $(1 \to 2$ or $2 \to 1)$; the coefficient reflects the multiplicity. Thus in particular this polynomial is not invariant.

In computing invariance under the circle action, all other indices remain fixed so

$$
\begin{aligned}
T_z^*(g_{34/11} + g_{34/22}) &= (a^2 g_{34/11} + b^2 g_{34/22} + 2ab g_{34/12} \\
&\quad + a^2 g_{34/22} + b^2 g_{34/11} - 2ab g_{34/12}) \\
&= (a^2 + b^2)(g_{34/11} + g_{34/22})
\end{aligned}
$$

is invariant. Similarly, it is easy to compute:

$$
T_z^*(g_{11/22} + g_{22/11} - 2g_{12/12}) = (a^2 + b^2)^2(g_{11/22} + g_{22/11} - 2g_{12/12})
$$

so this is invariant. We note this second example is homogeneous of degree 4 in the (a, b) variables since $\deg_1(A) + \deg_2(A) = 4$.

With these examples in mind, we begin the proof of Lemma 2.5.1. Let P be invariant under the action of the circle acting on the first two coordinates. We decompose $P = P_0 + P_1 + \cdots$ where each monomial A of P_j satisfies $\deg_1(A) + \deg_2(A) = j$. If A is such a monomial, then $T_z^*(A)$ is a sum of similar monomials. Therefore each of the P_j is invariant separately so we may assume $P = P_n$ for some n. By setting $a = b = -1$, we see n must be even. Decompose:

$$
T_z^*(P) = a^n P^{(0)} + ba^{n-1} P^{(1)} + \cdots + b^n P^{(n)}
$$

where $P = P^{(0)}$. We use the assumption $T_z^*(P) = P$ and replace b by $-b$ to see:

$$
0 = T_{(a,b)}^*(P) - T_{(a,-b)}^*(P) = 2ba^{n-1} P^{(1)} + 2b^3 a^{n-3} P^{(3)} + \cdots
$$

We divide this equation by b and take the limit as $b \to 0$ to show $P^{(1)} = 0$. (In fact, it is easy to show $P^{(2j+1)} = 0$ and $P^{(2j)} = \binom{n/2}{j} P$ but as we shall not need this fact, we omit the proof).

We let A_0 denote a variable monomial which will play the same role as the generic constant C of Chapter 1. We introduce additional notation we shall find useful. If A is a monomial, we decompose $T^*_{(a,b)}A = a^n A + a^{n-1}bA^{(1)} + \cdots$. If B is a monomial of $A^{(1)}$ then $\deg_1(B) = \deg_1(A) \pm 1$ and B can be constructed from the monomial A by changing exactly one index $1 \to 2$ or $2 \to 1$. If $\deg_1(B) = \deg_1(A) + 1$, then B is obtained from A by changing exactly one index $2 \to 1$; $c(B, A^{(1)})$ is a negative integer which reflects the multiplicity with which this change can be made. If $\deg_1(B) = \deg_1(A) - 1$, then B is obtained from A by changing exactly one index $1 \to 2$; $c(B, A^{(1)})$ is a positive integer. We define:

$$A(1 \to 2) = \sum c(B, A^{(1)})B \qquad \text{summed over } \deg_1(B) = \deg_1(A) - 1$$

$$A(2 \to 1) = \sum -c(B, A^{(1)})B \qquad \text{summed over } \deg_1(B) = \deg_1(A) + 1$$

$$A^{(1)} = A(1 \to 2) - A(2 \to 1).$$

For example, if $A = (g_{12/33})^2 g_{11/44}$ then $n = 6$ and:

$$A(1 \to 2) = 2g_{12/33}g_{22/33}g_{11/44} + 2(g_{12/33})^2 g_{12/44}$$

$$A(2 \to 1) = 2g_{12/33}g_{11/33}g_{11/44}.$$

It is immediate from the definition that:

$$\sum_B c(B, A(1 \to 2)) = \deg_1(A) \qquad \text{and} \qquad \sum_B c(B, A(2 \to 1)) = \deg_2(A).$$

Finally, it is clear that $c(B, A^{(1)}) \neq 0$ if and only if $c(A, B^{(1)}) \neq 0$, and that these two coefficients will be opposite in sign, and not necessarily equal in magnitude.

LEMMA 2.5.2. *Let P be invariant under the action of $\mathrm{SO}(2)$ on the first two coordinates and let A be a monomial of P. Let B be a monomial of $A^{(1)}$. Then there exists a monomial of A_1 different from A so that*

$$c(B, A^{(1)})c(A, P)c(B, A_1^{(1)})c(A_1, P) < 0.$$

PROOF: We know $P^{(1)} = 0$. We decompose

$$P^{(1)} = \sum_A c(A, P)A^{(1)} = \sum_B c(A, P)c(B, A^{(1)})B.$$

Therefore $c(B, P^{(1)}) = 0$ implies

$$\sum_A c(A, P)c(B, A^{(1)}) = 0.$$

for all monomials B. If we choose B so $c(B, A^{(1)}) \neq 0$ then there must be some other monomial A_1 of P which helps to cancel this contribution. The signs must be opposite which proves the lemma.

This lemma is somewhat technical and formidable looking, but it is exactly what we need about orthogonal invariance. We can now complete the proof of Lemma 2.5.1. Suppose first that $A = (g_{12/\alpha})^k A_0$ for $k > 0$ where $g_{12/\alpha}$ does not divide A_0. Assume $c(A, P) \neq 0$. Let $B = g_{11/\alpha}(g_{12/\alpha})^{k-1} A_0$ then $c(B, A^{(1)}) = -k \neq 0$. Choose $A_1 \neq A$ so $c(A_1, P) \neq 0$. Then $c(A_1, B^{(1)}) \neq 0$. If $c(A_1, B(2 \to 1)) \neq 0$ then A_1 is constructed from B by changing a $2 \to 1$ index so $A_1 = g_{11/\beta} \ldots$ has the desired form. If $c(A_1, B(1 \to 2)) \neq 0$, we expand $B(1 \to 2) = A+$ terms divisible by $g_{11/\beta}$ for some β. Since $A_1 \neq A$, again A_1 has the desired form which proves (a).

To prove (b), we choose $g_{ij/\alpha}$ dividing some monomial of P. Let β be chosen with $\beta(1)+\beta(2) = \alpha(1)+\alpha(2)$ and $\beta(k) = \alpha(k)$ for $k > 2$ with $\alpha(1)$ maximal so $g_{uv/\beta}$ divides some monomial of P for some (u, v). Suppose $\beta(2) \neq 0$, we argue for a contradiction. Set $\gamma = (\beta(1)+1, \beta(2)-1, \beta(3), \ldots, \beta(m))$. Expand $A = (g_{uv/\beta})^k A_0$ and define $B = g_{uv/\gamma}(g_{uv/\beta})^{k-1} A_0$ where $g_{uv/\beta}$ does not divide A_0. Then $c(B, A^{(1)}) = -\beta(2)k \neq 0$ so we may choose $A_1 \neq A$ so $c(A_1, B^{(1)}) \neq 0$. If $c(A_1, B(2 \to 1)) \neq 0$ then either A_1 is divisible by $g_{u'v'/\gamma}$ or by $g_{uv/\gamma'}$ where $\gamma'(1) = \gamma(1) + 1$, $\gamma'(2) = \gamma(2) - 1$, and $\gamma'(j) = \alpha(j)$ for $j > 2$. Either possibility contradicts the choice of β as maximal so $c(A_1, B(1 \to 2)) \neq 0$. However, $B(1 \to 2) = \beta(1)A+$ terms divisible by $g_{u'v'/\gamma}$ for some (u', v'). This again contradicts the maximality as $A \neq A_1$ and completes the proof of Lemma 2.5.1 and thereby of Theorems 2.4.7 and 2.4.8.

If $I = \{1 \leq i_1 \leq \cdots \leq i_p \leq m\}$, let $|I| = p$ and $dx^I = dx_{i_1} \wedge \cdots \wedge dx_{i_p}$. A p-form valued polynomial is a collection $\{P_I\} = P$ for $|I| = p$ of polynomials P_I in the derivatives of the metric. We will also sometimes write $P = \sum_{|I|=p} P_I \, dx^I$ as a formal sum to represent P. If all the $\{P_I\}$ are homogeneous of order n, we say P is homogeneous of order n. We define:

$$P(X, G)(x_0) = \sum_I P_I(X, G) \, dx^I \in \Lambda^p(T^*M)$$

to be the evaluation of such a polynomial. We say P is invariant if $P(X, G)(x_0) = P(Y, G)(x_0)$ for every normalized coordinate systems X and Y; as before we denote the common value by $P(G)(x_0)$. In analogy with Lemma 2.4.4 we have:

LEMMA 2.5.3. *Let P be p-form valued and invariant. Then P is homogeneous of order n if and only if $P(c^2 G)(x_0) = c^{p-n} P(G)(x_0)$ for every $c \neq 0$.*

The proof is exactly the same as that given for Lemma 2.4.4. The only new feature is that $dy^I = c^p \, dx^I$ which contributes the extra feature of c^p in this equation.

We define $P_{m,n,p}$ to be the vector space of all p-form valued invariants which are homogeneous of order n on a manifold of dimension m. If $P_{m,*,p}$ denotes the vector space of all p-form valued invariant polynomials, then we have a direct sum decomposition $P_{m,*,p} = \bigoplus_n P_{m,n,p}$ exactly as in the scalar case $p = 0$.

We define $\deg_k(I)$ to be 1 if k appears in I and 0 if k does not appear in I.

LEMMA 2.5.4. Let $P \in P_{m,n,p}$ with $P \neq 0$. Then $n - p$ is even. If A is a monomial then A is a monomial of at most one P_I. If $c(A, P_I) \neq 0$ then:

$$\deg_k(A) + \deg_k(I) \qquad \text{is always even.}$$

PROOF: Let T be the coordinate transformation defined by $T(x_k) = -x_k$ and $T(x_j) = x_j$ for $j \neq k$. Then

$$P = T^*(P) = \sum_{I,A} (-1)^{\deg_k(A) + \deg_k(I)} c(A, P_I) \, dx^I$$

which implies $\deg_k(A) + \deg_k(I)$ is even if $c(A, P_I) \neq 0$. Therefore

$$\sum_k \deg_k(A) + \deg_k(I) = 2\ell(A) + \text{ord}(A) + p = 2\ell(A) + n + p$$

must be even. This shows $n + p$ is even if $P \neq 0$. Furthermore, if $c(A, P_I) \neq 0$ then I is simply the ordered collection of indices k so $\deg_k(A)$ is odd which shows A is a monomial of at most one P_I and completes the proof.

We extend the Riemannian metric to a fiber metric on $\Lambda(T^*M) \otimes \mathbb{C}$. It is clear that $P \cdot P = \sum_I P_I \bar{P}_I$ since the $\{dx^I\}$ form an orthonormal basis at x_0. $P \cdot P$ is a scalar invariant. We use Lemma 2.5.1 to prove:

LEMMA 2.5.5. Let P be p-form valued and invariant under the action of $O(m)$. Let A be a monomial of P. Then there is a monomial A_1 of P with $\deg_k(A_1) = 0$ for $k > 2\ell(A)$.

PROOF: Let $r = \ell(A)$ and let $P'_r = \sum_{\ell(B)=r} c(B, P_I) \, dx^I \neq 0$. Since this is invariant under the action of $O(m)$, we may assume without loss of generality $P = P_r$. We construct a scalar invariant by taking the inner product $Q = (P, P)$. By applying Lemma 2.5.1(a) and making a coordinate permutation if necessary, we can assume g_{11/α_1} divides some monomial of Q. We apply 2.5.1(b) to the indices > 1 to assume $\alpha_1(k) = 0$ for $k > 2$. g_{11/α_1} must divide some monomial of P. Let

$$P_1 = \sum_{A_0, I} c(g_{11/\alpha_1} A_0, P_I) g_{11/\alpha_1} A_0 \, dx^I \neq 0.$$

This is invariant under the action of $O(m-2)$ on the last $m-2$ coordinates. We let $Q_1 = (P_1, P_1)$ and let $g_{i_2 j_2 / \alpha_2}$ divide some monomial of Q_1. If both i_2 and j_2 are ≤ 2 we leave this alone. If $i_2 \leq 2$ and $j_2 \geq 3$ we perform a coordinate permutation to assume $j_2 = 3$. In a similar fashion if $i_2 \geq 3$ and $j_2 \leq 2$, we choose this variable so $i_2 = 3$. Finally, if both indices ≥ 3, we apply Lemma 2.5.1(a) to choose this variable so $i_2 = j_2 = 3$. We apply Lemma 2.5.1(b) to the variables $k \geq 4$ to choose this variable so $\alpha_2(k) = 0$ for $k > 4$. If $A_2 = g_{11/\alpha_1} g_{i_2 j_2 / \alpha_2}$ then:

$$\deg_k(A_2) = 0 \text{ for } k > 4 \text{ and } A_2 \text{ divides some monomial of } P.$$

We continue inductively to construct $A_r = g_{11/\alpha_1} \cdots g_{i_r j_r / \alpha_r}$ so that

$$\deg_k(A_r) = 0 \text{ for } k > 2r \text{ and } A_r \text{ divides some monomial of } P.$$

Since every monomial of P has length r, this implies A_r itself is a monomial of P and completes the proof.

Let $P_j(G) = p_j(TM)$ be the jth Pontrjagin form computed relative to the curvature tensor of the Levi-Civita connection. If we expand p_j in terms of the curvature tensor, then p_j is homogeneous of order $2j$ in the $\{R_{ijkl}\}$ tensor so p_j is homogeneous of order $4j$ in the jets of the metric. It is clear p_j is an invariantly defined $4j$-form so $p_j \in \mathcal{P}_{m,4j,4j}$. The algebra generated by the p_j is called the algebra of the Pontrjagin forms. By Lemma 2.2.2, this is also the algebra of real characteristic forms of $T(M)$. If ρ is a partition of $k = i_1 + \cdots + i_j$ we define $p_\rho = p_{i_1} \ldots p_{i_j} \in \mathcal{P}_{m,4k,4k}$. The $\{p_\rho\}$ form a basis of the Pontrjagin $4k$ forms. By Lemma 2.3.4, these are linearly independent if $m = 4k$ since the matrix $\int_{M_r} p_\rho$ is non-singular. By considering products of these manifolds with flat tori T^{m-4k} we can easily show that the $\{p_\rho\}$ are linearly independent in $\mathcal{P}_{m,4k,4k}$ if $4k \leq m$. We let $\pi(k)$ be the number of partitions of k; this is the dimension of the Pontrjagin forms.

The axiomatic characterization of the real characteristic forms of the tangent space which is the analogue of the axiomatic characterization of the Euler class given in Theorem 2.4.7 is the following:

LEMMA 2.5.6.
(a) $\mathcal{P}_{m,n,p} = 0$ if $n < p$.
(b) $\mathcal{P}_{m,n,n}$ is spanned by the Pontrjagin forms—i.e.,

$$\mathcal{P}_{m,n,n} = 0 \text{ if } n \text{ is not divisible by } 4k,$$

$$\mathcal{P}_{m,4k,4k} = \text{span}\{p_\rho\} \text{ for } 4k \leq m \text{ has dimension } \pi(k).$$

PROOF: By decomposing P into its real and imaginary parts, it suffices to prove this lemma for polynomials with real coefficients. Let $0 \neq P \in \mathcal{P}_{m,n,p}$

and let A be a monomial of P. Use Lemma 2.5.5 to find a monomial A_1 of some P_I where $I = \{1 \le i_1 < \cdots < i_p \le m\}$ so $\deg_k(A) = 0$ for $k > 2\ell(A)$. Since $\deg_{i_p}(A)$ is odd (and hence non-zero) and since $2\ell(A) \le \sum |\alpha_\nu| = n$ as A is a polynomial in the jets of order 2 and higher, we estimate:

$$p \le i_p \le 2\ell(A) \le n.$$

This proves $P = 0$ if $n < p$, which proves (a).

If $n = p$, then all of these inequalities must have been equalities. This shows that the higher order jets of the metric do not appear in P so P is a polynomial in the $\{g_{ij/kl}\}$ variables. Furthermore, $i_p = p$ and there is some monomial A so that

$$\deg_k(A) = 0 \text{ for } k > p \quad \text{and} \quad A\,dx^1 \wedge \cdots \wedge dx^p \text{ appears in } P.$$

There is a natural restriction map $r: \mathcal{P}_{m,n,p} \to \mathcal{P}_{m-1,n,p}$ defined in the same way as the restriction map $r: \mathcal{P}_{m,n,0} \to \mathcal{P}_{m-1,n,0}$ discussed earlier. This argument shows $r: \mathcal{P}_{m,n,n} \to \mathcal{P}_{m-1,n,n}$ is injective for $n < m$ since $r(A\,dx^1 \wedge \cdots \wedge dx^p) = A\,dx^1 \wedge \cdots \wedge dx^p$ appears in $r(P)$. The Pontrjagin forms have dimension $\pi(k)$ for $n = 4k$. By induction, $r^{m-n}: \mathcal{P}_{m,n,n} \to \mathcal{P}_{n,n,n}$ is injective so $\dim \mathcal{P}_{m,n,n} \le \dim \mathcal{P}_{n,n,n}$.

We shall prove Lemma 2.5.6(b) for the special case $n = m$. If n is not divisible by $4k$, then $\dim \mathcal{P}_{n,n,n} = 0$ which implies $\dim \mathcal{P}_{m,n,n} = 0$. If $n = 4k$, then $\dim \mathcal{P}_{n,n,n} = \pi(k)$ implies that $\pi(k) \le \dim \mathcal{P}_{m,n,n} \le \dim \mathcal{P}_{n,n,n} \le \pi(k)$ so $\dim \mathcal{P}_{m,n,n} = \pi(k)$. Since the Pontrjagin forms span a subspace of exactly dimension $\pi(k)$ in $\mathcal{P}_{m,n,n}$ this will complete the proof of (b).

This lemma is at the heart of our discussion of the index theorem. We shall give two proofs for the case $n = m = p$. The first is based on H. Weyl's theorem for the orthogonal group and follows the basic lines of the proof given in Atiyah-Bott-Patodi. The second proof is purely combinatorial and follows the basic lines of the original proof first given in our thesis. The H. Weyl based proof has the advantage of being somewhat shorter but relies upon a deep theorem we have not proved here while the second proof although longer is entirely self-contained and has some additional features which are useful in other applications.

We review H. Weyl's theorem briefly. Let V be a real vector space with a fixed inner product. Let $O(V)$ denote the group of linear maps of $V \to V$ which preserve this inner product. Let $\bigotimes^k(V) = V \otimes \cdots \otimes V$ denote the kth tensor product of V. If $g \in O(V)$, we extend g to act orthogonally on $\bigotimes^k(V)$. We let $z \mapsto g(z)$ denote this action. Let $f: \bigotimes^k(V) \to \mathbf{R}$ be a multi-linear map, then we say f is $O(V)$ invariant if $f(g(z)) = f(z)$ for every $g \in O(V)$. By letting $g = -1$, it is easy to see there are no $O(V)$ invariant maps if k is odd. We let $k = 2j$ and construct a map

$f_0 : \bigotimes^k(V) = (V \otimes V) \otimes (V \otimes V) \otimes \cdots \otimes (V \otimes V) \to \mathbf{R}$ using the metric to map $(V \otimes V) \to \mathbf{R}$. More generally, if ρ is any permutation of the integers 1 through k, we define $z \mapsto z_\rho$ as a map from $\bigotimes^k(V) \to \bigotimes^k(V)$ and let $f_\rho(z) = f_0(z_\rho)$. This will be $O(V)$ invariant for any permutation ρ. H. Weyl's theorem states that the maps $\{f_\rho\}$ define a spanning set for the collection of $O(V)$ invariant maps.

For example, let $k = 4$. Let $\{v_i\}$ be an orthonormal basis for V and express any $z \in \bigotimes^4(V)$ in the form $a_{ijkl}v_i \otimes v_j \otimes v_k \otimes v_l$ summed over repeated indices. Then after weeding out duplications, the spanning set is given by:

$$f_0(z) = a_{iijj}, \qquad f_1(z) = a_{ijij}, \qquad f_2(z) = a_{ijji}$$

where we sum over repeated indices. f_0 corresponds to the identity permutation; f_1 corresponds to the permutation which interchanges the second and third factors; f_2 corresponds to the permutation which interchanges the second and fourth factors. We note that these need not be linearly independent; if $\dim V = 1$ then $\dim(\bigotimes^4 V) = 1$ and $f_1 = f_2 = f_3$. However, once $\dim V$ is large enough these become linearly independent.

We are interested in p-form valued invariants. We take $\bigotimes^k(V)$ where $k - p$ is even. Again, there is a natural map we denote by

$$f^p(z) = f_0(z_1) \wedge \Lambda(z_2)$$

where we decompose $\bigotimes^k(V) = \bigotimes^{k-p}(V) \otimes \bigotimes^p(V)$. We let f_0 act on the first $k - p$ factors and then use the natural map $\bigotimes^p(V) \xrightarrow{\Lambda} \Lambda^p(V)$ on the last p factors. If ρ is a permutation, we set $f_\rho^p(z) = f^p(z_\rho)$. These maps are equivariant in the sense that $f_\rho^p(gz) = g f_\rho^p(z)$ where we extend g to act on $\Lambda^p(V)$ as well. Again, these are a spanning set for the space of equivariant multi-linear maps from $\bigotimes^k(V)$ to $\Lambda^p(V)$. If $k = 4$ and $p = 2$, then after eliminating duplications this spanning set becomes:

$$f_1(z) = a_{iijk}v_j \wedge v_k, \quad f_2(z) = a_{ijik}v_j \wedge v_k, \quad f_3(z) = a_{ijki}v_j \wedge v_k$$
$$f_4(z) = a_{jiki}v_j \wedge v_k, \quad f_5(z) = a_{jiik}v_i \wedge v_k, \quad f_6(z) = a_{jkii}v_j \wedge v_k.$$

Again, these are linearly independent if $\dim V$ is large, but there are relations if $\dim V$ is small. Generally speaking, to construct a map from $\bigotimes^k(V) \to \Lambda^p(V)$ we must alternate p indices (the indices j, k in this example) and contract the remaining indices in pairs (there is only one pair i, i here).

THEOREM 2.5.7 (H. WEYL'S THEOREM ON THE INVARIANTS OF THE ORTHOGONAL GROUP). *The space of maps $\{f_\rho^p\}$ constructed above span the space of equivariant multi-linear maps from $\bigotimes^k V \to \Lambda^p V$.*

The proof of this theorem is beyond the scope of the book and will be omitted. We shall use it to give a proof of Lemma 2.5.6 along the lines of the proof given by Atiyah, Bott, and Patodi. We will then give an independent proof of Lemma 2.5.6 by other methods which does not rely on H. Weyl's theorem.

We apply H. Weyl's theorem to our situation as follows. Let $P \in \mathcal{P}_{n.n,n}$ then P is a polynomial in the 2-jets of the metric. If we let X be a system of geodesic polar coordinates centered at x_0, then the 2-jets of the the metric are expressible in terms of the curvature tensor so we can express P as a polynomial in the $\{R_{ijkl}\}$ variables which is homogeneous of order $n/2$. The curvature defines an element $R \in \bigotimes^4(T(M))$ since it has 4 indices. There are, however, relations among the curvature variables:

$$R_{ijkl} = R_{klij}, \qquad R_{ijkl} = -R_{jikl}, \qquad \text{and} \qquad R_{ijkl} + R_{iklj} + R_{iljk} = 0.$$

We let V be the sub-bundle of $\bigotimes^4(T(M))$ consisting of tensors satisfying these 3 relations.

If $n = 2$, then $P: V \to \Lambda^2(T(M))$ is equivariant while more generally, $P: \bigotimes^{n/2}(V) \to \Lambda^n(T(M))$ is equivariant under the action of $O(T(M))$. (We use the metric tensor to raise and lower indices and identify $T(M) = T^*(M)$). Since these relations define an $O(T(M))$ invariant subspace of $\bigotimes^{2n}(T(M))$, we extend P to be zero on the orthogonal complement of $\bigotimes^{n/2}(V)$ in $\bigotimes^{2n}(T(M))$ to extend P to an equivariant action on the whole tensor algebra. Consequently, we can use H. Weyl's theorem to span $\mathcal{P}_{n,n,n}$ by expressions in which we alternate n indices and contract in pairs the remaining n indices.

For example, we compute:

$$p_1 = C \cdot R_{ijab} R_{ijcd} \, dx^a \wedge dx^b \wedge dx^c \wedge dx^d$$

for a suitable normalizing constant C represents the first Pontrjagin form. In general, we will use letters a, b, $c,\ldots$ for indices to alternate on and indices i, j, $k,\ldots$ for indices to contract on. We let P be such an element given by H. Weyl's theorem. There are some possibilities we can eliminate on a priori grounds. The Bianchi identity states:

$$R_{iabc} \, dx^a \wedge dx^b \wedge dx^c = \frac{1}{3}(R_{iabc} + R_{ibca} + R_{icab}) \, dx^a \wedge dx^b \wedge dx^c = 0$$

so that three indices of alternation never appear in any $R_{\ldots}$ variable. Since there are $n/2$ R variables and n indices of alternation, this implies each R variable contains exactly two indices of alternation. We use the Bianchi identity again to express:

$$R_{iajb} \, dx^a \wedge dx^b = \frac{1}{2}(R_{iajb} - R_{ibja}) \, dx^a \wedge dx^b$$

$$= \frac{1}{2}(R_{iajb} + R_{ibaj}) \, dx^a \wedge dx^b$$

$$= -\frac{1}{2}R_{ijba} \, dx^a \wedge dx^b = \frac{1}{2}R_{ijab} \, dx^a \wedge dx^b.$$

This together with the other curvature identities means we can express P in terms of $R_{ijab}\,dx^a \wedge dx^b = \Omega_{ij}$ variables. Thus $P = P(\Omega_{ij})$ is a polynomial in the components of the curvature matrix where we regard $\Omega_{ij} \in \Lambda^2(T^*M)$. (This differs by various factors of 2 from our previous definitions, but this is irrelevant to the present argument). $P(\Omega)$ is an $O(n)$ invariant polynomial and thus is a real characteristic form. This completes the proof.

The remainder of this subsection is devoted to giving a combinatorial proof of this lemma in the case $n = m = p$ independent of H. Weyl's theorem. Since the Pontrjagin forms span a subspace of dimension $\pi(k)$ if $n = 4k$ we must show $P_{n,n,n} = 0$ if n is not divisible by 4 and that $\dim P_{4k,4k,4k} \leq \pi(k)$ since then equality must hold.

We showed $n = 2j$ is even and any polynomial depends on the 2-jets of the metric. We improve Lemma 2.5.1 as follows:

LEMMA 2.5.8. *Let P satisfy the hypothesis of Lemma 2.5.1 and be a polynomial in the 2-jets of the metric. Then:*
(a) Let $A = g_{12/\alpha}A_0$ be a monomial of P. Either by interchanging 1 and 2 indices or by changing two 2 indices to 1 indices we can construct a monomial A_1 of the form $A_1 = g_{11/\beta}A_0'$ which is a monomial of P.
(b) Let $A = g_{ij/12}A_0$ be a monomial of P. Either by interchanging 1 and 2 indices or by changing two 2 indices to 1 indices we can construct a monomial of A_1 of the form $A_1 = g_{i'j'/11}A_0'$ which is a monomial of P.
(c) The monomial $A_1 \neq A$. If $\deg_1(A_1) = \deg_1(A)+2$ then $c(A,P)c(A_1,P) > 0$. Otherwise $\deg_1(A_1) \neq \deg_1(A)$ and $c(A,P)c(A_1,P) < 0$.

PROOF: We shall prove (a) as (b) is the same; we will also verify (c). Let $B = g_{11/\alpha}A_0$ so $c(B, A(2 \to 1)) \neq 0$. Then $\deg_1(B) = \deg_1(A) + 1$. Apply Lemma 2.5.2 to find $A_1 \neq A$ so $c(B, A_1^{(1)}) \neq 0$. We noted earlier in the proof of 2.5.1 that A_1 must have the desired form. If $c(A_1, B(2 \to 1)) \neq 0$ then $\deg_1(A_1) = \deg_1(B)+1 = \deg_1(A)+2$ so A_1 is constructed from A by changing two 2 to 1 indices. Furthermore, $c(B, A_1^{(1)}) > 0$ and $c(B, A^{(1)}) < 0$ implies $c(A,P)c(A_1,P) > 0$. If, on the other hand, $c(A_1, B(1 \to 2)) \neq 0$ then $\deg_1(A_1) = \deg_1(B) - 1 = \deg_1(A)$ and A changes to A_1 by interchanging a 2 and a 1 index. Furthermore, $c(B, A_1^{(1)}) < 0$ and $c(B, A^{(1)}) < 0$ implies $c(A,P)c(A_1,P) < 0$ which completes the proof.

Let $P \in P_{n,n,n}$ and express $P = P' dx^1 \wedge \cdots \wedge dx^n$. $P' = *P$ is a scalar invariant which changes sign if the orientation of the local coordinate system is reversed. We identify P with P' for notational convenience and henceforth regard P as a skew-invariant scalar polynomial. Thus $\deg_k(A)$ is odd for every k and every monomial A of P.

The indices with $\deg_k(A) = 1$ play a particularly important role in our discussion. We say that an index i touches an index j in the monomial A

if A is divisible by a variable $g_{ij/..}$ or $g_{../ij}$ where ".." indicate two indices which are not of interest.

LEMMA 2.5.9. *Let $P \in \mathcal{P}_{n,n,n}$ with $P \neq 0$. Then there exists a monomial A of P so*
(a) $\deg_k(A) \leq 3$ *all* k.
(b) *If $\deg_k(A) = 1$ there exists an index $j(k)$ which touches k in A. The index $j(k)$ also touches itself in A.*
(c) *Let $\deg_j A + \deg_k A = 4$. Suppose the index j touches itself and the index k in A. There is a unique monomial A_1 different from A which can be formed from A by interchanging a j and k index. $\deg_j A_1 + \deg_k A_1 = 4$ and the index j touches itself and the index k in A_1. $c(A, P) + c(A_1, P) = 0$.*

PROOF: Choose A so the number of indices with $\deg_k(A) = 1$ is minimal. Among all such choices, we choose A so the number of indices which touch themselves is maximal. Let $\deg_k(A) = 1$ then k touches some index $j = j(k) \neq k$ in A. Suppose A has the form $A = g_{jk/..}A_0$ as the other case is similar. Suppose first that $\deg_j(A) \geq 5$. Use lemma 2.5.8 to find $A_1 = g_{kk/..}A_0$. Then $\deg_k(A_1) \neq \deg_k(A)$ implies $\deg_k(A_1) = \deg_k(A) + 2 = 3$. Also $\deg_j(A_1) = \deg_j(A) - 2 \geq 3$ so A_1 has one less index with degree 1 which contradicts the minimality of the choice of A. We suppose next $\deg_j(A) = 1$. If T is the coordinate transformation interchanging x_j and x_k, then T reverses the orientation so $T^*P = -P$. However $\deg_j(A) = \deg_k(A) = 1$ implies $T^*A = A$ which contradicts the assumption that A is a monomial of P. Thus $\deg_j(A) = 3$ which proves (a).

Suppose j does not touch itself in A. We use Lemma 2.5.8 to construct $A_1 = g_{kk/..}A_0'$. Then $\deg_k(A_1) = \deg_k(A) + 2 = 3$ and $\deg_j(A_1) = \deg_j(A) - 2 = 1$. This is a monomial with the same number of indices of degree 1 but which has one more index (namely k) which touches itself. This contradicts the maximality of A and completes the proof of (b).

Finally, let $\mathcal{A}_\nu = \{k : \deg_k(A) = \nu\}$ for $\nu = 1, 3$. The map $k \mapsto j(k)$ defines an injective map from $\mathcal{A}_1 \to \mathcal{A}_3$ since no index of degree 3 can touch two indices of degree 1 as well as touching itself. The equalities:

$$n = \operatorname{card}(\mathcal{A}_1) + \operatorname{card}(\mathcal{A}_3) \quad \text{and} \quad 2n = \sum_k \deg_k(A) = \operatorname{card}(\mathcal{A}_1) + 3\operatorname{card}(\mathcal{A}_3)$$

imply $2\operatorname{card}(\mathcal{A}_3) = n$. Thus $\operatorname{card}(\mathcal{A}_1) = \operatorname{card}(\mathcal{A}_3) = n/2$ and the map $k \mapsto j(k)$ is bijective in this situation.

(c) follows from Lemma 2.5.8 where $j = 1$ and $k = 2$. Since $\deg_k(A) = 1$, A_1 cannot be formed by transforming two k indices to j indices so A_1 must be the unique monomial different from A obtained by interchanging these indices. For example, if $A = g_{jj/ab}g_{jk/cd}A_0$, then $A_1 = g_{jk/ab}g_{jj/cd}A_0$. The multiplicities involved are all 1 so we can conclude $c(A, P) + c(A_1, P) = 0$ and not just $c(A, P)c(A_1, P) < 0$.

Before further normalizing the choice of the monomial A, we must prove a lemma which relies on cubic changes of coordinates:

LEMMA 2.5.10. *Let P be a polynomial in the 2-jets of the metric invariant under changes of coordinates of the form $y_i = x_i + c_{jkl}x_j x_k x_l$. Then:*
(a) $g_{12/11}$ and $g_{11/12}$ divide no monomial A of P.
(b) Let $A = g_{23/11}A_0$ and $B = g_{11/23}A_0$ then A is a monomial of P if and only if B is a monomial of P. Furthermore, $c(A,P)$ and $c(B,P)$ have the same sign.
(c) Let $A = g_{11/22}A_0$ and $B = g_{22/11}A_0$, then A is a monomial of P if and only if B is a monomial of P. Furthermore, $c(A,P)$ and $c(B,P)$ have the same sign.

PROOF: We remark the use of the indices 1, 2, 3 is for notational convenience only and this lemma holds true for any triple of distinct indices. Since $g_{ij}(X,G) = \delta_{ij} + O(x^2)$, $g^{ij}{}_{/kl}(X,G)(x_0) = -g_{ij/kl}(X,G)(x_0)$. Furthermore, under changes of this sort, $d_y^\alpha(x_0) = d_x^\alpha(x_0)$ if $|\alpha| = 1$. We consider the change of coordinates:

$$y_2 = x_2 + cx_1^3, \qquad\qquad y_k = k_x \quad \text{otherwise}$$
$$dy_2 = dx_2 + 3cx_1^2\, dx_1, \qquad dy_k = dx_k \quad \text{otherwise}$$

with
$$g^{12}(Y,G) = g^{12}(X,G) + 3cx_1^2 + O(x^4),$$
$$g^{ij}(Y,G) = g^{ij}(X,G) + O(x^4) \quad \text{otherwise}$$
$$g_{12/11}(Y,G)(x_0) = g_{12/11}(X,G)(x_0) - 6c$$
$$g_{ij/kl}(Y,G)(x_0) = g_{ij/kl}(X,G)(x_0) \quad \text{otherwise.}$$

We decompose $A = (g_{12/11})^\nu A_0$. If $\nu > 0$, $T^*(A) = A - 6\nu c(g_{12/11})^{\nu-1}A_0 + O(c^2)$. Since $T^*(P) = P$, and since there is no way to cancel this additional contribution, A cannot be a monomial of P so $g_{12/11}$ divides no monomial of P.

Next we consider the change of coordinates:

$$y_1 = x_1 + cx_1^2 x_2, \qquad\qquad y_k = x_k \quad \text{otherwise}$$
$$dy_1 = dx_1 + 2cx_1 x_2\, dx_1 + cx_1^2\, dx_2, \qquad dy_k = dx_k \quad \text{otherwise}$$

with
$$g_{11/12}(Y,G)(x_0) = g_{11/12}(X,G)(x_0) - 4c$$
$$g_{12/11}(Y,G)(x_0) = g_{12/11}(X,G)(x_0) - 2c$$
$$g_{ij/kl}(Y,G)(x_0) = g_{ij/kl}(X,G)(x_0) \qquad\qquad \text{otherwise.}$$

We noted $g_{12/11}$ divides no monomial of P. If $A = (g_{11/12})^\nu A_0$, then $\nu > 0$ implies $T^*(A) = A - 4cv(g_{11/12})^{\nu-1}A_0 + O(c^2)$. Since there would be no way to cancel such a contribution, A cannot be a monomial of P which completes the proof of (a).

Let A_0' be a monomial not divisible by any of the variables $g_{23/11}$, $g_{12/13}$, $g_{13/12}$, $g_{11/23}$ and let $A(p, q, r, s) = (g_{23/11})^p (g_{12/13})^q (g_{13/12})^r (g_{11/23})^s A_0'$. We set $c(p, q, r, s) = c(A(p, q, r, s), P)$ and prove (b) by establishing some relations among these coefficients. We first consider the change of coordinates,

$$y_2 = x_2 + cx_1^2 x_3, \qquad\qquad y_k = x_k \quad \text{otherwise}$$
$$dy_2 = dx_2 + 2cx_1 x_3\, dx_1 + cx_1^2\, dx_3, \qquad dy_k = dx_k \quad \text{otherwise}$$

with

$$g_{12/13}(Y, G)(x_0) = g_{12/13}(X, G)(x_0) - 2c$$
$$g_{23/11}(Y, G)(x_0) = g_{23/11}(X, G)(x_0) - 2c$$
$$g_{ij/kl}(Y, G)(x_0) = g_{ij/kl}(X, G)(x_0) \qquad\qquad \text{otherwise.}$$

We compute that:

$$T^*(A(p, q, r, s))$$
$$= A(p, q, r, s) + c\{-2qA(p, q-1, r, s) - 2pA(p-1, q, r, s)\} + O(c^2).$$

Since $T^*(P) = P$ is invariant, we conclude

$$pc(p, q, r, s) + (q+1)c(p-1, q+1, r, s) = 0.$$

By interchanging the roles of 2 and 3 in the argument we also conclude:

$$pc(p, q, r, s) + (r+1)c(p-1, q, r+1, s) = 0.$$

(We set $c(p, q, r, s) = 0$ if any of these integers is negative.)

Next we consider the change of coordinates:

$$y_1 = x_1 + cx_1 x_2 x_3, \qquad\qquad y_k = x_k \quad \text{otherwise}$$
$$dy_1 = dx_1 + cx_1 x_2\, dx_3 + cx_1 x_3\, dx_2 + cx_2 x_3\, dx_1, \qquad dy_k = dx_k \quad \text{otherwise}$$

with

$$g_{11/23}(Y, G)(x_0) = g_{11/23}(X, G)(x_0) - 2c$$
$$g_{12/13}(Y, G)(x_0) = g_{12/13}(X, G)(x_0) - c$$
$$g_{13/12}(Y, G)(x_0) = g_{13/12}(Y, G)(x_0) - c$$

so that

$$T^*(A(p, q, r, s)) = A(p, q, r, s) + c\{-2sA(p, q, r, s-1)$$
$$- rA(p, q, r-1, s) - qA(p, q-1, r, s)\} + O(c^2).$$

This yields the identities

$$(q+1)c(p, q+1, r, s-1) + (r+1)c(p, q, r+1, s-1) + 2sc(p, q, r, s) = 0.$$

Let $p \neq 0$ and let $A = g_{23/11}A_0'$ be a monomial of P. Then $c(p-1, q+1, r, s)$ and $c(p-1, q, r+1, s)$ are non-zero and have the opposite sign as $c(p, q, r, s)$. Therefore $(q+1)c(p-1, q+1, r, s) + (r+1)c(p-1, q, r+1, s)$ is non-zero which implies $c(p-1, q, r, s+1)(s+1)$ is non-zero and has the same sign as $c(p, q, r, s)$. This shows $g_{11/23}A_0'$ is a monomial of P. Conversely, if A is not a monomial of P, the same argument shows $g_{11/23}A_0'$ is not a monomial of P. This completes the proof of (b).

The proof of (c) is essentially the same. Let A_0' be a monomial not divisible by the variables $\{g_{11/22}, g_{12/12}, g_{22/11}\}$ and let

$$A(p, q, r) = (g_{11/22})^p (g_{12/12})^q (g_{22/11})^r.$$

Let $c(p, q, r) = c(A(p, q, r), P)$. Consider the change of coordinates:

$$y_1 = x_1 + cx_1 x_2^2, \qquad\qquad y_k = x_k \quad \text{otherwise}$$
$$dy_1 = dx_1 + cx_2^2\, dx_1 + 2cx_1 x_2\, dx_2, \quad dy_k = dx_k \quad \text{otherwise}$$

with

$$g_{11/22}(Y, g)(x_0) = g_{11/22}(X, G)(x_0) - 4c$$
$$g_{12/12}(Y, G)(x_0) = g_{12/12}(X, G)(x_0) - 2c$$
$$g_{ij/kl}(Y, G)(x_0) = g_{ij/kl}(X, G)(x_0) \qquad\qquad \text{otherwise.}$$

This yields the relation $2pc(p, q, r) + (q+1)c(p-1, q+1, r) = 0$. By interchanging the roles of 1 and 2 we obtain the relation $2rc(p, q, r) + (q+1)c(p, q+1, r-1) = 0$ from which (c) follows.

This step in the argument is functionally equivalent to the use made of the $\{R_{ijkl}\}$ variables in the argument given previously which used H. Weyl's formula. It makes use in an essential way of the invariance of P under a wider group than just first and second order transformations. For the Euler form, by contrast, we only needed first and second order coordinate transformations.

We can now construct classifying monomials using these lemmas. Fix $n = 2n_1$ and let A be the monomial of P given by Lemma 2.5.9. By making a coordinate permutation, we may assume $\deg_k(A) = 3$ for $k \leq n_1$ and $\deg_k(A) = 1$ for $k > n_1$. Let $x(i) = i + n_1$ for $1 \leq i \leq n_0$; we may assume the index I touches itself and $x(i)$ in A for $i \leq n_0$.

We further normalize the choice of A as follows. Either $g_{11/ij}$ or $g_{ij/11}$ divides A. Since $\deg_1(A) = 3$ this term is not $g_{11/11}$ and by Lemma 2.5.10 it is not $g_{11/1x(1)}$ nor $g_{1x(1)/11}$. By Lemma 2.5.10(b) or 2.5.10(c), we may assume $A = g_{11/ij} \ldots$ for $i, j \geq 2$. Since not both i and j can have degree

1 in A, by making a coordinate permutation if necessary we may assume that $i = 2$. If $j = 2$, we apply Lemma 2.5.9(c) to the indices 2 and $x(2)$ to perform an interchange and assume $A = g_{11/2x(2)}A_0$. The index 2 must touch itself elsewhere in A. We apply the same considerations to choose A in the form $A = g_{11/2x(2)}g_{22/ij}A_0$. If i or j is 1, the cycle closes and we express $A = g_{11/2x(2)}g_{11/1x(1)}A_0$ where $\deg_k(A_0) = 0$ for $k = 1, 2, 1 + n_1$, $2 + n_1$. If i, $j > 2$ we continue this argument until the cycle closes. This permits us to choose A to be a monomial of P in the form:

$$A = g_{11/2x(2)}g_{22/3x(3)} \cdots g_{j-1,j-1/jx(j)}g_{jj/1x(1)}A_0$$

where $\deg_k(A_0) = 0$ for $1 \le k \le j$ and $n_1 + 1 \le k \le n_1 + j$.

We wish to show that the length of the cycle involved is even. We apply Lemma 2.5.10 to show

$$B = g_{2x(2)/11}g_{3x(3)/22} \cdots g_{1x(1)/jj}A_0$$

satisfies $c(A, P)c(B, P) > 0$. We apply Lemma 2.5.9 a total of j times to see

$$C = g_{22,1x(1)}g_{33/2x(2)} \cdots g_{11/jx(j)}A_0$$

satisfies $c(B, P)c(C, P)(-1)^j > 0$. We now consider the even permutation:

$$\rho(1) = j$$
$$\rho(k) = \begin{cases} k - 1, & 2 \le k \le j \\ k, & j < k \le n_1 \end{cases}$$
$$\rho(k(j)) = x(\rho(j)) \qquad \text{for } 1 \le j \le n_1$$

to see that $c(C_\rho, P)c(C, P) > 0$. However $A = C_\rho$ so $c(A, P)^2(-1)^j > 0$ which shows j is necessarily even. (This step is formally equivalent to using the skew symmetry of the curvature tensor to show that only polynomials of even degree can appear to give non-zero real characteristic forms).

We decompose A_0 into cycles to construct A inductively so that A has the form:

$$A = \{g_{11/2x(2)} \cdots g_{i_1 i_1/1x(1)}\}$$
$$\{g_{i_1+1,i_1+1/i_1+2,x(i_1+2)} \cdots g_{i_1+i_2,i_1+i_2/i_1,x(i_1)}\} \cdots$$

where we decompose A into cycles of length $i_1, i_2, \ldots$ with $n_1 = i_1 + \cdots + i_j$. Since all the cycles must have even length, $\ell(A) = n/2$ is even so n is divisible by 4.

We let $n = 4k$ and let ρ be a partition of $k = k_1 + \cdots + k_j$. We let A_ρ be defined using the above equation where $i_1 = 2k_1$, $i_2 = 2k_2, \ldots$. By making a coordinate permutation we can assume $i_1 \ge i_2 \ge \cdots$. We have shown that if $P = 0$, then $c(A_\rho, P) = 0$ for some ρ. Since there are exactly $\pi(k)$ such partitions, we have constructed a family of $\pi(k)$ linear functionals on $\mathcal{P}_{n,n,n}$ which form a seperating family. This implies dim $\mathcal{P}_{n,n,n} < \pi(k)$ which completes the proof.

We conclude this subsection with a few remarks on the proofs we have given of Theorem 2.4.7 and Lemma 2.5.6. We know of no other proof of Theorem 2.4.7 other than the one we have given. H. Weyl's theorem is only used to prove the surjectivity of the restriction map r and is inessential in the axiomatic characterization of the Euler form. This theorem gives an immediate proof of the Gauss-Bonnet theorem using heat equation methods. It is also an essential step in settling Singer's conjecture for the Euler form as we shall discuss in the fourth chapter. The fact that $r(E_m) = 0$ is, of course, just an invariant statement of the fact E_m is an unstable characteristic class; this makes it difficult to get hold of axiomatically in contrast to the Pontrjagin forms which are stable characteristic classes.

We have discussed both of the proofs of Lemma 2.5.6 which exist in the literature. The proof based on H. Weyl's theorem and on geodesic normal coordinates is more elegant, but relies heavily on fairly sophisticated theorems. The second is the original proof and is more combinatorial. It is entirely self-contained and this is the proof which generalizes to Kaehler geometry to yield an axiomatic characterization of the Chern forms of $T_c(M)$ for a holomorphic Kaehler manifold. We shall discuss this case in more detail in section 3.7.

2.6. Invariance Theory and
Mixed Characteristic Classes
of the Tangent Space and of a Coefficient Bundle.

In the previous subsection, we gave in Lemma 2.5.6 an axiomatic characterization of the Pontrjagin forms in terms of functorial properties. In discussing the Hirzebruch signature formula in the next chapter, it will be convenient to have a generalization of this result to include invariants which also depend on the derivatives of the connection form on an auxilary bundle.

Let V be a complex vector bundle. We assume V is equipped with a Hermitian fiber metric and let ∇ be a Riemannian or unitary connection on V. We let $\vec{s} = (s_1, \ldots, s_a, \ldots, s_v)$ be a local orthonormal frame for V and introduce variables ω_{abi} for the connection 1-form;

$$\nabla(s_a) = \omega_{abi}\, dx^i \otimes s_b, \qquad \text{i.e., } \nabla \vec{s} = \omega \otimes \vec{s}.$$

We introduce variables $\omega_{abi/\alpha} = d_x^\alpha(\omega_{abi})$ for the partial derivatives of the connection 1-form. We shall also use the notation $\omega_{abi/jk...}$. We use indices $1 \leq a, b, \cdots \leq v$ to index the frame for V and indices $1 \leq i, j, k \leq m$ for the tangent space variables. We define:

$$\operatorname{ord}(\omega_{abi/\alpha}) = 1 + |\alpha| \qquad \text{and} \qquad \deg_k(\omega_{abi/\alpha}) = \delta_{i,k} + \alpha(k).$$

We let $\mathcal{Q}$ be the polynomial algebra in the $\{\omega_{abi/\alpha}\}$ variables for $|\alpha| \geq 1$; if $Q \in \mathcal{Q}$ we define the evaluation $Q(X, \vec{s}, \nabla)(x_0)$. We normalize the choice of frame $\vec{s}$ by requiring $\nabla(\vec{s})(x_0) = 0$. We also normalize the coordinate system X as before so $X(x_0) = 0$, $g_{ij}(X, G)(x_0) = \delta_{ij}$, and $g_{ij/k}(X, G)(x_0) = 0$. We say Q is invariant if $Q(X, \vec{s}, \nabla)(x_0) = Q(Y, \vec{s}', \nabla)(x_0)$ for any normalized frames $\vec{s}$, $\vec{s}'$ and normalized coordinate systems X, Y; we denote this common value by $Q(\nabla)$ (although it also depends in principle on the metric tensor and the 1-jets of the metric tensor through our normalization of the coordinate system X). We let $\mathcal{Q}_{m,p,v}$ denote the space of all invariant p-form valued polynomials in the $\{\omega_{abi/\alpha}\}$ variables for $|\alpha| \geq 1$ defined on a manifold of dimension m and for a vector bundle of complex fiber dimension v. We let $\mathcal{Q}_{m,n,p,v}$ denote the subspace of invariant polynomials homogeneous of order n in the jets of the connection form. Exactly as was done for the $\mathcal{P}_*$ algebra in the jets of the metric, we can show there is a direct sum decomposition

$$\mathcal{Q}_{m,p,v} = \bigoplus_n \mathcal{Q}_{m,n,p,v} \qquad \text{and} \qquad \mathcal{Q}_{m,n,p,v} = 0 \qquad \text{for } n - p \text{ odd.}$$

Let $Q(gAg^{-1}) = Q(A)$ be an invariant polynomial of order q in the components of a $v \times v$ matrix. Then $Q(\Omega)$ defines an element of $\mathcal{Q}_{m,2q,q,v}$ for any $m \geq 2q$. By taking $Q \cdot \bar{Q}$ we can define scalar valued invariants and by taking $\delta(Q)$ we can define other form valued invariants in $\mathcal{Q}_{m,2q+1,2q-1,v}$. Thus there are a great many such form valued invariants.

In addition to this algebra, we let $\mathcal{R}_{m,n,p,v}$ denote the space of p-form valued invariants which are homogeneous of order n in the $\{g_{ij/\alpha}, \omega_{abk/\beta}\}$ variables for $|\alpha| \geq 2$ and $|\beta| \geq 1$. The spaces $\mathcal{P}_{m,n,p}$ and $\mathcal{Q}_{m,n,p,v}$ are both subspaces of $\mathcal{R}_{m,n,p,v}$. Furthermore wedge product gives a natural map $\mathcal{P}_{m,n,p} \otimes \mathcal{Q}_{m,n',p',v} \to \mathcal{R}_{m,n+n',p+p',v}$. We say that $R \in \mathcal{R}_{m,p,p,v}$ is a characteristic form if it is in the linear span of wedge products of Pontrjagin forms of $T(M)$ and Chern forms of V. The characteristic forms are characterized abstractly by the following Theorem. This is the generalization of Lemma 2.5.6 which we shall need in discussing the signature and spin complexes.

THEOREM 2.6.1.
(a) $\mathcal{R}_{m,n,p,v} = 0$ if $n < p$ or if $n = p$ and n is odd.
(b) If $R \in \mathcal{R}_{m,n,n,v}$ then R is a characteristic form.

PROOF: The proof of this fact relies heavily on Lemma 2.5.6 but is much easier. We first need the following generalization of Lemma 2.5.1:

LEMMA 2.6.2. Using the notation of Lemma 2.5.1 we define $T_z^*(R)$ if R is a scalar invariant in the $\{g_{ij/\alpha}, \omega_{abj/\beta}\}$ variables.
(a) Let $g_{12/\alpha}$ divide some monomial of R, then $g_{../\beta}$ divides some other monomial of R.
(b) Let $g_{../\alpha}$ divide some monomial of R, then $g_{../\beta}$ divides some other monomial of R where $\beta(1) = \alpha(1) + \alpha(2)$ and $\beta(2) = 0$.
(c) Let $\omega_{abi/\alpha}$ divide some monomial of R for $i > 2$, then $\omega_{abi/\beta}$ divides some other monomial of R where $\beta(1) = \alpha(1) + \alpha(2)$ and $\beta(2) = 0$.

The proof of this is exactly the same as that given for Lemma 2.5.1 and is therefore omitted.

The proof of Theorem 2.6.1 parallels the proof of Lemma 2.5.6 for a while so we summarize the argument briefly. Let $0 \neq R \in \mathcal{R}_{m,n,p,v}$. The same argument given in Lemma 2.5.3 shows an invariant polynomial is homogeneous of order n if $R(c^2G, \nabla) = c^{p-n}R(G, \nabla)$ which gives a invariant definition of the order of a polynomial. The same argument as given in Lemma 2.5.4 shows $n - p$ must be even and that if A is a monomial of R, A is a monomial of exactly one of the R_I. If $c(A, R_I) = 0$ then $\deg_k(A) + \deg_k(I)$ is always even. We decompose A in the form:

$$A = g_{i_1 j_1/\alpha_1} \cdots g_{i_q j_q/\alpha_q} \omega_{a_1 b_1 k_1/\beta_1} \cdots \omega_{a_r b_r k_r/\beta_r} = A^g A^\omega$$

and define $\ell(A) = q + r$ to the length of A. We argue using Lemma 2.6.2 to choose A so $\deg_k(A^g) = 0$ for $k > 2q$. By making a coordinate permutation we can assume that the $k_\nu \leq 2q + r$ for $1 \leq \nu \leq r$. We apply Lemma 2.6.2(c) a total of r times to choose the β_i so $\beta_1(k) = 0$ for $k > 2q + r + 1$, $\beta_2(k) = 0$ for $k > 2q + r + 2, \ldots$, $\beta_r(k) = 0$ for $k > 2q + 2r$. This chooses A so $\deg_k(A) = 0$ for $k > 2\ell(A)$. If A is a monomial of R_I for $I = \{1 \le i_1 < \cdots < i_p \le m\}$ then $\deg_{i_p}(A)$ is odd. We estimate $p \le i_p \le 2\ell(A) \le \sum |\alpha_\nu| + \sum (|\beta_\mu| + 1) = n$ so that $P_{m,n,p,v} = \{0\}$ if $n < p$ or if $n - p$ is odd which proves (a) of Theorem 2.6.1.

In the limiting case, we must have equalities so $|\alpha_\nu| = 2$ and $|\beta_\mu| = 1$. Furthermore, $i_p = p$ so there is some monomial A so $\deg_k(A) = 0$ for $k > p = n$ and $A\, dx^1 \wedge \cdots \wedge dx^p$ appears in R. There is a natural restriction map

$$r: \mathcal{R}_{m,n,p,v} \to \mathcal{R}_{m-1,n,p,v}$$

and our argument shows $r: \mathcal{R}_{m,n,n,v} \to \mathcal{R}_{m-1,n,n,v}$ is injective for $n < m$. Since the restriction of a characteristic form is a characteristic form, it suffices to prove (b) of Theorem 2.6.1 for the case $m = n = p$.

Let $0 \ne R \in \mathcal{R}_{n,n,n,v}$ then R is a polynomial in the $\{g_{ij/kl}, \omega_{abi/j}\}$ variables. The restriction map r was defined by considering products $M_1 \times S^1$ but there are other functorial constructions which give rise to useful projections. Fix non-negative integers (s,t) so that $n = s + t$. Let M_1 be a Riemannian manifold of dimension s. Let M_2 be the flat torus of dimension t and let V_2 be a vector bundle with connection ∇_2 over M_2. Let $M = M_1 \times M_2$ with the product metric and let V be the natural extension of V_2 to M which is flat in the M_1 variables. More exactly, if $\pi_2: M \to M_2$ is a projection on the second factor, then $(V, \nabla) = \pi_2^*(V_2, \nabla_2)$ is the pull back bundle with the pull back connection. We define

$$\pi_{(s,t)}(R)(G_1, \nabla_2) = R(G_1 \times 1, \nabla).$$

Using the fact that $P_{s,n_1,p_1} = 0$ for $s < p_1$ or $n_1 < p_1$ and the fact $\mathcal{Q}_{t,n_2,p_2,k} = 0$ for $t < p_2$ or $n_2 < p_2$ it follows that $\pi_{(s,t)}$ defines a map

$$\pi_{(s,t)}: \mathcal{R}_{n,n,v} \to \mathcal{P}_{s,s,s} \otimes \mathcal{Q}_{t,t,t,v}.$$

More algebraically, let $A = A^g A^\omega$ be a monomial, then we define:

$$\pi_{(s,t)}(A) = \begin{cases} 0 & \text{if } \deg_k(A^g) > 0 \text{ for } k > s \text{ or } \deg_k(A^\omega) > 0 \text{ for } k \le s \\ A & \text{otherwise.} \end{cases}$$

The only additional relations imposed are to set $g_{ij/kl} = 0$ if any of these indices exceeds s and to set $\omega_{abi/j} = 0$ if either i or j is less than or equal to s.

We use these projections to reduce the proof of Theorem 2.6.1 to the case in which $R \in \mathcal{Q}_{t,t,t,v}$. Let $0 \ne R \in \mathcal{R}_{n,n,n,v}$ and let $A = A^g A^\omega$ be a monomial of R. Let $s = 2\ell(A^g) = \text{ord}(A^g)$ and let $t = n - s = 2\ell(A^\omega) =$

ord(A^ω). We choose A so $\deg_k(A^g) = 0$ for $k > s$. Since $\deg_k(A) \geq 1$ must be odd for each index k, we can estimate:

$$t \leq \sum_{k>s} \deg_k(A) = \sum_{k>s} \deg_k(A^\omega) \leq \sum_k \deg_k(A^\omega) = \text{ord}(A^\omega) = t.$$

As all these inequalities must be equalities, we conclude $\deg_k(A^\omega) = 0$ for $k \leq s$ and $\deg_k(A^\omega) = 1$ for $k > s$. This shows in particular that $\pi_{(s,t)}(R) \neq 0$ for some (s,t) so that

$$\bigoplus_{s+t=n} \pi_{(s,t)} \colon R_{n,n,n,v} \to \bigoplus_{s+t=n} P_{s,s,s} \otimes \mathcal{Q}_{t,t,t,v}$$

is injective.

We shall prove that $\mathcal{Q}_{t,t,t,v}$ consists of characteristic forms of V. We showed earlier that $P_{s,s,s}$ consists of Pontrjagin forms of $T(M)$. The characteristic forms generated by the Pontrjagin forms of $T(M)$ and of V are elements of $R_{n,n,n,v}$ and $\pi_{(s,t)}$ just decomposes such products. Thererfore π is surjective when restricted to the subspace of characteristic forms. This proves π is bijective and also that $R_{n,n,n,v}$ is the space of characteristic forms. This will complete the proof of Theorem 2.6.1.

We have reduced the proof of Theorem 2.6.1 to showing $\mathcal{Q}_{t,t,t,v}$ consists of the characteristic forms of V. We noted that $0 \neq Q \in \mathcal{Q}_{t,t,t,v}$ is a polynomial in the $\{\omega_{abi/j}\}$ variables and that if A is a monomial of Q, then $\deg_k(A) = 1$ for $1 \leq k \leq t$. Since $\text{ord}(A) = t$ is even, we conclude $\mathcal{Q}_{t,t,t,v} = 0$ if t is odd.

The components of the curvature tensor are given by:

$$\Omega_{abij} = \omega_{abi/j} - \omega_{abi/j} \quad \text{and} \quad \Omega_{ab} = \sum \Omega_{abij}\, dx_i \wedge dx_j$$

up to a possible sign convention and factor of $\frac{1}{2}$ which play no role in this discussion. If A is a monomial of P, we decompose:

$$A = \omega_{a_1 b_1 i_1/i_2} \cdots \omega_{a_u b_u i_{t-1}/i_t} \qquad \text{where } 2u = t.$$

All the indices i_ν are distinct. If ρ is a permutation of these indices, then $c(A, P) = \text{sign}(\rho) c(A^\rho, P)$. This implies we can express P in terms of the expressions:

$$\bar{A} = (\omega_{a_1 b_1 i_1/i_2} - \omega_{a_1 b_1 i_2/i_1}) \cdots (\omega_{a_u b_u i_{t-1}/i_t} - \omega_{a_u b_u i_t/i_{t-1}})$$
$$dx_{i_1} \wedge \cdots \wedge dx_{i_t}$$
$$= \Omega_{a_1 b_1 i_1 i_2} \cdots \Omega_{a_u b_u i_{t-1} i_t}\, dx_{i_1} \wedge \cdots \wedge dx_{i_t}$$

Again, using the alternating nature of these expression, we can express P in terms of expressions of the form:

$$\Omega_{a_1 b_1} \wedge \cdots \wedge \Omega_{a_u b_u}$$

so that $Q = Q(\Omega)$ is a polynomial in the components Ω_{ab} of the curvature. Since the value of Q is independent of the frame chosen, Q is the invariant under the action of $U(v)$. Using the same argument as that given in the proof of Lemma 2.1.3 we see that in fact Q is a characteristic form which completes the proof.

We conclude this subsection with some uniqueness theorems regarding the local formulas we have been considering. If M is a holomorphic manifold of real dimension m and if V is a complex vector bundle of fiber dimension v, we let $\mathcal{R}^{ch}_{m,m,p,v}$ denote the space of p-form valued invariants generated by the Chern forms of V and of $T_c(M)$. There is a suitable axiomatic characterization of these spaces using invariance theory for Kaehler manifolds which we shall discuss in section 3.7. The uniqueness result we shall need in proving the Hirzebruch signature theorem and the Riemann-Roch theorem is the following:

LEMMA 2.6.3.
(a) Let $0 \neq R \in \mathcal{R}_{m,m,m,v}$ then there exists (M,V) so M is oriented and $\int_M R(G,\nabla) \neq 0$.
(b) Let $0 \neq R \in \mathcal{R}^{ch}_{m,m,m,v}$ then there exists (M,V) so M is a holomorphic manifold and $\int_M R(G,\nabla) \neq 0$.

PROOF: We prove (a) first. Let $\rho = \{1 \leq i_1 \leq \cdots \leq i_\nu\}$ be a partition $k(\rho) = i_1 + \cdots + i_\nu$ for $4k = s \leq m$. Let $\{M^r_\rho\}$ be the collection of manifolds of dimension s discussed in Lemma 2.3.4. Let P_ρ be the corresponding real characteristic form so

$$\int_{M^r_\rho} P_\tau(G) = \delta_{\rho,\tau}.$$

The $\{P_\rho\}$ forms a basis for $\mathcal{P}_{m,s,s}$ for any $s \leq m$. We decompose

$$R = \sum_\rho P_\rho Q_\rho \qquad \text{for } Q_\rho \in \mathcal{Q}_{m,t,t,k}, \text{ where } t + s = m.$$

This decomposition is, of course, nothing but the decomposition defined by the projections $\pi_{(s,t)}$ discussed in the proof of the previous lemma.

Since $R \neq 0$, at least one of the $Q_\rho \neq 0$. We choose ρ so $k(\rho)$ is maximal with $Q_\rho \neq 0$. We consider $M = M^r_\rho \times M_2$ and $(V,\nabla) = \pi^*_2(V_2,\nabla_2)$ where (V_2,∇_2) is a bundle over M_2 which will be specified later. Then we compute:

$$\int_M P_\tau(G)Q_\tau(\nabla) = 0$$

unless $\mathrm{ord}(Q_\tau) \leq \dim(M_2)$ since ∇ is flat along M^r_ρ. This implies $k(\tau) \geq k(\rho)$ so if this integral is non-zero $k(\tau) = k(\rho)$ by the maximality of ρ. This implies

$$\int_M P_\tau(G)Q_\tau(\nabla) = \int_{M^r_\rho} P_\tau(G) \cdot \int_{M_2} Q_\tau(\nabla_2) = \delta_{\rho,\tau} \int_{M_2} Q_\tau(\nabla_2).$$

This shows that

$$\int_M R(G, \nabla) = \int_{M_2} Q_\rho(\nabla_2)$$

and reduces the proof of this lemma to the special case $Q \in \mathcal{Q}_{m,m,m,v}$.

Let A be a $v \times v$ complex matrix and let $\{x_1, \ldots, x_v\}$ be the normalized eigenvalues of A. If $2k = m$ and if ρ is a partition of k we define

$$x_\rho = x_1^{i_1} \ldots x_\nu^{i_\nu},$$

then x_ρ is a monomial of $Q(A)$ for some ρ. We let $M = M_1 \times \cdots \times M_\nu$ with $\dim(M_j) = 2i_j$ and let $V = L_1 \oplus \cdots \oplus L_\nu \oplus 1^{v-\nu}$ where the L_j are line bundles over M_j. If $c(x_\rho, Q)$ is the coefficient of x in Q, then:

$$\int_M Q(\nabla) = c(x_\rho, Q) \prod_{j=1}^{\nu} \int_{M_j} c_1(L_j)^{i_j}.$$

This is, of course, nothing but an application of the splitting principle. This reduces the proof of this lemma to the special case $Q \in \mathcal{Q}_{m,m,m,1}$.

If $Q = c_1^k$ we take $M = S^2 \times \cdots \times S^2$ to be the k-fold product of two dimensional spheres. We let L_j be a line bundle over the jth factor of S^2 and let $L = L_1 \otimes \cdots \otimes L_k$. Then $c_1(L) = c_1(L_j)$ so

$$\int_M c_1(L)^k = k! \prod_{j=1}^{k} \int_{S^2} c_1(L_j)$$

which reduces the proof to the case $m = 2$ and $k = 1$. We gave an example in Lemma 2.1.5 of a line bundle over S^2 so $\int_{S^2} c_1(L) = 1$. Alternatively, if we use the Gauss-Bonnet theorem with $L = T_c(S^2)$, then $\int_{S^2} c_1(T_c(S^2)) = \int_{S^2} e_2(T(S^2)) = \chi(S^2) = 2 \neq 0$ completes the proof of (a). The proof of (b) is the same where we replace the real manifolds M_ρ^r by the corresponding manifolds M_ρ^c and the real basis P_ρ by the corresponding basis P_ρ^c of characteristic forms of $T_c(M)$. The remainder of the proof is the same and relies on Lemma 2.3.4 exactly as for (a) and is therefore omitted in the interests of brevity.

CHAPTER 3

THE INDEX THEOREM

Introduction

In this the third chapter, we complete the proof of the index theorem for the four classical elliptic complexes. We give a proof of the Aityah-Singer theorem in general based on the Chern isomorphism between K-theory and cohomology (which is not proved). Our approach is to use the results of the first chapter to show there exists a suitable formula with the appropriate functorial properties. The results of the second chapter imply it must be a characteristic class. The normalizing constants are then determined using the method of universal examples.

In section 3.1, we define the twisted signature complex and prove the Hirzebruch signature theorem. We shall postpone until section 3.4 the determination of all the normalizing constants if we take coefficients in an auxilary bundle. In section 3.2, we introduce spinors as a means of connecting the de Rham, signature and Dolbeault complexes. In section 3.3, we discuss the obstruction to putting a spin structure on a real vector bundle in terms of Stieffel-Whitney classes. We compute the characteristic classes of spin bundles.

In section 3.4, we discuss the spin complex and the $\hat{A}$ genus. In section 3.5, we use the spin complex together with the spin_c representation to discuss the Dolbeault complex and to prove the Riemann-Roch theorem for almost complex manifolds. In sections 3.6 and 3.7 we give another treatment of the Riemann-Roch theorem based on a direct approach for Kaehler manifolds. For Kaehler manifolds, the integrands arising from the heat equation can be studied directly using an invariant characterization of the Chern forms similar to that obtained for the Euler form. These two subsections may be deleted by a reader not interested in Kaehler geometry.

In section 3.8, we give the preliminaries we shall need to prove the Atiyah-Singer index theorem in general. The only technical tool we will use which we do not prove is the Chern isomorphism between rational cohomology and K-theory. We give a discussion of Bott periodicity using Clifford algebras. In section 3.9, we show that the index can be treated as a formula in rational K-theory. We use constructions based on Clifford algebras to determine the normalizing constants involved. For these two subsections, some familarity with K-theory is helpful, but not essential.

Theorems 3.1.4 and 3.6.10 were also derived by V. K. Patodi using a complicated cancellation argument as a replacement of the invariance theory presented in Chapter 2. A similar although less detailed discussion may also be found in the paper of Atiyah, Bott and Patodi.

3.1. The Hirzebruch Signature Formula.

The signature complex is best described using Clifford algebras as these provide a unified framework in which to discuss (and avoid) many of the $\pm$ signs which arise in dealing with the exterior algebra directly. The reader will note that we are choosing the opposite sign convention for our discussion of Clifford algebras from that adopted in the example of Lemma 2.1.5. This change in sign convention is caused by the $\sqrt{-1}$ present in discussing the symbol of a first order operator.

Let V be a real vector with a positive definite inner product. The Clifford algebra $\text{CLIF}(V)$ is the universal algebra generated by V subject to the relations

$$v * v + (v, v) = 0 \text{ for } v \in V.$$

If the $\{e_i\}$ are an orthonormal basis for V and if $I = \{1 \leq i_1 < \cdots < i_p \leq \dim(V)\}$ then $e_i * e_j + e_j * e_i = -2\delta_{ij}$ is the Kronecker symbol and

$$e_I = e_{i_1} * \cdots * e_{i_p}$$

is an element of the Clifford algebra. $\text{CLIF}(V)$ inherits a natural inner product from V and the $\{e_I\}$ form an orthonormal basis for V.

If $\Lambda(V)$ denotes the exterior algebra of V and if $\text{END}(\Lambda(V))$ is the algebra of linear endomorphisms of $\Lambda(V)$, there is a natural representation of $\text{CLIF}(V)$ into $\text{END}(\Lambda(V))$ given by Clifford multiplication. Let $\text{ext}: V \rightarrow \text{END}(\Lambda(V))$ be exterior multiplication on the left and let $\text{int}(v)$ be interior multiplication, the adjoint. For example:

$$\text{ext}(e_1)(e_{i_1} \wedge \cdots \wedge e_{i_p}) = \begin{cases} e_1 \wedge e_{i_1} \wedge \cdots \wedge e_{i_p} & \text{if } i_1 > 1 \\ 0 & \text{if } i_1 = 1 \end{cases}$$

$$\text{int}(e_1)(e_{i_1} \wedge \cdots \wedge e_{i_p}) = \begin{cases} e_{i_2} \wedge \cdots \wedge e_{i_p} & \text{if } i_1 = 1 \\ 0 & \text{if } i_1 > 1. \end{cases}$$

We define

$$c(v) = \text{ext}(v) - \text{int}(v): V \rightarrow \text{END}(\Lambda(V)).$$

It is immediate from the definition that

$$c(v)^2 = -(\text{ext}(v)\,\text{int}(v) + \text{int}(v)\,\text{ext}(v)) = -|v|^2 I$$

so that c extends to define an algebra morphism

$$c: \text{CLIF}(V) \rightarrow \text{END}(\Lambda(V)).$$

Furthermore:

$$c(e_I)1 = e_{i_1} \wedge \cdots \wedge e_{i_p}$$

so the map $w \mapsto c(w)1$ defines a vector space isomorphism (which is not, of course an algebra morphism) between $\mathrm{CLIF}(V)$ and $\Lambda(V)$. Relative to an orthonormal frame, we simply replace Clifford multiplication by exterior multiplication.

Since these constructions are all independent of the basis chosen, they extend to the case in which V is a real vector bundle over M with a fiber metric which is positive definite. We define $\mathrm{CLIF}(V)$, $\Lambda(V)$, and $c: \mathrm{CLIF}(V) \rightarrow \mathrm{END}(\Lambda(V))$ as above. Since we only want to deal with complex bundles, we tensor with $\mathbf{C}$ at the end to enable us to view these bundles as being complex. We emphasize, however, that the underlying constructions are all real.

Clifford algebras provide a convenient way to describe both the de Rham and the signature complexes. Let $(d + \delta): C^\infty(\Lambda(T^*M)) \rightarrow C^\infty(\Lambda(T^*M))$ be exterior differentiation plus its adjoint as discussed earlier. The leading symbol of $(d+\delta)$ is $\sqrt{-1}(\mathrm{ext}(\xi) - \mathrm{int}(\xi)) = \sqrt{-1}c(\xi)$. We use the following diagram to define an operator A; let ∇ be covariant differentiation. Then:

$$A: C^\infty(\Lambda(T^*M)) \xrightarrow{\nabla} C^\infty(T^*M \otimes \Lambda(T^*M)) \xrightarrow{c} C^\infty(\Lambda(T^*M)).$$

A is invariantly defined. If $\{e_i\}$ is a local orthonormal frame for $T^*(M)$ which we identify with $T(M)$, then:

$$A(\omega) = \sum_i (\mathrm{ext}(e_i) - \mathrm{int}(e_i))\nabla_{e_i}(\omega).$$

Since the leading symbol of A is $\sqrt{-1}c(\xi)$, these two operators have the same leading symbol so $(d+\delta) - A = A_0$ is an invariantly defined 0th order operator. Relative to a coordinate frame, we can express A_0 as a linear combination of the 1-jets of the metric with coefficients which are smooth in the $\{g_{ij}\}$ variables. Given any point x_0, we can always choose a frame so the $g_{ij/k}$ variables vanish at x_0 so $A_0(x_0) = 0$ so $A_0 \equiv 0$. This proves $A = (d + \delta)$ is defined by this diagram which gives a convenient way of describing the operator $(d + \delta)$ in terms of Clifford multiplication.

This trick will be useful in what follows. If A and B are natural first order differential operators with the same leading symbol, then $A = B$ since $A - B$ is a 0th order operator which is linear in the 1-jets of the metric. This trick does not work in the holomorphic category unless we impose the additional hypothesis that M is Kaehler. This makes the study of the Riemann-Roch theorem more complicated as we shall see later since there are many natural operators with the same leading symbol.

We let $\alpha \in \mathrm{END}(\Lambda(T^*M))$ be defined by:

$$\alpha(\omega_p) = (-1)^p \omega_p \qquad \text{for } \omega_p \in \Lambda^p(T^*M).$$

It is immediate that

$$\mathrm{ext}(\xi)\alpha = -\alpha\,\mathrm{ext}(\xi) \qquad \text{and} \qquad \mathrm{int}(\xi)\alpha = -\alpha\,\mathrm{int}(\xi)$$

so that $-\alpha(d+\delta)\alpha$ and $(d+\delta)$ have the same leading symbol. This implies

$$\alpha(d+\delta) = -(d+\delta)\alpha.$$

We decompose $\Lambda(T^*M) = \Lambda^e(T^*M) \oplus \Lambda^o(T^*M)$ into the differential forms of even and odd degree. This decomposes $\Lambda(T^*M)$ into the ± 1 eigenspaces of α. Since $(d+\delta)$ anti-commutes with α, we decompose

$$(d+\delta)_{\mathrm{e,o}}\colon C^\infty(\Lambda^{e,o}(T^*M)) \to C^\infty(\Lambda^{e,o}(T^*M))$$

where the adjoint of $(d+\delta)_e$ is $(d+\delta)_o$. This is, of course, just the de Rham complex, and the index of this elliptic operator is $\chi(M)$.

If $\dim(M) = m$ is even and if M is oriented, there is another natural endomorphism $\tau \in \mathrm{END}(\Lambda(T^*M))$. It can be used to define an elliptic complex over M called the signature complex in just the same way that the de Rham complex was defined. Let $\mathrm{dvol} \in \Lambda^m(T^*M)$ be the volume form. If $\{e_i\}$ is an oriented local orthonormal frame for T^*M, then $\mathrm{dvol} = e_1 \wedge \cdots \wedge e_m$. We can also regard $\mathrm{dvol} = e_1 * \cdots * e_m \in \mathrm{CLIF}(T^*M)$ and we define

$$\tau = (\sqrt{-1})^{m/2}c(\mathrm{dvol}) = (\sqrt{-1})^{m/2}c(e_1)\ldots c(e_m).$$

We compute:

$$\begin{aligned}
\tau^2 &= (-1)^{m/2}c(e_1 * \cdots * e_m * e_1 * \cdots * e_m) \\
&= (-1)^{m/2}(-1)^{m+(m-1)+\cdots+1} \\
&= (-1)^{(m+(m+1)m)/2} = 1.
\end{aligned}$$

Because m is even, $c(\xi)\tau = -\tau c(\xi)$ so τ anti-commutes with the symbol of $(d+\delta)$. If we decompose $\Lambda(T^*M) = \Lambda^+(T^*M) \oplus \Lambda^-(T^*M)$ into the ± 1 eigenvalues of τ, then $(d+\delta)$ decomposes to define:

$$(d+\delta)_\pm\colon C^\infty(\Lambda^\pm(T^*M)) \to C^\infty(\Lambda^\mp(T^*M))$$

where the adjoint of $(d+\delta)_+$ is $(d+\delta)_-$. We define:

$$\mathrm{signature}(M) = \mathrm{index}(d+\delta)_+$$

to be the signature of M. (This is also often refered to as the index of M, but we shall not use this notation as it might be a source of some confusion).

We decompose the Laplacian $\Delta = \Delta^+ \oplus \Delta^-$ so

$$\text{signature}(M) = \dim \mathrm{N}(\Delta^+) - \dim \mathrm{N}(\Delta^-).$$

Let $a_n^s(x, \text{orn}) = a_n(x, \Delta^+) - a_n(x, \Delta^-)$ be the invariants of the heat equation; they depend on the orientation orn chosen. Although we have complexified the bundles $\Lambda(T^*M)$ and $\mathrm{CLIF}(T^*M)$, the operator $(d+\delta)$ is real. If $m \equiv 2$ (4), then τ is pure imaginary. Complex conjugation defines an isomorphism

$$\Lambda^+(T^*M) \xrightarrow{\simeq} \Lambda^-(T^*M) \qquad \text{and} \qquad \Delta^+ \xrightarrow{\simeq} \Delta^-.$$

This implies signature$(M) = 0$ and $a_n^s(x, \text{orn}) = 0$ in this case. We can get a non-zero index if $m \equiv 2$ (4) if we take coefficients in some auxiliary bundle as we shall discuss shortly.

If $m \equiv 0$ (4), then τ is a real endomorphism. In general, we compute:

$$\begin{aligned}
\tau(e_1 \wedge \cdots \wedge e_p) &= (\sqrt{-1})^{m/2} e_1 * \cdots * e_m * e_1 * \cdots * e_p \\
&= (\sqrt{-1})^{m/2}(-1)^{p(p-1)/2} e_{p+1} \wedge \cdots \wedge e_m.
\end{aligned}$$

If "$*$" is the Hodge star operator discussed in the first sections, then

$$\tau_p = (\sqrt{-1})^{m/2+p(p-1)} *_p$$

acting on p-forms. The spaces $\Lambda^p(T^*M) \oplus \Lambda^{m-p}(T^*M)$ are invariant under τ. If $p \neq m - p$ there is a natural isomorphism

$$\Lambda^p(T^*M) \xrightarrow{\simeq} (\Lambda^p(T^*M) \oplus \Lambda^{m-p}(T^*M))^\pm \qquad \text{by} \qquad \omega_p \mapsto \tfrac{1}{2}(\omega_p \pm \omega_p).$$

This induces a natural isomorphism from

$$\Delta_p \xrightarrow{\simeq} \Delta^\pm \text{ on } (\Lambda^p(T^*M) \oplus \Lambda^{m-p}(T^*M))^\pm$$

so these terms all cancel off in the alternating sum and the only contribution is made in the middle dimension $p = m - p$.

If $m = 4k$ and $p = 2k$ then $\tau = *$. We decompose $\mathrm{N}(\Delta_p) = \mathrm{N}(\Delta_p^+) \oplus \mathrm{N}(\Delta_p^-)$ so signature$(M) = \dim \mathrm{N}(\Delta_p^+) - \dim \mathrm{N}(\Delta_p^-)$. There is a natural symmetric bilinear form on $H^{2k}(T^*M; \mathbf{C}) = \mathrm{N}(\Delta_p)$ defined by

$$I(\alpha_1, \alpha_2) = \int_M \alpha_1 \wedge \alpha_2.$$

If we use the de Rham isomorphism to identify de Rham and simplicial cohomology, then this bilinear form is just the evaluation of the cup product of two cohomology classes on the top dimensional cycle. This shows I can be defined in purely topological terms.

The index of a real quadratic form is just the number of $+1$ eigenvalues minus the number of -1 eigenvalues when it is diagonalized over $\mathbf{R}$. Since

$$I(\alpha, \beta) = \int_M \alpha \wedge \beta = \int_M (\alpha, *\beta)\, \mathrm{dvol} = (\alpha, *\beta)_{L^2},$$

we see

$$\begin{aligned}
\mathrm{index}(I) &= \dim\{+1 \text{ eigenspace of } * \text{ on } H^p\} \\
&\quad - \dim\{-1 \text{ eigenspace of } * \text{ on } H^p\} \\
&= \dim N(\Delta^+) - \dim N(\Delta^-) = \mathrm{signature}(M).
\end{aligned}$$

This gives a purely topological definition of the signature of M in terms of cup product. We note that if we reverse the orientation of M, then the signature changes sign.

Example: Let $M = \mathbf{C}P_{2k}$ be complex projective space. Let $x \in H^2(M; \mathbf{C})$ be the generator. Since x^k is the generator of $H^{2k}(M; \mathbf{C})$ and since $x^k \wedge x^k = x^{2k}$ is the generator of $H^{4k}(M; \mathbf{C})$, we conclude that $*x^k = x^k$. $\dim N(\Delta_{2k}^+) = 1$ and $\dim N(\Delta_{2k}^-) = 0$ so $\mathrm{signature}(\mathbf{C}P_{2k}) = 1$.

An important tool in the study of the de Rham complex was its multiplicative properties under products. Let M_i be oriented even dimensional manifolds and let $M = M_1 \times M_2$ with the induced orientation. Decompose:

$$\begin{aligned}
\Lambda(T^*M) &= \Lambda(T^*M_1) \otimes \Lambda(T^*M_2) \\
\mathrm{CLIF}(T^*M) &= \mathrm{CLIF}(T^*M_1) \otimes \mathrm{CLIF}(T^*M_2)
\end{aligned}$$

as graded non-commutative algebras—i.e.,

$$(\omega_1 \otimes \omega_2) \circ (\omega_1' \otimes \omega_2') = (-1)^{\deg \omega_2 \cdot \deg \omega_1'} (\omega_1 \circ \omega_1') \otimes (\omega_2 \circ \omega_2')$$

for $\circ = $ either $\wedge$ or $*$. Relative to this decomposition, we have:

$$\tau = \tau_1 \otimes \tau_2 \text{ where the } \tau_i \text{ commute.}$$

This implies that:

$$\begin{aligned}
\Lambda^+(M) &= \Lambda^+(T^*M_1) \otimes \Lambda^+(T^*M_2) \oplus \Lambda^-(T^*M_1) \otimes \Lambda^-(T^*M_2) \\
\Lambda^-(M) &= \Lambda^-(T^*M_1) \otimes \Lambda^+(T^*M_2) \oplus \Lambda^+(T^*M_1) \otimes \Lambda^-(T^*M_2) \\
N(\Delta^+) &= N(\Delta_1^+) \otimes N(\Delta_2^+) \oplus N(\Delta_1^-) \otimes N(\Delta_2^-) \\
N(\Delta^-) &= N(\Delta_1^-) \otimes N(\Delta_2^+) \oplus N(\Delta_1^+) \otimes N(\Delta_2^-) \\
\end{aligned}$$
$$\mathrm{signature}(M) = \mathrm{signature}(M_1)\,\mathrm{signature}(M_2).$$

Example: Let ρ be a partition of $k = i_1 + \cdots + i_j$ and $M_\rho^r = \mathbb{CP}_{2i_1} \times \cdots \times \mathbb{CP}_{2i_j}$. Then signature$(M_\rho^r) = 1$. Therefore if L_k is the Hirzebruch L-polynomial,

$$\text{signature}(M_\rho^r) = \int_{M_\rho^p} L_k(T(M_\rho^r))$$

by Lemma 2.3.5.

We can now begin the proof of the Hirzebruch signature theorem. We shall use the same argument as we used to prove the Gauss-Bonnet theorem with suitable modifications. Let $m = 4k$ and let $a_n^s(x, \text{orn}) = a_n(x, \Delta^+) - a_n(x, \Delta^-)$ be the invariants of the heat equation. By Lemma 1.7.6:

$$\int_M a_n^s(x, \text{orn}) = \begin{cases} 0 & \text{if } n \neq m \\ \text{signature}(M) & \text{if } n = m \end{cases}$$

so this gives a local formula for signature(M). We can express τ functorially in terms of the metric tensor. We can find functorial local frames for $\Lambda^\pm$ relative to any oriented coordinate system in terms of the coordinate frames for $\Lambda(T^*M)$. Relative to such a frame, we express the symbol of $\Delta^\pm$ functorially in terms of the metric. The leading symbol is $|\xi|^2 I$; the first order symbol is linear in the 1-jets of the metric with coefficients which depend smoothly on the $\{g_{ij}\}$ variables; the 0th order symbol is linear in the 2-jets of the metric and quadratic in the 1-jets of the metric with coefficients which depend smoothly on the $\{g_{ij}\}$ variables. By Lemma 2.4.2, we conclude $a_n^s(x, \text{orn})$ is homogeneous of order n in the jets of the metric.

It is worth noting that if we replace the metric G by $c^2 G$ for $c > 0$, then the spaces $\Lambda^\pm$ are not invariant. On Λ^p we have:

$$\tau(c^2 G)(\omega_p) = c^{2p-m} \tau(G)(\omega_p).$$

However, in the middle dimension we have τ is invariant as are the spaces $\Lambda_p^\pm$ for $2p = m$. Clearly $\Delta_p^\pm(c^2 G) = c^{-2} \Delta_p^\pm(G)$. Since $a_n^s(x, \text{orn})$ only depends on the middle dimension, this provides another proof that a_n^s is homogeneous of order n in the derivatives of the metric since

$$a_n^s(x, \text{orn})(c^2 G) = a_n(x, c^{-2} \Delta_p^+) - a_n(x, c^{-2} \Delta_p^-)$$
$$= c^{-n} a_n(x, \Delta_p^+) - c^{-n} a_n(x, \Delta_p^-) = c^{-n} a_n^s(x, \text{orn})(G).$$

If we reverse the orientation, we interchange the roles of Δ^+ and Δ^- so a_n^s changes sign if we reverse the orientation. This implies a_n^s can be regarded as an invariantly defined m-form; $a_n^s(x) = a_n^s(x, \text{orn}) \, \text{dvol} \in \mathcal{P}_{m,n,m}$.

THEOREM 3.1.1. *Let* $a_n^s = \{a_n(x, \Delta^+) - a_n(x, \Delta^-)\} \, \mathrm{dvol} \in \mathcal{P}_{m,n,m}$ *then:*
(a) $a_n^s = 0$ *if either* $m \equiv 2$ (4) *or if* $n < m$.
(b) If $m = 4k$, *then* $a_{4k}^s = L_k$ *is the Hirzebruch polynomial so*

$$\mathrm{signature}(M) = \int_M L_k.$$

(Hirzebruch Signature Theorem).

PROOF: We already noted that Δ^+ is naturally isomorphic to Δ^- if $m \equiv 2$ (4) so $a_n^s = 0$ in that case. Lemma 2.5.6 implies $a_n^s = 0$ for $n < m$. If $m = 4k$, then a_m^s is a characteristic form of $T(M)$ by Lemma 2.5.6. We know

$$\int_{M_\rho^r} a_m^s = \mathrm{signature}(M_\rho^r) = 1 = \int_{M_\rho^r} L_k$$

so Lemma 2.3.4 implies $a_m^s = L_k$. Since $\mathrm{signature}(M) = \int_M a_m^s$ for any manifold M, we conclude $\mathrm{signature}(M) = \int_M L_k$ in general which completes the proof of (b).

If $\omega \in \Lambda(T^*M)$, we define $\int_M \omega = \int_M \omega_m$ of the top degree form. With this notational convention, we can also express

$$\mathrm{signature}(M) = \int_M L$$

which is a common form in which the Hirzebruch signature theorem appears.

It is worth making a few remarks about the proof of this result. Just as in the case of the de Rham complex, the heat equation furnishes us with the a priori local formula for the signature of M. The invariance theory of the second chapter identifies this local formula as a characteristic class. We evaluate this local formula on a sufficient number of classifying examples to determine the normalizing constants to prove $a_m^s = L_k$.

There are a great many consequences of this theorem and of the Gauss-Bonnet theorem. We present just a few to illustrate some of the applications:

CORROLARY 3.1.2.
(a) Let $F \to M_1 \to M_2$ *be a finite covering projection. Then* $\chi(M_1) = \chi(M_2)|F|$. *If* M_2 *is orientable, then* M_1 *is orientable and we give it the natural orientation inherited from* M_2. *Then* $\mathrm{signature}(M_1) = \mathrm{signature}(M_2)|F|$.
(b) If M_1 *and* M_2 *are manifolds of dimension* m, *we let* $M_1 \# M_2$ *be the connected sum. This is defined by punching out disks in both manifolds and gluing along the common resulting boundaries. Then* $\chi(M_1 \# M_2) + \chi(S^m) = \chi(M_1) + \chi(M_2)$. *If* M_1 *and* M_2 *are oriented by some orientation*

on $M_1 \# M_2$ then signature$(M_1 \# M_2)$ = signature(M_1) + signature(M_2), if $m \equiv 0$ (4).

PROOF: (a) is an immediate consequence of the fact we have local formulas for the Euler characteristic and for the signature. To prove (b), we note that the two disks we are punching out glue together to form a sphere. We use the additivity of local formulas to prove the assertion about X. The second assertion follows similarly if we note signature$(S^m) = 0$.

This corollary has topological consequences. Again, we present just one to illustrate the methods involved:

COROLLARY 3.1.3. Let $F \to \mathbf{C}P_{2j} \to M$ be a finite covering. Then $|F| = 1$ and $M = \mathbf{C}P_{2j}$.

PROOF: $\chi(\mathbf{C}P_{2j}) = 2j + 1$ so as $2j + 1 = |F|\chi(M)$, we conclude $|F|$ must be odd. Therefore, this covering projection is orientation preserving so M is orientable. The identity $1 = \text{signature}(\mathbf{C}P_{2j}) = |F| \text{signature}(M)$ implies $|F| = 1$ and completes the proof.

If $m \equiv 2$ (4), the signature complex does not give a non-trivial index. We twist by taking coefficients in an auxiliary complex vector bundle V to get a non-trivial index problem in any even dimension m.

Let V be a smooth complex vector bundle of dimension v equipped with a Riemannian connection ∇. We take the Levi-Civita connection on $T^*(M)$ and on $\Lambda(T^*M)$ and let ∇ be the tensor product connection on $\Lambda(T^*M) \otimes V$. We define the operator $(d + \delta)_V$ on $C^\infty(\Lambda(T^*M) \otimes V)$ using the diagram:

$$(d + \delta)_V : C^\infty(\Lambda(T^*M) \otimes V) \xrightarrow{\nabla} C(T^*M \otimes \Lambda(T^*M) \otimes V)$$
$$\xrightarrow{c \otimes 1} C^\infty(\Lambda(T^*M) \otimes V).$$

We have already noted that if $V = 1$ is the trivial bundle with flat connection, then the resulting operator is $(d + \delta)$.

We define $\tau_V = \tau \otimes 1$, then a similar argument to that given for the signature complex shows $\tau_V^2 = 1$ and τ_V anti-commutes with $(d+\delta)_V$. The ± 1 eigenspaces of τ_V are $\Lambda^\pm(T^*M) \otimes V$ and the twisted signature complex is defined by the diagram:

$$(d + \delta)_V^\pm : C^\infty(\Lambda^\pm(T^*M) \otimes V) \to C^\infty(\Lambda^\mp(T^*M) \otimes V)$$

where as before $(d + \delta)_V^-$ is the adjoint of $(d + \delta)_V^+$. We let $\Delta_V^\pm$ be the associated Laplacians and define:

$$\text{signature}(M, V) = \text{index}((d + \delta)_V^+) = \dim N(\Delta_V^+) - \dim N(\Delta_V^-)$$
$$a_n^s(x, V) = \{a_n(x, \Delta_V^+) - a_n(x, \Delta_V^-)\}\, \text{dvol} \in \Lambda^m$$
$$\int_M a_n^s(x, V) = \text{signature}(M, V).$$

The invariance of the index under homotopies shows signature(M,V) is independent of the metric on M, of the fiber metric on V, and of the Riemannian connection on V. If we do *not* choose a Riemannian connection on V, we can still compute signature$(M,V) = $ index$((d+\delta)^+_V)$, but then $(d+\delta)^-_V$ is not the adjoint of $(d+\delta)^+_V$ and a^s_n is not an invariantly defined m form.

Relative to a functorial coordinate frame for $\Lambda^\pm$, the leading symbol of $\Delta^\pm$ is $|\xi|^2 I$. The first order symbol is linear in the 1-jets of the metric and the connection form on V. The 0th order symbol is linear in the 2-jets of the metric and the connection form and quadratic in the 1-jets of the metric and the connection form. Thus $a^s_n(X,V) \in \mathcal{R}_{m,n,m,v}$. Theorem 2.6.1 implies $a^s_n = 0$ for $n < m$ while a^s_m is a characteristic form of $T(M)$ and of V.

If $m = 2$ and if $v = 1$, then $\mathcal{R}_{2,2,2,1}$ is one dimensional and is spanned by the first Chern class $c_1(V) = ch(V) = \frac{i}{2\pi}\Omega$. Consequently $a^s_2 = cc_1$ in this case. We shall show later that this normalizing constant $c = 2$—i.e.,

LEMMA 3.1.4. *Let $m = 2$ and let V be a line bundle over M_2. Then:*

$$\text{signature}(M,V) = 2 \int_M c_1(L).$$

We postpone the proof of this lemma until later in this chapter.

With this normalizing constant established, we can compute a formula for signature(M,V) in general:

THEOREM 3.1.5. *Let L be the total L-polynomial and let $ch(V)$ be the Chern character. Then:*
(a) $a^s_n(x,V) = 0$ for $n < m$.
(b) $a^s_m(x,V) = \sum_{4s+2t=m} L_s(TM) \wedge 2^t ch_t(V)$ so that:

$$\text{signature}(M,V) = \sum_{4s+2t=m} \int_M L_s(TM) \wedge 2^t ch_t(V).$$

The factors of 2^t are perhaps a bit mysterious at this point. They arise from the normalizing constant of Lemma 3.1.4 and will be explained when we discuss the spin and Dolbeault complexes.

PROOF: We have already proved (a). We know $a^s_m(x,V)$ is a characteristic form which integrates to signature(M,V) so it suffices to verify the formula of (b). If V_1 and V_2 are bundles, we let $V = V_1 \oplus V_2$ with the direct sum connection. Since $\Delta^\pm_V = \Delta^\pm_{V_1} \oplus \Delta^\pm_{V_2}$ we conclude

$$\text{signature}(M,V_1 \oplus V_2) = \text{signature}(M,V_1) + \text{signature}(M,V_2).$$

Since the integrals are additive, we apply the uniqueness of Lemma 2.6.3 to conclude the local formulas must be additive. This also follows from Lemma 1.7.5 so that:

$$a_n^s(x, V_1 \oplus V_2) = a_n^s(x, V_1) + a_n^s(x, V_2).$$

Let $\{P_\rho\}_{|\rho|=s}$ be the basis for $\mathcal{P}_{m,4s,4s}$ and expand:

$$a_m^s(x, V) = \sum_{4|\rho|+2t=m} P_\rho \wedge Q_{m,t,v,\rho} \qquad \text{for } Q_{m,t,v,\rho} \in \mathcal{Q}_{m,2t,2t,v}$$

a characteristic form of V. Then the additivity under direct sum implies:

$$Q_{m,t,v,\rho}(V_1 \oplus V_2) = Q_{m,t,v_1,\rho}(V_1) + Q_{m,t,v_2,\rho}(V_2).$$

If $v = 1$, then $Q_{m,t,1,\rho}(V_1) = c \cdot c_1(V)^t$ since $\mathcal{Q}_{m,2t,2t,1}$ is one dimensional. If A is diagonal matrix, then the additivity implies:

$$Q_{m,t,v,\rho}(A) = Q_{m,t,v,\rho}(\lambda) = c \cdot \sum_j \lambda_j^t = c \cdot ch_t(A).$$

Since Q is determined by its values on diagonal matrices, we conclude:

$$Q_{m,t,v,\rho}(V) = c(m, t, \rho)ch_t(V)$$

where the normalizing constant does not depend on the dimension v. Therefore, we expand a_m^s in terms of $ch_t(V)$ to express:

$$a_m^s(x, V) = \sum_{4s+2t=m} P_{m,s} \wedge 2^t ch_t(V) \qquad \text{for } P_{m,s} \in \mathcal{P}_{m,4s,4s}.$$

We complete the proof of the theorem by identifying $P_{m,s} = L_s$; we have reduced the proof of the theorem to the case $v = 1$.

We proceed by induction on m; Lemma 3.1.4 establishes this theorem if $m = 2$. Suppose $m \equiv 0 \ (4)$. If we take V to be the trivial bundle, then if $4k = m$,

$$a_m^s(x, 1) = L_k = P_{m,k}$$

follows from Theorem 3.1.1. We may therefore assume $4s < m$ in computing $P_{m,s}$. Let $M = M_1 \times S^2$ and let $V = V_1 \otimes V_2$ where V_1 is a line bundle over M_1 and where V_2 is a line bundle over S^2 so $\int_{S^2} c_1(V_2) = 1$. (We constructed such a line bundle in section 2.1 using Clifford matrices). We take the product connection on $V_1 \otimes V_2$ and decompose:

$$\Lambda^+(V) = \Lambda^+(V_1) \otimes \Lambda^+(V_2) \ \oplus \ \Lambda^-(V_1) \otimes \Lambda^-(V_2)$$
$$\Lambda^-(V) = \Lambda^-(V_1) \otimes \Lambda^+(V_2) \ \oplus \ \Lambda^+(V_1) \otimes \Lambda^-(V_2).$$

A similar decomposition of the Laplacians yields:

$$\text{signature}(M,V) = \text{signature}(M_1,V_1)\,\text{signature}(M_2,V_2)$$
$$= 2\,\text{signature}(M_1,V_1)$$

by Lemma 3.1.4. Since the signatures are multiplicative, the local formulas are multiplicative by the uniqueness assertion of Lemma 2.6.3. This also follows by using Lemma 1.7.5 that

$$a_m^s(x,V) = \sum_{p+q=m} a_p^s(x_1,V_1)a_q^s(x_2,V_2)$$

and the fact $a_p = 0$ for $p < m_1$ and $a_q = 0$ for $q < m_2$. Thus we conclude:

$$a_m^s(x,V) = a_{m_1}^s(x_1,V_1)a_{m_2}^s(x_2,V_2)$$

where, of course, $m_2 = 2$ and $m_1 = m - 2$.

We use the identity:

$$ch(V_1 \otimes V_2) = ch(V_1)ch(V_2)$$

to conclude therefore:

$$\text{signature}(M_1,V_1)$$
$$= \frac{1}{2}\,\text{signature}(M,V)$$
$$= \frac{1}{2}\left\{ \sum_{4s+2t=m-2} \int_{M_1} P_{m,s} \wedge 2^t ch_t(V_1) \right\} \int_{M_2} 2ch_1(V_2)$$
$$= \sum_{4s+2t=m-2} \int_{M_1} P_{m,s} \wedge 2^t ch_t(V_1).$$

We apply the uniqueness assertion of Lemma 2.6.3 to conclude $P_{m,s} = P_{m-2,s}$ for $4s \le m - 2$. Since by induction, $P_{m-2,s} = L_s$ this completes the proof of the theorem.

We note that the formula is non-zero, so Lemma 2.6.3 implies that in any even dimension m, there always exist (M,V) so signature$(M,V) \neq 0$. In fact, much more is true. Given any orientable manifold M, we can find V over M so signature$(M,V) \neq 0$ if $\dim(M)$ is even. Since the proof of this assertion relies on the fact $ch\colon K(M) \otimes \mathbf{Q} \xrightarrow{\sim} H^{2*}(M;\mathbf{Q})$ we postpone a discussion of this fact until we discuss the index theorem in general.

3.2. Spinors and their Representations.

To define the signature complex, we needed to orient the manifold M. For any Riemannian manifold, by restricting to local orthonormal frames, we can always assume the transition functions of $T(M)$ are maps $g_{\alpha\beta}: U_\alpha \cap U_\beta \to O(m)$. If M is oriented, by restricting to local orthonormal frames, we can choose the transition functions of $T(M)$ to lie in $SO(m)$ and to reduce the structure group from $GL(m, R)$ to $SO(m)$. The signature complex results from the represntation $\Lambda^\pm$ of $SO(m)$; it cannot be defined in terms of $GL(m, R)$ or $O(m)$. By contrast, the de Rham complex results from the representation $\Lambda^{e,o}$ which is a representation of $GL(m, R)$ so the de Rham complex is defined for non-orientable manifolds.

To define the spin complex, which is in some sense a more fundamental elliptic complex than is either the de Rham or signature complex, we will have to lift the transition functions from $SO(m)$ to $SPIN(m)$. Just as every manifold is not orientable, in a similar fashion there is an obstruction to defining a spin structure.

If $m \geq 3$, then $\pi_1(SO(m)) = \mathbf{Z}_2$. Abstractly, we define $SPIN(m)$ to be the universal cover of $SO(m)$. (If $m = 2$, then $SO(2) = S^1$ and we let $SPIN(2) = S^1$ with the natural double cover $\mathbf{Z}_2 \to SPIN(2) \to SO(2)$ given by $\theta \mapsto 2\theta$). To discuss the representations of $SPIN(m)$, it is convenient to obtain a more concrete representation of $SPIN(m)$ in terms of Clifford algebras.

Let V be a real vector space of dimension $v \equiv 0$ (2). Let

$$\bigotimes V = \mathbf{R} \oplus V \oplus (V \otimes V) \oplus \cdots \oplus \bigotimes^k V \oplus \cdots$$

be the complete tensor algebra of V. We assume V is equipped with a symmetric positive definite bilinear form. Let I be the two-sided ideal of $\bigotimes V$ generated by $\{v \otimes v + |v|^2\}_{v \in V}$, then the real Clifford algebra $CLIF(V) = \bigotimes V$ mod I. (Of course, we will always construct the corresponding complex Clifford algebra by tensoring $CLIF(V)$ with the complex numbers). There is a natural transpose defined on $\bigotimes V$ by:

$$(v_1 \otimes \cdots \otimes v_k)^t = v_k \otimes \cdots \otimes v_1.$$

Since this preserves the ideal I, it extends to $CLIF(V)$. If $\{e_1, \ldots, e_v\}$ are the orthonormal basis for V, then $e_i * e_j + e_j * e_i = -2\delta_{ij}$ and

$$(e_{i_1} * \cdots * e_{i_p})^t = (-1)^{p(p-1)/2} e_{i_1} * \cdots * e_{i_p}.$$

If V is oriented, we let $\{e_i\}$ be an oriented orthonormal basis and define:

$$\tau = (\sqrt{-1})^{v/2} e_1 * \cdots * e_v.$$

We already computed that:

$$\tau^2 = (-1)^{v/2} e_1 * \cdots * e_v * e_1 * \ldots e_v$$
$$= (-1)^{v(v-1)/2} e_1 * \cdots * e_v * e_1 * \cdots * e_v$$
$$= e_1 * \cdots * e_v * e_v * \cdots * e_1 = (-1)^v = 1.$$

We let $\mathrm{SPIN}(V)$ be the set of all $w \in \mathrm{CLIF}(V)$ such that w can be decomposed as a formal product $w = v_1 * \cdots * v_{2j}$ for some j where the $v_i \in V$ are elements of length 1. It is clear that w^t is such an element and that $ww^t = 1$ so that $\mathrm{SPIN}(V)$ forms a group under Clifford multiplication. We define

$$\rho(w)x = wxw^t \qquad \text{for } x \in \mathrm{CLIF}, w \in \mathrm{CLIF}(V).$$

For example, if $v_1 = e_1$ is the first element of our orthonormal basis, then:

$$e_1 e_i e_1 = \begin{cases} -e_1 & i = 1 \\ e_i & i \neq 1. \end{cases}$$

The natural inclusion of V in $\bigotimes V$ induces an inclusion of V in $\mathrm{CLIF}(V)$. $\rho(e_1)$ preserves V and is reflection in the hyperplane defined by e_1. If $w \in \mathrm{SPIN}(V)$, then $\rho(w): V \to V$ is a product of an even number of hyperplane reflections. It is therefore in $\mathrm{SO}(V)$ so

$$\rho: \mathrm{SPIN}(V) \to \mathrm{SO}(V)$$

is a group homomorphism. Since any orthogonal transformation of determinant one can be decomposed as a product of an even number of hyperplane reflections, ρ is subjective.

If $w \in \mathrm{CLIF}(V)$ is such that

$$wvw^t = v \text{ all } v \in V \text{ and } ww^t = 1$$

then $wv = vw$ for all $v \in V$ so w must be in the center of $\mathrm{CLIF}(V)$.

LEMMA 3.2.1. *If* $\dim V = v$ *is even, then the center of* $\mathrm{CLIF}(V)$ *is one dimensional and consists of the scalars.*

PROOF: Let $\{e_i\}$ be an orthonormal basis for V and let $\{e_I\}$ be the corresponding orthonormal basis for $\mathrm{CLIF}(V)$. We compute:

$$e_i * e_{i_1} * \cdots * e_{i_p} * e_i = e_{i_1} * \cdots * e_{i_p} \qquad \text{for } p \geq 1,$$

if (a) p is even and i is one of the i_j or (b) p is odd and i is not one of the i_j. Thus given I we can choose i so $e_i * e_I = -e_I * e_i$. Thus

$e_i * (\sum_I c_I e_I) = \sum_I c_I e_I * e_i$ for all i implies $c_I = 0$ for $|I| > 0$ which completes the proof.

(We note that this lemma fails if $\dim V$ is odd since the center in that case consists of the elements $a + b e_1 * \cdots * e_v$ and the center is two dimensional).

If $\rho(w) = 1$ and $w \in \text{SPIN}(V)$, this implies w is scalar so $w = \pm 1$. By considering the arc in $\text{SPIN}(V)$ given by $w(\theta) = ((\cos\theta)e_1 + (\sin\theta)e_2) * (-(\cos\theta)e_1 + (\sin\theta)e_2)$, we note $w(0) = 1$ and $w(\frac{\pi}{2}) = -1$ so $\text{SPIN}(V)$ is connected. This proves we have an exact sequence of groups in the form:

$$\mathbf{Z}_2 \to \text{SPIN}(V) \to \text{SO}(V)$$

and shows that $\text{SPIN}(V)$ is the universal cover of $\text{SO}(V)$ for $v > 2$.

We note that is it possible to define $\text{SPIN}(V)$ using Clifford algebras even if v is odd. Since the center of $\text{CLIF}(V)$ is two dimensional for v odd, more care must be used with the relevant signs which arise. As we shall not need that case, we refer to Atiyah-Bott-Shapiro (see bibliography) for further details.

$\text{SPIN}(V)$ acts on $\text{CLIF}(V)$ from the left. This is an orthogonal action.

LEMMA 3.2.2. *Let* $\dim V = 2v_1$. *We complexify* $\text{CLIF}(V)$. *As a left* $\text{SPIN}(V)$ *module, this is not irreducible. We can decompose* $\text{CLIF}(V)$ *as a direct sum of left* $\text{SPIN}(V)$ *modules in the form*

$$\text{CLIF}(V) = 2^{v_1} \Delta.$$

This representation is called the spin representation. It is not irreducible but further decomposes in the form

$$\Delta = \Delta^+ \oplus \Delta^-.$$

If we orient V *and let* τ *be the orientation form discussed earlier, then left multiplication by* τ *is* ± 1 *on* $\Delta^\pm$ *so these are inequivalent representations. They are irreducible and act on a representation space of dimension* $2^{v_1 - 1}$ *and are called the half-spin representations.*

PROOF: Fix an oriented orthonormal basis $\{e_i\}$ for V and define:

$$\alpha_1 = \sqrt{-1}\, e_1 e_2, \quad \alpha_2 = \sqrt{-1}\, e_3 e_4, \quad \ldots, \quad \alpha_{v_1} = \sqrt{-1}\, e_{v-1} e_v$$

as elements of $\text{CLIF}(V)$. It is immediate that $\tau = \alpha_1 \ldots \alpha_{v_1}$ and:

$$\alpha_i^2 = 1 \quad \text{and} \quad \alpha_i \alpha_j = \alpha_j \alpha_i.$$

We let the $\{\alpha_i\}$ act on $\text{CLIF}(V)$ from the right and decompose $\text{CLIF}(V)$ into the 2^{v_1} simultaneous eigenspaces of this action. Since right and left multiplication commute, each eigenspace is invariant as a left $\text{SPIN}(V)$

module. Each eigenspace corresponds to one of the 2^{v_1} possible sequences of $+$ and $-$ signs. Let ε be such a string and let Δ_ε be the corresponding representation space.

Since $e_1 \alpha_1 = -\alpha_1 e_1$ and $e_1 \alpha_i = \alpha_i e_1$ for $i > 1$, multiplication on the right by e_1 transforms $\Delta_\varepsilon \to \Delta_{\varepsilon'}$, where $\varepsilon(1) = -\varepsilon'(1)$ and $\varepsilon(i) = \varepsilon'(i)$ for $i > 1$. Since this map commutes with left multiplication by SPIN(V), we see these two representations are isomorphic. A similar argument on the other indices shows all the representation spaces Δ_ε are equivalent so we may decompose CLIF$(V) = 2^{v_1} \Delta$ as a direct sum of 2^{v_1} equivalent representation spaces Δ. Since $\dim(CLIF(V)) = 2^v = 4^{v_1}$, the representation space Δ has dimension 2^{v_1}. We decompose Δ into ± 1 eigenspaces under the action of τ to define $\Delta^\pm$. Since $e_1 * \cdots * e_v$ is in the center of SPIN(V), these spaces are invariant under the left action of SPIN(V). We note that elements of V anti-commute with τ so Clifford multiplication on the left defines a map:

$$cl: V \otimes \Delta^\pm \to \Delta^\mp$$

so both of the half-spin representations have the same dimension. They are clearly inequivalent. We leave the proof that they are irreducible to the reader as we shall not need this fact.

We shall use the notation $\Delta^\pm$ to denote both the representations and the corresponding representation spaces. Form the construction we gave, it is clear the CLIF(V) acts on the left to preserve the space Δ so Δ is a representation space for the left action by the whole Clifford algebra. Since $v\tau = -\tau v$ for $v \in V$, Clifford multiplication on the left by an element of V interchanges Δ^+ and Δ^-.

LEMMA 3.2.3. *There is a natural map given by Clifford multiplication of $V \otimes \Delta^\pm \to \Delta^\mp$. This map induces a map on the representations involved: $\rho \otimes \Delta^\pm \mapsto \Delta^\mp$. If this map is denoted by $v * w$ then $v * v * w = -|v|^2 w$.*

PROOF: We already checked the map on the spaces. We check:

$$wvw^t \otimes wx \mapsto wvw^t wx = wvx$$

to see that the map preserves the relevant representation. We emphasize that $\Delta^\pm$ are complex representations since we must complexify CLIF(V) to define these representation spaces ($ww^t = 1$ for $w \in$ SPIN).

There is a natural map $\rho: $ SPIN$(V) \to$ SO(V). There are natural representations Λ, $\Lambda^\pm$, $\Lambda^{e,o}$ of SO(V) on the subspaces of $\Lambda(V) =$ CLIF(V). We use ρ to extend these representations of SPIN(V) as well. They are related to the half-spin representations as follows:

LEMMA 3.2.4.
(a) $\Lambda = \Delta \otimes \Delta$.
(b) $(\Lambda^+ - \Lambda^-) = (\Delta^+ - \Delta^-) \otimes (\Delta^+ + \Delta^-)$.

(c) $(\Lambda^e - \Lambda^o) = (\Delta^+ - \Delta^-) \otimes (\Delta^+ - \Delta^-)(-1)^{v/2}$.

These identities are to be understood formally (in the sense of K-theory). They are shorthand for the identities:

$$\Lambda^+ = (\Delta^+ \otimes \Delta^+) \oplus (\Delta^+ \otimes \Delta^-) \quad \text{and} \quad \Delta^- = (\Delta^- \otimes \Delta^+) \oplus (\Delta^- \otimes \Delta^-)$$

and so forth.

PROOF: If we let SPIN(V) act on CLIF(V) by right multiplication by $w^t = w^{-1}$, then this representation is equivalent to left multiplication so we get the same decomposition of CLIF$(V) = 2^{v_1} \Delta$ as right spin modules. Since $\rho(w)x = wxw^t$, it is clear that $\Lambda = \Delta \otimes \Delta$ where one factor is viewed as a left and the other as a right spin module. Since $\Lambda^\pm$ is the decomposition of SPIN(V) under the action of τ from the left, (b) is immediate. (c) follows similarly once the appropriate signs are taken into consideration; $w \in$ CLIF$(V)^{\text{even}}$ if and only if

$$\tau w \tau^t = (-1)^{v/2} w$$

which proves (c).

It is helpful to illustrate this for the case $\dim V = 2$. Let $\{e_1, e_2\}$ be an oriented orthonormal basis for V. We compute that:

$$\big((\cos \alpha)e_1 + (\sin \alpha)e_2\big)\big((\cos \beta)e_1 + (\sin \beta)e_2\big)$$
$$= \big(-\cos \alpha \cos \beta - \sin \alpha \sin \beta\big)$$
$$+ \big(\cos \alpha \sin \beta - \sin \alpha \cos \beta\big)e_1 e_2$$

so elements of the form $\cos \gamma + (\sin \gamma)e_1 e_2$ belong to SPIN(V). We compute:

$$\big(\cos \alpha + (\sin \alpha)e_1 e_2\big)\big(\cos \beta + (\sin \beta)e_1 e_2\big) = \cos(\alpha + \beta) + \sin(\alpha + \beta)e_1 e_2$$

so spin(V) is the set of all elements of this form and is naturally isomorphic to the circle $S^1 = [0, 2\pi]$ with the endpoints identified. We compute that

$$\rho(w)(e_1) = \big(\cos \theta + (\sin \theta)e_1 e_2\big)e_1\big(\cos \theta + (\sin \theta)e_2 e_1\big)$$
$$= (\cos^2 \theta - \sin^2 \theta)e_1 + 2(\cos \theta \sin \theta)e_2$$
$$= \cos(2\theta)e_1 + \sin(2\theta)e_2$$
$$\rho(w)(e_2) = \cos(2\theta)e_2 - \sin(2\theta)e_1$$

so the map $\rho: S^1 \to S^1$ is the double cover $\theta \mapsto 2\theta$.

We construct the one dimensional subspaces V_i generated by the elements:

$$v_1 = 1 + \tau, \quad v_2 = 1 - \tau, \quad v_3 = (1 + \tau)e_1, \quad v_4 = (1 - \tau)e_1,$$
$$\tau v_1 = v_1, \quad \tau v_2 = -v_2, \quad \tau v_3 = v_3, \quad \tau v_4 = -v_4,$$
$$v_1 \tau = v_1, \quad v_2 \tau = -v_2, \quad v_3 \tau = -v_3, \quad v_4 \tau = v_4.$$

When we decompose $\mathrm{CLIF}(V)$ under the left action of $\mathrm{SPIN}(V)$,

$$V_1 \simeq V_3 \simeq \Delta^+ \qquad \text{and} \qquad V_2 \simeq V_4 \simeq \Delta^-.$$

When we decompose $\mathrm{CLIF}(V)$ under the right action of $\mathrm{SPIN}(V)$, and replace τ by $-\tau = \tau^t$ acting on the right:

$$V_2 \simeq V_3 \simeq \Delta^+ \qquad \text{and} \qquad V_1 \simeq V_4 \simeq \Delta^-.$$

From this it follows that as $\mathrm{SO}(V)$ modules we have:

$$V_1 = \Delta^+ \otimes \Delta^-, \quad V_2 = \Delta^- \otimes \Delta^+, \quad V_3 = \Delta^+ \otimes \Delta^+, \quad V_4 = \Delta^- \otimes \Delta^-$$

from which it is immediate that:

$$\Lambda^e = V_1 \oplus V_2 = \Delta^+ \otimes \Delta^- \ \oplus \ \Delta^- \otimes \Delta^+$$
$$\Lambda^o = V_3 \oplus V_4 = \Delta^+ \otimes \Delta^+ \ \oplus \ \Delta^- \otimes \Delta^-$$
$$\Lambda^+ = V_1 \oplus V_3 = \Delta^+ \otimes \Delta^- \ \oplus \ \Delta^+ \otimes \Delta^+$$
$$\Lambda^- = V_2 \oplus V_4 = \Delta^- \otimes \Delta^+ \ \oplus \ \Delta^- \otimes \Delta^-.$$

If V is a one-dimensional complex vector space, we let V_r be the underlying real vector space. This defines a natural inclusion of $\mathrm{U}(1) \to \mathrm{SO}(2)$. If we let $J : V \to V$ be complex multiplication by $\sqrt{-1}$ then $J(e_1) = e_2$ and $J(e_2) = -e_1$ so J is equivalent to Clifford multiplication by $e_1 e_2$ on the left. We define a complex linear map from V to $\Delta^- \otimes \Delta^-$ by:

$$T(v) = v - iJ(v) \in V_4$$

by computing $T(e_1) = (e_1 - ie_2) = (1 - ie_1 e_2)(e_1)$ and $T(e_2) = e_2 + ie_1 = (i)(e_1 - ie_2)$.

LEMMA 3.2.5. *Let $\mathrm{U}(1)$ be identified with $\mathrm{SO}(2)$ in the usual manner. If V is the underlying complex 1-dimensional space corresponding to V_r then $V \simeq \Delta^- \otimes \Delta^-$ and $V^* \simeq \Delta^+ \otimes \Delta^+$ as representation spaces of $\mathrm{SPIN}(2)$.*

PROOF: We have already verified the first assertion. The second follows from the fact that V^* was made into a complex space using the map $-J$ instead of J on V_r. This takes us into V_3 instead of into V_4. It is also clear that $\Delta^- = (\Delta^+)^*$ if $\dim V = 2$. Of course, all these statements are to be interpreted as statements about representations since they are trivial as statements about vector spaces (since any vector spaces of the same dimension are isomorphic).

3.3. Spin Structures on Vector Bundles.

We wish to apply the constructions of 3.2 to vector bundles over manifolds. We first review some facts regarding principal bundles and Stieffel-Whitney classes which we shall need.

Principal bundles are an extremely convenient bookkeeping device. If G is a Lie group, a principal G-bundle is a fiber space $\pi: P_G \to M$ with fiber G such that the transition functions are elements of G acting on G by left multiplication in the group. Since left and right multiplication commute, we can define a right action of G on P_G which is fiber preserving. For example, let SO($2k$) and SPIN($2k$) be the groups defined by $\mathbf{R}^{2k}$ with the cannonical inner product. Let V be an oriented Riemannian vector bundle of dimension $2k$ over M and let P_{SO} be the bundle of oriented frames of V. P_{SO} is an SO($2k$) bundle and the natural action of SO($2k$) from the right which sends an oriented orthonormal frame $s = (s_1, \ldots, s_{2k})$ to $s \cdot g = (s_1', \ldots, s_{2k}')$ is defined by:

$$s_i' = s_1 g_{1,i} + \cdots + s_{2k} g_{2k,i}.$$

The fiber of P_{SO} is SO(V_x) where V_x is the fiber of V over the point x. This isomorphism is not natural but depends upon the choice of a basis.

It is possible to define the theory of characteristic classes using principal bundles rather than vector bundles. In this approach, a connection is a splitting of $T(P_G)$ into vertical and horizontal subspaces in an equivariant fashion. The curvature becomes a Lie algebra valued endomorphism of $T(P_G)$ which is equivariant under the right action of the group. We refer to Euguchi, Gilkey, Hanson for further details.

Let $\{U_\alpha\}$ be a cover of M so V is trivial over U_α and let $\vec{s}_\alpha$ be local oriented orthonormal frames over U_α. On the overlap, we express $\vec{s}_\alpha = g_{\alpha\beta} \vec{s}_\beta$ where $g_{\alpha\beta}: U_\alpha \cap U_\beta \to$ SO($2k$). These satisfy the cocycle condition:

$$g_{\alpha\beta} g_{\beta\gamma} g_{\gamma\alpha} = I \qquad \text{and} \qquad g_{\alpha\alpha} = I.$$

The principal bundle P_{SO} of oriented orthonormal frames has transition functions $g_{\alpha\beta}$ acting on SO($2k$) from the left.

A spin structure on V is a lifting of the transition functions to SPIN($2k$) preserving the cocyle condition. If the lifting is denoted by g', then we assume:

$$\rho(g_{\alpha\beta}') = g_{\alpha\beta}, \qquad g_{\alpha\beta}' g_{\beta\gamma}' g_{\gamma\alpha}' = I, \qquad \text{and} \qquad g_{\alpha\alpha}' = I.$$

This is equivalent to constructing a principal SPIN($2k$) bundle P_{SPIN} together with a double covering map $\rho: P_{\mathrm{SPIN}} \to P_{\mathrm{SO}}$ which preserves the group action—i.e.,

$$\rho(x \cdot g') = \rho(x) \cdot \rho(g').$$

The transition functions of P_{SPIN} are just the $g'_{\alpha\beta}$ acting on the left.

Attempting to find a spin structure on V is equivalent to taking a square root in a certain sense which we will make clear later. There is an obstruction to defining a spin structure which is similar to that to defining an orientation on V. These obstructions are $\mathbf{Z}_2$ characteristic classes called the Stieffel-Whitney classes and are most easily defined in terms of Čech cohomology. To understand these obstructions better, we review the construction briefly.

We fix a Riemannian structure on $T(M)$ to define a notion of distance. Geodesics on M are curves which locally minimize distance. A set U is said to be geodesically convex if (a) given $x, y \in U$ there exists a unique geodesic in M joining x to y with $d(x,y) = \text{length}(\gamma)$ and (b) if γ is any such geodesic then γ is actually contained in U. It is immediate that the intersection of geodesically convex sets is again geodesically convex and that every geodesically convex set is contractible.

It is a basic theorem of Riemannian geometry that there exist open covers of M by geodesically convex sets. A cover $\{U_\alpha\}$ is said to be *simple* if the intersection of any number of sets of the cover is either empty or is contractible. A cover of M by open geodesically convex sets is a simple cover.

We fix such a simple cover hence forth. Since U_α is contractible, any vector bundle over M is trivial over U_α. Let $\mathbf{Z}_2$ be the multiplicative group $\{\pm 1\}$. A Čech j-cochain is a function $f(\alpha_0, \ldots, \alpha_j) \in \mathbf{Z}_2$ defined for $j + 1$-tuples of indices where $U_{\alpha_0} \cap \cdots \cap U_{\alpha_j} \neq \emptyset$ which is totally symmetric—i.e.,

$$f(\alpha_{\sigma(0)}, \ldots, \alpha_{\sigma(j)}) = f(\alpha_0, \ldots, \alpha_j)$$

for any permutation σ. If $C^j(M; \mathbf{Z}_2)$ denotes the multiplicative group of all such functions, the coboundary $\delta: C^j(M, \mathbf{Z}_2) \to C^{j+1}(M; \mathbf{Z}_2)$ is defined by:

$$(\delta f)(\alpha_0, \ldots, \alpha_{j+1}) = \prod_{i=0}^{j+1} f(\alpha_0, \ldots, \widehat{\alpha_i}, \ldots, \alpha_{j+1}).$$

The multiplicative identity of $C^j(M; \mathbf{Z}_2)$ is the function 1 and it is an easy combinatorial exercise that $\delta^2 f = 1$. For example, if $f_0 \in C^0(M; \mathbf{Z}_2)$ and if $f_1 \in C^1(M; \mathbf{Z}_2)$, then:

$$\delta(f_0)(\alpha_0, \alpha_1) = f(\alpha_1)f(\alpha_0)$$
$$\delta(f_1)(\alpha_0, \alpha_1, \alpha_2) = f(\alpha_1, \alpha_2)f(\alpha_0, \alpha_1)f(\alpha_o, \alpha_2).$$

$\mathbf{Z}_2$ is a particularly simple coefficient group to work with since every element is its own inverse; in defining the Čech cohomology with coefficients in other abelian groups, more care must be taken with the signs which arise.

Let $H^j(M; \mathbf{Z}_2) = \mathrm{N}(\delta_j)/\mathrm{R}(\delta_{j-1})$ be the cohomology group; it is an easy exercise to show these groups are independent of the particular simple cover chosen. There is a ring structure on $H^*(M; \mathbf{Z}_2)$, but we shall only use the "additive" group structure (which we shall write multiplicatively).

Let V be a real vector bundle, not necessarily orientable. Since the U_α are contractible, V is trivial over the U and we can find a local orthonormal frame $\vec{s}_\alpha$ for V over U_α. We let $\vec{s}_\alpha = g_{\alpha\beta}\vec{s}_\beta$ and define the 1-cochain:

$$f(\alpha\beta) = \det(g_{\alpha\beta}) = \pm 1.$$

This is well defined since $U_\alpha \cap U_\beta$ is contractible and hence connected. Since $f(\alpha, \beta) = f(\beta, \alpha)$, this defines an element of $C^1(M; \mathbf{Z}_2)$. Since the $\{g_{\alpha\beta}\}$ satisfy the cocycle condition, we compute:

$$\delta f(\alpha, \beta, \gamma) = \det(g_{\alpha\beta}g_{\beta\gamma}g_{\gamma\alpha}) = \det(I) = 1$$

so $\delta(f) = 1$ and f defines an element of $H^1(M; \mathbf{Z}_2)$. If we replace $\vec{s}_\alpha$ by $\vec{s}'_\alpha = h_\alpha \vec{s}_\alpha$ the new transition functions become $g'_{\alpha\beta} = h_\alpha g_{\alpha\beta} h_\beta^{-1}$ so if $f_0(\alpha) = \det(h_\alpha)$,

$$f'(\alpha, \beta) = \det(h_\alpha g_{\alpha\beta} h_\beta^{-1}) = \det(h_\alpha)f(\alpha, \beta)\det(h_\beta) = \delta(f_0)f$$

and f changes by a coboundary. This proves the element in cohomology defined by f is independent of the particular frame chosen and we shall denote this element by $w_1(V) \in H^1(M; \mathbf{Z}_2)$.

If V is orientable, we can choose frames so $\det(g_{\alpha\beta}) = 1$ and thus $w_1(V) = 1$ represents the trivial element in cohomology. Conversely, if $w_1(V)$ is trivial, then $f = \delta f_0$. If we choose h_α so $\det(h_\alpha) = f_0(\alpha)$, then the new frames $\vec{s}'_\alpha = h_\alpha \vec{s}_\alpha$ will have transition functions with $\det(g'_{\alpha\beta}) = 1$ and define an orientation of V. Thus V is orientable if and only if $w_1(V)$ is trivial and $w_1(V)$, which is called the first Stieffel-Whitney class, measures the obstruction to orientability.

If V is orientable, we restrict henceforth to oriented frames. Let $\dim V = 2k$ be even and let $g_{\alpha\beta} \in \mathrm{SO}(2k)$ be the transition functions. We choose any lifting $\tilde{g}_{\alpha\beta}$ to $\mathrm{SPIN}(2k)$ so that:

$$\rho(\tilde{g}_{\alpha\beta}) = g_{\alpha\beta} \qquad \text{and} \qquad \tilde{g}_{\alpha\beta}\tilde{g}_{\beta\alpha} = I;$$

since the U_α are contractible such lifts always exist. We have $g_{\alpha\beta}g_{\beta\gamma}g_{\gamma\alpha} = I$ so $\rho(\tilde{g}_{\alpha\beta}\tilde{g}_{\beta\gamma}\tilde{g}_{\gamma\alpha}) = I$ and hence $\tilde{g}_{\alpha\beta}\tilde{g}_{\beta\gamma}\tilde{g}_{\gamma\alpha} = \pm I = f(\alpha, \beta, \gamma)I$ where $f(\alpha, \beta, \gamma) \in \mathbf{Z}_2$. V admits a spin structure if and only if we can choose the lifting so $f(\alpha, \beta, \gamma) = 1$. It is an easy combinatorial exercise to show that f is symmetric and that $\delta f = 0$. Furthermore, if we change the choice of the $\vec{s}_\alpha$ or change the choice of lifts, then f changes by a coboundary. This implies f defines an element $w_2(V) \in H^2(M; \mathbf{Z}_2)$ independent of the choices made. w_2 is called the second Stieffel-Whitney class and is trivial if and only if V admits a spin structure.

We suppose $w_1(V) = w_2(V) = 1$ are trivial so V is orientable and admits a spin structure. Just as there are two orientations which can be chosen on V, there can be several possible inequivalent spin structures (i.e., several non-isomorphic principal bundles P_{SPIN}). It is not difficult to see that inequivalent spin structures are parametrized by representations of the fundamental group $\pi_1(M) \to \mathbf{Z}_2$ just as inequivalent orientations are parametrized by maps of the components of M into $\mathbf{Z}_2$.

To illustrate the existence and non-existence of spin structures on bundles, we take $M = S^2$. $S^2 = \mathbf{CP}_1$ is the Riemann sphere. There is a natural projection from $\mathbf{C}^2 - 0$ to S^2 given by sending $(x, w) \mapsto z/w$; S^2 is obtained by identifying $(z, w) = (\lambda z, \lambda w)$ for $(z, w) \neq (0, 0)$ and $\lambda \neq 0$. There are two natural charts for S^2:

$$U_1 = \{z : |z| \le 1\} \quad \text{and} \quad U_2 = \{z : |z| \ge 1\} \cup \{\infty\}$$

where $w = 1/z$ gives coordinates on U_2.

$\mathbf{CP}_1$ is the set of lines in $\mathbf{C}^2$; we let L be the natural line bundle over S^2; this is also refered to as the tautological line bundle. We have natural sections to L over U_i defined by:

$$s_1 = (z, 1) \text{ over } U_1 \quad \text{and} \quad s_2 = (1, w) \text{ over } U_2.$$

these are related by the transition function:

$$s_1 = z s_2$$

so on $U_1 \cap U_2$ we have $g_{12} = e^{i\theta}$. The double cover $\text{SPIN}(2) \to \text{SO}(2)$ is defined by $\theta \mapsto 2\theta$ so this bundle does not have a spin structure since this transition function cannot be lifted to $\text{SPIN}(2)$.

This cover is not a simple cover of S^2 so we construct the cover given in the following

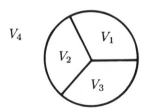

(where we should "fatten up" the picture to have an open cover). If we have sections $\vec{s}_i$ to L over V_i, then we can choose the transition functions so $\vec{s}_j = e^{i\theta}\vec{s}_4$ for $1 \leq j \leq 3$ and $\vec{s}_1 = \vec{s}_2 = \vec{s}_3$. We let

$$
\begin{aligned}
0 \leq \theta \leq 2\pi/3 \qquad &\text{parametrize} \quad V_1 \cap V_4 \\
2\pi/3 \leq \theta \leq 4\pi/3 \qquad &\text{parametrize} \quad V_2 \cap V_4 \\
4\pi/3 \leq \theta \leq 2\pi \qquad &\text{parametrize} \quad V_3 \cap V_4
\end{aligned}
$$

and define:

$$
\tilde{g}_{14} = e^{i\theta/2}, \quad \tilde{g}_{24} = e^{i\theta/2}, \quad \tilde{g}_{34} = e^{i\theta/2}
$$

and all the $\tilde{g}_{jk} = 1$ for $1 \leq j, k \leq 3$ then we compute:

$$
\tilde{g}_{13}\tilde{g}_{34}\tilde{g}_{41} = e^0 e^{i\pi} = -1
$$

so the cochain defining w_2 satisfies:

$$
f(1,3,4) = -1 \quad \text{with} \quad f(i,j,k) = 1 \quad \text{if } i \text{ or } j \text{ or } k=2.
$$

We compute this is non-trivial in cohomology as follows: suppose $f = \delta f_1$. Then:

$$
\begin{aligned}
1 = f(1,2,3) = f_1(1,2)f_1(1,3)f_1(2,3) \\
1 = f(1,2,4) = f_1(1,2)f_1(1,4)f_1(2,4) \\
1 = f(2,3,4) = f_1(2,3)f_1(2,4)f_1(3,4)
\end{aligned}
$$

and multiplying together:

$$
\begin{aligned}
1 &= f_1(1,2)^2 f_1(2,3)^2 f_1(2,4)^2 f_1(1,4)f_1(1,3)f_1(3,4) \\
&= f_1(1,4)f_1(1,3)f_1(3,4) = f(1,3,4) = -1
\end{aligned}
$$

which is a contradiction. Thus we have computed combinatorially that $w_2 \neq 1$ is non-trivial.

Next we let $V = T(M)$ be the real tangent bundle. We identify $U(1)$ with $SO(2)$ to identify $T(M)$ with $T_C(M)$. Since $w = 1/z$ we have:

$$
\frac{d}{dz} = \frac{dw}{dz}\frac{d}{dw} = -z^{-2}\frac{d}{dw}
$$

so the transition function on the overlap is $-e^{-2i\theta}$. The minus sign can be eliminated by using $-\dfrac{d}{dw}$ instead of $\dfrac{d}{dw}$ as a section over U_2 to make the transition function be $e^{-2i\theta}$. Since the double cover of $SPIN(2) \to SO(2)$ is given by $\theta \mapsto 2\theta$, this transition function lifts and $T(M)$ has a spin structure. Since S^2 is simply connected, the spin structure is unique. This also proves $T_C(M) = L^* \otimes L^*$ since the two bundles have the same transition functions.

There is a natural inclusion of $\mathrm{SPIN}(V)$ and $\mathrm{SPIN}(W)$ into subgroups of $\mathrm{SPIN}(V \oplus W)$ which commute. This induces a map from $\mathrm{SPIN}(V) \times \mathrm{SPIN}(W) \to \mathrm{SPIN}(V \oplus W)$. Using this map, it is easy to compute $w_2(V \oplus W) = w_2(V)w_2(W)$. To define $w_2(V)$, we needed to choose a fixed orientation of V. It is not difficult to show that $w_2(V)$ is independent of the orientation chosen and of the fiber metric on V. (Our definition does depend upon the fact that V is orientable. Although it is possible to define $w_2(V)$ more generally, the identity $w_2(V \oplus W) = w_2(V)w_2(W)$ fails if V and W are not orientable in general). We summarize the properties of w_2 we have derived:

LEMMA 3.3.1. Let V be a real oriented vector bundle and let $w_2(V) \in H^2(M; \mathbf{Z}_2)$ be the second Stieffel-Whitney class. Then:
(a) $w_2(V) = 1$ represents the trivial cohomology class if and only if V admits a spin structure.
(b) $w_2(V \oplus W) = w_2(V)w_2(W)$.
(c) If L is the tautological bundle over S^2 then $w_2(L)$ is non-trivial.

We emphasize that (b) is written in multiplicative notation since we have chosen the multiplicative version of $\mathbf{Z}_2$. w_2 is also functorial under pull-backs, but as we have not defined the Čech cohomology as a functor, we shall not discuss this property. We also note that the Stieffel-Whitney classes can be defined in general; $w(V) = 1 + w_1(V) + \cdots \in H^*(M; \mathbf{Z}_2)$ is defined for any real vector bundle and has many of the same properties that the Chern class has for complex bundles.

We can use Lemma 3.3.1 to obtain some other results on the existence of spin structures:

LEMMA 3.3.2.
(a) If $M = S^m$ then any bundle over S^m admits a spin structure for $m \neq 2$.
(b) If $M = \mathbf{CP}_k$ and if L is the tautological bundle over M, then L does not admit a spin structure.
(c) If $M = \mathbf{CP}_k$ and if $V = T(M)$ is the tangent space, then V admits a spin structure if and only if k is odd.
(d) If $M = \mathbf{QP}_k$ is quaternionic projective space, then any bundle over M admits a spin structure.

PROOF: $H^2(M; \mathbf{Z}_2) = \{1\}$ in (a) and (d) so w_2 must represent the trivial element. To prove (b), we suppose L admits a spin structure. S^2 is embedded in $\mathbf{CP}_k$ for $k \geq 2$ and the restriction of L to S^2 is the tautological bundle over S^2. This would imply L admits a spin structure over S^2 which is false. Finally, we use the representation:

$$T_c(\mathbf{CP}_j) \oplus 1 = L^* \oplus \cdots \oplus L^*$$

so that

$$T(\mathbf{CP}_j) \oplus 1^2 = \underbrace{(L_r^*) \oplus \cdots \oplus (L_r^*)}_{j+1 \text{ times}}$$

where L_r^* denotes the real vector bundle obtained from L^* by forgetting the complex structure. Then $w_2(L_r^*) = w_2(L_r)$ since these bundles are isomorphic as real bundles. Furthermore:

$$w_2(T(\mathbf{CP}_j)) = w_2(L_r)^{j+1}$$

and this is the trivial class if $j + 1$ is even (i.e., j is odd) while it is $w_2(L_r)$ and non-trivial if $j + 1$ is odd (i.e., j is even).

Let M be a manifold and let V be a real vector bundle over M. If V admits a spin structure, we shall say V is spin and not worry for the moment about the lack of uniqueness of the spin structure. This will not become important until we discuss Lefschetz fixed point formulas.

If V is spin, we define the bundles $\Delta^{\pm}(V)$ to have transition functions $\Delta^{\pm}(\tilde{g}_{\alpha\beta})$ where we apply the representation $\Delta^{\pm}$ to the lifted transition functions. Alternatively, if P_{SPIN} is the principal spin bundle defining the spin structure, we define:

$$\Delta^{\pm}(V) = P_{\text{SPIN}} \otimes_{\Delta^{\pm}} \Delta^{\pm}$$

where the tensor product across a representation is defined to be $P_{\text{SPIN}} \times \Delta^{\pm}$ module the identification $(p \cdot g) \times z = p \times (\Delta^{\pm}(g)z)$ for $p \in P_{\text{SPIN}}$, $g \in P_{\text{SPIN}}(2k)$, $z \in \Delta^{\pm}$. (The slight confusion of notation is caused by our convention of using the same symbol for the representation and the representation space.)

Let $b_{\pm}$ be a fixed unitary frame for $\Delta^{\pm}$. If $\vec{s}$ is a local oriented orthonormal frame for V, we let $\tilde{s}_i$ be the two lifts of $\vec{s} \in P_{\text{SO}}$ to P_{SPIN}. $\tilde{s}_1 = -\tilde{s}_2$ but there is no natural way to distinguish these lifts, although the pair is cannonically defined. If ∇ is a Riemannian connection on V, let $\nabla_{\vec{s}} = \omega\vec{s}$ be the connection 1-form. ω is a skew symmetric matrix of 1-forms and is a 1-form valued element of the Lie algebra of $SO(2k)$. Since the Lie algebra of $SO(2k)$ and $SPIN(2k)$ coincide, we can also regard ω as an element which is 1-form valued of the Lie algebra of $SPIN(2k)$ and let $\Delta^{\pm}(\omega)$ act on $\Delta^{\pm}(V)$. We define bases $\tilde{s}_i \otimes b_{\pm}$ for $\Delta^{\pm}(V)$ and define:

$$\nabla(\tilde{s}_i \otimes b_{\pm}) = \tilde{s}_i \otimes \Delta^{\pm}(\omega)b_{\pm}$$

to define a natural connection on $\Delta^{\pm}(V)$. Since the same connection is defined whether $\tilde{s}_1$ or $\tilde{s}_2 = -\tilde{s}_1$ is chosen, the $\mathbf{Z}_2$ ambiguity is irrelevant and ∇ is well defined.

Lemma 3.2.4 and 3.2.5 extend to:

LEMMA 3.3.3. *Let V be a real oriented Riemannian vector bundle of even fiber dimension v. Suppose V admits a spin structure and let $\Delta^{\pm}(V)$ be the half-spin bundles. Then:*

(a) $\Lambda(V) \simeq \Delta(V) \otimes \Delta(V)$.

(b) $(\Lambda^+ - \Lambda^-)(V) \simeq (\Delta^+ - \Delta^-) \otimes (\Delta^+ - \Delta^-)(V)$.

(c) $(\Lambda^e - \Lambda^o)(V) \simeq (-1)^{v/2}(\Delta^+ - \Delta^-) \otimes (\Delta^+ - \Delta^-)(V)$.

(d) If $v = 2$ and if V is the underlying real bundle of a complex bundle V_c then $V_c \simeq \Delta^- \otimes \Delta^-$ and $V_c^* \simeq \Delta^+ \otimes \Delta^+$.

(e) If ∇ is a Riemannian connection on V and if we extend ∇ to $\Delta^{\pm}(V)$, then the isomorphisms given are unitary isomorphisms which preserve ∇.

Spinors are multiplicative with respect to products. It is immediate from the definitions we have given that:

LEMMA 3.3.4. *Let V_i be real oriented Riemannian vector bundles of even fiber dimensions v_i with given spin structures. Let $V_1 \oplus V_2$ have the natural orientation and spin structure. Then:*

$$(\Delta^+ - \Delta^-)(V_1 \oplus V_2) \simeq \{(\Delta^+ - \Delta^-)(V_1)\} \otimes \{(\Delta^+ - \Delta^-)(V_2)\}.$$

If ∇_i are Riemannian connections on V_i and if we define the natural induced connections on these bundles, then the isomorphism is unitary and preserves the connections.

A spin structure always exists locally since the obstruction to a spin structure is global. Given any real oriented Riemannian bundle V of even fiber dimension with a fixed Riemannian connection ∇, we define $\Lambda(V)$ and $\Delta(V) = \Delta^+(V) \oplus \Delta^-(V)$. With the natural metrics and connections, we relate the connection 1-forms and curvatures of these 3 bundles:

LEMMA 3.3.5. *Let $\{e_i\}$ be a local oriented orthonormal frame for V. We let $\nabla e_j = \omega_{jk} s_k$ represent the connection 1-form and $\Omega e_j = \Omega_{jk} e_k$ be the curvature matrix for $\Omega_{jk} = d\omega_{jk} - \omega_{jl} \wedge \omega_{lk}$. Let Λe and Δe denote the natural orthonormal frames on $\Lambda(V)$ and $\Delta(V)$. Relative to these frames we compute the connection 1-forms and curvature matrices of $\Lambda(V)$ and $\Delta(V)$ by:*

(a) $\omega_\Lambda = \omega_{jk} \operatorname{ext}(e_k) \operatorname{int}(e_j)$ $\Omega_\Lambda = \Omega_{jk} \operatorname{ext}(e_k) \operatorname{int}(e_j)$

(b) $\omega_\Delta = \frac{1}{4}\omega_{jk} e_j * e_k$ $\Omega_\Delta = \frac{1}{4}\Omega_{jk} e_j * e_k.$

PROOF: We sum over repeated indices in these expressions. We note (a) is true by definition on $\Lambda^1(V) = V$. Since both ω and Ω extend to act as derivations on the exterior algebra, this implies (a) is true on forms of all degree.

Let $so(n) = \{A \in n \times n$ real matrices with $A + A^t = 0\}$ be the Lie algebra of $SO(n)$. This is also the Lie algebra of $SPIN(n)$ so we must identify this with an appropriate subset of the Clifford algebra in order to prove (b). We choose a representative element of so(n) and define

$$Ae_1 = e_2, \quad Ae_2 = -e_1, \quad Ae_j = 0 \qquad \text{for } j > 2;$$

i.e.,

$$A_{12} = 1, \quad A_{21} = -1, \quad A_{jk} = 0 \qquad \text{otherwise.}$$

If we let $g(t) \in$ SO(n) be defined by

$$g(t)e_1 = (\cos t)e_1 + (\sin t)e_2,$$
$$g(t)e_2 = (\cos t)e_2 - (\sin t)e_1,$$
$$g(t)e_j = e_j \qquad j > 2,$$

then $g(0) = I$ and $g'(0) = A$. We must lift $g(t)$ from SO(n) to SPIN(n). Define:

$$h(t) = \big(\cos(t/4)e_1 + \sin(t/4)e_2\big)\big(-\cos(t/4)e_1 + \sin(t/4)e_2\big)$$
$$= \cos(t/2) + \sin(t/2)e_1e_2 \in \text{SPIN}(n)$$

Then $\rho(h) \in$ SO(n) is defined by:

$$\rho(h)e_j = \big(\cos(t/2) + \sin(t/2)e_1e_2\big)e_j\big(\cos(t/2) - \sin(t/2)e_1e_2\big)$$

so that $\rho(h)e_j = e_j$ for $j > 2$. We compute:

$$\rho(h)e_1 = \big(\cos(t/2)e_1 + \sin(t/2)e_2\big)\big(\cos(t/2) - \sin(t/2)e_1e_2\big)$$
$$= \big(\cos^2(t/2) - \sin^2(t/2)\big)e_1 + 2\sin(t/2)\cos(t/2)e_2$$
$$\rho(h)e_2 = \big(\cos(t/2)e_2 - \sin(t/2)e_1\big)\big(\cos(t/2) - \sin(t/2)e_1e_2\big)$$
$$= \big(\cos^2(t/2) - \sin^2(t/2)\big)e_2 - 2\sin(t/2)\cos(t/2)e_1$$

so that $\rho(h) = g$. This gives the desired lift from SO(n) to SPIN(n). We differentiate to get

$$h'(0) = \frac{1}{2}e_1e_2 = \frac{1}{4}A_{jk}e_j * e_k.$$

This gives the lift of a matrix in this particular form. Since the whole Lie algebra is generated by elements of this form, it proves that the lift of A_{jk} in general is given by $\frac{1}{4}A_{jk}e_j * e_k$. Since SPIN($n$) acts on the Clifford algebra by Clifford multiplication on the left, this gives the action of the curvature and connection 1-form on the spin representations and completes the proof.

We can define $ch(\Delta^\pm(V))$ as SO characteristic forms. They can be expressed in terms of the Pontrjagin and Euler forms. We could use the explicit representation given by Lemma 3.3.5 to compute these forms; it is most easy, however, to compute in terms of generating functions. We introduce formal variables $\{x_j\}$ for $1 \le j \le (\dim V)/2$ so that

$$p(V) = \prod_j (1 + x_j^2) \qquad \text{and} \qquad e(V) = \prod_j x_j.$$

All these computations are really induced from the corresponding matrix identities and the $\{x_j\}$ arise from putting a skew-symmetric matrix A in block form.

LEMMA 3.3.6. *Let V be an oriented real Riemannian vector bundle of dimension $n \equiv 0\ (2)$. Let $ch(\Delta^\pm(V))$ be a real characteristic form; these are well defined even if V does not admit a global spin structure. Then:*
(a) $ch(\Delta^+(V)) + ch(\Delta^-(V)) = ch(\Delta(V)) = \prod_j \{e^{x_j/2} + e^{-x_j/2}\}.$
(b)

$$(-1)^{n/2}\{ch(\Delta^+(V)) - ch(\Delta^-(V))\} = \prod_j \{e^{x_j/2} - e^{-x_j/2}\}$$

$$= e(V)(1 + \text{ higher order terms}).$$

PROOF: This is an identity among invariant polynomials in the Lie algebra $so(n)$. We may therefore restrict to elements of the Lie algebra which split in block diagonal form. Using the multiplicative properties of the Chern character and Lemma 3.3.4, it suffices to prove this lemma for the special case that $n = 2$ so $\Delta^\pm(V)$ are complex line bundles. Let V_c be a complex line bundle and let V be the underlying real bundle. Then $x_1 = x = c_1(V_c) = e(V)$. Since:

$$V_c = \Delta^- \otimes \Delta^- \qquad \text{and} \qquad V_c^* = \Delta^+ \otimes \Delta^+$$

we conclude:

$$x = 2c_1(\Delta^-) \qquad \text{and} \qquad -x = 2c_1(\Delta^+)$$

which shows (a) and the first part of (b). We expand:

$$e^{x/2} - e^{-x/2} = x + \frac{1}{24}x^3 + \cdots$$

to see $ch(\Delta^-) - ch(\Delta^+) = e(V)(1 + \frac{1}{24}p_1(V) + \cdots)$ to complete the proof of (b).

There is one final calculation in characteristic classes which will prove helpful. We defined L and $\hat{A}$ by the generating functions:

$$L(x) = \prod_j \frac{x_j}{\tanh x_j}$$

$$= \prod_j x_j \frac{e^{x_j} + e^{-x_j}}{e^{x_j} - e^{-x_j}}$$

$$\hat{A}(x) = \prod_j \frac{x_j}{\sinh(x_j/2)} = \prod_j \frac{x_j}{e^{x_j/2} - e^{-x_j/2}}.$$

LEMMA 3.3.7. Let V be an oriented Riemannian real vector bundle of dimension $n = 2n_0$. If we compute the component of the differential form which is in $\Lambda^n(T^*M)$ then:

$$\{L(V))\}_n = \{ch(\Delta(V)) \wedge \hat{A}(V)\}_n.$$

PROOF: It suffices to compute the component of the corresponding symmetric functions which are homogeneous of order n_0. If we replace x_j by $x_j/2$ then:

$$\{L(x_j)\}_{n_0} = \{2^{n_0} L(x_j/2)\}_{n_0}$$

$$= \left\{ \prod_j x_j \tanh(x_j/2) \right\}_{n_0}$$

$$= \left\{ \prod_j x_j \frac{e^{x/2} + e^{-x/2}}{e^{x_j/2} - e^{-x_j/2}} \right\}_{n_0}$$

$$= \{ch(\Delta(V)) \wedge \hat{A}(V)\}_{n_0}$$

which completes the proof.

3.4. The Spin Complex.

We shall use the spin complex chiefly as a formal construction to link the de Rham, signature, and Dolbeault complexes. Let M be a Riemannian manifold of even dimension m. Let $T(M)$ be the real tangent space. We assume that M is orientable and that $T(M)$ admits a spin structure. We let $\Delta^{\pm}(M)$ be the half-spin representations. There is a natural map given by Lemma 3.2.3 from the representations of

$$T^*(M) \to \mathrm{HOM}(\Delta^{\pm}(M), \Delta^{\mp}(M))$$

which we will call $c(\xi)$ (since it is essentially Clifford multiplication) such that $c(\xi)^2 = -|\xi|^2 I$. We extend the Levi-Civita connection to act naturally on these bundles and define the spin complex by the diagram:

$$A^{\pm}: C^{\infty}(\Delta^{\pm}(M)) \xrightarrow{\nabla} C^{\infty}(T^*M \otimes \Delta^{\pm}(M)) \xrightarrow{c} C^{\infty}(\Delta^{\mp}(M))$$

to be the operator with leading symbol c. $(A^+)^* = A^-$ and $A^{\pm}$ is elliptic since $c(\xi)^2 = -|\xi|^2 I$; this operator is called the Dirac operator.

Let V be a complex bundle with a Riemannian connection ∇. We define the spin complex with coefficients in ∇ by using the diagram:

$$A_V^{\pm}: C^{\infty}(\Delta^{\pm}(M) \otimes V) \to C^{\infty}(T^*M \otimes \Delta^{\pm}(M) \otimes V)$$
$$\xrightarrow{c \otimes 1} C^{\infty}(\Delta^{\mp}(M) \otimes V).$$

This is completely analogous to the signature complex with coefficients in V. We define:

$$\mathrm{index}(V, \mathrm{spin}) = \mathrm{index}(A_V^+);$$

a priori this depends on the particular spin structure chosen on V, but we shall show shortly that it does not depend on the particular spin structure, although it does depend on the orientation. We let

$$a_n^{\mathrm{spin}}(x, V)$$

denote the invariants of the heat equation. If we reverse the orientation of M, we interchange the roles of Δ^+ and of Δ^- so a_n^{spin} changes sign. This implies a_n^{spin} can be regarded as an invariantly defined m-form since the scalar invariant changes sign if we reverse the orientation.

Let X be an oriented coordinate system. We apply the Gramm-Schmidt process to the coordinate frame to construct a functorial orthonormal frame $\vec{s}(X)$ for $T(M) = T^*(M)$. We lift this to define two local sections $\vec{s}_i(X)$ to the principal bundle P_{SPIN} with $\vec{s}_1(X) = -\vec{s}_2(X)$. There is, of course, no cannonical way to prefer one over the other, but the pair is invariantly defined. Let $b^{\pm}$ be fixed bases for the representation space $\Delta^{\pm}$ and let

$b_i^\pm(X) = \vec{s}_i(X) \otimes b^\pm$ provide local frames for $\Delta^\pm(M)$. If we fix $i = 1$ or $i = 2$, the symbol of the spin complex with coefficients in V can be functorially expressed in terms of the 1-jets of the metric on M and in terms of the connection 1-form on V. The leading symbol is given by Clifford multiplication; the 0th order term is linear in the 1-jets of the metric on M and in the connection 1-form on V with coefficients which are smooth functions of the metric on M. If we replace $b_1^\pm$ by $b_2^\pm = -b_1^\pm$ then the local representation of the symbol is unchanged since multiplication by -1 commutes with differential operators. Thus we may regard $a_n^{\text{spin}}(x, V) \in \mathcal{R}_{m,n,m,\dim V}$ as an invariantly defined polynomial which is homogeneous of order n in the jets of the metric and of the connection form on V which is m-form valued.

This interpretation defines $a_n^{\text{spin}}(x, V)$ even if the base manifold M is not spin. We can always define the spin complex locally as the $\mathbf{Z}_2$ indeterminacy in the choice of a spin structure will not affect the symbol of the operator. Of course, $\int_M a_m^{\text{spin}}(x, V)$ can only be given the interpretation of index(V, spin) if M admits a spin structure. In particular, if this integral is not an integer, M cannot admit a spin structure.

$a_n^{\text{spin}}(x, V)$ is a local invariant which is not affected by the particular global spin structure chosen. Thus index(V, spin) is independent of the particular spin structure chosen.

We use exactly the same arguments based on the theorem of the second chapter and the multiplicative nature of the twisted spin complex as were used to prove the Hirzebruch signature theorem to establish:

LEMMA 3.4.1.
(a) $a_n^{\text{spin}} = 0$ if $n < m$ and a_m^{spin} is a characteristic form of $T(M)$ and of V.
(b) $\int_M a_m^{\text{spin}}(x, V) = \text{index}(V, \text{spin})$.
(c) There exists a characteristic form $\hat{A}'$ of $T(M)$ in the form:

$$\hat{A}' = 1 + \hat{A}_1 + \cdots$$

which does not depend on the dimension of M together with a universal constant c such that

$$a_m^{\text{spin}} = \sum_{4s+2t=m} \hat{A}'_s \wedge c^t ch(V)_t.$$

We use the notation $\hat{A}'$ since we have not yet shown it is the A-roof genus defined earlier. In proving the formula splits into this form, we do not rely on the uniqueness property of the second chapter, but rather on the multiplicative properties of the invariants of the heat equation discussed in the first chapter.

The spin complex has an intimate relation with both the de Rham and signature complexes:

LEMMA 3.4.2. *Let U be an open contractible subset of M. Over U, we define the signature, de Rham, and spin complexes. Then:*
(a) $(\Lambda^e - \Lambda^o) \otimes V \simeq (-1)^{m/2}(\Delta^+ - \Delta^-) \otimes (\Delta^+ - \Delta^-) \otimes V$.
(b) $(\Lambda^+ - \Lambda^-) \otimes V \simeq (\Delta^+ - \Delta^-) \otimes (\Delta^+ + \Delta^-) \otimes V$.
(c) These two isomorphisms preserve the unitary structures and the connections. They also commute with Clifford multiplication on the left.
(d) The natural operators on these complexes agree under this isomorphism.

PROOF: (a)–(c) follow from previous results. The natural operators on these complexes have the same leading symbol and therefore must be the same since they are natural first order operators. This proves (d).

We apply this lemma in dimension $m = 2$ with $V = $ the trivial bundle and $M = S^2$:

$$\chi(S^2) = 2 = \text{index}(d + \delta) = \text{index}(\Delta^- - \Delta^+, \text{spin})$$
$$= \int_{S^2} c\{c_1(\Delta^-) - c_1(\Delta^+)\}$$
$$= c \int_{S^2} e(TM) = 2c$$

by Lemmas 2.3.1, 3.3.5, and 3.3.4. This establishes that the normalizing constant of Lemma 3.4.1 must be 1 so that:

$$a_m^{\text{spin}} = \sum_{4s+2t=m} \hat{A}' \wedge ch_t(V).$$

We apply this lemma to the twisted signature complex in dimension $m = 2$ with V a non-trivial line bundle over S^2 to conclude:

$$\text{signature}(S^2, V) = \text{index}((\Delta^+ \oplus \Delta^-) \otimes V, \text{spin})$$
$$= \int_{S^2} ch((\Delta^+ \oplus \Delta^-) \otimes V) = \int_{S^2} 2 \cdot c_1(V).$$

This shows that the normalizing constant of Lemma 3.1.4 for the twisted signature complex is 2 and completes the proof of Lemma 3.1.4. This therefore completes the proof of the Hirzebruch signature theorem in general.

The de Rham complex with coeffiecients in V is defined by the diagram:

$$C^\infty(\Lambda^{e,o}(M) \otimes V) \xrightarrow{\nabla} C^\infty(T^*M \otimes \Lambda^{e,o}(M) \otimes V) \xrightarrow{c \otimes 1} C^\infty(\Lambda^{o,e}(M) \otimes V)$$

and we shall denote the operator by $(d + \delta)^e_V$. The relations given by Lemma 3.4.2 give rise to relations among the local formulas:

$$a_n(x, (d+\delta)^e_V) = a_n(x, A^+_{(-1)^{m/2}(\Delta^+ - \Delta^-) \otimes V})$$
$$a_n(x, (d+\delta)^+_V) = a_n(x, A^+_{(\Delta^+ - \Delta^-) \otimes V})$$

where $(d + \delta)^+_V$ is the operator of the twisted signature complex. These relations are well defined regardless of whether or not M admits a spin structrue on $T(M)$.

We deal first with the de Rham complex. Using Lemma 3.4.1 and the fact that the normalizing constant c is 1, we conclude:

$$a_n(x, (d + \delta)^e_V) = \hat{A}' \wedge ch((-1)^{m/2}(\Delta^+ - \Delta^-))) \wedge ch(V).$$

Since $ch((-1)^{m/2}(\Delta^+ - \Delta^-)) = e(M)$ is already a top dimensional form by Lemma 3.3.5(b), we conclude $a_n(x, (d + \delta)^e_V) = (\dim V)e(M)$. This proves:

THEOREM 3.4.3. Let $(d + \delta)^e_V$ be the the de Rham complex with coefficients in the bundle V and let $a_n(x, (d + \delta)^e_V)$ be the invariants of the heat equation. Then:
(a) $a_n(x, (d + \delta)^e_V) = 0$ if $n < m$.
(b) $a_m(x, (d + \delta)^e_V) = (\dim V)e(M)$ where $e(M)$ is the Euler form of $T(M)$.
(c)

$$\text{index}((d + \delta)^e_V) = (\dim V)\chi(M) = \int_m (\dim V)e(M).$$

This shows that no information about V (except its dimension) is obtained by considering the de Rham complex with coefficients in V and it is for this reason we did not introduce this complex earlier. This gives a second proof of the Gauss-Bonnet theorem independent of the proof we gave earlier.

Next, we study the signature complex in order to compute $\hat{A}$. Using Lemma 3.4.1 with the bundle $V = 1$, we conclude for $m = 4k$,

$$a_m(x, (d + \delta)^+_V) = \{\hat{A}' \wedge ch(\Delta)\}_m.$$

Using Theorem 3.1.1 and Lemma 3.3.6, we compute therefore:

$$L_k = \{ch(\Delta) \wedge \hat{A}\}_m = \{ch(\Delta) \wedge \hat{A}'\}_m = a_m(x, (d + \delta)^+_V)$$

so as the Chern character is formally invertible; $\hat{A} = \hat{A}'$ is given by the generating function $z_j / \sinh(z_j/2)$ by Lemma 3.3.7. We can now improve Lemma 3.4.1 and determine all the relevant normalizing constants.

THEOREM 3.4.4. Let $T(M)$ admit a spin structure, then:
(a) $a_n^{\text{spin}} = 0$ if $n < m$.
(b) $a_m^{\text{spin}}(x, V) = \sum_{4s+2t=m} \hat{A}_s \wedge ch_t(V)$.
(c) $\text{index}(V, \text{spin}) = \int_M a_m^{\text{spin}}(x, V)$.

3.5. The Riemann-Roch Theorem
For Almost Complex Manifolds.

So far, we have discussed three of the four classical elliptic complexes, the de Rham, the signature, and the spin complexes. In this subsection, we define the Dolbeault complex for an almost complex manifold and relate it to the spin complex.

Let M be a Riemannian manifold of dimension $m = 2n$. An almost complex structure on M is a linear map $J : T(M) \to T(M)$ with $J^2 = -1$. The Riemannian metric G is unitary if $G(X, Y) = G(JX, JY)$ for all $X, Y \in T(M)$; we can always construct unitary metrics by averaging over the action of J. Henceforth we assume G is unitary. We extend G to be Hermitian on $T(M) \otimes \mathbf{C}$, $T^*(M) \otimes \mathbf{C}$, and $\Lambda(M) \otimes \mathbf{C}$.

Since $J^2 = -1$, we decompose $T(M) \otimes \mathbf{C} = T'(M) \oplus T''(M)$ into the $\pm i$ eigenspaces of J. This direct sum is orthogonal with respect to the metric G. Let $\Lambda^{1,0}(M)$ and $\Lambda^{0,1}(M)$ be the dual spaces in $T^*(M) \otimes \mathbf{C}$ to T' and T''. We choose a local frame $\{e_j\}$ for $T(M)$ so that $J(e_j) = e_{j+n}$ for $1 \leq i \leq n$. Let $\{e^j\}$ be the corresponding dual frame for $T^*(M)$. Then:

$$T'(M) = \operatorname{span}_{\mathbf{C}} \{e_j - i e_{j+n}\}_{j=1}^n \qquad T''(M) = \operatorname{span}_{\mathbf{C}} \{e_j + i e_{j+n}\}_{j=1}^n$$
$$\Lambda^{1,0}(M) = \operatorname{span}_{\mathbf{C}} \{e^j + i e^{j+n}\}_{j=1}^n \qquad \Lambda^{0,1}(M) = \operatorname{span}_{\mathbf{C}} \{e^j - i e^{j+n}\}_{j=1}^n.$$

These four vector bundles are all complex vector bundles over M. The metric gives rise to natural isomorphisms $T'(M) \simeq \Lambda^{0,1}(M)$ and $T''(M) = \Lambda^{1,0}(M)$. We will use this isomorphism to identify e_j with e^j for much of what follows.

If we forget the complex structure on $T'(M)$, then the underlying real vector bundle is naturally isomorphic to $T(M)$. Complex multiplication by i on $T'(M)$ is equivalent to the endomorphism J under this identification. Thus we may regard J as giving a complex structure to $T(M)$.

The decomposition:

$$T^*(M) \otimes \mathbf{C} = \Lambda^{1,0}(M) \oplus \Lambda^{0,1}(M)$$

gives rise to a decomposition:

$$\Lambda(T^* M) \otimes \mathbf{C} = \bigoplus_{p,q} \Lambda^{p,q}(M)$$

for

$$\Lambda^{p,q}(M) = \Lambda^p(\Lambda^{1,0}(M)) \otimes \Lambda^q(\Lambda^{0,1}(M)).$$

Each of the bundles $\Lambda^{p,q}$ is a complex bundle over M and this decomposition of $\Lambda(T^* M) \otimes \mathbf{C}$ is orthogonal. Henceforth we will denote these bundles by T', T'', and $\Lambda^{p,q}$ when no confusion will arise.

If M is a holomorphic manifold, we let $z = (z_1, \ldots, z_n)$ be a local holomorphic coordinate chart. We expand $z_j = x_j + iy_j$ and define:

$$\frac{\partial}{\partial z_j} = \frac{1}{2}\left(\frac{\partial}{\partial x_j} - i\frac{\partial}{\partial y_j}\right), \qquad \frac{\partial}{\partial \bar{z}_j} = \frac{1}{2}\left(\frac{\partial}{\partial x_j} + i\frac{\partial}{\partial y_j}\right)$$
$$dz_j = dx_j + idy_j, \qquad\qquad d\bar{z}_j = dx_j - idy_j.$$

We define:

$$\partial(f) = \sum_j \frac{\partial f}{\partial z_j} dz_j \qquad \text{and} \qquad \bar{\partial}(f) = \sum_j \frac{\partial f}{\partial \bar{z}_j} d\bar{z}_j.$$

The Cauchy-Riemann equations imply that a function f is holomorphic if and only if $\bar{\partial}(f) = 0$. If $w = (w_1, \ldots, w_n)$ is another holomorphic coordinate system, then:

$$\frac{\partial}{\partial w_j} = \sum_k \frac{\partial z_k}{\partial w_j}\frac{\partial}{\partial z_k}, \qquad \frac{\partial}{\partial \bar{w}_j} = \sum_k \frac{\partial \bar{z}_k}{\partial \bar{w}_j}\frac{\partial}{\partial \bar{z}_k}$$
$$dw_j = \sum_k \frac{\partial w_j}{\partial z_k} dz_k, \qquad d\bar{w}_j = \frac{\partial \bar{w}_j}{\partial \bar{z}_k} d\bar{z}_k.$$

We define:

$$T'(M) = \text{span}\left\{\frac{\partial}{\partial z_j}\right\}_{j=1}^n, \qquad T''(M) = \text{span}\left\{\frac{\partial}{\partial \bar{z}_j}\right\}_{j=1}^n$$
$$\Lambda^{1,0}(M) = \text{span}\{dz_j\}_{j=1}^n, \qquad \Lambda^{0,1}(M) = \text{span}\{d\bar{z}_j\}_{j=1}^n$$

then these complex bundles are invariantly defined independent of the choice of the coordinate system. We also note

$$\partial\colon C^\infty(M) \to C^\infty(\Lambda^{1,0}(M)) \qquad \text{and} \qquad \bar{\partial}\colon C^\infty(M) \to C^\infty(\Lambda^{0,1}(M))$$

are invariantly defined and decompose $d = \partial + \bar{\partial}$.

There is a natural isomorphism of $T'(M)$ with $T(M)$ as real bundles in this example and we let J be complex multiplication by i under this isomorphism. Equivalently:

$$J\left(\frac{\partial}{\partial x_j}\right) = \left(\frac{\partial}{\partial y_j}\right) \qquad \text{and} \qquad J\left(\frac{\partial}{\partial y_j}\right) = -\frac{\partial}{\partial x_j}.$$

$T'(M) = T_c(M)$ is the complex tangent bundle in this example; we shall reserve the notation $T_c(M)$ for the holomorphic case.

Not every almost complex structure arises from a complex structure; there is an integrability condition. If J is an almost complex structure, we decompose the action of exterior differentiation d on $C^\infty(\Lambda)$ with respect to the bigrading (p, q) to define:

$$\partial: C^\infty(\Lambda^{p,q}) \to C^\infty(\Lambda^{p+1,q}) \qquad \text{and} \qquad \bar{\partial}: C^\infty(\Lambda^{p,q}) \to C^\infty(\Lambda^{p,q+1}).$$

THEOREM 3.5.1 (NIRENBERG-NEULANDER). *The following are equivalent and define the notion of an integrable almost complex structure:*
(a) The almost complex structure J arises from a holomorphic structure on M.
(b) $d = \partial + \bar{\partial}$.
(c) $\bar{\partial}\bar{\partial} = 0$.
(d) $T'(M)$ is integrable—i.e., given $X, Y \in C^\infty(T'(M))$, then the Lie bracket $[X, Y] \in C^\infty(T'(M))$ where we extend $[\ ,\ ]$ to complex vector fields in the obvious fashion.

PROOF: This is a fairly deep result and we shall not give complete details. It is worth, however, giving a partial proof of some of the implications to illustrate the concepts we will be working with. Suppose first M is holomorphic and let $\{z_j\}$ be local holomorphic coordinates on M. Define:

$$dz^I = dz_{i_1} \wedge \cdots \wedge dz_{i_p} \qquad \text{and} \qquad d\bar{z}^J = d\bar{z}_{j_1} \wedge \cdots \wedge d\bar{z}_{j_q}$$

then the collection $\{dz^I \wedge d\bar{z}^J\}$ gives a local frame for $\Lambda^{p,q}$ which is closed. If $\omega \in C^\infty(\Lambda^{p,q})$, we decompose $\omega = \sum f_{I,J} \, dz^I \wedge d\bar{z}^J$ and compute:

$$d\omega = d\left\{ \sum f_{I,J} \, dz^I \wedge d\bar{z}^J \right\} = \sum df_{I,J} \wedge dz^I \wedge d\bar{z}^J.$$

On functions, we decompose $d = \partial + \bar{\partial}$. Thus $d\omega \in C^\infty(\Lambda^{p+1,q} \oplus \Lambda^{p,q+1})$ has no other components. Therefore $d\omega = \partial\omega + \bar{\partial}\omega$ so (a) implies (b).

We use the identity $d^2 = 0$ to compute $(\partial + \bar{\partial})^2 = (\partial^2) + (\partial\bar{\partial} + \bar{\partial}\partial) + (\bar{\partial}^2) = 0$. Using the bigrading and decomposing this we conclude $(\partial^2) = (\partial\bar{\partial} + \bar{\partial}\partial) = (\bar{\partial}^2) = 0$ so (b) implies (c). Conversely, suppose that $\bar{\partial}\bar{\partial} = 0$ on $C^\infty(M)$. We must show $d: C^\infty(\Lambda^{p,q}) \to C^\infty(\Lambda^{p+1,q} \oplus \Lambda^{p,q+1})$ has no other components. Let $\{e_j\}$ be a local frame for $\Lambda^{1,0}$ and let $\{\bar{e}_j\}$ be the corresponding local frame for $\Lambda^{0,1}$. We decompose:

$$de_j = \partial e_j + \bar{\partial} e_j + A_j \qquad \text{for } A_j \in \Lambda^{0,2}$$
$$d\bar{e}_j = \bar{\partial}\bar{e}_j + \partial\bar{e}_j + \bar{A}_j \qquad \text{for } \bar{A}_j \in \Lambda^{2,0}.$$

Then we compute:

$$d\left(\sum f_j e_j \right) = \sum df_j \wedge e_j + f_j de_j$$
$$= \sum \{\partial f_j \wedge e_j + \bar{\partial} f_j \wedge e_j + f_j \partial e_j + f_j \bar{\partial} e_j + f_j A_j\}$$
$$= \partial\left(\sum f_j e_j \right) + \bar{\partial}\left(\sum f_j e_j \right) + \sum f_j A_j.$$

Similarly

$$d\left(\sum f_j \bar{e}_j\right) = \partial\left(\sum f_j \bar{e}_j\right) + \bar{\partial}\left(\sum f_j \bar{e}_j\right) + \sum f_j \bar{A}_j.$$

If A and $\bar{A}$ denote the 0th order operators mapping $\Lambda^{1,0} \to \Lambda^{0,2}$ and $\Lambda^{0,1} \to \Lambda^{2,0}$ then we compute:

$$d(\omega) = \partial(\omega) + \bar{\partial}(\omega) + A\omega \qquad \text{for } \omega \in C^\infty(\Lambda^{1,0})$$
$$d(\bar{\omega}) = \partial(\bar{\omega}) + \bar{\partial}(\bar{\omega}) + \bar{A}\bar{\omega} \qquad \text{for } \bar{\omega} \in C^\infty(\Lambda^{0,1}).$$

Let $f \in C^\infty(M)$ and compute:

$$0 = d^2 f = d(\partial f + \bar{\partial} f) = (\partial^2 + \bar{A}\bar{\partial})f + (\partial\bar{\partial} + \bar{\partial}\partial)f + (\bar{\partial}^2 + A\partial)f$$

where we have decomposed the sum using the bigrading. This implies that $(\bar{\partial}^2 + A\partial)f = 0$ so $A\partial f = 0$. Since $\{\partial f\}$ spans $\Lambda^{1,0}$, this implies $A = \bar{A} = 0$ since A is a 0th order operator. Thus $de_j = \partial e_j + \bar{\partial} e_j$ and $d\bar{e}_j = \partial\bar{e}_j + \bar{\partial}\bar{e}_j$. We compute $d(e^I \wedge \bar{e}^J)$ has only $(p+1, q)$ and $(p, q+1)$ components so $d = \partial + \bar{\partial}$. Thus (b) and (c) are equivalent.

It is immediate from the definition that $X \in T'(M)$ if and only if $\omega(X) = 0$ for all $\omega \in \Lambda^{0,1}$. If $X, Y \in C^\infty(T^*M)$, then Cartan's identity implies:

$$d\omega(X, Y) = X(\omega Y) - Y(\omega X) - \omega([X, Y]) = -\omega([X, Y]).$$

If (b) is true then $d\omega$ has no component in $\Lambda^{2,0}$ so $d\omega(X, Y) = 0$ which implies $\omega([X, Y]) = 0$ which implies $[X, Y] \in C^\infty(T'M)$ which implies (d). Conversely, if (d) is true, then $d\omega(X, Y) = 0$ so $d\omega$ has no component in $\Lambda^{2,0}$ so $d = \partial + \bar{\partial}$ on $\Lambda^{0,1}$. By taking conjugates, $d = \partial + \bar{\partial}$ on $\Lambda^{1,0}$ as well which implies as noted above that $d = \partial + \bar{\partial}$ in general which implies (c).

We have proved that (b)–(d) are equivalent and that (a) implies (b). The hard part of the theorem is showing (b) implies (a). We shall not give this proof as it is quite lengthy and as we shall not need this implication of the theorem.

As part of the previous proof, we computed that $d - (\partial + \bar{\partial})$ is a 0th order operator (which vanishes if and only if M is holomorphic). We now compute the symbol of both ∂ and $\bar{\partial}$. We use the metric to identify $T(M) = T^*(M)$. Let $\{e_j\}$ be a local orthonormal frame for $T(M)$ such that $J(e_j) = e_{j+n}$ for $1 \le j \le n$. We extend ext and int to be complex linear maps from $T^*(M) \otimes \mathbf{C} \to \text{END}(\Lambda(T^*M) \otimes \mathbf{C})$.

LEMMA 3.5.2. Let ∂ and $\bar{\partial}$ be defined as before and let δ' and δ'' be the formal adjoints. Then:

(a)

$$\partial: C^\infty(\Lambda^{p,q}) \to C^\infty(\Lambda^{p+1,q})$$

$$\text{and} \quad \sigma_L(\partial)(x,\xi) = \frac{i}{2}\sum_{j\leq n}(\xi_j - i\xi_{j+n})\,\text{ext}(e_j + ie_{j+n})$$

$$\bar{\partial}: C^\infty(\Lambda^{p,q}) \to C^\infty(\Lambda^{p,q+1})$$

$$\text{and} \quad \sigma_L(\bar{\partial})(x,\xi) = \frac{i}{2}\sum_{j\leq n}(\xi_j + i\xi_{j+n})\,\text{ext}(e_j - ie_{j+n})$$

$$\delta': C^\infty(\Lambda^{p,q}) \to C^\infty(\Lambda^{p-1,q})$$

$$\text{and} \quad \sigma_L(\delta')(x,\xi) = -\frac{i}{2}\sum_{j\leq n}(\xi_j + i\xi_{j+n})\,\text{int}(e_j - ie_{j+n})$$

$$\delta'': C^\infty(\Lambda^{p,q}) \to C^\infty(\Lambda^{p,q-1})$$

$$\text{and} \quad \sigma_L(\delta'')(x,\xi) = -\frac{i}{2}\sum_{j\leq n}(\xi_j - i\xi_{j+n})\,\text{int}(e_j + ie_{j+n}).$$

(b) If $\Delta_c' = (\partial + \delta')^2$ and $\Delta_c'' = (\bar{\partial} + \delta'')^2$ then these are elliptic on Λ with $\sigma_L(\Delta_c') = \sigma_l(\Delta_c'') = \frac{1}{2}|\xi|^2 I$.

PROOF: We know $\sigma_L(d)(x,\xi) = i\sum_j \xi_j\,\text{ext}(e_j)$. We define

$$A(\xi) = \frac{1}{2}\sum_{j\leq n}(\xi_{j+n})\,\text{ext}(e_j + ie_{j+n})$$

then $A(\xi): \Lambda^{p,q} \to \Lambda^{p+1,q}$ and $\bar{A}(\xi): \Lambda^{p,q} \to \Lambda^{p,q+1}$. Since $iA(\xi) + i\bar{A}(\xi) = \sigma_L(d)$, we conclude that iA and $i\bar{A}$ represent the decomposition of $\sigma_L(d)$ under the bigrading and thus define the symbols of ∂ and $\bar{\partial}$. The symbol of the adjoint is the adjoint of the symbol and this proves (a). (b) is an immediate consequence of (a).

If M is holomorphic, the Dolbeault complex is the complex $\{\bar{\partial}, \Lambda^{0,q}\}$ and the index of this complex is called the arithmetic genus of M. If M is not holomorphic, but only has an almost complex structure, then $\bar{\partial}^2 \neq 0$ so we can not define the arithmetic genus in this way. Instead, we use a trick called "rolling up" the complex. We define:

$$\Lambda^{0,+} = \bigoplus_q \Lambda^{0,2q} \quad \text{and} \quad \Lambda^{0,-} = \bigoplus_q \Lambda^{0,2q+1}$$

to define a $\mathbf{Z}_2$ grading on the Dolbeault bundles. (These are also often denoted by $\Lambda^{0,\text{even}}$ and $\Lambda^{0,\text{odd}}$ in the literature). We consider the two term elliptic complex:

$$(\bar{\partial} + \delta'')_\pm: C^\infty(\Lambda^{0,\pm}) \to C^\infty(\Lambda^{0,\mp})$$

and define the arithmetic genus of M to be the index of $(\bar{\partial} + \delta'')_+$. The adjoint of $(\bar{\partial} + \delta'')_+$ is $(\bar{\partial} + \delta'')_-$ and the associated Laplacian is just Δ_c'' restricted to $\Lambda^{0,\pm}$ so this is an elliptic complex. If M is holomorphic, then the index of this elliptic complex is equal to the index of the complex $(\bar{\partial}, \Lambda^{0,q})$ by the Hodge decomposition theorem.

We define:

$$c'(\xi) = (\sqrt{2})^{-1} \sum_{j \leq n} \{(\xi_j + i\xi_{j+n}) \, \mathrm{ext}(e_j - ie_{j+n})$$
$$- (\xi_j - i\xi_{j+n}) \, \mathrm{int}(e_j + ie_{j+n})\}$$

then it is immediate that:

$$c'(\xi)c'(\xi) = -|\xi|^2 \qquad \text{and} \qquad \sigma_L(\bar{\partial} + \delta'') = ic'(\xi)/\sqrt{2}.$$

Let ∇ be a connection on $\Lambda^{0,\pm}$, then we define the operator $A^\pm(\nabla)$ by the diagram:

$$A^\pm(\nabla): C^\infty(\Lambda^{0,\pm}) \xrightarrow{\nabla} C^\infty(T^*M \otimes \Lambda^{0,\pm}) \xrightarrow{c'/\sqrt{2}} C^\infty(\Lambda^{0,\mp})$$

This will have the same leading symbol as $(\bar{\partial} + \delta'')_\pm$. There exists a unique connection so $A^\pm(\nabla) = (\bar{\partial} + \delta'')_\pm$ but as the index is constant under lower order perturbations, we shall not need this fact as the index of $A^+(\nabla) = \mathrm{index}(\bar{\partial} + \delta'')_+$ for any ∇. We shall return to this point in the next subsection.

We now let V be an arbitrary coefficient bundle with a connection ∇ and define the Dolbeault complex with coefficients in V using the diagram:

$$A^\pm: C^\infty(\Lambda^{0,\pm} \otimes V) \xrightarrow{\nabla} C^\infty(T^*M \otimes \Lambda^{0,\pm} \otimes V) \xrightarrow{c'/\sqrt{2} \otimes 1} C^\infty(\Lambda^{0,\mp} \otimes V)$$

and we define index$(V, \text{Dolbeault})$ to be the index of this elliptic complex. The index is independent of the connections chosen, of the fiber metrics chosen, and is constant under perturbations of the almost complex structure.

If M is holomorphic and if V is a holomorphic vector bundle, then we can extend $\bar{\partial}: C^\infty(\Lambda^{0,q} \otimes V) \to C^\infty(\Lambda^{0,q+1} \otimes V)$ with $\bar{\partial}\bar{\partial} = 0$. Exactly as was true for the arithmetic genus, the index of this elliptic complex is equal to the index of the rolled up elliptic complex so our definitions generalize the usual definitions from the holomorphic category to the almost complex category.

We will compute a formula for index$(V, \text{Dolbeault})$ using the spin complex. There is a natural inclusion from $U(\frac{m}{2}) = U(n)$ into $SO(m)$, but this does not lift in general to $\mathrm{SPIN}(m)$. We saw earlier that $T(CP_k)$

does not admit a spin structure if k is even, even though it does admit a unitary structure. We define $\mathrm{SPIN}_c(m) = \mathrm{SPIN}(m) \times S^1/\mathbf{Z}_2$ where we choose the $\mathbf{Z}_2$ identification $(g, \lambda) = (-g, -\lambda)$. There is a natural map $\rho_c \colon \mathrm{SPIN}_c(m) \to SO(m) \times S^1$ induced by the map which sends $(g, \lambda) \mapsto (\rho(g), \lambda^2)$. This map is a $\mathbf{Z}_2$ double cover and is a group homomorphism.

There is a natural map of $\mathrm{U}(\frac{m}{2})$ into $SO(m) \times S^1$ defined by sending $g \mapsto (g, \det(g))$. The interesting thing is that this inclusion does lift; we can define $f \colon \mathrm{U}(\frac{m}{2}) \to \mathrm{SPIN}_c(m)$ so the following diagram commutes:

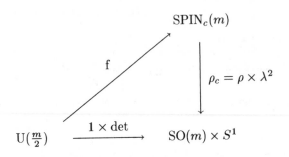

We define the lifting as follows. For $U \in \mathrm{U}(\frac{m}{2})$, we choose a unitary basis $\{e_j\}_{j=1}^n$ so that $U(e_j) = \lambda_j e_j$. We define $e_{j+n} = ie_j$ so $\{e_j\}_{j=1}^{2n}$ is an orthogonal basis for $\mathbf{R}^m = \mathbf{C}^n$. Express $\lambda_j = e^{i\theta_j}$ and define:

$$f(U) = \prod_{j=1}^n \{\cos(\theta_j/2) + \sin(\theta_j/2)e_j e_{j+n}\} \times \prod_{j=1}^n e^{i\theta_j/2} \in \mathrm{SPIN}_c(m).$$

We note first that $e_j e_{j+n}$ is an invariant of the one-dimensional complex subspace spanned by e_j and does not change if we replace e_j by ze_j for $|z| = 1$. Since all the factors commute, the order in which the eigenvalues is taken does not affect the product. If there is a multiple eigenvalue, this product is independent of the particular basis which is chosen. Finally, if we replace θ_j by $\theta_j + 2\pi$ then both the first product and the second product change sign. Since $(g, \lambda) = (-g, -\lambda)$ in $\mathrm{SPIN}_c(m)$, this element is invariantly defined. It is clear that $f(I) = I$ and that f is continuous. It is easily verified that $\rho_c f(U) = i(U) \times \det(U)$ where $i(U)$ denotes the matrix U viewed as an element of $SO(m)$ where we have forgotten the complex structure on $\mathbf{C}^n$. This proves that f is a group homomorphism near the identity and consequently f is a group homomorphism in general. ρ_c is a covering projection.

The cannonical bundle K is given by:

$$K = \Lambda^{n,0} = \Lambda^n(T'M)^*$$

so $K^* = \Lambda^n(T'M)$ is the bundle with transition functions $\det(U_{\alpha\beta})$ where the $U_{\alpha\beta}$ are the transitions for $T'M$. If we have a spin structure on M, we can split:

$$P_{\text{SPIN}_c} = P_{\text{SPIN}} \times P_{S^1}/\mathbf{Z}_2$$

where P_{S^1} represents a line bundle L_1 over M. From this description, it is clear $K^* = L_1 \otimes L_1$. Conversely, if we can take the square root of K (or equivalently of K^*), then M admits a spin structure so the obstruction to constructing a spin structure on an almost complex manifold is the obstruction to finding a square root of the cannonical bundle.

Let $V = \mathbf{C}^n = \mathbf{R}^m$ with the natural structures. We extend $\Delta^\pm$ to representations $\Delta_c^\pm$ of $\text{SPIN}_c(m)$ in the natural way. We relate this representation to the Dolbeault representation as follows:

LEMMA 3.5.3. *There is a natural isomorphism between $\Lambda^{0,\pm}$ and $\Delta_c^\pm$ which defines an equivalence of these two representations of $\mathrm{U}(\frac{m}{2})$. Under this isomorphism, the action of V by Clifford multiplication on the left is preserved.*

PROOF: Let $\{e_j\}$ be an orthonormal basis for $\mathbf{R}^m$ with $Je_j = e_{j+n}$ for $1 \leq j \leq n$. Define:

$$\alpha_j = ie_je_{j+n}, \qquad \beta_j = e_j + ie_{j+n}, \qquad \bar{\beta}_j = e_j - ie_{j+n} \qquad \text{for } 1 \leq j \leq n.$$

We compute:

$$\beta_j\alpha_j = -\beta_j \quad \text{and} \quad \beta_j\alpha_k = \alpha_k\beta_j \qquad \text{for } k \neq j.$$

We define $\gamma = \beta_1 \ldots \beta_n$ then γ spans $\Lambda^{n,0}$ and $\gamma\alpha_j = -\gamma$, $1 \leq j \leq n$. We define:

$$\Lambda^{0,*} = \bigoplus_q \Lambda^{0,q} \qquad \text{and} \qquad \Lambda^{n,*} = \bigoplus_q \Lambda^{n,q} = \Lambda^{0,*}\gamma.$$

Since $\dim(\Lambda^{0,*}) = 2^n$, we conclude that $\Lambda^{0,*}\gamma$ is the simultaneous -1 eigenspace for all the α_j and that therefore:

$$\Delta_c = \Lambda^{0,*}\gamma$$

as a left representation space for $\text{SPIN}(m)$. Again, we compute:

$$\alpha_j\beta_j = \beta_j, \quad \alpha_j\bar{\beta}_j = -\bar{\beta}_j, \quad \alpha_j\beta_k = \beta_k\alpha_j, \quad \alpha_j\bar{\beta}_k = \bar{\beta}_k\alpha_j \qquad \text{for } j \neq k$$

so that if $x \in \Lambda^{0,q}\gamma$ then since $\tau = \alpha_1 \ldots \alpha_n$ we have $\tau x = (-1)^q x$ so that

$$\Delta_c^\pm = \Lambda^{0,\pm}\gamma.$$

We now study the induced representation of $U(n)$. Let

$$g = \big(\cos(\theta/2) + \sin(\theta/2)e_1 J e_1\big) \cdot e^{i\theta/2} \in \mathrm{SPIN}_c(m).$$

We compute:

$$g\beta_1 = \{\cos(\theta/2) - i\sin(\theta/2)\}e^{i\theta/2}\beta_1 = \beta_1$$
$$g\bar{\beta}_1 = \{\cos(\theta/2) + i\sin(\theta/2)\}e^{i\theta/2}\bar{\beta}_1 = e^{i\theta}\bar{\beta}_1$$
$$g\beta_k = \beta_k g \quad \text{and} \quad g\bar{\beta}_k = \bar{\beta}_k g \quad \text{for } k > 1.$$

Consequently:

$$g\bar{\beta}_J\gamma = \begin{cases} \bar{\beta}_J\gamma & \text{if } j_1 > 1 \\ e^{i\theta}\bar{\beta}_J\gamma & \text{if } j_1 = 1. \end{cases}$$

A similar computation goes for all the other indices and thus we compute that if $Ue_j = e^{i\theta_j}e_j$ is unitary that:

$$f(U)\bar{\beta}_J\gamma = e^{i\theta_{j_1}}\ldots e^{i\theta_{j_q}}\bar{\beta}_J\gamma$$

which is of course the natural represenation of $U(n)$ on $\Lambda^{0,q}$.

Finally we compare the two actions of V by Clifford multiplication. We assume without loss of generality that $\xi = (1, 0, \ldots, 0)$ so we must study Clifford multiplication by e_1 on Δ_c and

$$\{\mathrm{ext}(e_1 - ie_{1+n}) - \mathrm{int}(e_1 + ie_{1+n})\}/\sqrt{2} = c'(e_1)$$

on $\Lambda^{0,*}$. We compute:

$$e_1\beta_1 = -1 + ie_1e_2 = (e_1 - ie_2)(e_1 + ie_2)/2 = \bar{\beta}_1\beta_1/2,$$
$$e_1\bar{\beta}_1\beta_1 = -2\beta_1,$$
$$e_1\bar{\beta}_k = -\bar{\beta}_k e_1 \quad \text{for } k > 1.$$

From this it follows immediately that:

$$e_1\bar{\beta}_J\gamma = (-1)^q \begin{cases} \bar{\beta}_1\bar{\beta}_J\gamma/2 & \text{if } j_1 > 1 \\ 2\bar{\beta}_{J'}\gamma & \text{where } J' = \{j_2, \ldots, j_q\} \text{ if } j_1 = 1. \end{cases}$$

Similarly, we compute:

$$c'(e_1)\bar{\beta}_J = \begin{cases} \bar{\beta}_1\bar{\beta}_J/\sqrt{2} & \text{if } j_1 > 1 \\ -\sqrt{2}\bar{\beta}_{J'} & \text{for } J' = \{j_2, \ldots, j_q\} \text{ if } j_1 = 1. \end{cases}$$

From these equations, it is immediate that if we define $T(\bar{\beta}_J) = \bar{\beta}_J\gamma$, then T will not preserve Clifford multiplication. We let $a(q)$ be a sequence of non-zero constants and define $T: \Lambda^{0,\pm} \to \Delta_c^{\pm}$ by:

$$T(\bar{\beta}_J) = a(|J|)\bar{\beta}_J\gamma.$$

Since the spaces $\Lambda^{0,q}$ are $U(n)$ invariant, this defines an equivalence between these two representations of $U(n)$. T will induce an equivalence between Clifford multiplication if and only if we have the relations:

$$(-1)^q a(q)/2 = a(q+1)/\sqrt{2} \qquad \text{and} \qquad \sqrt{2}a(q-1) = (-1)^q \cdot 2 \cdot a(q).$$

These give rise to the inductive relations:

$$a(q+1) = (-1)^q a(q)/\sqrt{2} \qquad \text{and} \qquad a(q) = (-1)^{q-1} a(q-1)/\sqrt{2}.$$

These relations are consistent and we set $a(q) = (\sqrt{2})^{-q}(-1)^{q(q-1)/2}$ to define the equivalence T and complete the proof of the Lemma.

From this lemma, we conclude:

LEMMA 3.5.4. *There is a natural isomorphism of elliptic complexes* $(\Delta_c^+ - \Delta_c^-) \otimes V \simeq (\Lambda^{0,+} - \Lambda^{0,-}) \otimes V$ *which takes the operator of the* SPIN$_c$ *complex to an operator which has the same leading symbol as the Dolbeault complex (and thus has the same index). Furthermore, we can represent the* SPIN$_c$ *complex locally in terms of the* SPIN *complex in the form:* $(\Delta_c^+ - \Delta_c^-) \otimes V = (\Delta^+ - \Delta^-) \otimes L_1 \otimes V$ *where* L_1 *is a local square root of* $\Lambda^n(T'M)$.

We use this sequence of isomorphisms to define an operator on the Dolbeault complex with the same leading symbol as the operator $(\bar{\partial} + \partial'')$ which is locally isomorphic to the natural operator of the SPIN complex. The $\mathbf{Z}_2$ ambiguity in the definition of L_1 does not affect this construction. (This is equivalent to choosing an appropriate connection called the spin connection on $\Lambda^{0,\pm}$.) We can compute index$(V, \text{Dolbeault})$ using this operator. The local invariants of the heat equation for this operator are the local invariants of the twisted spin complex and therefore arguing exactly as we did for the signature complex, we compute:

$$\text{index}(V, \text{Dolbeault}) = \int_M \hat{A}(TM) \wedge ch(L_1) \wedge ch(V)$$

where $ch(L_1)$ is to be understood as a complex characteristic class of $T'(M)$.

THEOREM 3.5.5 (RIEMANN-ROCH). *Let* $Td(T'M)$ *be the Todd class defined earlier by the generating function* $Td(A) = \prod_\nu x_\nu/(1 - e^{-x_\nu})$. *Then*

$$\text{index}(V, \text{Dolbeault}) = \int_M Td(T'M) \wedge ch(V).$$

PROOF: We must simply identify $\hat{A}(TM) \wedge ch(L_1)$ with $Td(T'M)$. We perform a computation in characteristic classes using the splitting principal.

We formally decompose $T'M = \tilde{L}_1 \oplus \cdots \oplus \tilde{L}_n$ as the direct sum of line bundles. Then $\Lambda^n(T'M) = \tilde{L}_1 \otimes \cdots \otimes \tilde{L}_n$ so $c_1(\Lambda^n T'M) = c_1(\tilde{L}_1) + \cdots + c_1(\tilde{L}_n) = x_1 + \cdots + x_n$. Since $L_1 \otimes L_1 = \Lambda^n T'M$, we conclude $c_1(L_1) = \frac{1}{2}(x_1 + \cdots + x_n)$ so that:

$$ch(L_1) = \prod_\nu e^{x_\nu/2}.$$

Therefore:

$$\hat{A}(M) \wedge ch(L_1) = \prod_\nu x_\nu \frac{e^{x_\nu/2}}{e^{x_\nu/2} - e^{-x_\nu/2}}$$

$$= \prod_\nu \frac{x_\nu}{1 - e^{-x_\nu}} = Td(T'M).$$

Of course, this procedure is only valid if the given bundle does in fact split as the direct sum of line bundles. We can make this procedure a correct way of calculating characteristic classes by using flag manifolds or by actually calculating on the group representaton and using the fact that the diagonizable matrices are dense.

It is worth giving another proof of the Riemann-Roch formula to ensure that we have not made a mistake of sign somewhere in all our calculations. Using exactly the same multiplicative considerations as we used earlier and using the qualitative form of the formala for index$(V, \text{Dolbeault})$ given by the SPIN_c complex, it is immediate that there is some formula of the form:

$$\text{index}(V, \text{Dolbeault}) = \int_N \sum_{s+t=n} Td'_s(T'M) \wedge c^t ch_t(V)$$

where c is some universal constant to be determined and where Td'_s is some characteristic form of $T'M$.

If $M = \mathbf{CP}_j$, then we shall show in Lemma 3.6.8 that the arithmetic genus of $\mathbf{CP}_j$ is 1. Using the multiplicative property of the Dolbeault complex, we conclude the arithmetic genus of $\mathbf{CP}_{j_1} \times \cdots \times \mathbf{CP}_{j_k}$ is 1 as well. If M^c_ρ are the manifolds of Lemma 2.3.4 then if we take $V = 1$ we conclude:

$$1 = \text{index}(1, \text{Dolbeault}) = \text{ arithmetic genus of } \mathbf{CP}_{j_1} \times \cdots \times \mathbf{CP}_{j_k}$$

$$= \int_{M^c_\rho} Td'(T'M).$$

We verified in Lemma 2.3.5 that $Td(T'M)$ also has this property so by the uniqueness assertion of Lemma 2.3.4 we conclude $Td = Td'$. We take $m = 2$ and decompose:

$$\Lambda^0 \oplus \Lambda^2 = \Lambda^{0,0} \oplus \Lambda^{1,1} \qquad \text{and} \qquad \Lambda^1 = \Lambda^{1,0} \oplus \Lambda^{0,1}$$

so that
$$\Lambda^e - \Lambda^o = (\Lambda^{0,0} - \Lambda^{0,1}) \otimes (\Lambda^{0,0} - \Lambda^{1,0}).$$

In dimension 2 (and more generally if M is Kaehler), it is an easy exercise to compute that $\Delta = 2(\bar{\partial}\delta'' + \delta''\bar{\partial})$ so that the harmonic spaces are the same therefore:

$$\chi(M) = \text{index}(\Lambda^{0,0}, \text{Dolbeault}) - \text{index}(\Lambda^{1,0}, \text{Dolbeault})$$
$$= -c \int_M ch(\Lambda^{1,0}) = c \int_M ch(T'M) = c \int_M e(M).$$

The Gauss-Bonnet theorem (or the normalization of Lemma 2.3.3) implies that the normalizing constant $c = 1$ and gives another equivalent proof of the Riemann-Roch formula.

There are many applications of the Riemann-Roch theorem. We present a few in dimension 4 to illustrate some of the techniques involved. If $\dim M = 4$, then we showed earlier that:

$$c_2(T'M) = e_2(TM) \qquad \text{and} \qquad (2c_2 - c_1^2)(T'M) = p_1(T'M).$$

The Riemann-Roch formula expresses the arithmetic genus of M in terms of c_2 and c_1^2. Consequently, there is a formula:

$$\text{arithmetic genus} = a_1 \int_M e_2(TM) + a_2 \int_M p_1(TM)/3$$
$$= a_1 \chi(M) + a_2 \text{ signature}(M)$$

where a_1 and a_2 are universal constants. If we consider the manifolds $S^2 \times S^2$ and $\mathbf{CP}_2$ we derive the equations:

$$1 = a_1 \cdot 4 + a_2 \cdot 0 \qquad \text{and} \qquad 1 = a_1 \cdot 3 + a_2 \cdot 1$$

so that $a_1 = a_2 = 1/4$ which proves:

LEMMA 3.5.6. *If M is an almost complex manifold of real dimension 4, then:*
$$\text{arithmetic genus}(M) = \{\chi(M) + \text{signature}(M)\}/4.$$

Since the arithmetic genus is always an integer, we can use this result to obtain some non-integrability results:

COROLLARY 3.5.7. *The following manifolds do not admit almost complex structures:*
(a) S^4 (the four dimensional sphere).
(b) $\mathbf{CP}_2$ with the reversed orientation.

(c) $M_1 \# M_2$ where the M_i are 4-dimensional manifolds admitting almost complex structures (# denotes connected sum).

PROOF: We assume the contrary in each of these cases and attempt to compute the arithmetic genus:

$$\text{a.g.}(S^4) = \tfrac{1}{4}(2+0) = \tfrac{1}{2}$$
$$\text{a.g.}(-\mathbf{C}P_2) = \tfrac{1}{4}(3-1) = \tfrac{1}{2}$$
$$\text{a.g.}(M_1 \# M_2) = \tfrac{1}{4}(\chi(M_1 \# dM_2) + \text{sign}(M_1 \# M_2)$$
$$= \tfrac{1}{4}(\chi(M_1) + \chi(M_2) - 2 + \text{sign}(M_1) + \text{sign}(M_2))$$
$$= \text{a.g.}(M_1) + \text{a.g.}(M_2) - \tfrac{1}{2}.$$

In none of these examples is the arithmetic genus an integer which shows the impossibility of constructing the desired almost complex structure.

3.6. A Review of Kaehler Geometry.

In the previous subsection, we proved the Riemann-Roch theorem using the SPIN_c complex. This was an essential step in the proof even if the manifold was holomorphic.

Let M be holomorphic and let the coefficient bundle V be holomorphic (i.e., the transition functions are holomorphic). We can define the Dolbeault complex directly by defining:

$$\bar{\partial}_V : C^\infty(\Lambda^{0,q} \otimes V) \to C^\infty(\Lambda^{0,q+1} \otimes V) \qquad \text{by } \bar{\partial}_V(\omega \otimes s) = \bar{\partial}\omega \otimes s$$

where s is a local holomorphic section to V. It is immediate $\bar{\partial}_V \bar{\partial}_V = 0$ and that this is an elliptic complex; the index of this elliptic complex is just $\text{index}(V, \text{Dolbeault})$ as defined previously.

We let $a_n(x, V, \text{Dolbeault})$ be the invariant of the heat equation for this elliptic complex; we use the notation $a_j(x, \text{Dolbeault})$ when V is the trivial bundle. Then:

REMARK 3.6.1. *Let $m = 2n > 2$. Then there exists a unitary Riemannian metric on the m-torus (with its usual complex structure) and a point x such that:*
(a) $a_j(x, \text{Dolbeault}) \neq 0$ for j even and $j \geq n$,
(b) $a_m(x, \text{Dolbeault}) \neq Td_n$ where both are viewed as scalar invariants.

The proof of this is quite long and combinatorial and is explained elsewhere; we simply present the result to demonstrate that it is not in general possible to prove the Riemann-Roch theorem directly by heat equation methods.

The difficulty is that the metric and the complex structure do not fit together properly. Choose local holomorphic coordinates $z = (z_1, \ldots, z_n)$ and extend the metric G to be Hermitian on $T(M) \otimes \mathbf{C}$ so that T' and T'' are orthogonal. We define:

$$g_{j\bar{k}} = G(\partial/\partial z_j, \partial/\partial z_k)$$

then the matrix $g_{j\bar{k}}$ is a positive definite Hermitian matrix which determines the original metric on $T(M)$:

$$ds^2 = 2 \sum_{j,k} g_{j\bar{k}} dz^j \cdot d\bar{z}^k$$

$$G(\partial/\partial x_j, \partial/\partial x_k) = G(\partial/\partial y_j, \partial/\partial y_k) = (g_{j\bar{k}} + g_{k\bar{j}})$$

$$G(\partial/\partial x_j, \partial/\partial y_k) = -G(\partial/\partial y_j, \partial/\partial x_k) = \frac{1}{i}(g_{j\bar{k}} - g_{k\bar{j}})$$

We use this tensor to define the Kaehler 2-form:

$$\Omega = i \sum_{j,k} g_{j\bar{k}} dz^j \wedge d\bar{z}^k.$$

This is a real 2-form which is defined by the identity:

$$\Omega[X,Y] = -G(X,JY).$$

(This is a slightly different sign convention from that sometimes followed.) The manifold M is said to be Kaehler if Ω is closed and the heat equation gives a direct proof of the Riemann-Roch theorem for Kaehler manifolds.
We introduce variables:

$$g_{j\bar{k}/l} \quad \text{and} \quad g_{j\bar{k}/\bar{l}}$$

for the jets of the metric. On a Riemannian manifold, we can always find a coordinate system in which all the 1-jets of the metric vanish at a point; this concept generalizes as follows:

LEMMA 3.6.2. *Let M be a holomorphic manifold and let G be a unitary metric on M. The following statements are equivalent:*
(a) The metric is Kaeher (i.e., $d\Omega = 0$).
(b) For every $z_0 \in M$ there is a holomorphic coordinate system Z centred at z_0 so that $g_{j\bar{k}}(Z,G)(z_0) = \delta_{jk}$ and $g_{j\bar{k}/l}(Z,G)(z_0) = 0$.
(c) For every $z_0 \in M$ there is a holomorphic coordinate system Z centred at z_0 so $g_{j\bar{k}}(Z,G)(z_0) = \delta_{jk}$ and so that all the 1-jets of the metric vanish at z_0.

PROOF: We suppose first the metric is Kaehler and compute:

$$d\Omega = i\sum_{j,k}\{g_{j\bar{k}/l}dz^l \wedge dz^j \wedge d\bar{z}^k + g_{j\bar{k}/\bar{l}}d\bar{z}^l \wedge dz^j \wedge d\bar{z}^k\}.$$

Thus Kaehler is equivalent to the conditions:

$$g_{j\bar{k}/l} - g_{l\bar{k}/j} = g_{j\bar{k}/\bar{l}} - g_{j\bar{l}/\bar{k}} = 0.$$

By making a linear change of coordinates, we can assume that the holomorphic coordinate system is chosen to be orthogonal at the center z_0. We let

$$z'_j = z_j + \sum c^j_{kl}z_k z_l \qquad \text{(where } c^j_{kl} = c^j_{lk})$$

and compute therefore:

$$dz'_j = dz_j + 2\sum c^j_{kl}z_l dz_k$$

$$\partial/\partial z'_j = \partial/\partial z_j - 2\sum c^k_{jl}z_l \partial/\partial z_k + O(z^2)$$

$$g'_{j\bar{k}} = g_{j\bar{k}} - 2\sum c^k_{jl}z_l + \text{ terms in } \bar{z} + O(|z|^2)$$

$$g'_{j\bar{k}/l} = g_{j\bar{k}/l} - 2c^k_{jl} \text{ at } z_0.$$

We define $c_{jl}^k = \frac{1}{2}g_{j\bar{k}/l}$ and observe that the Kaehler condition shows this is symmetric in the indices (j, l) so (a) implies (b). To show (b) implies (c) we simply note that $g_{j\bar{k}/\bar{l}} = \overline{g_{k\bar{j}/l}} = 0$ at z_0. To prove (c) implies (a) we observe $d\Omega(z_0) = 0$ so since z_0 was arbitrary, $d\Omega \equiv 0$.

If M is Kaehler, then $\delta\Omega$ is linear in the 1-jets of the metric when we compute with respect to a holomorphic coordinate system. This implies that $\delta\Omega = 0$ so Ω is harmonic. We compute that Ω^k is also harmonic for $1 \leq k \leq n$ and that $\Omega^n = c \cdot \text{dvol}$ is a multiple of the volume form (and in particular is non-zero). This implies that if x is the element in $H^2(M; \mathbf{C})$ defined by Ω using the Hodge decomposition theorem, then $1, x, \ldots, x^n$ all represent non-zero elements in the cohomology ring of M. This can be used to show that there are topological obstructions to constructing Kaehler metrics:

REMARK 3.6.3. If $M = S^1 \times S^{m-1}$ for m even and $m \geq 4$ then this admits a holomorphic structure. No holomorphic structure on M admits a Kaehler metric.

PROOF: Since $H^2(M; \mathbf{C}) = 0$ it is clear M cannot admit a Kaehler metric. We construct holomorphic structures on M as follows: let $\lambda \in \mathbf{C}$ with $|\lambda| > 1$ and let $M_\lambda = \{\mathbf{C}^n - 0\}/\lambda$ (where we identify z and w if $z = \lambda^k w$ for some $k \in \mathbf{Z}$). The M_λ are all topologically $S^1 \times S^{m-1}$; for example if $\lambda \in \mathbf{R}$ and we introduce spherical coordinates (r, θ) on $\mathbf{C}^n = \mathbf{R}^m$ then we are identifying (r, θ) and $(\lambda^k r, \theta)$. If we let $t = \log r$, then we are identifying t with $t + k \log \lambda$ so the manifold is just $[1, \log \lambda] \times S^{m-1}$ where we identify the endpoints of the interval. The topological identification of M_λ for other λ is similar.

We now turn to the problem of constructing a Kaehler metric on $\mathbf{C}P_n$. Let L be the tautological line bundle over $\mathbf{C}P_n$ and let $x = -c_1(L) \in \Lambda^{1,1}(\mathbf{C}P_n)$ be the generator of $H^2(\mathbf{C}P_n; \mathbf{Z})$ discussed earlier. We expand x in local coordinates in the form:

$$x = \frac{i}{2\pi} \sum_{j,k} g_{j\bar{k}} dz^j \wedge d\bar{z}^k$$

and define $G(\partial/\partial z_k, \partial/\partial z_k) = g_{j\bar{k}}$. This gives an invariantly defined form called the Fubini-Study metric on $T_c(\mathbf{C}P_n)$. We will show G is positive definite and defines a unitary metric on the real tangent space. Since the Kaehler form of G, $\Omega = 2\pi x$, is a multiple of x, we conclude $d\Omega = 0$ so G will be a Kaehler metric.

The 2-form x is invariant under the action of $U(n+1)$. Since $U(n+1)$ acts transitively on $\mathbf{C}P_n$, it suffices to show G is positive definite and symmetric at a single point. Using the notation of section 2.3, let $U_n = \{z \in \mathbf{C}P_n : z_{n+1}(z) \neq 0\}$. We identify U_n with $\mathbf{C}^n$ by identifying $z \in \mathbf{C}^n$ with the line in $\mathbf{C}^{n+1}$ through the point $(z,1)$. Then:

$$x = \frac{i}{2\pi}\partial\bar{\partial}\log(1+|z|^2)$$

so at the center $z = 0$ we compute that:

$$x = \frac{i}{2\pi}\sum_j dz^j \wedge d\bar{z}^j \quad \text{so that} \quad g_{j\bar{k}} = \delta_{j\bar{k}}.$$

If G' is any other unitary metric on $\mathbf{CP}_n$ which is invariant under the action of $U(n+1)$, then G' is determined by its value on $T_c(\mathbf{CP}_n)$ at a single point. Since $U(n)$ preserves the origin of $U_n = \mathbf{C}^n$, we conclude $G' = cG$ for some $c > 0$. This proves:

LEMMA 3.6.4.
(a) There exists a unitary metric G on $\mathbf{CP}_n$ which is invariant under the action of $U(n+1)$ such that the Kaehler form of G is given by $\Omega = 2\pi x$ where $x = -c_1(L)$ (L is the tautological bundle over $\mathbf{CP}_n$). G is a Kaehler metric.
(b) $\int_{\mathbf{CP}_n} x^n = 1$.
(c) If G' is any other unitary metric on $\mathbf{CP}_n$ which is invariant under the action of $U(n+1)$ then $G' = cG$ for some constant $c > 0$.

A holomorphic manifold M is said to be Hodge if it admits a Kaehler metric G such that the Kaehler form Ω has the form $\Omega = c_1(L)$ for some holomorphic line bundle L over M. Lemma 3.6.4 shows $\mathbf{CP}_n$ is a Hodge manifold. Any submanifold of a Hodge manifold is again Hodge where the metric is just the restriction of the given Hodge metric. Thus every algebraic variety is Hodge. The somewhat amazing fact is that the converse is true:

REMARK 3.6.5. A holomorphic manifold M is an algebraic variety (i.e., is holomorphically equivalent to a manifold defined by algebraic equations in $\mathbf{CP}_n$ for some n) if and only if it admits a Hodge metric.

We shall not prove this remark but simply include it for the sake of completeness. We also note in passing that there are many Kaehler manifolds which are not Hodge. The Riemann period relations give obstructions on complex tori to those tori being algebraic.

We now return to our study of Kaehler geometry. We wish to relate the $\bar{\partial}$ cohomology of M to the ordinary cohomology. Define:

$$H^{p,q}(M) = \ker\bar{\partial}/\operatorname{im}\bar{\partial} \quad \text{in bi-degree } (p,q).$$

Since the Dolbeault complex is elliptic, these groups are finite dimensional using the Hodge decomposition theorem.

LEMMA 3.6.6. *Let M be Kaehler and let $\Omega = i \sum g_{j\bar{k}} dz^j \wedge d\bar{z}^k$ be the Kaehler form. Let $d = \partial + \bar{\partial}$ and let $\delta = \delta' + \delta''$. Then we have the identities:*

(a) $\bar{\partial} \operatorname{int}(\Omega) - \operatorname{int}(\Omega)\bar{\partial} = i\delta'$.

(b) $\bar{\partial}\delta' + \delta'\bar{\partial} = \partial\delta'' + \delta''\partial = 0$.

(c) $d\delta + \delta d = 2(\partial\delta' + \delta'\partial) = 2(\bar{\partial}\delta'' + \delta''\bar{\partial})$.

(d) We can decompose the de Rham cohomology in terms of the Dolbeault cohomology:

$$H^n(M; \mathbf{C}) = \bigoplus_{p+q=n} H^{p,q}(M; \mathbf{C}).$$

(e) There are isomorphisms $H^{p,q} \simeq H^{q,p} \simeq H^{n-p,n-q} \simeq H^{n-q,n-p}$.

PROOF: We first prove (a). We define:

$$A = \bar{\partial} \operatorname{int}(\Omega) - \operatorname{int}(\Omega)\bar{\partial} - i\delta'.$$

If we can show that A is a 0th order operator, then A will be a functorially defined endomorphism linear in the 1-jets of the metric. This implies $A = 0$ by Lemma 3.6.2. Thus we must check $\sigma_L(A) = 0$. We choose an orthonormal basis $\{e_j\}$ for $T(M) = T^*(M)$ so $Je_j = e_{j+n}$, $1 \leq j \leq n$. By Lemma 3.5.2, it suffices to show that:

$$\operatorname{ext}(e_j - ie_{j+n}) \operatorname{int}(\Omega) - \operatorname{int}(\Omega) \operatorname{ext}(e_j - ie_{j+n}) = -i \cdot \operatorname{int}(e_j - ie_{j+n})$$

for $1 \leq j \leq n$. Since $g_{j\bar{k}} = \frac{1}{2}\delta_{j\bar{k}}$, Ω is given by

$$\Omega = \frac{i}{2} \sum_j (e_j + ie_{j+n}) \wedge (e_j - ie_{j+n}) = \sum_j e_j \wedge e_{j+n}.$$

The commutation relations:

$$\operatorname{int}(e_k) \operatorname{ext}(e_l) + \operatorname{ext}(e_l) \operatorname{int}(e_k) = \delta_{k,l}$$

imply that $\operatorname{int}(e_k \wedge e_{k+n}) = -\operatorname{int}(e_k)\operatorname{int}(e_{k+n})$ commutes with $\operatorname{ext}(e_j - ie_{j+n})$ for $k \neq j$ so these terms disappear from the commutator. We must show:

$$\operatorname{ext}(e_j - ie_{j+n}) \operatorname{int}(e_j \wedge e_{j+n}) - \operatorname{int}(e_j \wedge e_{j+n}) \operatorname{ext}(e_j - ie_{j+n})$$
$$= -i \cdot \operatorname{int}(e_j - ie_{j+n}).$$

This is an immediate consequence of the previous commutation relations for int and ext together with the identity $\operatorname{int}(e_j) \operatorname{int}(e_k) + \operatorname{int}(e_k) \operatorname{int}(e_j) = 0$. This proves (a).

From (a) we compute:

$$\bar{\partial}\delta' + \delta'\bar{\partial} = i(-\bar{\partial}\operatorname{int}(\Omega)\bar{\partial}) + i(\bar{\partial}\operatorname{int}(\Omega)\bar{\partial}) = 0$$

and by taking complex conjugates $\partial\delta'' + \delta''\partial = 0$. This shows (b). Thus

$$\Delta = (d+\delta)^2 = d\delta + \delta d = \{\partial\delta' + \delta'\partial\} + \{\bar{\partial}\delta'' + \delta''\partial\} = \{\partial + \delta'\}^2 + \{\bar{\partial} + \delta''\}^2$$

since all the cross terms cancel. We apply (a) again to compute:

$$\partial\bar{\partial}\operatorname{int}(\Omega) - \partial\operatorname{int}(\Omega)\bar{\partial} = i\partial\delta'$$
$$\bar{\partial}\operatorname{int}(\Omega)\partial - \operatorname{int}(\Omega)\bar{\partial}\partial = i\delta'\partial$$

which yields the identity:

$$i(\partial\delta' + \delta'\partial) = \partial\bar{\partial}\operatorname{int}(\Omega) - \partial\operatorname{int}(\Omega)\bar{\partial} + \bar{\partial}\operatorname{int}(\Omega)\partial - \operatorname{int}(\Omega)\bar{\partial}\partial.$$

Since Ω is real, we take complex conjugate to conclude:

$$-i(\bar{\partial}\delta'' + \delta''\bar{\partial}) = \bar{\partial}\partial\operatorname{int}(\Omega) - \bar{\partial}\operatorname{int}(\Omega)\partial + \partial\operatorname{int}(\Omega)\bar{\partial} - \operatorname{int}(\Omega)\partial\bar{\partial}$$
$$= -i(\partial\delta' + \delta'\partial).$$

This shows $(\bar{\partial}\delta'' + \delta''\bar{\partial}) = (\partial\delta' + \delta'\partial)$ so $\Delta = 2(\bar{\partial}\delta'' + \delta''\bar{\partial})$ which proves (c).

$H^n(M; \mathbf{C})$ denotes the de Rham cohomology groups of M. The Hodge decomposition theorem identifies these groups with the null-space of $(d+\delta)$. Since $N(d + \delta) = N(\bar{\partial} + \delta'')$, this proves (d). Finally, taking complex conjugate and applying $*$ induces the isomorphisms:

$$H^{p,q} \simeq H^{q,p} \simeq H^{n-q,n-p} \simeq H^{n-p,n-q}.$$

In particular $\dim H^1(M; \mathbf{C})$ is even if M admits a Kaehler metric. This gives another proof that $S^1 \times S^{2n-1}$ does not admit a Kaehler metric for $n > 1$.

The duality operations of (e) can be extended to the Dolbeault complex with coefficients in a holomorphic bundle V. If V is holomorphic, the transition functions are holomorphic and hence commute with $\bar{\partial}$. We define:

$$\bar{\partial} : C^\infty(\Lambda^{p,q} \otimes V) \to (\Lambda^{p,q+1} \otimes V)$$

by defining $\bar{\partial}(\omega \otimes s) = \bar{\partial}\omega \otimes s$ relative to any local holomorphic frame for V. This complex is equivalent to the one defined in section 3.5 and defines cohomology classes:

$$H^{p,q}(V) = \ker \bar{\partial}/\operatorname{im}\bar{\partial} \qquad \text{on } C^\infty(\Lambda^{p,q} \otimes V).$$

It is immediate from the definition that:

$$H^{p,q}(V) = H^{0,q}(\Lambda^{p,0} \otimes V).$$

We define

$$\text{index}(V, \bar{\partial}) = \text{index}(V, \text{Dolbeault}) = \sum (-1)^q \dim H^{0,q}(V)$$

to be the index of this elliptic complex.

There is a natural duality (called Serre duality) which is conjugate linear $\bar{*}: H^{p,q}(V) \rightarrow H^{n-p,n-q}(V^*)$, where V^* is the dual bundle. We put a unitary fiber metric on V and define:

$$T(s_1)(s_2) = s_2 \cdot s_1$$

so $T: V \rightarrow V^*$ is conjugate linear. If $*: \Lambda^k \rightarrow \Lambda^{m-k}$ is the ordinary Hodge operator, then it extends to define $*: \Lambda^{p,q} \rightarrow \Lambda^{n-q,n-p}$. The complex conjugate operator defines $\bar{*}: \Lambda^{p,q} \rightarrow \Lambda^{n-p,n-q}$. We extend this to have coefficients in V by defining:

$$\bar{*}(\omega \otimes s) = \bar{*}\omega \otimes Ts$$

so $\bar{*}: \Lambda^{p,q} \otimes V \rightarrow \Lambda^{n-p,n-q} \otimes V^*$ is conjugate linear. Then:

LEMMA 3.6.7. Let δ_V'' be the adjoint of $\bar{\partial}_V$ on $C^\infty(\Lambda^{p,*} \otimes V)$ with respect to a unitary metric on M and a Hermitian metric on V. Then
(a) $\delta'' = -\bar{*}\bar{\partial}_V\bar{*}$.
(b) $\bar{*}$ induces a conjugate linear isomorphism between the groups $H^{p,q}(V)$ and $H^{n-p,n-q}(V^*)$.

PROOF: It is important to note that this lemma, unlike the previous one, does not depend upon having a Kaehler metric. We suppose first V is holomorphically trivial and the metric on V is flat. Then $\Lambda \otimes V = \Lambda \otimes 1^k$ for some k and we may suppose $k = 1$ for the sake of simplicity. We noted

$$\delta = - * d* = -\bar{*}d\bar{*}$$

in section 1.5. This decomposes

$$\delta = \delta' + \delta'' = -\bar{*}\partial\bar{*} - \bar{*}\bar{\partial}\bar{*}.$$

Since $\bar{*}: \Lambda^{p,q} \rightarrow \Lambda^{n-p,n-q}$, we conclude $\delta' = -\bar{*}\partial\bar{*}$ and $\delta'' = -\bar{*}\bar{\partial}\bar{*}$. This completes the proof if V is trivial. More generally, we define

$$A = \delta_V'' + \bar{*}\bar{\partial}\bar{*}.$$

This must be linear in the 1-jets of the metric on V. We can always normalize the choice of local holomorphic frame so the 1-jets vanish at a basepoint z_0 and thus $A = 0$ in general which proves (a). We identify $H^{p,q}(V)$ with $\mathrm{N}(\bar{\partial}_V) \cap \mathrm{N}(\delta_V'')$ using the Hodge decomposition theorem. Thus $\bar{*} \colon H^{p,q}(V) \to H^{n-p,n-q}(V^*)$. Since $\bar{*}^2 = \pm 1$, this completes the proof.

We defined $x = -c_1(L) \in \Lambda^{1,1}(\mathbf{CP}_n)$ for the standard metric on the tautological line bundle over $\mathbf{CP}_n$. In section 2.3 we showed that x was harmonic and generates the cohomology ring of $\mathbf{CP}_n$. Since x defines a Kaehler metric, called the Fubini-Study metric, Lemma 3.6.6 lets us decompose $H^n(\mathbf{CP}_n; \mathbf{C}) = \bigoplus_{p+q=n} H^{p,q}(\mathbf{CP}_n)$. Since $x^k \in H^{k,k}$ this shows:

LEMMA 3.6.8. *Let $\mathbf{CP}_n$ be given the Fubini-Study metric. Let $x = -c_1(L)$. Then $H^{k,k}(M) \simeq \mathbf{C}$ is generated by x^k for $1 \le k \le n$. $H^{p,q}(M) = 0$ for $p \ne q$. Thus in particular,*

$$\mathrm{index}(\bar{\partial}) = \text{arithmetic genus} = \sum (-1)^k \dim H^{0,k} = 1 \qquad \text{for } \mathbf{CP}_n.$$

Since the arithmetic genus is multiplicative with respect to products,

$$\mathrm{index}(\bar{\partial}) = 1 \qquad \text{for } \mathbf{CP}_{n_1} \times \cdots \times \mathbf{CP}_{n_k}.$$

We used this fact in the previous subsection to derive the normalizing constants in the Riemann-Roch formula.

We can now present another proof of the Riemann-Roch theorem for Kaehler manifolds which is based directly on the Dolbeault complex and not on the SPIN_c complex. Let Z be a holomorphic coordinate system and let

$$g_{u\bar{v}} = G(\partial/\partial z_u, \partial/\partial z_v)$$

represent the components of the metric tensor. We introduce additional variables:

$$g_{u\bar{v}/\alpha\bar{\beta}} = d_z^\alpha d_{\bar{z}}^\beta g_{u\bar{v}}$$
$$\text{for } 1 \le u, v \le n, \ \alpha = (\alpha_1, \ldots, \alpha_n), \text{ and } \beta = (\beta_1, \ldots, \beta_n).$$

These variables represent the formal derivatives of the metric.

We must also consider the Dolbeault complex with coefficients in a holomorphic vector bundle V. We choose a local fiber metric H for V. If $s = (s_1, \ldots, s_k)$ is a local holomorphic frame for V, we introduce variables:

$$h_{p,\bar{q}} = H(s_p, s_q), \qquad h_{p,\bar{q}/\alpha\bar{\beta}} = d_z^\alpha d_{\bar{z}}^\beta h_{p\bar{q}} \qquad \text{for } 1 \le p, q \le k = \dim V.$$

If Z is a holomorphic coordinate system centered at z_0 and if s is a local holomorphic frame, we normalize the choice so that:

$$g_{u\bar{v}}(Z, G)(z_0) = \delta_{u,v} \qquad \text{and} \qquad h_{p\bar{q}}(s, H)(z_0) = \delta_{p,q}.$$

The coordinate frame $\{\partial/\partial z_u\}$ and the holomorphic frame $\{s_p\}$ for V are orthogonal at z_0. Let $\mathcal{R}$ be the polynomial algebra in these variables. If $P \in \mathcal{R}$ we can evaluate $P(Z, G, s, H)(z_0)$ once a holomorphic frame Z is chosen and a holomorphic frame s is chosen. We say P is invariant if $P(Z, G, s, H)(z_0) = P(G, H)(z_0)$ is independent of the particular (Z, s) chosen; let $\mathcal{R}_{n,k}^c$ denote the sub-algebra of all invariant polynomials. As before, we define:

$$\operatorname{ord}(g_{u,\bar{v}/\alpha\bar{\beta}}) = \operatorname{ord}(s_{p\bar{q}/\alpha\bar{\beta}}) = |\alpha| + |\beta|.$$

We need only consider those variables of positive order as $g_{u,\bar{v}} = \delta_{j\bar{v}}$ and $s_{p,\bar{q}} = \delta_{p,q}$ at z_0. If $\mathcal{R}_{n,\nu,k}^c$ denotes the subspace of invariant polynomials homogenous of order ν in the jets of the metrics (G, H), then there is a direct sum decomposition:

$$\mathcal{R}_{n,k}^c = \bigoplus \mathcal{R}_{n,\nu,k}^c.$$

In the real case in studying the Euler form, we considered the restriction map from manifolds of dimension m to manifolds of dimension $m - 1$. In the complex case, there is a natural restriction map

$$r \colon \mathcal{R}_{n,\nu,k}^c \to \mathcal{R}_{n-1,\nu,k}^c$$

which lowers the complex dimension by one (and the real dimension by two). Algebraically, r is defined as follows: let

$$\deg_k(g_{u\bar{v}/\alpha\bar{\beta}}) = \delta_{k,u} + \alpha(k) \qquad \deg_{\bar{k}}(g_{u\bar{v}/\alpha\bar{\beta}}) = \delta_{k,v} + \beta(k)$$

$$\deg_k(h_{p\bar{q}/\alpha\bar{\beta}}) = \alpha(k) \qquad \deg_{\bar{k}}(h_{p\bar{q}/\alpha\bar{\beta}}) = \beta(k).$$

We define:

$$r(g_{u\bar{v}/\alpha\bar{\beta}}) = \begin{cases} g_{u\bar{v}/\alpha\bar{\beta}} & \text{if } \deg_n(g_{u\bar{v}/\alpha\bar{\beta}}) + \deg_{\bar{n}}(g_{u\bar{v}/\alpha\bar{\beta}}) = 0 \\ 0 & \text{if } \deg_n(g_{u\bar{v}/\alpha\bar{\beta}}) + \deg_{\bar{n}}(g_{u\bar{v}/\alpha\bar{\beta}}) > 0 \end{cases}$$

$$r(h_{p\bar{q}/\alpha\bar{\beta}}) = \begin{cases} h_{p\bar{q}/\alpha\bar{\beta}} & \text{if } \deg_n(h_{p\bar{q}/\alpha\bar{\beta}}) + \deg_{\bar{n}}(h_{p\bar{q}/\alpha\bar{\beta}}) = 0 \\ 0 & \text{if } \deg_n(h_{p\bar{q}/\alpha\bar{\beta}}) + \deg_{\bar{n}}(h_{p\bar{q}/\alpha\bar{\beta}}) > 0. \end{cases}$$

This defines a map $r \colon \mathcal{R}_n \to \mathcal{R}_{n-1}$ which is an algebra morphism; we simply set to zero those variables which do not belong to $\mathcal{R}_{n-1}$. It is immediate that r preserves both invariance and the order of a polynomial. Geometrically, we are considering manifolds of the form $M^n = M^{n-1} \times T_2$ where T_2 is the flat two torus, just as in the real case we considered manifolds $M^m = M^{m-1} \times S^1$.

Let $\mathcal{R}_{n,p,k}^{ch}$ be the space of p-forms generated by the Chern forms of $T_c(M)$ and by the Chern forms of V. We take the holomorphic connection defined by the metrics G and H on $T_c(M)$ and on V. If $P \in \mathcal{R}_{n,m,k}^{ch}$ then $*P$ is a scalar invariant. Since P vanishes on product metrics of the form $M^{n-1} \times T_2$ which are flat in one holomorphic direction, $r(*P) = 0$. This is the axiomatic characterization of the Chern forms which we shall need.

THEOREM 3.6.9. *Let the complex dimension of M be n and let the fiber dimension of V be k. Let $P \in \mathcal{R}^c_{n,\nu,k}$ and suppose $r(P) = 0$. Then:*
(a) If $\nu < 2n$ then $P(G, H) = 0$ for all Kaehler metrics G.
(b) If $\nu = 2n$ then there exists a unique $Q \in \mathcal{R}^{ch}_{n,m,k}$ so $P(G, H) = *Q(G, H)$ for all Kaehler metrics G.

As the proof of this theorem is somewhat technical, we shall postpone the proof until section 3.7. This theorem is false if we don't restrict to Kaehler metrics. There is a suitable generalization to form valued invariants. We can apply Theorem 3.6.9 to give a second proof of the Riemann-Roch theorem:

THEOREM 3.6.10. *Let V be a holomorphic bundle over M with fiber metric H and let G be a Kaehler metric on M. Let $\bar{\partial}_V : C^\infty(\Lambda^{0,q} \otimes V) \to C^\infty(\Lambda^{0,q+1} \otimes V)$ denote the Dolbeault complex with coefficients in V and let $a_\nu(x, \bar{\partial}_V)$ be the invariants of the heat equation. $a_\nu \in \mathcal{R}^c_{n,\nu,k}$ and*

$$\int_M a_\nu(x, \bar{\partial}_V) \, \mathrm{dvol}(x) = \begin{cases} 0 & \text{if } \nu \neq 2n \\ \mathrm{index}(\bar{\partial}_V) & \text{if } \nu = 2n. \end{cases}$$

Then
(a) $a_\nu(x, \bar{\partial}_V) = 0$ for $\nu < 2n$.
(b) $a_{2n}(x, \bar{\partial}_V) = *\{Td(T_cM) \wedge ch(V)\}_m$.

PROOF: We note this implies the Riemann-Roch formula. By remark 3.6.1, we note this theorem is false in general if the metric is not assumed to be Kaehler. The first assertions of the theorem (including the homogeneity) follow from the results of Chapter 1, so it suffices to prove (a) and (b). The Dolbeault complex is multiplicative under products $M = M^{n-1} \times T_2$. If we take the product metric and assume V is the pull-back of a bundle over M_{n-1} then the natural decomposition:

$$\Lambda^{0,q} \otimes V \simeq \Lambda^{0,q}(M^{n-1}) \otimes V \; \oplus \; \Lambda^{0,q-1}(M^{n-1}) \otimes V$$

shows that $r(a_\nu) = 0$. By Theorem 3.6.9, this shows $a_\nu = 0$ for $\nu < 2n$ proving (a).

In the limiting case $\nu = 2n$, we conclude $*a_{2n}$ is a characteristic $2n$ form. We first suppose $m = 1$. Any one dimensional complex manifold is Kaehler. Then

$$\mathrm{index}(\bar{\partial}_V) = a_1 \int_M c_1(T_c(M)) + a_2 \int_M c_1(V)$$

where a_1 and a_2 are universal constants to be determined. Lemma 3.6.6 implies that $\dim H^{0,0} = \dim H^0 = 1 = \dim H^2 = \dim H^{1,1}$ while $\dim H^{1,0} = \dim H^{0,1} = g$ where g is the genus of the manifold. Consequently:

$$\chi(M) = 2 - 2g = 2 \, \mathrm{index}(\bar{\partial}).$$

We specialize to the case $M = S^2$ and $V = 1$ is the trivial bundle. Then $g = 0$ and

$$1 = \text{index}(\bar{\partial}_V) = a_1 \int_M c_1(T_c(M)) = 2a_1$$

so $a_1 = \frac{1}{2}$. We now take the line bundle $V = \Lambda^{1,0}$ so

$$\text{index}(\bar{\partial}_V) = g - 1 = -1 = 1 + a_2 \int_M c_1(T_c^*) = 1 - 2a_2$$

so that $a_2 = 1$. Thus the Riemann-Roch formula in dimension 1 becomes:

$$\text{index}(\bar{\partial}_V) = \frac{1}{2} \int_M c_1(T_c(M)) + \int_M c_1(V).$$

We now use the additive nature of the Dolbeault complex with respect to V and the multiplicative nature with respect to products $M_1 \times M_2$ to conclude that the characteristic m-form $*a_m$ must have the form:

$$\{T_d'(T_cM) \wedge ch(V)\}$$

where we use the fact the normalizing constant for $c_1(V)$ is 1 if $m = 1$. We now use Lemma 2.3.5(a) together with the fact that $\text{index}(\bar{\partial}_V) = 1$ for products of complex projective spaces to show $Td' = Td$. This completes the proof.

3.7. An Axiomatic Characterization
Of the Characteristic Forms for
Holomorphic Manifolds with Kaehler Metrics.

In subsection 3.6 we gave a proof of the Riemann-Roch formula for Kaehler manifolds based on Theorem 3.6.9 which gave an axiomatic characterization of the Chern forms. This subsection will be devoted to giving the proof of Theorem 3.6.9. The proof is somewhat long and technical so we break it up into a number of steps to describe the various ideas which are involved.

We introduce the notation:

$$g_{u_0 v_0 / u_1 \ldots u_j \bar{v}_1 \ldots \bar{v}_k}$$

for the jets of the metric on M. Indices (u, v, w) will refer to $T_c(M)$ and will run from 1 thru $n = \frac{m}{2}$. We use $(\bar{u}, \bar{v}, \bar{w})$ for anti-holomorphic indices. The symbol "$*$" will refer to indices which are not of interest in some particular argument. We let A_0 denote a generic monomial. The Kaehler condition is simply the identity:

$$g_{u\bar{v}/w} = g_{w\bar{v}/u} \qquad \text{and} \qquad g_{u\bar{v}/\bar{w}} = g_{u\bar{w}/\bar{v}}.$$

We introduce new variables:

$$g(u_0, \ldots, u_j; \bar{v}_0, \ldots, \bar{v}_k) = g_{u_0 v_0 / u_1 \ldots u_j \bar{v}_1 \ldots \bar{v}_k}.$$

If we differentiate the Kaehler identity, then we conclude $g(\vec{u}; \vec{v})$ is symmetric in $\vec{u} = (u_0, \ldots, u_j)$ and $\vec{v} = (v_0, \ldots, v_j)$. Consequently, we may also use the multi-index notation $g(\alpha; \bar{\beta})$ to denote these variables. We define:

$$\text{ord}(g(\alpha; \bar{\beta})) = |\alpha| + |\beta| - 2, \quad \deg_u g(\alpha; \bar{\beta}) = \alpha(u), \quad \deg_{\bar{u}} g(\alpha; \bar{\beta}) = \beta(u).$$

We already noted in Lemma 3.6.2 that it was possible to normalize the coordinates so $g_{u\bar{v}/w}(z_0) = g_{u\bar{v}/\bar{w}}(z_0) = 0$. The next lemma will permit us to normalize coordinates to arbitrarily high order modulo the action of the unitary group:

LEMMA 3.7.1. *Let G be a Kaehler metric and let $z_0 \in M$. Let $\nu \geq 2$ be given. Then there exists a holomorphic coordinate system Z centered at z_0 so that*
(a) $g_{u\bar{v}}(Z, G)(z_0) = \delta_{uv}$.
(b) $g_{u\bar{v}/\alpha}(Z, G)(z_0) = g_{u\bar{v}/\bar{\alpha}}(Z, G)(z_0) = 0$ for $1 \leq |\alpha| \leq \nu - 1$.
(c) *Z is unique modulo the action of the unitary group* U(n) *and modulo coordinate transformations of order $\nu + 1$ in z.*

PROOF: We proceed by induction. It is clear U(n) preserves such coordinate systems. The case $\nu = 2$ is just Lemma 3.6.2 and the uniqueness is clear. We now consider W given which satisfies (a) and (b) for

$1 \le |\alpha| \le \nu - 2$ and define a new coordinate system Z by setting:

$$w_u = z_u + \sum_{|\alpha|=\nu} C_{u,\alpha} z^\alpha$$

where the constants $C_{u,\alpha}$ remain to be determined. Since this is the identity transformation up to order ν, the $\nu - 2$ jets of the metric are unchanged so conditions (a) and (b) are preserved for $|\beta| < \nu - 1$. We compute:

$$\frac{\partial}{\partial z_u} = \frac{\partial}{\partial w_u} + \sum_{v,|\alpha|=\nu} C_{v,\alpha} \alpha(u) z^{\alpha_u} \frac{\partial}{\partial w_v}$$

where α_u is the multi-index defined by the identity $z_u z^{\alpha_u} = z^\alpha$ (and which is undefined if $\alpha(u) = 0$.) This implies immediately that:

$$g_{u\bar{v}}(Z,G) = g_{u\bar{v}}(W,G) + \sum_{|\alpha|=n} C_{v,\alpha} \alpha(u) z^{\alpha_u} + \text{ terms in } \bar{z} + O(z^{\nu+1}).$$

The $\partial/\partial z_u$ and $\partial/\partial w_u$ agree to first order. Consequently:

$$g_{u\bar{v}/\alpha_u}(Z,G) = g_{u\bar{v}/\alpha_u}(W,G) + \alpha! \, C_{v,\alpha} + O(z,\bar{z}).$$

Therefore the symmetric derivatives are given by:

$$g(\alpha; \bar{v})(Z,G)(z_0) = g(\alpha; \bar{v})(W,G)(z_0) + \alpha! \, C_{v,\alpha}$$

The identity $g(\alpha; \bar{v})(Z,G)(z_0) = 0$ determines the $C_{v,\alpha}$ uniquely. We take complex conjugate to conclude $g(v, \bar{\alpha})(Z,G)(z_0) = 0$ as well.

This permits us to choose the coordinate system so all the purely holomorphic and anti-holomorphic derivatives vanish at a single point. We restrict henceforth to variables $g(\alpha; \bar{\beta})$ so that $|\alpha| \ge 2, |\beta| \ge 2$.

We introduced the notation $h_{p\bar{q}/\alpha\bar{\beta}}$ for the jets of the metric H on the auxilary coefficient bundle V.

LEMMA 3.7.2. *Let H be a fiber metric on a holomorphic bundle V. Then given $v \ge 1$ there exists a holomorphic frame s near z_0 for V so that:*
(a) $h_{p\bar{q}}(s,H)(z_0) = \delta_{pq}$.
(b) $h_{p\bar{q}/\alpha}(s,H)(z_0) = h_{p\bar{q}/\bar{\alpha}}(s,H)(z_0) = 0$ for $1 \le |\alpha| \le \nu$.
(c) The choice of s is unique modulo the action of the unitary group $U(k)$ (where k is the fiber dimension of V) and modulo transformations of order $\nu + 1$.

PROOF: If $\nu = 0$, we make a linear change to assume (a). We proceed by induction assuming s' chosen for $\nu - 1$. We define:

$$s_p = s'_p + \sum_{q,|\alpha|=\nu} C_{q,\alpha} s'_q z^\alpha.$$

We adopt the notational conventions that indices (p,q) run from 1 through k and index a frame for V. We compute:

$$h_{p\bar{q}}(s,H) = h_{p\bar{q}}(s',H) + \sum_{|\alpha|=\nu} C_{q,\alpha} z^\alpha$$

$$+ \text{ terms in } \bar{z} + \text{ terms vanishing to order } \nu+1$$

$$h_{p\bar{q}/\alpha}(x,H)(z_0) = h_{p\bar{q}/\alpha}(s',H)(z_0) + \alpha! C_{q,\alpha} \qquad \text{for } |\alpha| = \nu$$

and derivatives of lower order are not disturbed. This determines the $C_{q,\alpha}$ uniquely so $h_{p\bar{q}/\alpha}(s,H)(z_0) = 0$ and taking complex conjugate yields $h_{q\bar{p}/\bar{\alpha}}(s,H)(z_0) = 0$ which completes the proof of the lemma.

Using these two normalizing lemmas, we restrict henceforth to polynomials in the variables $\{g(\alpha;\bar{\beta}), h_{p\bar{q}/\alpha_1\bar{\beta}_1}\}$ where $|\alpha| \geq 2$, $|\beta| \geq 2$, $|\alpha_1| \geq 1$, $|\beta_1| \geq 1$. We also restrict to unitary transformations of the coordinate system and of the fiber of V.

If P is a polynomial in these variables and if A is a monomial, let $c(A,P)$ be the coefficient of A in P. A is a monomial of P if $c(A,P) \neq 0$. Lemma 2.5.1 exploited invariance under the group SO(2) and was central to our axiomatic characterization of real Pontrjagin forms. The natural groups to study here are U(1), SU(2) and the coordinate permutations. If we set $\partial/\partial w_1 = a\partial/\partial z_1$, $\partial/\partial \bar{w}_1 = \bar{a}\partial/\partial \bar{z}_1$ for $a\bar{a} = 1$ and if we leave the other indices unchanged, then we compute:

$$A(W,*) = a^{\deg_1(A)} \bar{a}^{\deg_{\bar{1}}(A)} A(Z,*).$$

If A is a monomial of an invariant polynomial P, then $\deg_1(A) = \deg_{\bar{1}}(A)$ follows from this identity and consequently $\deg_u(A) = \deg_{\bar{u}}(A)$ for all $1 \leq u \leq n$.

This is the only conclusion which follows from U(1) invariance so we now study the group SU(2). We consider the coordinate transformation:

$$\partial/\partial w_1 = a\partial/\partial z_1 + b\partial/\partial z_2, \qquad \partial/\partial w_2 = -\bar{b}\partial/\partial z_1 + \bar{a}\partial/\partial z_2,$$

$$\partial/\partial \bar{w}_1 = \bar{a}\partial/\partial \bar{z}_1 + \bar{b}\partial/\partial \bar{z}_2, \qquad \partial/\partial \bar{w}_2 = -b\partial/\partial \bar{z}_1 + a\partial/\partial \bar{z}_2,$$

$$\partial/\partial w_u = \partial/\partial z_u \qquad \text{for } u > 2,$$

$$\partial/\partial \bar{w}_u = \partial/\partial \bar{z}_u \qquad \text{for } u > 2,$$

$$a\bar{a} + b\bar{b} = 1.$$

We let $j = \deg_1(A) + \deg_2(A) = \deg_{\bar{1}}(A) + \deg_{\bar{2}}(A)$ and expand

$$A(W,*) = a^j \bar{a}^j A(Z,*) + a^{j-1} \bar{a}^j b A(1 \to 2 \text{ or } \bar{2} \to \bar{1})(Z,*)$$

$$+ a^j \bar{a}^{j-1} \bar{b} A(2 \to 1 \text{ or } \bar{1} \to \bar{2})(Z,*)$$

$$+ \text{ other terms.}$$

The notation "$A(1 \to 2$ or $\bar{2} \to \bar{1})$" indicates all the monomials (with multiplicity) of this polynomial constructed by either changing a single index $1 \to 2$ or a single index $\bar{2} \to \bar{1}$. This plays the same role as the polynomial $A^{(1)}$ of section 2.5 where we must now also consider holomorphic and anti-holomorphic indices.

Let P be $U(2)$ invariant. Without loss of generality, we may assume $\deg_1(A) + \deg_2(A) = j$ is constant for all monomials A of P since this condition is $U(2)$ invariant. (Of course, the use of the indices 1 and 2 is for notational convenience only as similar statements will hold true for any pairs of indices). We expand:

$$P(W, *) = a^j \bar{a}^j P(Z, *) + a^{j-1} \bar{a}^j b P(1 \to 2 \text{ or } \bar{2} \to \bar{1})$$
$$+ a^j \bar{a}^{j-1} \bar{b} P(2 \to 1 \text{ or } \bar{1} \to \bar{2}) + O(b, \bar{b})^2.$$

Since P is invariant, we conclude:

$$P(1 \to 2 \text{ or } \bar{2} \to \bar{1}) = P(2 \to 1 \text{ or } \bar{1} \to \bar{2}) = 0.$$

We can now study these relations. Let B be an arbitrary monomial and expand:

$$B(1 \to 2) = c_0 A_0 + \cdots + c_k A_k \qquad B(\bar{2} \to \bar{1}) = -(d_0 A_0' + \cdots + d_k A_k')$$

where the c's and d's are positive integers with $\sum c_\nu = \deg_1(B)$ and $\sum d_\nu = \deg_{\bar{2}}(B)$. Then it is immediate that the $\{A_j\}$ and the $\{A_j'\}$ denote disjoint collections of monomials since $\deg_1(A)_\nu = \deg_1(B) - 1$ while $\deg_1(A)_\nu' = \deg_1(B)$. We compute:

$$A_\nu(2 \to 1) = -c_\nu' B + \text{other terms} \qquad A_\nu'(\bar{1} \to \bar{2}) = d_\nu' B + \text{other terms}$$

where again the c's and d's are positive integers (related to certain multiplicities). Since $P(2 \to 1$ or $\bar{1} \to \bar{2})$ is zero, if P is invariant, we conclude an identity:

$$c(B, P(2 \to 1 \text{ or } \bar{1} \to \bar{2})) = \sum_\nu -c_\nu' c(A_\nu, P) + \sum_\nu d_\nu' c(A_\nu', P) = 0.$$

By varying the creating monomial B (and also interchanging the indices 1 and 2), we can construct many linear equations among the coefficients. We note that in practice B will never be a monomial of P since $\deg_1(B) \neq \deg_{\bar{1}}(B)$. We use this principle to prove the following generalization of Lemmas 2.5.1 and 2.5.5:

LEMMA 3.7.3. *Let P be invariant under the action of $U(2)$ and let A be a monomial of P.*
(a) *If $A = g(\alpha; \bar{\beta})A_0'$, then by changing only 1 and 2 indices and $\bar{1}$ and $\bar{2}$ indices we can construct a new monomial A_1 of P which has the form $A_1 = g(\alpha_1; \bar{\beta}_1)A_0''$ where $\alpha(1) + \alpha(2) = \alpha_1(1)$ and where $\alpha_1(2) = 0$.*
(b) *If $A = h_{p\bar{q}/\alpha\bar{\beta}}A_0'$ then by changing only 1 and 2 indices and $\bar{1}$ and $\bar{2}$ indices we can construct a new monomial A_1 of P which has the form $A_1 = h_{p\bar{q}/\alpha_1\bar{\beta}_1}A_0''$ where $\alpha(1) + \alpha(2) = \alpha_1(1)$ and where $\alpha_1(2) = 0$.*

PROOF: We prove (a) as the proof of (b) is the same. Choose A of this form so $\alpha(1)$ is maximal. If $\alpha(2)$ is zero, we are done. Suppose the contrary. Let $B = g(\alpha_1; \bar{\beta})A_0'$ where $\alpha_1 = (\alpha(1) + 1, \alpha(2) - 1, \alpha(3), \dots, \alpha(n))$. We expand:

$$B(1 \to 2) = \alpha_1(1)A + \text{ monomials divisible by } g(\alpha_1; \bar{\beta})$$

$$B(\bar{2} \to \bar{1}) = \text{ monomials divisible by } g(\alpha_1; \bar{\beta}_1) \text{ for some } \beta_1.$$

Since A is a monomial of P, we use the principle described previously to conclude there is some monomial of P divisible by $g(\alpha_1; \bar{\beta}_1)$ for some β_1. This contradicts the maximality of α and shows $\alpha(2) = 0$ completing the proof.

We now begin the proof of Theorem 3.6.9. Let $0 \neq P \in \mathcal{R}^c_{n,\nu,k}$ be a scalar valued invariant homogeneous of order ν with $r(P) = 0$. Let A be a monomial of P. Decompose

$$A = g(\alpha_1; \bar{\beta}_1) \dots g(\alpha_r; \bar{\beta}_r)h_{p_1 q_1/\alpha_{r+1}\bar{\beta}_{r+1}} \cdots h_{p_s\bar{q}_s/\alpha_{r+s}\bar{\beta}_{r+s}}.$$

Let $\ell(A) = r + s$ be the length of A. We show $\ell(A) \geq n$ as follows. Without changing $(r, s, |\alpha_\nu|, |\beta_\nu|)$ we can choose A in the same form so that $\alpha_1(u) = 0$ for $u > 1$. We now fix the index 1 and apply Lemma 3.7.3 to the remaining indices to choose A so $\alpha_2(u) = 0$ for $\nu > 2$. We continue in this fashion to construct such an A so that $\deg_u(A) = 0$ for $u > r + s$. Since $\deg_n(A) = \deg_{\bar{n}}(A) > 0$, we conclude therefore that $r + s \geq n$. We estimate:

$$\nu = \text{ord}(P) = \sum_{\mu \leq r}\{|\alpha_\mu| + |\beta_\mu| - 2\} + \sum_{r < \mu \leq r+s}\{|\alpha_\mu| + |\beta_u|\}$$

$$\geq 2r + 2s \geq 2n = m$$

since $\text{ord}(g(\alpha, \beta)) \geq 2$ and $\text{ord}(h_{p\bar{q}/\alpha\bar{\beta}}) \geq 2$. This proves $\nu \geq m \geq 2$. Consequently, if $\nu < m$ we conclude $P = 0$ which proves the first assertion of Theorem 3.6.

We assume henceforth that we are in the limiting case $\nu = m$. In this case, all the inequalities must have been equalities. This implies $r + s = n$ and

$$|\alpha_\mu| + |\beta_\mu| = 4 \quad \text{for } \mu \leq r \quad \text{and} \quad |\alpha_\mu| + |\beta_\mu| = 2 \quad \text{for } r < \mu \leq s + r.$$

This holds true for every monomial of P since the construction of A_1 from A involved the use of Lemma 3.7.3 and does not change any of the orders involved. By Lemma 3.7.1 and 3.7.2 we assumed all the purely holomorphic and purely anti-holomorphic derivatives vanished at z_0 and consequently:

$$|\alpha_\mu| = |\beta_\mu| = 2 \quad \text{for } \mu \leq r \quad \text{and} \quad |\alpha_\mu| = |\beta_\mu| = 1 \quad \text{for } r < \mu \leq s + r.$$

Consequently, P is a polynomial in the $\{g(i_1 i_2; \bar{\jmath}_1 \bar{\jmath}_2) h_{p\bar{q}/i\bar{\jmath}}\}$ variables; it only involves the mixed 2-jets involved.

We wish to choose a monomial of P in normal form to begin counting the number of possible such P. We begin this process with:

LEMMA 3.7.4. *Let P satisfy the hypothesis of Theorem 3.6.8(b). Then there exists a monomial A of P which has the form:*

$$A = g(11; \bar{\jmath}_1 \bar{\jmath}_1') \dots g(tt; \bar{\jmath}_t \bar{\jmath}_t') h_{p_1 q_1/t+1, \bar{\jmath}_{t+1}} \dots h_{p_s q_s/m, \bar{\jmath}_n}$$

where $t + s = n$.

PROOF: We let $*$ denote indices which are otherwise unspecified and let A_0 be a generic monomial. We have $\deg_n(A) > 0$ for every monomial A of P so $\deg_u(A) > 0$ for every index u as well. We apply Lemma 3.7.3 to choose A of the form:

$$A = g(11; \bar{\jmath}_1 \bar{\jmath}_1') A_0.$$

Suppose $r > 1$. If $A = g(11; *)g(11 : *)A_0$, then we could argue as before using Lemma 3.7.3 that we could choose a monomial A of P so $\deg_k(A) = 0$ for $k > 1 + \ell(A_0) = r + s - 1 = n - 1$ which would be false. Thus $A = g(11; *)g(jk; *)A_0$ where not both j and k are 1. We may apply a coordinate permutation to choose A in the form $A = g(11; *)g(2j; *)A_0$. If $j \geq 2$ we can apply Lemma 3.7.3 to choose $A = g(11; *)g(22; *)A_0$. Otherwise we suppose $A = g(11; *)g(12; *)A_0$. We let $B = g(11; *)g(22; *)A_0$ and compute:

$$B(2 \to 1) = A + \text{ terms divisible by } g(11; *)g(22; *)$$
$$B(\bar{1} \to \bar{2}) = \text{ terms divisible by } g(11; *)g(22; *)$$

so that we conclude in any event we can choose $A = g(11; *)g(22; *)A_0$. We continue this argument with the remaining indices to construct:

$$A = g(11; *)g(22; *) \dots g(tt; *) h_{p_1 \bar{q}_1/u_1 \bar{v}_1} \dots h_{p_s \bar{q}_s/u_s \bar{v}_s}.$$

We have $\deg_u(A) \neq 0$ for all u. Consequently, the indices $\{t+1, \dots, t+s\}$ must appear among the indices $\{u_1, \dots, u_s\}$. Since these two sets have s elements, they must coincide. Thus by rearranging the indices we can assume $u_\nu = \nu + t$ which completes the proof. We note $\deg_u(A) = 2$ for $u \leq t$ and $\deg_u(A) = 1$ for $u > t$.

This lemma does not control the anti-holomorphic indices, we further normalize the choice of A in the following:

Lemma 3.7.5. *Let P satisfy the hypothesis of Theorem 3.6.8(b). Then there exists a monomial A of P which has the form:*

$$A = g(11; \bar{\jmath}_1 \bar{\jmath}_1') \dots g(tt; \bar{\jmath}_t \bar{\jmath}_t') h_{p_1 \bar{q}_1 / t+1, \overline{t+1}} \dots h_{p_s \bar{q}_s / n \bar{n}}$$

for

$$1 \le j_\mu, j_\mu' \le t.$$

Proof: If A has this form, then $\deg_j(A) = 1$ for $j > t$. Therefore $\deg_{\bar{\jmath}}(A) = 1$ for $j > t$ which implies $1 \le j_\mu, j_\mu' \le t$ automatically. We say that $\bar{\jmath}$ touches itself in A if A is divisible by $g(*; \bar{\jmath}\bar{\jmath})$ for some $*$. We say that j touches $\bar{\jmath}$ in A if A is divisible by $h_{pq/j\bar{\jmath}}$ for some (p,q). Choose A of the form given in Lemma 3.7.4 so the number of indices $j > t$ which touch $\bar{\jmath}$ in A is maximal. Among all such A, choose A so the number of $\bar{\jmath} \le \bar{t}$ which touch themselves in A is maximal. Suppose A does not satisfy the conditions of the Lemma. Thus A must be divisible by $h_{p\bar{q}/u\bar{v}}$ for $u \ne v$ and $u > t$. We suppose first $v > t$. Since $\deg_{\bar{v}}(A) = \deg_v(A) = 1$, v does not touch $\bar{v}$ in A. We let $A = h_{p\bar{q}/u\bar{v}} A_0$ and $B = h_{p\bar{q}/u\bar{u}} A_0$ then $\deg_{\bar{v}}(B) = 0$. We compute:

$$B(\bar{u} \to \bar{v}) = A + A_1 \qquad \text{and} \qquad B(v \to u) = A_2$$

where A_1 is defined by interchanging $\bar{u}$ and $\bar{v}$ and where A_2 is defined by replacing both v and $\bar{v}$ by u and $\bar{u}$ in A. Thus $\deg_v(A)_2 = 0$ so A_2 is not a monomial of A. Thus $A_1 = h_{p\bar{q}/u\bar{u}} A_0'$ must be a monomial of P. One more index (namely u) touches its holomorphic conjugate in A_1 than in A. This contradicts the maximality of A and consequently $A = h_{p\bar{q}/u\bar{v}} A_0'$ for $u > t$, $v \le t$. (This shows $h_{p_0 q_0 / u_0 \bar{v}_0}$ does not divide A for any $u_0 \ne v_0$ and $\bar{v}_0 > \bar{t}$.) We have $\deg_{\bar{u}}(A) = 1$ so the anti-holomorphic index $\bar{u}$ must appear somewhere in A. It cannot appear in a h variable and consequently A has the form $A = g(*; \bar{u}\bar{w}) h_{p\bar{q}/u\bar{v}} A_0''$. We define $B = g(*; \bar{w}\bar{w}) h_{p\bar{q}/u\bar{v}} A_0''$ and compute:

$$B(\bar{w} \to \bar{u}) = A + \text{ terms divisible by } g(*; \bar{w}\bar{w})$$
$$B(u \to w) = g(*; \bar{w}\bar{w}) h_{p\bar{q}/w\bar{v}} A_0''.$$

Since $g(*; \bar{w}\bar{w}) h_{p\bar{q}/w\bar{v}} A_0''$ does not have the index u, it cannot be a monomial of P. Thus terms divisible by $g(*; \bar{w}\bar{w})$ must appear in P. We construct these terms by interchanging a $\bar{w}$ and $\bar{u}$ index in P so the maximality of indices u_0 touching $\bar{u}_0$ for $u_0 > t$ is unchanged. Since $\bar{w}$ does not touch $\bar{w}$ in A, we are adding one additional index of this form which again contradicts the maximality of A. This final contradiction completes the proof.

This constructs a monomial A of P which has the form

$$A = A_0 A_1 \qquad \text{for } \begin{cases} \deg_u(A_0) = \deg_{\bar{u}}(A_0) = 0, & u > t, \\ \deg_u(A_1) = \deg_{\bar{u}}(A_1) = 0, & u \le t. \end{cases}$$

A_0 involves only the derivatives of the metric g and A_1 involves only the derivatives of h. We use this splitting in exactly the same way we used a similar splitting in the proof of Theorem 2.6.1 to reduce the proof of Theorem 3.6.9 to the following assertions:

LEMMA 3.7.6. *Let P satisfy the hypothesis of Theorem 3.6.9(b).*
(a) If P is a polynomial in the $\{h_{p\bar{q}/u\bar{v}}\}$ variables, then $P = *Q$ for Q a Chern m form of V.
(b) If P is a polynomial in the $\{g(u_1 u_2; \bar{v}_1 \bar{v}_2)\}$ variables, then $P = *Q$ for Q a Chern m form of $T_c M$.

PROOF: (a). If A is a monomial of P, then $\deg_u(A) = \deg_{\bar{u}}(A) = 1$ for all u so we can express

$$A = h_{p_1 q_1/1\bar{u}_1} \cdots h_{p_n q_n/n\bar{u}_n}$$

where the $\{u_\nu\}$ are a permutation of the indices i through n. We let $A = h_{*/1\bar{u}_1} h_{*/2\bar{u}_2} A_0'$ and $B = h_{*/1\bar{u}_1} h_{*/2\bar{u}_1} A_0'$. Then:

$$B(\bar{u}_1 \to \bar{u}_2) = A + A_1 \qquad \text{for } A_1 = h_{*/1\bar{u}_2} h_{*/2\bar{u}_1} A_0'$$
$$B(u_2 \to u_1) = A_2 \qquad \text{for } \deg_{u_2}(A_2) = 0.$$

Therefore A_2 is not a monomial of P. Thus A_1 is a monomial of P and furthermore $c(A, P) + c(A_1, P) = 0$. This implies when we interchange $\bar{u}_1$ and $\bar{u}_2$ that we change the sign of the coefficient involved. This implies immediately we can express P is terms of expressions:

$$*(\Omega^V_{p_1\bar{q}_1} \wedge \cdots \wedge \Omega^V_{p_n\bar{q}_n}) = \sum_{\vec{u},\vec{v}} *(h_{p_1\bar{q}_1/u_1\bar{v}_1} \, du^1 \wedge d\bar{v}_1 \cdots$$
$$\wedge h_{p_n\bar{q}_n/u_n\bar{v}_n} \, du^n \wedge dv^n)$$
$$= n! \sum_\rho \text{sign}(\rho) h_{p_1 q_1/1\bar{\rho}(1)} \cdots h_{p_n q_n/n\bar{\rho}(n)} + \cdots$$

where ρ is a permutation. This implies $*P$ can be expressed as an invariant polynomial in terms of curvature which implies it must be a Chern form as previously computed.

The remainder of this section is devoted to the proof of (b).

LEMMA 3.7.7. *Let P satisfy the hypothesis of Lemma 3.7.6(b). Then we can choose a monomial A of P which has the form:*

$$A = g(11; \bar{u}_1 \bar{u}_1) \ldots g(nn; \bar{u}_n \bar{u}_n).$$

This gives a normal form for a monomial. Before proving Lemma 3.7.7, we use this lemma to complete the proof of Lemma 3.7.6(b). By making a coordinate permutation if necessary we can assume A has either the form $g(11;\bar{1}\bar{1})A_0'$ or $g(11;\bar{2}\bar{2})A_0'$. In the latter case, we continue inductively to express $A = g(11;\bar{2}\bar{2})g(22;\bar{3}\bar{3})\ldots g(u-1,u-1;\bar{u}\bar{u})g(uu;\bar{1}\bar{1})A_0'$ until the cycle closes. If we permit $u = 1$ in this decomposition, we can also include the first case. Since the indices 1 through u appear exactly twice in A they do not appear in A_0'. Thus we can continue to play the same game to decompose A into cycles. Clearly A is determined by the length of the cycles involved (up to coordinate permutations); the number of such classifying monomials is $\pi(n)$, the number of partitions of n. This shows that the dimension of the space of polynomials P satisfying the hypothesis of Lemma 3.7.6(b) is $\leq \pi(n)$. Since there are exactly $\pi(n)$ Chern forms the dimension must be exactly $\pi(n)$ and every such P must be a Chern m form as claimed.

We give an indirect proof to complete the proof of Lemma 3.7.7. Choose A of the form given by Lemma 3.7.5 so the number of anti-holomorphic indices which touch themselves is maximal. If every anti-holomorphic index touches itself, then A has the form of Lemma 3.7.7 and we are done. We suppose the contrary. Since every index appears exactly twice, every anti-holomorphic index which does not touch itself touches another index which also does not touch itself. Every holomorphic index touches only itself. We may choose the notation so $A = g(*;\bar{1}\bar{2})A_0'$. Suppose first $\bar{1}$ does not touch $\bar{2}$ in A_0'. Then we can assume A has the form:

$$A = g(*;\bar{1}\bar{2})g(*;\bar{1}\bar{3})g(*;\bar{2}\bar{k})A_0'$$

where possibly $k = 3$ in this expression. The index 1 touches itself in A. The generic case will be:

$$A = g(11;*)g(*;\bar{1}\bar{2})g(*;\bar{1}\bar{3})g(*;\bar{2}\bar{k})A_0'.$$

The other cases in which perhaps $A = g(11;\bar{1}\bar{2})\ldots$ or $g(11,\bar{1}\bar{3})\ldots$ or $g(11;\bar{2}\bar{k})\ldots$ are handled similarly. The holomorphic and anti-holomorphic indices do not interact. In exactly which variable they appear does not matter. This can also be expressed as a lemma in tensor algebras.

We suppose $k \neq 3$; we will never let the 2 and 3 variables interact so the case in which $k = 3$ is exactly analogous. Thus A has the form:

$$A = g(11;*)g(*;\bar{1}\bar{2})g(*;\bar{1}\bar{3})g(*;\bar{2}\bar{k})g(*;\bar{3}\bar{\jmath})A_0'$$

where possibly $j = k$. Set $B = (11;*)g(*;\bar{2}\bar{2})g(*;\bar{1}\bar{3})g(*;\bar{2}\bar{k})g(*;\bar{3}\bar{\jmath})A_0'$ then:

$$B(\bar{2} \to \bar{1}) = 2A + A''$$
$$B(1 \to 2) = 2A_1$$

for

$$A'' = g(11;*)g(*;\bar{2}\bar{2})g(*;\bar{1}\bar{3})g(*;\bar{1}\bar{k})g(*;\bar{3}\bar{j})A_0'$$
$$A_1 = g(12;*)g(*;\bar{2}\bar{2})g(*;\bar{1}\bar{3})g(*;\bar{2}\bar{k})g(*;\bar{3}\bar{j})A_0'.$$

A'' is also a monomial of the form given by Lemma 3.7.5. Since one more anti-holomorphic index touches itself in A'' then does in A, the maximality of A shows A'' is not a monomial of P. Consequently A_1 is a monomial of P. Set $B_1 = g(12;*)g(*;\bar{2}\bar{2})g(*;\bar{3}\bar{3})g(*;\bar{2}\bar{k})g(*;\bar{3}\bar{j})A_0'$ then:

$$B_1(\bar{3} \to \bar{1}) = 2A_1 + A_2$$
$$B_1(1 \to 3) = A'''$$

for

$$A_2 = g(12;*)g(*;\bar{2}\bar{2})g(*;\bar{3}\bar{3})g(*;\bar{2}\bar{k})g(*;\bar{1}\bar{j})A_0'$$
$$A''' = g(32;*)g(*;\bar{2}\bar{2})g(*;\bar{3}\bar{3})g(*;\bar{2}\bar{k})g(*;\bar{3}\bar{j})A_0'.$$

However, $\deg_1(A''') = 0$. Since $r(P) = 0$, A''' cannot be a monomial of P so A_2 is a monomial of P. Finally we set

$$B_2 = g(11;*)g(*;\bar{2}\bar{2})g(*;\bar{3}\bar{3})g(*;\bar{2}\bar{k})g(*;\bar{1}\bar{j})A_0'$$

so

$$B_2(1 \to 2) = A_2$$
$$B_2(\bar{2} \to \bar{1}) = A_3 + 2A_4$$

for

$$A_3 = g(11;*)g(*;\bar{2}\bar{2})g(*;\bar{3}\bar{3})g(*;\bar{1}\bar{k})g(*;\bar{1}\bar{j})A_0'$$
$$A_4 = g(11;*)g(*;\bar{1}\bar{2})g(*;\bar{3}\bar{3})g(*;\bar{2}\bar{k})g(*;\bar{1}\bar{j})A_0'.$$

This implies either A_3 or A_4 is monomial of P. Both these have every holomorphic index touching itself. Furthermore, one more anti-holomorphic index (namely $\bar{3}$) touches itself. This contradicts the maximality of A.

In this argument it was very important that $\bar{2} \neq \bar{3}$ as we let the index $\bar{1}$ interact with each of these indices separately. Thus the final case we must consider is the case in which A has the form:

$$A = g(11;*)g(22;*)g(*;\bar{1}\bar{2})g(*;\bar{1}\bar{2})A_0.$$

So far we have not had to take into account multiplicities or signs in computing $A(1 \to 2)$ etc; we have been content to conclude certain coefficients are non-zero. In studying this case, we must be more careful in our analysis as the signs involved are crucial. We clear the previous notation and define:

$$A_1 = g(12;*)g(22;*)g(*;\bar{1}\bar{2})g(*;\bar{2}\bar{2})A_0$$
$$A_2 = g(12;*)g(22;*)g(*;\bar{2}\bar{2})g(*;\bar{1}\bar{2})A_0$$
$$A_3 = g(11;*)g(22;*)g(*;\bar{1}\bar{1})g(*;\bar{2}\bar{2})A_0$$
$$A_4 = g(22;*)g(22;*)g(*;\bar{2}\bar{2})g(*;\bar{2}\bar{2})A_0.$$

We note that A_3 is not a monomial of P by the maximality of A. A_4 is not a monomial of P as $\deg_1(A_4) = 0$ and $r(P) = 0$. We let $B = g(11; *)g(22; *)g(*; \bar{1}\bar{2})g(*; \bar{2}\bar{2})A_0$. Then

$$B(\bar{2} \to \bar{1}) = 2A + A_3, \qquad B(1 \to 2) = 2A_1$$

so that A_1 must be a monomial of P since A is a monomial of P and A_4 is not. We now pay more careful attention to the multiplicities and signs:

$$A(\bar{1} \to \bar{2}) = B + \cdots \qquad \text{and} \qquad A_1(2 \to 1) = B + \cdots$$

However $\bar{1} \to \bar{2}$ introduces a $\bar{b}$ while $2 \to 1$ introduces a $-\bar{b}$ so using the argument discussed earlier we conclude not only $c(A_1, P) \neq 0$ but that $c(A, P) - c(A_1, P) = 0$ so $c(A_1, P) = c(A, P)$. A_2 behaves similarly so the analogous argument using $B_1 = g(11; *)g(22; *)g(*; \bar{2}\bar{2})g(*; \bar{1}\bar{2})$ shows $c(A_2, P) = c(A, P)$. We now study $B_2 = g(12; *)g(22; *)g(*, \bar{2}\bar{2})g(*; \bar{2}\bar{2})$ and compute:

$$B_2(1 \to 2) = A_4 \qquad B_2(\bar{2} \to \bar{1}) = 2A_1 + 2A_2$$

A_4 is not a monomial of P as noted above. We again pay careful attention to the signs:

$$A_1(\bar{1} \to \bar{2}) = B_2 \qquad \text{and} \qquad A_2(\bar{1} \to \bar{2}) = B_2$$

This implies $c(A_1, P) + c(A_2, P) = 0$. Since $c(A_1, P) = c(A_2, P) = c(A, P)$ this implies $2c(A, P) = 0$ so A was not a monomial of P. This final contradiction completes the proof.

This proof was long and technical. However, it is not a theorem based on unitary invariance alone as the restriction axiom plays an important role in the development. We know of no proof of Theorem 3.6.9 which is based only on H. Weyl's theorem; in the real case, the corresponding characterization of the Pontrjagin classes was based only on orthogonal invariance and we gave a proof based on H. Weyl's theorem in that case.

3.8. The Chern Isomorphism and Bott Periodicity.

In section 3.9 we shall discuss the Atiyah-Singer index theorem in general using the results of section 3.1. The index theorem gives a topological formula for the index of an arbitrary elliptic operator. Before begining the proof of that theorem, we must first review briefly the Bott periodicity theorem from the point of Clifford modules. We continue our consideration of the bundles over S^n constructed in the second chapter. Let

$$\text{GL}(k, \mathbf{C}) = \{\, A : A \text{ is a } k \times k \text{ complex matrix with } \det(A) \neq 0 \,\}$$
$$\text{GL}'(k, \mathbf{C}) = \{\, A \in \text{GL}(k, \mathbf{C}) : \det(A - it) \neq 0 \text{ for all } t \in \mathbf{R} \,\}$$
$$\text{U}(k) = \{\, A \in \text{GL}(k, \mathbf{C}) : A \cdot A^* = I \,\}$$
$$\text{S}(k) = \{\, A \in \text{GL}(k, \mathbf{C}) : A^2 = I \text{ and } A = A^* \,\}$$
$$\text{S}_0(k) = \{\, A \in \text{S}(k) : \text{Tr}(A) = 0 \,\}.$$

We note that $\text{S}_0(k)$ is empty if k is odd. $\text{U}(k)$ is compact and is a deformation retract of $\text{GL}(k, \mathbf{C})$; $\text{S}(k)$ is compact and is a deformation retract of $\text{GL}'(k, \mathbf{C})$. $\text{S}_0(k)$ is one of the components of $\text{S}(k)$.

Let X be a finite simplicial complex. The suspension ΣX is defined by identifying $X \times \{\frac{\pi}{2}\}$ to a single point N and $X \times \{-\frac{\pi}{2}\}$ to single point S in the product $X \times \left[-\frac{\pi}{2}, \frac{\pi}{2}\right]$. Let $D_{\pm}(X)$ denote the northern and southern "hemispheres" in the suspension; the intersection $D_+(X) \cap D_-(X) = X$.

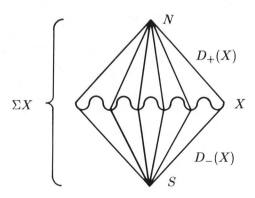

We note both $D_+(X)$ and $D_-(X)$ are contractible. $\Sigma(X)$ is a finite simplicial complex. If S^n is the unit sphere in $\mathbf{R}^{n+1}$, then $\Sigma(S^n) = S^{n+1}$. Finally, if W is a vector bundle over some base space Y, then we choose a fiber metric on W and let $S(W)$ be the unit sphere bundle. $\Sigma(W)$ is the fiberwise suspension of $S(W)$ over Y. This can be identified with $S(W \oplus 1)$.

It is beyond the scope of this book to develop in detail the theory of vector bundles so we shall simply state relevant facts as needed. We let $\text{Vect}_k(X)$ denote the set of isomorphism classes of complex vector bundles over X of fiber dimension k. We let $\text{Vect}(X) = \bigcup_k \text{Vect}_k(X)$ be the set of isomorphism classes of complex vector bundles over X of all fiber dimensions. We assume X is connected so that the dimension of a vector bundle over X is constant.

There is a natural inclusion map $\text{Vect}_k(X) \to \text{Vect}_{k+1}(X)$ defined by sending $V \mapsto V \oplus 1$ where 1 denotes the trivial line bundle over X.

LEMMA 3.8.1. *If* $2k \geq \dim X$ *then the map* $\text{Vect}_k(X) \to \text{Vect}_{k+1}(X)$ *is bijective.*

We let $F(\text{Vect}(X))$ be the free abelian group on these generators. We shall let (V) denote the element of $K(X)$ defined by $V \in \text{Vect}(X)$. $K(X)$ is the quotient modulo the relation $(V \oplus W) = (V) + (W)$ for $V, W \in \text{Vect}(X)$. $K(X)$ is an abelian group. The natural map $\dim: \text{Vect}(X) \to \mathbf{Z}^+$ extends to $\dim: K(X) \to \mathbf{Z}$; $\tilde{K}(X)$ is the kernel of this map. We shall say that an element of $K(X)$ is a virtual bundle; $\tilde{K}(X)$ is the subgroup of virtual bundles of virtual dimension zero.

It is possible to define a kind of inverse in $\text{Vect}(X)$:

LEMMA 3.8.2. *Given* $V \in \text{Vect}(X)$, *there exists* $W \in \text{Vect}(X)$ *so* $V \oplus W = 1^j$ *is isomorphic to a trivial bundle of dimension* $j = \dim V + \dim W$.

We combine these two lemmas to give a group structure to $\text{Vect}_k(X)$ for $2k \geq \dim X$. Given $V, W \in \text{Vect}_k(X)$, then $V \oplus W \in \text{Vect}_{2k}(X)$. By Lemma 3.8.1, the map $\text{Vect}_k(X) \to \text{Vect}_{2k}(X)$ defined by sending $U \mapsto U \oplus 1^k$ is bijective. Thus there is a unique element we shall denote by $V + W \in \text{Vect}_k(X)$ so that $(V + W) \oplus 1^k = V \oplus W$. It is immediate that this is an associative and commutative operation and that the trivial bundle 1^k functions as the unit. We use Lemma 3.8.2 to construct an inverse. Given $V \in \text{Vect}_k(X)$ there exists j and $W \in \text{Vect}_j(X)$ so $V \oplus W = 1^{j+k}$. We assume without loss of generality that $j \geq k$. By Lemma 3.8.1, we choose $\overline{W} \in \text{Vect}_k(X)$ so that $V \oplus \overline{W} \oplus 1^{j-k} = V \oplus \overline{W} = 1^{j+k}$. Then the bijectivity implies $V \oplus \overline{W} = 1^{2k}$ so $\overline{W}$ is the inverse of V. This shows $\text{Vect}_k(X)$ is a group under this operation.

There is a natural map $\text{Vect}_k(X) \to \tilde{K}(X)$ defined by sending $V \mapsto (V) - (1^k)$. It is immediate that:

$$(V + W) - 1^k = (V + W) + (1^k) - (1^{2k}) = ((V + W) \oplus 1^k) - (1^{2k})$$
$$= (V \oplus W) - (1^{2k})$$
$$= (V) + (W) - (1^{2k}) = (V) - (1^k) + (W) - (1^k)$$

so the map is a group homomorphism. $\tilde{K}(X)$ is generated by elements of the form $(V) - (W)$ for $V, W \in \text{Vect}_j X$ for some j. If we choose $\overline{W}$ so $W \oplus \overline{W} = 1^v$ then $(V) - (W) = (V \oplus \overline{W}) - (1^v)$ so $\tilde{K}(X)$ is generated by elements of the form $(V) - (1^v)$ for $V \in \text{Vect}_v(X)$. Again, by adding trivial factors, we may assume $v \geq k$ so by Lemma 3.8.1 $V = V_1 \oplus 1^{j-k}$ and $(V) - (1^j) = (V_1) - (1^k)$ for $V_1 \in \mathrm{Vect}_k(X)$. This implies the map $\mathrm{Vect}_k(X) \to \tilde{K}(X)$ is subjective. Finally, we note that that in fact $K(X)$ is generated by $\mathrm{Vect}_k(X)$ subject to the relation $(V) + (W) = (V \oplus W) = (V + W) + (1^k)$ so that this map is injective and we may identify $\tilde{K}(X) = \mathrm{Vect}_k(X)$ and $K(X) = \mathbf{Z} \oplus \tilde{K}(X) = \mathbf{Z} \oplus \mathrm{Vect}_k(X)$ for any k such that $2k \geq \dim X$.

Tensor product defines a ring structure on $K(X)$. We define $(V) \otimes (W) = (V \otimes W)$ for $V, W \in \mathrm{Vect}(X)$. Since $(V_1 \oplus V_2) \otimes (W) = (V_1 \otimes W \oplus V_2 \otimes W)$, this extends from $F(\mathrm{Vect}(X))$ to define a ring structure on $K(X)$ in which the trivial line bundle 1 functions as a multiplicative identity. $\tilde{K}(X)$ is an ideal of $K(X)$.

$K(X)$ is a $\mathbf{Z}$-module. It is convenient to change the coefficient group and define:

$$K(X; \mathbf{C}) = K(X) \otimes_{\mathbf{Z}} \mathbf{C}$$

to permit complex coefficients. $K(X; \mathbf{C})$ is the free $\mathbf{C}$-vector space generated by $\mathrm{Vect}(X)$ subject to the relations $V \oplus W = V + W$. By using complex coefficients, we eliminate torsion which makes calculations much simpler. The Chern character is a morphism:

$$ch \colon \mathrm{Vect}(X) \to H^{\mathrm{even}}(X; \mathbf{C}) = \bigoplus_q H^{2q}(X; \mathbf{C}).$$

We define ch using characteristic classes in the second section if X is a smooth manifold; it is possible to extend this definition using topological methods to more general topological settings. The identities:

$$ch(V \oplus W) = ch(V) + ch(W) \qquad \text{and} \qquad ch(V \otimes W) = ch(V) \, ch(W)$$

imply that we can extend:

$$ch \colon K(X) \to H^{\mathrm{even}}(X; \mathbf{C})$$

to be a ring homomorphism. We tensor this $\mathbf{Z}$-linear map with $\mathbf{C}$ to get

$$ch \colon K(X; \mathbf{C}) \to H^{\mathrm{even}}(X; \mathbf{C}).$$

LEMMA 3.8.3 (CHERN ISOMORPHISM). $ch \colon K(X; \mathbf{C}) \to H^{\mathrm{even}}(X; \mathbf{C})$ is a ring isomorphism.

If $\tilde{H}^{\mathrm{even}}(X; \mathbf{C}) = \bigoplus_{q>o} H^{2q}(X; \mathbf{C})$ is the reduced even dimensional cohomology, then $ch \colon \tilde{K}(X; \mathbf{C}) \to \tilde{H}^{\mathrm{even}}(X; \mathbf{C})$ is a ring isomorphism. For this reason, $\tilde{K}(X)$ is often refered to as reduced K-theory. We emphasize that in this isomorphism we are ignoring torsion and that torsion is crucial to understanding K-theory in general. Fortunately, the index is $\mathbf{Z}$-valued and we can ignore torsion in K-theory for an understanding of the index theorem.

We now return to studying the relation between $K(X)$ and $K(\Sigma X)$. We shall let $[X, Y]$ denote the set of homotopy classes of maps from X to Y. We shall always assume that X and Y are equipped with base points and that all maps are basepoint preserving. We fix $2k \geq \dim X$ and let $f: X \to S_0(2k)$. Since $f(x)$ is self-adjoint, and $f(x)^2 = I$, the eigenvalues of $f(x)$ are ± 1. Since $\operatorname{Tr} f(x) = 0$, each eigenvalue appears with multiplicity k. We let $\Pi_{\pm}(f)$ be the bundles over X which are sub-bundles of $X \times 1^k$ so that the fiber of $\Pi_{\pm}(f)$ at x is just the ± 1 eigenspace of $f(x)$. If we define $\pi_{\pm}(f)(x) = \frac{1}{2}(1 \pm f(x))$ then these are projections of constant rank k with range $\Pi_{\pm}(f)$. If f and f_1 are homotopic maps, then they determine isomorphic vector bundles. Thus the assignment $f \mapsto \Pi_+(f) \in \operatorname{Vect}_k(X)$ defines a map $\Pi_+ \colon [X, S_0(2k)] \to \operatorname{Vect}_k(X)$.

LEMMA 3.8.4. *The natural map* $[X, S_0(2k)] \to \operatorname{Vect}_k(X)$ *is bijective for* $2k \geq \dim X$.

PROOF: Given $V \in \operatorname{Vect}_k(X)$ we choose $W \in \operatorname{Vect}_k(X)$ so $V \oplus W = 1^{2k}$. We choose fiber metrics on V and on W and make this sum orthogonal. By applying the Gram-Schmidt process to the given global frame, we can assume that there is an orthonormal global frame and consequently that $V \oplus W = 1^{2k}$ is an orthogonal direct sum. We let $\pi_+(x)$ be orthogonal projection on the fiber V_x of 1^{2k} and $f(x) = 2\pi_+(x) - I$. Then it is immediate that $f: X \to S_0(2k)$ and $\pi_+(f) = V$. This proves the map is subjective. The injectivity comes from the same sorts of considerations as were used to prove Lemma 3.8.1 and is therefore omitted.

If $f: X \to \operatorname{GL}'(k, \mathbf{C})$, we can extend the definition to let $\Pi_{\pm}(f)(x)$ be the span of the generalized eigenvectors of f corresponding to eigenvalues with positive/negative real part. Since there are no purely imaginary eigenvalues, $\Pi_{\pm}(f)$ has constant rank and gives a vector bundle over X.

In a similar fashion, we can classify $\operatorname{Vect}_k \Sigma X = [X, \operatorname{U}(k)]$. Since $\operatorname{U}(k)$ is a deformation retract of $\operatorname{GL}(k, \mathbf{C})$, we identify $[X, \operatorname{U}(k)] = [X, \operatorname{GL}(k, \mathbf{C})]$. If $g: X \to \operatorname{GL}(k, \mathbf{C})$, we use g as a clutching function to define a bundle over ΣX. Over $D_{\pm}(X)$, we take the bundles $D_{\pm}(X) \times \mathbf{C}^k$. $D_+(X) \cap D_-(X) = X$. On the overlap, we identify $(x, z)_+ = (x, z')_-$ if $z \cdot g(x) = z'$. If we let s^+ and s^- be the usual frames for C^k over $D_{\pm}$ then $\sum z_i^+ g_{ij} s_j^+ = \sum z_i^- s_i^-$ so that we identify the frames using the identity

$$s^- = gs^+.$$

We denote this bundle by V_g. Homotopic maps define isomorphic bundles so we have a map $[X, \operatorname{GL}(k, \mathbf{C})] \to \operatorname{Vect}_k \Sigma X$. Conversely, given a vector bundle V over ΣX we can always choose local trivializations for V over

$D_\pm(X)$ since these spaces are contractible. The transition function $s^- = gs^+$ gives a map $g: X \to \mathrm{GL}(k, \mathbf{C})$. It is convenient to assume X has a base point x_0 and to choose $s^- = s^+$ at x_0. Thus $g(x_0) = I$. This shows the map $[X, \mathrm{GL}(k, \mathbf{C})] \to \mathrm{Vect}_k \Sigma X$ is surjective. If we had chosen different trivializations $\tilde{s}^+ = h^+ s^+$ and $\tilde{s}^- = h^- s^-$ then we would have obtained a new clutching function $\tilde{g} = h^- g(h^+)^{-1}$. Since $h^\pm$ are defined on contractible sets, they are null homotopic so $\tilde{g}$ is homotopic to g. This proves:

LEMMA 3.8.5. *The map $[X, \mathrm{GL}(k, \mathbf{C})] = [X, \mathrm{U}(k)] \to \mathrm{Vect}_k(\Sigma X)$, given by associating to a map g the bundle defined by the clutching function g, is bijective.*

It is always somewhat confusing to try to work directly with this definition. It is always a temptation to confuse the roles of g and of its inverse as well as the transposes involved. There is another definition which avoids this difficulty and which will be very useful in computing specific examples. If $g: X \to \mathrm{GL}(k, \mathbf{C})$, we shall let $g(x)z$ denote matrix multiplication. We shall regard $\mathbf{C}^k$ as consisting of column vectors and let g act as a matrix from the left. This is, of course, the opposite convention from that used previously.

Let $\theta \in [-\frac{\pi}{2}, \frac{\pi}{2}]$ be the suspension parameter. We define:

$$\Sigma: [X, \mathrm{GL}(k, \mathbf{C})] \to [\Sigma X, \mathrm{GL}'(2k, \mathbf{C})]$$
$$\Sigma: [X, \mathrm{GL}'(k, \mathbf{C})] \to [\Sigma X, \mathrm{GL}(k, \mathbf{C})]$$

by

$$\Sigma g(x, \theta) = \begin{pmatrix} (\sin\theta)I_k & (\cos\theta)g^*(x) \\ (\cos\theta)g(x) & -(\sin\theta)I_k \end{pmatrix}$$
$$\Sigma f(x, \theta) = (\cos\theta)f(x) - i(\sin\theta)I_k.$$

We check that Σ has the desired ranges as follows. If $g: X \to \mathrm{GL}(k, \mathbf{C})$, then it is immediate that Σg is self-adjoint. We compute:

$$(\Sigma g)^2 = \begin{pmatrix} (\sin^2\theta)I_k + (\cos^2\theta)g^*g & 0 \\ 0 & (\sin^2\theta)I_k + (\cos^2\theta)gg^* \end{pmatrix}.$$

This is non-singular since g is invertible. Therefore Σg is invertible. If $\theta = \pi/2$, then $\Sigma g = \begin{pmatrix} I & 0 \\ 0 & -I \end{pmatrix}$ is independent of x. If g is unitary, $\Sigma g \in S_0(2k)$. If $f \in \mathrm{GL}'(k, \mathbf{C})$, then f has no purely imaginary eigenvalues so Σf is non-singular.

LEMMA 3.8.6. *Let $g: X \to \mathrm{GL}(k, \mathbf{C})$ and construct the bundle $\Pi_+(\Sigma g)$ over ΣX. Then this bundle is defined by the clutching function g.*

PROOF: We may replace g by a homotopic map without changing the isomorphism class of the bundle $\Pi_+(\Sigma g)$. Consequently, we may assume without loss of generality that $g: X \to \mathrm{U}(k)$ so $g^*g = gg^* = I_k$. Consequently $(\Sigma g)^2 = I_{2k}$ and $\pi_+(\Sigma g) = \frac{1}{2}(I_{2k} + \Sigma g)$. If $z \in \mathbf{C}^k$, then:

$$\pi_+(\Sigma g)(x, \theta) \begin{pmatrix} z \\ 0 \end{pmatrix} = \frac{1}{2} \begin{pmatrix} z + (\sin\theta)z \\ (\cos\theta)g(x)z \end{pmatrix}.$$

This projection from $\mathbf{C}^k$ to $\Pi_+(\Sigma g)$ is non-singular away from the south pole S and can be used to give a trivialization of $\Pi_+(\Sigma g)$ on $\Sigma X - S$.

From this description, it is clear that $\Pi_+(\Sigma g)$ is spanned by vectors of the form $\dfrac{1}{2} \begin{pmatrix} (1 + (\sin\theta))z \\ (\cos\theta)g(x)z \end{pmatrix}$ away from the south pole. At the south pole, $\Pi_+(\Sigma g)$ consists of all vectors of the form $\begin{pmatrix} 0 \\ w \end{pmatrix}$. Consequently, projection on the second factor $\pi_+(S): \begin{pmatrix} a \\ b \end{pmatrix} \to \begin{pmatrix} 0 \\ b \end{pmatrix}$ is non-singular away from the north pole N and gives a trivialization of $\Pi_+(\Sigma g)$ on $\Sigma X - N$. We restrict to the equator X and compute the composite of these two maps to determine the clutching function:

$$\begin{pmatrix} z \\ 0 \end{pmatrix} \mapsto \frac{1}{2} \begin{pmatrix} z \\ g(x)z \end{pmatrix} \mapsto \frac{1}{2} \begin{pmatrix} 0 \\ g(x)z \end{pmatrix}.$$

The function $g(x)/2$ is homotopic to g which completes the proof.

This is a very concrete description of the bundle defined by the clutching function g. In the examples we shall be considering, it will come equipped with a natural connection which will make computing characteristic classes much easier.

We now compute the double suspension. Fix $f: X \to S_0(2k)$ and $g: X \to \mathrm{GL}(k, \mathbf{C})$.

$$\Sigma^2 f(x, \theta, \phi)$$
$$= \begin{pmatrix} \sin\phi & \cos\phi\{(\cos\theta)f^* + i\sin\theta\} \\ \cos\phi\{(\cos\theta)f(x) - i\sin\theta\} & -\sin\phi \end{pmatrix}$$

$$\Sigma^2 g(x, \theta, \phi)$$
$$= \begin{pmatrix} \cos\phi\sin\theta - i\sin\phi & \cos\phi\cos\theta g^*(x) \\ \cos\phi\cos\theta g(x) & -\cos\phi\cos\theta - i\sin\phi \end{pmatrix}$$

We let $U(\infty)$ be the direct limit of the inclusions $U(k) \to U(k+1) \cdots$ and $S_0(\infty)$ be the direct limit of the inclusions $S_0(2k) \to S_0(2k+2) \to \cdots$. Then we have identified:

$$\widetilde{K}(X) = [X, S_0(\infty)] \qquad \text{and} \qquad \widetilde{K}(\Sigma X) = [\Sigma X, S_0(\infty)] = [S, U(\infty)]$$

We can now state:

THEOREM 3.8.7 (BOTT PERIODICITY). *The map*

$$\Sigma^2 \colon \widetilde{K}(X) = [X, S_0(\infty)] \to \widetilde{K}(\Sigma^2 X) = [\Sigma^2 X, S_0(\infty)]$$

induces a ring isomorphism. Similarly, the map

$$\Sigma^2 \colon \widetilde{K}(\Sigma X) = [X, U(\infty)] \to \widetilde{K}(\Sigma^2 X) = [\Sigma^2 X, U(\infty)]$$

is a ring isomorphism.

We note that $[X, U(\infty)]$ inherits a natural additive structure from the group structure on $U(\infty)$ by letting $g \oplus g'$ be the direct sum of these two maps. This group structure is compatible with the additive structure on $\widetilde{K}(\Sigma X)$ since the clutching function of the direct sum is the direct sum of the clutching functions. Similarly, we can put a ring structure on $[X, U(\infty)]$ using tensor products to be compatible with the ring structure on $\widetilde{K}(\Sigma X)$.

The Chern character identifies $\widetilde{K}(X; \mathbf{C})$ with $\widetilde{H}^{even}(X; \mathbf{C})$. We may identify $\Sigma^2 X$ with a certain quotient of $X \times S^2$. Bott periodicity in this context becomes the assertion $K(X \times S^2) = K(X) \otimes K(S^2)$ which is the Kunneth formula in cohomology.

We now consider the case of a sphere $X = S^n$. The unitary group $U(2)$ decomposes as $U(2) = U(1) \times SU(2) = S^1 \times S^3$ topologically so $\pi_1(U(2)) = \mathbf{Z}$, $\pi_2(U(2)) = 0$ and $\pi_3(U(2)) = \mathbf{Z}$. This implies $\widetilde{K}(S^1) = \widetilde{K}(S^3) = 0$ and $\widetilde{K}(S^2) = \widetilde{K}(S^4) = \mathbf{Z}$. Using Bott periodicity, we know more generally that:

LEMMA 3.8.8 (BOTT PERIODICITY). *If n is odd, then*

$$\widetilde{K}(S^n) \simeq \pi_{n-1}(U(\infty)) = 0.$$

If n is even, then

$$\widetilde{K}(S^n) \simeq \pi_{n-1}(U(\infty)) = \mathbf{Z}.$$

It is useful to construct explicit generators of these groups. Let $x = (x_0, \ldots, x_n) \in S^n$, let $y = (x, x_{n+1}) \in S^{n+1}$, and let $z = (y, x_{n+2}) \in S^{n+2}$. If $f \colon S^n \to GL'(k, \mathbf{C})$ and $g \colon X \to GL(k, \mathbf{C})$ we extend these to be homogeneous of degree 1 with values in the $k \times k$ matrices. Then we compute:

$$\Sigma f(y) = f(x) - ix_{n+1}$$

$$\Sigma^2 f(z) = \begin{pmatrix} x_{n+2} & f^*(x) + ix_{n+1} \\ f(x) - ix_{n+1} & -x_{n+2} \end{pmatrix}$$

$$\Sigma g(y) = \begin{pmatrix} x_{n+1} & g^*(x) \\ g(x) & -x_{n+1} \end{pmatrix}$$

$$\Sigma^2 g(z) = \begin{pmatrix} x_{n+1} - ix_{n+2} & g^*(x) \\ g(x) & -x_{n+1} - ix_{n+2} \end{pmatrix}.$$

If we suppose that f is self-adjoint, then we can express:

$$\Sigma^2 f(z) = x_{n+2} \begin{pmatrix} 1 & 0 \\ 0 & -1 \end{pmatrix} \otimes I_k + x_{n+1} \begin{pmatrix} 0 & i \\ -i & 0 \end{pmatrix} \otimes I_k + \begin{pmatrix} 0 & 1 \\ 1 & 0 \end{pmatrix} \otimes f(x).$$

We can now construct generators for $\pi_{n-1} U(\infty)$ and $\widetilde{K}(S^n)$ using Clifford algebras. Let $g(x_0, x_1) = x_0 - ix_1$ generate $\pi_1(S^1) = \mathbf{Z}$ then

$$\Sigma g(x_0, x_1, x_2) = x_0 e_0 + x_1 e_1 + x_2 e_2$$

for

$$e_0 = \begin{pmatrix} 0 & 1 \\ 1 & 0 \end{pmatrix}, \qquad e_1 = \begin{pmatrix} 0 & i \\ -i & 0 \end{pmatrix}, \qquad e_2 = \begin{pmatrix} 1 & 0 \\ 0 & -1 \end{pmatrix}.$$

The $\{e_j\}$ satisfy the relations $e_j e_k + e_k e_j = 2\delta_{jk}$ and $e_0 e_1 e_2 = -iI_2$. $\Pi_+(\Sigma g)$ is a line bundle over S^2 which generates $\widetilde{K}(S^2)$. We compute:

$$\Sigma^2 g(x_0, x_1, x_2, x_3) = x_0 e_0 + x_1 e_1 + x_2 e_2 - ix_3 I_2.$$

If we introduce $z_1 = x_0 + ix_1$ and $z_2 = x_2 + ix_3$ then

$$\Sigma^2 g(z_1, z_2) = \begin{pmatrix} \bar{z}_2 & z_1 \\ \bar{z}_1 & -z_2 \end{pmatrix}.$$

Consequently $\Sigma^2 g$ generates $\pi_3(U(\infty))$. We suspend once to construct:

$$\Sigma^3 g(x_0, x_1, x_2, x_3, x_4)$$
$$= x_0 e_0 \otimes e_0 + x_1 e_0 \otimes e_1 + x_2 e_0 \otimes e_2 + x_3 e_1 \otimes I + x_4 e_2 \otimes I.$$

The bundle $\Pi_+(\Sigma^3 g)$ is a 2-plane bundle over S^4 which generates $\widetilde{K}(S^4)$. We express $\Sigma^3 g = x_0 e_0^4 + \cdots + x_4 e_4^4$ then these matrices satisfy the commutation relations $e_j^4 e_j^4 + e_k^4 e_j^4 = 2\delta_{jk}$ and $e_0^4 e_1^4 e_2^4 e_3^4 e_4^4 = -I$.

We proceed inductively to define matrices e_j^{2k} for $0 \leq j \leq 2k$ so that $e_i e_j + e_j e_i = 2\delta_{ij}$ and $e_0^{2k} \ldots e_{2k}^{2k} = (-i)^k I$. These are matrics of shape $2^k \times 2^k$ such that $\Sigma^{2k-1}(g)(x) = \sum x_j e_j^{2k}$. In Lemma 2.1.5 we computed that:

$$\int_{S^{2k}} ch_k(\Pi_+ \Sigma^{2k-1} g) = i^k 2^{-k} \operatorname{Tr}(e_0^{2k} \ldots e_{2k}^{2k}) = 1.$$

These bundles generate $\widetilde{K}(S^{2k})$ and $\Sigma^{2k}(g)$ generates $\pi_{2k+1}(U(\infty))$. We summarize these calculations as follows:

LEMMA 3.8.9. Let $\{e_0, \ldots, e_{2k}\}$ be a collection of self-adjoint matrices of shape $2^k \times 2^k$ such that $e_i e_j + e_j e_i = 2\delta_{ij}$ and such that $e_0 \ldots e_{2k} = (-i)^k I$. We define $e(x) = x_0 e_0 + \cdots + x_{2k} e_{2k}$ for $x \in S^{2k}$. Let $\Pi_+(e)$ be the bundle of $+1$ eigenvectors of e over S^{2k}. Then $\Pi_+(e)$ generates $\widetilde{K}(S^{2k}) \simeq \mathbf{Z}$. $\Sigma e(y) = e(x) - i x_{2k+1}$ generates $\pi_{2k+1}(U(\infty)) \simeq \mathbf{Z}$. $\int_{S^{2k}} ch_k(V) : \widetilde{K}(S^{2k}) \to \mathbf{Z}$ is an isomorphism and $\int_{S^{2k}} ch_k(\Pi_+ e) = 1$.

PROOF: We note that $\int_{S^{2k}} ch_k(V)$ is the index of the spin complex with coefficients in V since all the Pontrjagin forms of $T(S^{2k})$ vanish except for p_0. Thus this integral, called the topological charge, is always an integer. We have checked that the integral is 1 on a generator and hence the map is surjective. Since $\widetilde{K}(S^{2k}) = \mathbf{Z}$, it must be bijective.

It was extremely convenient to have the bundle with clutching function $\Sigma^{2k-2} g$ so concretely given so that we could apply Lemma 2.1.5 to compute the topological charge. This will also be important in the next chapter.

3.9. The Atiyah-Singer Index Theorem.

In this section, we shall discuss the Atiyah-Singer theorem for a general elliptic complex by interpreting the index as a map in K-theory. Let M be smooth, compact, and without boundary. For the moment we make no assumptions regarding the parity of the dimension m. We do not assume M is orientable. Let $P: C^\infty(V_1) \to C^\infty(V_2)$ be an elliptic complex with leading symbol $p(x, \xi): S(T^*M) \to \mathrm{HOM}(V_1, V_2)$. We let $\Sigma(T^*M)$ be the fiberwise suspension of the unit sphere bundle $S(T^*M)$. We identify $\Sigma(T^*M) = S(T^*M \oplus 1)$. We generalize the construction of section 3.8 to define $\Sigma p: \Sigma(T^*M) \to \mathrm{END}(V_1 \oplus V_2)$ by:

$$\Sigma p(x, \xi, \theta) = \begin{pmatrix} (\sin \theta) I_{V_1} & (\cos \theta) p^*(x, \xi) \\ (\cos \theta) p(x, \xi) & -(\sin \theta) I_{V_2} \end{pmatrix}.$$

This is a self-adjoint invertible endomorphism. We let $\Pi_\pm(\Sigma p)$ be the sub-bundle of $V_1 \oplus V_2$ over $\Sigma(T^*M)$ corresponding to the span of the eigenvectors of Σp with positive/negative eigenvalues. If we have given connections on V_1 and V_2, we can project these connections to define natural connections on $\Pi_\pm(\Sigma p)$. The clutching function of $\Pi_+(\Sigma p)$ is p in a sense we explain as follows:

We form the disk bundles $D_\pm(M)$ over M corresponding to the northern and southern hemispheres of the fiber spheres of $\Sigma(T^*M)$. Lemma 3.8.6 generalizes immediately to let us identify $\Pi_+(\Sigma p)$ with the bundle $V_1^+ \cup V_2^-$ over the disjoint union $D_+(M) \cup D_-(M)$ attached using the clutching function p over their common boundary $S(T^*M)$. If $\dim V_1 = k$, then $\Pi_+(\Sigma p) \in \mathrm{Vect}_k(\Sigma(T^*M))$. Conversely, we suppose given a bundle $V \in \mathrm{Vect}_k(\Sigma(T^*M))$. Let $N: M \to \Sigma(T^*M)$ and $S: M \to \Sigma(T^*M)$ be the natural sections mapping M to the northern and southern poles of the fiber spheres; $N(x) = (x, 0, 1)$ and $S(x) = (x, 0, -1)$ in $S(T^*M \oplus 1)$. N and S are the centers of the disk bundles $D_\pm(M)$. We let $N^*(V) = V_1$ and $S^*(V) = V_2$ be the induced vector bundles over M. $D_\pm(M)$ deformation retracts to $M \times \{N\}$ and $M \times \{S\}$. Thus V restricted to $D_\pm(M)$ is cannonically isomorphic to the pull back of V_1 and V_2. On the intersection $S(T^*M) = D_+(M) \cap D_-(M)$ we have a clutching or glueing function relating the two decompositions of V. This gives rise to a map $p: S(T^*M) \to \mathrm{HOM}(V_1, V_2)$ which is non-singular. The same argument as that given in the proof of Lemma 3.8.5 shows that V is completely determined by the isomorphism class of V_1 and of V_2 together with the homotopy class of the map $p: S(T^*M) \to \mathrm{HOM}(V_1, V_2)$.

Given an order ν we can recover the leading symbol p by extending p from $S(T^*M)$ to T^*M to be homogeneous of order ν. We use this to define an elliptic operator $P_\nu: C^\infty(V_1) \to C^\infty(V_2)$ with leading symbol p_ν. If Q_ν is another operator with the same leading symbol, then we define

$P_\nu(t) = tP_\nu + (1-t)Q_\nu$. This is a 1-parameter family of such operators with the same leading symbol. Consequently by Lemma 1.4.4 index(P_ν) = index(Q_ν). Similarly, if we replace p by a homotopic symbol, then the index is unchanged. Finally, suppose we give two orders of homgeneity $\nu_1 > \nu_2$. We can choose a self-adjoint pseudo-differential operator R on $C^\infty(V_2)$ with leading symbol $|\xi|^{\nu_1-\nu_2}I_{V_2}$. Then we can let $P_{\nu_1} = RP_{\nu_2}$ so index(P_{ν_1}) = index(R)+index(P_{ν_2}). Since R is self-adjoint, its index is zero so index(P_{ν_1}) = index(P_{ν_2}). This shows that the index only depends on the homotopy class of the clutching map over $S(T^*M)$ and is independent of the order of homogeneity and the extension and the particular operator chosen. Consequently, we can regard index: Vect$(\Sigma(T^*M)) \to \mathbf{Z}$ so that index$(\Pi_+(\Sigma p))$ = index(P) if P is an elliptic operator. (We can always "roll up" an elliptic complex to give a 2-term elliptic complex in computing the index so it suffices to consider this case).

It is clear from our definition that $\Sigma(p \oplus q) = \Sigma(p) \oplus \Sigma(q)$ and therefore $\Pi_+(\Sigma(p \oplus q)) = \Pi_+(\Sigma(p)) \oplus \Pi_+(\Sigma(q))$. Since index$(P \oplus Q)$ = index(P) + index(Q) we conclude:

$$\text{index}(V \oplus W) = \text{index}(V) + \text{index}(W) \qquad \text{for } V, W \in \text{Vect}(\Sigma(T^*M)).$$

This permits us to extend index: $K(\Sigma(T^*M)) \to \mathbf{Z}$ to be $\mathbf{Z}$-linear. We tensor with the complex numbers to extend index: $K(\Sigma(T^*M); \mathbf{C}) \to \mathbf{C}$, so that:

LEMMA 3.9.1. *There is a natural map* index: $K(\Sigma(T^*M); \mathbf{C}) \to \mathbf{C}$ *which is linear so that* index(P) = index$(\Pi_+(\Sigma p))$ *if* $P: C^\infty V_1 \to C^\infty V_2$ *is an elliptic complex over* M *with symbol* p.

There is a natural projection map $\pi: \Sigma(T^*M) \to M$. This gives a natural map $\pi^*: K(M; \mathbf{C}) \to K(\Sigma(T^*M); \mathbf{C})$. If N denotes the north pole section, then $\pi N = 1_M$ so $N^*\pi^* = 1$ and consequently π^* is injective. This permits us to regard $K(M; \mathbf{C})$ as a subspace of $K(\Sigma(T^*M); \mathbf{C})$. If $V = \pi^*V_1$, then the clutching function defining V is just the identity map. Consequently, the corresponding elliptic operator P can be taken to be a self-adjoint operator on $C^\infty(V)$ which has index zero. This proves:

LEMMA 3.9.2. *If* $V \in K(\Sigma(T^*M); \mathbf{C})$ *can be written as* π^*V_1 *for* $V_1 \in K(M; \mathbf{C})$ *then* index$(V) = 0$. *Thus* index: $K(\Sigma(T^*M); \mathbf{C})/K(M; \mathbf{C}) \to \mathbf{C}$.

These two lemmas show that all the information contained in an elliptic complex from the point of view of computing its index is contained in the corresponding description in K-theory. The Chern character gives an isomorphism of $K(X; \mathbf{C})$ to the even dimensional cohomology. We will exploit this isomorphism to give a formula for the index in terms of cohomology.

In addition to the additive structure on $K(X; \mathbf{C})$, there is also a ring structure. This ring structure also has its analogue with respect to elliptic operators as we have discussed previously. The multiplicative nature of the four classical elliptic complexes played a fundamental role in determining the normalizing constants in the formula for their index.

We give $\Sigma(T^*M)$ the simplectic orientation. If $x = (x_1, \ldots, x_m)$ is a system of local coordinates on M, let $\xi = (\xi_1, \ldots, \xi_m)$ be the fiber coordinates on T^*M. Let u be the additional fiber coordinate on $T^*M \oplus 1$. We orient $T^*M \oplus 1$ using the form:

$$\omega_{2m+1} = dx_1 \wedge d\xi_1 \wedge \cdots \wedge dx_m \wedge d\xi_m \wedge du.$$

Let $\vec{N}$ be the outward normal on $S(T^*M \oplus 1)$ and define ω_{2m} on $S(T^*M \oplus 1)$ by:

$$\omega_{2m+1} = \vec{N} \wedge \omega_{2m} = \omega_{2m} \wedge \vec{N}.$$

This gives the orientation of Stokes theorem.

LEMMA 3.9.3.
(a) Let $P: C^\infty(V_1) \to C^\infty(V_2)$ be an elliptic complex over M_1 and let $Q: C^\infty(W_1) \to C^\infty(W_2)$ be an elliptic complex over M_2. We assume P and Q are partial differential operators of the same order and we let $M = M_1 \times M_2$ and define over M

$$R = (P \otimes 1 + 1 \otimes Q) \oplus (P^* \otimes 1 - 1 \otimes Q^*): C^\infty(V_1 \otimes W_1 \oplus V_2 \otimes W_2)$$
$$\to C^\infty(V_2 \otimes W_1 \oplus V_1 \otimes W_2).$$

Then R is elliptic and $\text{index}(R) = \text{index}(P) \cdot \text{index}(Q)$.
(b) The four classical elliptic complexes discussed earlier can be decomposed in this fashion over product manifolds.
(c) Let p and q be arbitrary elliptic symbols over M_1 and M_2 and define

$$r = \begin{pmatrix} p \otimes 1 & -1 \otimes q^* \\ 1 \otimes q & p^* \otimes 1 \end{pmatrix} \qquad \text{over } M = M_1 \times M_2.$$

Let $\theta_i \in H^{\text{even}}(M_i; \mathbf{C})$ for $i = 1, 2$, then:

$$\int_{\Sigma(T^*M)} \theta_1 \wedge \theta_2 \wedge ch(\Pi_+(\Sigma r))$$
$$= \int_{\Sigma(T^*M_1)} \theta_1 \wedge ch(\Pi_+(\Sigma p)) \cdot \int_{\Sigma(T^*M_2)} \theta_2 \wedge ch(\Pi_+(\Sigma q)).$$

(d) Let q be an elliptic symbol over M_1 and let p be a self-adjoint elliptic symbol over M_2. Over $M = M_1 \times M_2$ define the self-adjoint elliptic symbol r by:

$$r = \begin{pmatrix} 1 \otimes p & q^* \otimes 1 \\ q \otimes 1 & -1 \otimes p \end{pmatrix}.$$

Let $\theta_1 \in \Lambda^e(M_1)$ and $\theta_2 \in \Lambda^o(M_2)$ be *closed differential forms. Give* $S(T^*M)$, $\Sigma(T^*M_1)$ and $S(T^*M_2)$ *the orientations induced from the simplectic orientations. Then:*

$$\int_{S(T^*M)} \theta_1 \wedge \theta_2 \wedge ch(\Pi_+ r)$$

$$= \int_{\Sigma(T^*M_1)} \theta_1 \wedge ch(\Pi_+(\Sigma q)) \cdot \int_{S(T^*M_2)} \theta_1 \wedge ch(\Pi_+(p)).$$

Remark: We will use (d) to discuss the eta invariant in Chapter 4; we include this integral at this point since the proof is similar to that of (c).

PROOF: We let (p, q, r) be the symbols of the operators involved. Then:

$$r = \begin{pmatrix} p \otimes 1 & -1 \otimes q^* \\ 1 \otimes q & p^* \otimes 1 \end{pmatrix} \qquad r^* = \begin{pmatrix} p^* \otimes 1 & 1 \otimes q^* \\ -1 \otimes q & p \otimes 1 \end{pmatrix}$$

so that:

$$r^* r = \begin{pmatrix} p^* p \otimes 1 + 1 \otimes q^* q & 0 \\ 0 & pp^* \otimes 1 + 1 \otimes qq^* \end{pmatrix}$$

$$rr^* = \begin{pmatrix} pp^* \otimes 1 + 1 \otimes q^* q & 0 \\ 0 & p^* p \otimes 1 + 1 \otimes qq^* \end{pmatrix}$$

$r^* r$ and rr^* are positive self-adjoint matrices if $(\xi^1, \xi^2) \neq (0, 0)$. This verifies the ellipticity. We note that if (P, Q) are pseudo-differential, R will still be formally elliptic, but the symbol will not in general be smooth at $\xi^1 = 0$ or $\xi^2 = 0$ and hence R will not be a pseudo-differential operator in that case. We compute:

$$R^* R = \begin{pmatrix} P^* P \otimes 1 + 1 \otimes Q^* Q & 0 \\ 0 & PP^* \otimes 1 + 1 \otimes QQ^* \end{pmatrix}$$

$$RR^* = \begin{pmatrix} PP^* \otimes 1 + 1 \otimes Q^* Q & 0 \\ 0 & P^* P \otimes 1 + 1 \otimes QQ^* \end{pmatrix}$$

$$N(R^* R) = N(P^* P) \otimes N(Q^* Q) \oplus N(PP^*) \otimes N(QQ^*)$$

$$N(RR^*) = N(PP^*) \otimes N(Q^* Q) \oplus N(P^* P) \otimes N(QQ^*)$$

$$\mathrm{index}(R) = \{\dim N(P^* P) - \dim N(PP^*)\}$$
$$\times \{\dim N(Q^* Q) - \dim N(QQ^*)\}$$
$$= \mathrm{index}(P)\,\mathrm{index}(Q)$$

which completes the proof of (a).

We verify (b) on the symbol level. First consider the de Rham complex and decompose:

$$\Lambda(T^*M_1) = \Lambda_1^e \oplus \Lambda_1^o, \qquad \Lambda(T^*M_2) = \Lambda_2^e \oplus \Lambda_2^o,$$
$$\Lambda(T^*M) = (\Lambda_1^e \otimes \Lambda_2^e \ \oplus \ \Lambda_1^o \otimes \Lambda_2^o) \ \oplus \ (\Lambda_1^o \otimes \Lambda_2^e \ \oplus \ \Lambda_1^e \otimes \Lambda_2^o).$$

Under this decomposition:

$$\sigma_L((d+\delta)_M)(\xi^1, \xi^2) = \begin{pmatrix} c(\xi^1) \otimes 1 & -1 \otimes c(\xi^2) \\ 1 \otimes c(\xi^2) & c(\xi^1) \otimes 1 \end{pmatrix}.$$

$c(\cdot)$ denotes Clifford multiplication. This verifies the de Rham complex decomposes properly.

The signature complex is more complicated. We decompose

$$\Lambda^\pm(T^*M_1) = \Lambda_1^{\pm,e} \oplus \Lambda_1^{\pm,o}$$

to decompose the signature complex into two complexes:

$$(d+\delta): C^\infty(\Lambda_1^{+,e}) \to C^\infty(\Lambda_1^{-,o})$$
$$(d+\delta): C^\infty(\Lambda_1^{+,o}) \to C^\infty(\Lambda_1^{-,e}).$$

Under this decomposition, the signature complex of M decomposes into four complexes. If, for example, we consider the complex:

$$(d+\delta): C^\infty(\Lambda_1^{+,e} \otimes \Lambda_2^{+,e} \ \oplus \ \Lambda_1^{-,o} \otimes \Lambda_2^{-,o}) \to$$
$$C^\infty(\Lambda_1^{-,o} \otimes \Lambda_2^{+,e} \ \oplus \ \Lambda_1^{+,e} \otimes \Lambda_2^{-,o})$$

then the same argument as that given for the de Rham complex applies to show the symbol is:

$$\begin{pmatrix} c(\xi^1) \otimes 1 & -1 \otimes c(\xi^2) \\ 1 \otimes c(\xi^2) & c(\xi^1) \otimes 1 \end{pmatrix}.$$

If we consider the complex:

$$(d+\delta): C^\infty(\Lambda_1^{+,o} \otimes \Lambda_2^{+e} \ \oplus \ \Lambda_1^{-,e} \otimes \Lambda_2^{-,o})$$
$$\to C^\infty(\Lambda_1^{-,e} \otimes \Lambda_2^{+,e} \ \oplus \ \Lambda_1^{+,o} \otimes \Lambda_2^{-,o})$$

then we conclude the symbol is

$$\begin{pmatrix} c(\xi^1) \otimes 1 & 1 \otimes c(\xi^2) \\ -1 \otimes (\xi^2) & c(\xi^1) \otimes 1 \end{pmatrix}$$

which isn't right. We can adjust the sign problem for ξ^2 either by changing one of the identifications with $\Lambda^+(M)$ or by changing the sign of Q (which won't affect the index). The remaining two cases are similar. The spin and Dolbeault complexes are also similar. If we take coefficients in an auxiliary bundle V, the symbols involved are unchanged and the same arguments hold. This proves (b).

The proof of (c) is more complicated. We assume without loss of generality that p and q are homogeneous of degree 1. Let (ξ^1, ξ^2, u) parametrize the fibers of $T^*M \oplus 1$. At $\xi^1 = (1/\sqrt{2}, 0, \ldots, 0)$, $\xi^2 = (1/\sqrt{2}, 0, \ldots, 0)$, $u = 0$ the orientation is given by:

$$- dx_1^1 \wedge d\xi_1^1 \wedge dx_2^1 \wedge d\xi_2^1 \wedge \cdots$$
$$\wedge dx_{m_1}^1 \wedge d\xi_{m_1}^1 \wedge dx_1^2 \wedge dx_2^2 \wedge d\xi_2^2 \wedge \cdots \wedge dx_{m_2}^2 \wedge d\xi_{m_2}^2 \wedge du.$$

We have omitted $d\xi_2^1$ (and changed the sign) since it points outward at this point of the sphere to get the orientation on $\Sigma(T^*M)$.

In matrix form we have on $(V_1 \otimes W_1) \oplus (V_2 \otimes W_2) \oplus (V_2 \otimes W_1) \oplus (V_1 \otimes W_2)$ that:

$$\Sigma r = \begin{pmatrix} u\begin{pmatrix} 1 & 0 \\ 0 & 1 \end{pmatrix} & \begin{pmatrix} p^* \otimes 1 & 1 \otimes q^* \\ -1 \otimes q & p \otimes 1 \end{pmatrix} \\ \begin{pmatrix} p \otimes 1 & -1 \otimes q^* \\ 1 \otimes q & p^* \otimes 1 \end{pmatrix} & -u\begin{pmatrix} 1 & 0 \\ 0 & 1 \end{pmatrix} \end{pmatrix}.$$

This is not a very convenient form to work with. We define

$$\gamma_1 = \begin{pmatrix} 1 & 0 \\ 0 & -1 \end{pmatrix} \quad \text{and} \quad \alpha = \begin{pmatrix} 0 & p^* \\ p & 0 \end{pmatrix} \qquad \text{on } V_1 \oplus V_2 = V$$

$$\gamma_2 = \begin{pmatrix} 1 & 0 \\ 0 & -1 \end{pmatrix} \quad \text{and} \quad \beta = \begin{pmatrix} 0 & q^* \\ q & 0 \end{pmatrix} \qquad \text{on } W_1 \oplus W_2 = W$$

and compute that $\Sigma r = u\gamma_1 \otimes \gamma_2 + \alpha \otimes I_W + \gamma_1 \otimes \beta$. We replace p and q by homotopic symbols so that:

$$p^*p = |\xi^1|^2 I_{V_1}, \quad pp^* = |\xi^1|^2 I_{V_2}, \quad q^*q = |\xi^2|^2 I_{W_1}, \quad qq^* = |\xi^2|^2 I_{W_2}.$$

Since $\{\gamma_1 \otimes \gamma_2, \alpha \otimes I_W, \gamma_1 \otimes \beta\}$ all anti-commute and are self-adjoint,

$$(\Sigma r)^2 = (u^2 + |\xi^1|^2 + |\xi^2|^2)I \qquad \text{on } V \otimes W.$$

We parametrize $\Sigma(T^*M)$ by $S(T^*M_1) \times \left[0, \frac{\pi}{2}\right] \times \Sigma(T^*M_2)$ in the form:

$$\left(\xi^1 \cos\theta, \, \xi^2 \sin\theta, \, u \sin\theta\right).$$

We compute the orientation by studying $\theta = \frac{\pi}{4}$, $\xi^1 = (1, 0, \ldots, 0)$, $\xi^2 = (1, 0, \ldots, 0)$, $u = 0$. Since $d\xi_1^1 = -\sin\theta\, d\theta$, the orientation is given by:

$$dx_1^1 \wedge d\theta \wedge dx_2^1 \wedge d\xi_2^1 \wedge \cdots$$
$$\wedge\, dx_{m_1}^1 \wedge d\xi_{m_1}^1 \wedge dx_1^2 \wedge dx_2^2 \wedge d\xi_2^2 \wedge \cdots \wedge dx_{m_2}^2 \wedge d\xi_{m_2}^2 \wedge du.$$

If we identify $S(T^*M_1) \times \left[0, \frac{\pi}{2}\right]$ with $D_+(T^*M_1)$ then this orientation is:

$$\omega_{2m_1}^1 \wedge \omega_{2m_2}^2$$

so the orientations are compatible.

In this parametrization, we compute $\Sigma r(\xi^1, \theta, \xi^2, u) = (\sin\theta)\gamma_1 \otimes (u\gamma_2 + \beta(\xi^2)) + (\cos\theta)\alpha(\xi^1) \otimes I_W$. Since $\Sigma q = u\gamma_2 + \beta(\xi^2)$ satisfies $(\Sigma q)^2 = |\xi^2|^2 + u_2^2$ on W, we may decompose $W = \Pi_+(\Sigma q) \oplus \Pi_-(\Sigma q)$. Then:

$$\Sigma r = \{(\sin\theta)\gamma_1 + (\cos\theta)\alpha(\xi^1)\} \otimes I = \Sigma p(\xi^1, \theta) \otimes I \qquad \text{on } V \otimes \Pi_+(\Sigma q)$$
$$\Sigma r = \{\sin(-\theta)\gamma_1 + (\cos\theta)\alpha(\xi^1)\} \otimes I = \Sigma p(\xi^1, -\theta) \otimes I \quad \text{on } V \otimes \Pi_-(\Sigma q).$$

Consequently:

$$\Pi_+(\Sigma r) = \{\Pi_+(\Sigma p)(\xi^1, \theta) \otimes \Pi_+(\Sigma q)\} \oplus \{\Pi_+(\Sigma p)(\xi^1, -\theta) \otimes \Pi_-(\Sigma q)\}$$
$$\text{over } (D_+M_1) \times \Sigma(T^*M_2).$$

If we replace $-\theta$ by θ in the second factor, we may replace $\Pi_+(\Sigma p)(\xi^1, -\theta) \otimes \Pi_-(\Sigma q)$ by $\Pi_+(\Sigma p)(\xi^1, \theta) \otimes \Pi_-(\Sigma q)$. Since we have changed the orientation, we must change the sign. Therefore:

$$\int_{\Sigma(T^*M)} \theta_1 \wedge \theta_2 \wedge ch(\Pi_+(\Sigma r))$$
$$= \int_{D_+M_1} \theta_1 \wedge ch(\Pi_+(\Sigma p)) \cdot \int_{\Sigma(T^*M_2)} \theta_2 \wedge ch(\Pi_+(\Sigma q))$$
$$- \int_{D_-M_1} \theta_1 \wedge ch(\Pi_+(\Sigma p)) \cdot \int_{\Sigma(T^*M_2)} \theta_2 \wedge ch(\Pi_-(\Sigma q)).$$

$ch(\Pi_+(\Sigma q)) + ch(\Pi_-(\Sigma q)) = ch(V_2)$ does not involve the fiber coordinates of $\Sigma(T^*M_2)$ and thus

$$\int_{\Sigma(T^*M_2)} \theta_2 \wedge ch(V_2) = 0.$$

We may therefore replace $-ch(\Pi_-(\Sigma q))$ by $ch(\Pi_+(\Sigma q))$ in evaluating the integral over $D_-(M_1) \times \Sigma(T^*M_2)$ to complete the proof of (c).

We prove (d) in a similar fashion. We suppose without loss of generality that p and q are homogeneous of degree 1. We parameterize $S(T^*M) = S(T^*M_1) \times [0, \frac{\pi}{2}] \times S(T^*M_2)$ in the form $(\xi^1 \cos\theta, \xi^2 \sin\theta)$. Then

$$r = \begin{pmatrix} \sin\theta & 0 \\ 0 & -\sin\theta \end{pmatrix} \otimes p(\xi^2) + \begin{pmatrix} 0 & (\cos\theta)q^*(\xi^1) \\ (\cos\theta)q(\xi^1) & 0 \end{pmatrix}.$$

Again we decompose $V_2 = \Pi_+(p) \oplus \Pi_-(p)$ so that

$$\begin{aligned} \Sigma r = \big((\sin\theta)\gamma_1 + (\cos\theta)\alpha^1(\xi^1)\big) && \text{on } V \otimes \Pi_+(p) \\ \Sigma r = \big(\sin(-\theta)\gamma_1 + (\cos\theta)\alpha^1(\xi^1)\big) && \text{on } V \otimes \Pi_-(p) \end{aligned}$$

where

$$\gamma_1 = \begin{pmatrix} 1 & 0 \\ 0 & -1 \end{pmatrix} \qquad \alpha^1 = \begin{pmatrix} 0 & q^* \\ q & 0 \end{pmatrix}.$$

The remainder of the argument is exactly as before (with the appropriate change in notation from (c)) and is therefore omitted.

We now check a specific example to verify some normalizing constants:

LEMMA 3.9.4. *Let* $M = S^1$ *be the unit circle. Define* $P: C^\infty(M) \to C^\infty(M)$ *by:*

$$P(e^{in\theta}) = \begin{cases} ne^{i(n-1)\theta} & \text{for } n \geq 0 \\ ne^{in\theta} & \text{for } n \leq 0 \end{cases}$$

then P *is an elliptic pseudo-differential operator with* index$(P) = 1$. *Furthermore:*

$$\int_{\Sigma(T^*S^1)} ch(\Pi_+(\Sigma p)) = -1.$$

PROOF: Let $P_0 = -i\partial/\partial\theta$ and let $P_1 = \{-\partial^2/\partial\theta^2\}^{1/2}$. P_0 is a differential operator while P_1 is a pseudo-differential operator by the results of section 1.10. It is immediate that:

$$\begin{aligned} \sigma_L(P_0) = \xi, && P_0(e^{in\theta}) = ne^{in\theta} \\ \sigma_L(P_1) = |\xi|, && P_1(e^{in\theta}) = |n|e^{in\theta}. \end{aligned}$$

We define:

$$Q_0 = \frac{1}{2}e^{-i\theta}(P_0 + P_1)$$

$$\sigma_L Q_0 = \begin{cases} \xi e^{-i\theta} & \xi \geq 0 \\ 0 & \xi \leq 0 \end{cases}$$

$$Q_0(e^{in\theta}) = \begin{cases} ne^{i(n-1)\theta} & n \geq 0 \\ 0 & n \leq 0 \end{cases}$$

and

$$Q_1 = \frac{1}{2}(P_0 - P_1)$$

$$\sigma_L Q_1 = \begin{cases} 0 & \xi \geq 0 \\ \xi & \xi \leq 0 \end{cases}$$

$$Q_1(e^{in\theta}) = \begin{cases} 0 & n \geq 0 \\ ne^{in\theta} & n \leq 0. \end{cases}$$

Consequently, $P = Q_0 + Q_1$ is a pseudo-differential operator and

$$\sigma_L P = p = \begin{cases} \xi e^{-i\theta} & \xi > 0 \\ \xi & \xi < 0. \end{cases}$$

It is clear P is surjective so $N(P^*) = 0$. Since $N(P)$ is the space of constant functions, $N(P)$ is one dimensional so index$(P) = 1$. We compute:

$$\Sigma p(\theta, \xi, t) = \begin{cases} \begin{pmatrix} t & \xi e^{+i\theta} \\ \xi e^{-i\theta} & -t \end{pmatrix} & \text{if } \xi \geq 0 \\ \begin{pmatrix} t & \xi \\ \xi & -t \end{pmatrix} & \text{if } \xi \leq 0. \end{cases}$$

Since Σp does not depend on θ for $\xi \leq 0$, we may restrict to the region $\xi \geq 0$ in computing the integral. (We must smooth out the symbol to be smooth where $\xi = 0$ but suppress these details to avoid undue technical complications).

It is convenient to introduce the parameters:

$$u = t, \qquad v = \xi \cos\theta, \qquad w = \xi \sin\theta \qquad \text{for } u^2 + v^2 + w^2 = 1$$

then this parametrizes the region of $\Sigma(T^*S^1)$ where $\xi \geq 0$ in a 1-1 fashion except where $\xi = 0$. Since $du \wedge dv \wedge dw = -\xi \, d\theta \wedge d\xi \wedge dt$, S^2 inherits the reversed orientation from its natural one. Let

$$e_0 = \begin{pmatrix} 1 & 0 \\ 0 & -1 \end{pmatrix}, \qquad e_1 = \begin{pmatrix} 0 & 1 \\ 1 & 0 \end{pmatrix}, \qquad e_2 = \begin{pmatrix} 0 & i \\ -i & 0 \end{pmatrix}$$

so $\Sigma p(u, v, w) = ue_0 + ve_1 + we_2$. Then by Lemma 2.1.5 we have

$$-\int_{S^2} ch(\Pi_+(\Sigma p)) = (-1) \cdot \left(\frac{i}{2}\right) \cdot \text{Tr}(e_0 e_1 e_2) = -1$$

which completes the proof.

We can now state the Atiyah-Singer index theorem:

THEOREM 3.9.5 (THE INDEX THEOREM). *Let* $P: C^\infty(V_1) \to C^\infty(V_2)$
be an elliptic pseudo-differential operator. Let $Td_r(M) = Td(TM \otimes \mathbf{C})$ *be
the Todd class of the complexification of the real tangent bundle. Then:*

$$\text{index}(P) = (-1)^{\dim M} \int_{\Sigma(T^*M)} Td_r(M) \wedge ch(\Pi_+(\Sigma p)).$$

We remark that the additional factor of $(-1)^{\dim M}$ could have been avoided
if we changed the orientation of $\Sigma(T^*M)$.

We begin the proof by reducing to the case $\dim M = m$ even and M
orientable. Suppose first that m is odd. We take $Q: C^\infty(S^1) \to C^\infty(S^1)$
to be the operator defined in Lemma 3.9.4 with index $+1$. We then form
the operator R with $\text{index}(R) = \text{index}(P)\,\text{index}(Q) = \text{index}(P)$ defined
in Lemma 3.9.3. Although R is not a pseudo-differential operator, it can
be uniformly approximated by pseudo-differential operators in the natural
Fredholm topology (once the order of Q is adjusted suitably). (This process
does not work when discussing the twisted eta invariant and will involve
us in additional technical complications in the next chapter). Therefore:

$$(-1)^{m+1} \int_{\Sigma(T^*(M \times S^1))} Td_r(M \times S^1) \wedge ch(\Pi_+(\Sigma r))$$

$$= (-1)^m \int_{\Sigma(T^*M)} Td_r(M) \wedge ch(\Pi_+(\Sigma p)) \cdot (-1) \int_{\Sigma(T^*S^1)} ch(\Pi_+(\Sigma q))$$

$$= (-1)^m \int_{\Sigma(T^*M)} Td_r(M) \wedge ch(\Pi_+(\Sigma p)).$$

To show the last integral gives $\text{index}(P)$ it suffices to show the top integral
gives $\text{index}(R)$ and therefore reduces the proof of Theorem 3.9.3 to the case
$\dim M = m$ even.

If M is not orientable, we let M' be the orientable double cover of M.
It is clear the formula on the right hand side of the equation multiplies
by two. More careful attention to the methods of the heat equation for
pseudo-differential operators gives a local formula for the index even in
this case as the left hand side is also multiplied by two under this double
cover.

This reduces the proof of Theorem 3.9.3 to the case $\dim M = m$ even
and M orientable. We fix an orientation on M henceforth.

LEMMA 3.9.6. *Let* $P: C^\infty(\Lambda^+) \to C^\infty(\Lambda^-)$ *be the operator of the signature complex. Let* $\omega = ch_{m/2}(\Pi_+(\Sigma p)) \in H^m(\Sigma(T^*M); \mathbf{C})$. *Then if* ω_M
is the orientation class of M
(a) $\omega_M \wedge \omega$ *gives the orientation of* $\Sigma(T^*M)$.

(b) If S^m is a fiber sphere of $\Sigma(T^*M)$, then $\int_{S^m} \omega = 2^{m/2}$.

PROOF: Let $(x_1, \ldots, x_m)$ be an oriented local coordinate system on M so that the $\{dx_j\}$ are orthonormal at $x_0 \in M$. If $\xi = (\xi, \ldots, \xi_m)$ are the dual fiber coordinates for T^*M), then:

$$p(\xi) = \sum_j i\xi_j(c(dx_j)) = \sum_j i\xi_j(\text{ext}(dx_j) - \text{int}(\xi_j))$$

gives the symbol of $(d + \delta)$; $c(\cdot)$ denotes Clifford multiplication as defined previously. We let $e_j = ic(dx_j)$; these are self-adjoint matrices such that $e_j e_k + e_k e_j = 2\delta_{jk}$. The orientation class is defined by:

$$e_0 = i^{m/2}c(dx_1)\ldots c(dx_m) = (-i)^{m/2}e_1 \ldots e_m.$$

The bundles $\Lambda^\pm$ are defined as the ± 1 eigenspaces of e_0. Consequently,

$$\Sigma p(\xi, t) = te_0 + \sum \xi_j e_j.$$

Therefore by Lemma 2.1.5 when S^m is given its natural orientation,

$$\begin{aligned}
\int_{S^m} ch_{m/2} \Pi_+(\Sigma p) &= i^{m/2}2^{-m/2} \text{Tr}(e_0 e_1 \ldots e_m) \\
&= i^{m/2}2^{-m/2} \text{Tr}(e_0 i^{m/2} e_0) \\
&= (-1)^{m/2}2^{-m/2} \text{Tr}(I) = (-1)^{m/2}2^{-m/2}2^m \\
&= (-1)^{m/2}2^{m/2}.
\end{aligned}$$

However, S^m is in fact given the orientation induced from the orientation on $\Sigma(T^*M)$ and on M. At the point $(x, 0, \ldots, 0, 1)$ in $T^*M \oplus \mathbf{R}$ the natural orientations are:

$$\begin{aligned}
\text{of } X: \quad & dx_1 \wedge \cdots \wedge dx_m, \\
\text{of } \Sigma(T^*M): \quad & dx_1 \wedge d\xi_1 \wedge \cdots \wedge dx_m \wedge d\xi_m \\
& = (-1)^{m/2}dx_1 \wedge \cdots \wedge dx_m \wedge d\xi_1 \wedge \cdots \wedge d\xi_m \\
\text{of } S^m: \quad & (-1)^{m/2}d\xi_1 \wedge \cdots \wedge d\xi_m.
\end{aligned}$$

Thus with the induced orientation, the integral becomes $2^{m/2}$ and the lemma is proved.

Consequently, ω provides a cohomology extension and:

LEMMA 3.9.7. Let $\rho: \Sigma(T^*M) \to M$ where M is orientable and even dimensional. Then
(a) $\rho^*: H^*(M; \mathbf{C}) \to H^*(\Sigma(T^*M); \mathbf{C})$ is injective.

(b) If ω is as defined in Lemma 3.9.6, then we can express any $\alpha \in H^*(\Sigma(T^*M);\mathbf{C})$ uniquely as $\alpha = \rho^*\alpha_1 + \rho^*\alpha_2 \wedge \omega$ for $\alpha_i \in H^*(M;\mathbf{C})$.

Since ρ^* is injective, we shall drop it henceforth and regard $H^*(M;\mathbf{C})$ as being a subspace of $H^*(\Sigma(T^*M);\mathbf{C})$.

The Chern character gives an isomorphism $K(X;\mathbf{C}) \simeq H^e(X;\mathbf{C})$. When we interpret Lemma 3.9.7 in K-theory, we conclude that we can decompose $K(\Sigma(T^*M);\mathbf{C}) = K(M;\mathbf{C}) \oplus K(M;\mathbf{C}) \otimes \Pi_+(\Sigma p)$; $ch(V)$ generates $H^e(M;\mathbf{C})$ as V ranges over $K(M;\mathbf{C})$. Therefore $K(\Sigma(T^*M);\mathbf{C})/K(M;\mathbf{C})$ is generated as an additive module by the twisted signature complex with coefficients in bundles over M. $\Pi_+(\Sigma p_V) = V \otimes \Pi_+(\Sigma p)$ if p_V is the symbol of the signature complex with coefficients in V.

In Lemma 3.9.2, we interpreted the index as a map in K-theory. Since it is linear, it suffices to compute on the generators given by the signature complex with coefficients in V. This proves:

LEMMA 3.9.8. *Assume M is orientable and of even dimension m. Let P_V be the operator of the signature complex with coefficients in V. The bundles $\{\Pi_+(\Sigma p)_V\}_{V \in \mathrm{Vect}(M)}$ generate $K(\Sigma(T^*M);\mathbf{C})/K(M;\mathbf{C})$ additively. It suffices to prove Theorem 3.9.5 in general by checking it on the special case of the operators P_V.*

We will integrate out the fiber coordinates to reduce the integral of Theorem 3.9.5 from $\Sigma(T^*M)$ to M. We proceed as follows. Let W be an oriented real vector bundle of fiber dimension $k + 1$ over M equipped with a Riemannian inner product. Let $S(W)$ be the unit sphere bundle of W. Let $\rho \colon S(W) \to M$ be the natural projection map. We define a map $\mathcal{I} \colon C^\infty(\Lambda(S(W))) \to C^\infty(\Lambda(M))$ which is a $C^\infty(\Lambda(M))$ module homomorphism and which commutes with integration—i.e., if $\alpha \in C^\infty(\Lambda(S(W)))$ and $\beta \in C^\infty(\Lambda(M))$, we require the map $\alpha \mapsto \mathcal{I}(\alpha)$ to be linear so that $\mathcal{I}(\rho^*\beta \wedge \alpha) = \beta \wedge \mathcal{I}(\alpha)$ and $\int_{S(W)} \alpha = \int_M \mathcal{I}(\alpha)$.

We construct $\mathcal{I}$ as follows. Choose a local orthonormal frame for W to define fiber coordinates $u = (u_0, \ldots, u_k)$ on W. This gives a local representation of $S(W) = V \times S^k$ over the coordinate patch $\mathcal{U}$ on M. If $\alpha \in C^\infty(\Lambda(S(W)))$ has support over V, we can decompose $\alpha = \sum_\nu \beta_\nu \wedge \alpha_\nu$ for $\beta_\nu \in C^\infty(\Lambda(\mathcal{U}))$ and $\alpha_\nu \in C^\infty(\Lambda(S^k))$. We permit the α_ν to have coefficients which depend upon $x \in \mathcal{U}$. This expression is, of course, not unique. Then $\mathcal{I}(\alpha)$ is necessarily defined by:

$$\mathcal{I}(\alpha)(x) = \sum \beta_\nu \int_{S^k} \alpha_\nu(x).$$

It is clear this is independent of the particular way we have decomposed α. If we can show $\mathcal{I}$ is independent of the frame chosen, then this will define $\mathcal{I}$ in general using a partition of unity argument.

Let $u_i' = a_{ij}(x)u_j$ be a change of fiber coordinates. Then we compute:

$$du_i' = a_{ij}(x)\,du_j + da_{ij}(x)u_j.$$

Clearly if a is a constant matrix, we are just reparamatrizing S^k so the integral is unchanged. We fix x_0 and suppose $a(x_0) = I$. Then over x_0,

$$du_I = dv_I + \sum_{|I|<|J|} c_{I,J} \wedge dv_J \qquad \text{where } c_{I,J} \in \Lambda^{|I|-|J|}(M).$$

To integrate and get an answer different from 0 over S^k, we must have $|I| = k$ so these error terms integrate to zero and I is invariantly defined.

We specialize to the case $W = T^*M \oplus 1$. The orientation of M induces a natural orientation of $T^*M \oplus 1$ as a bundle in such a way as to agree with the orientation of $T^*M \oplus 1$ as a topological space. Let $\alpha = ch(\Pi_+(\Sigma p))$, so $I(\alpha) \in C^\infty(\Lambda(M))$. If we reverse the orientation of M, then we interchange the roles of Λ^+ and of Λ^-. This has the effect of replacing the parameter u by $-u$ which is equivalent to reversing the orientation of $T^*M \oplus 1$ as a topological space. Since both orientations have been reversed, the orientation of the fiber is unchanged so $I(\alpha)$ is invariantly defined independent of any local orientation of M. It is clear from the definition that $I(\alpha)$ is a polynomial in the jets of the metric and is invariant under changes of the metric by a constant factor. Therefore Theorems 2.5.6 and Lemma 2.5.3 imply $I(\alpha)$ is a real characteristic form. By Lemma 3.9.6, we can expand $I(\alpha) = 2^{m/2} + \cdots$.

We solve the equation:

$$\{I(ch(\Pi_+(\Sigma p))) \wedge Todd(m)\}_{m-4s} = 2^{(m-4s)/2} L_s$$

recursively to define a real characteristic class we shall call $Todd(m)$ for the moment. It is clear $Todd(m) = 1 + \cdots$. In this equation, L_s is the Hirzebruch genus.

LEMMA 3.9.9. *Let $Todd(m)$ be the real characteristic class defined above. Then if P is any elliptic pseudo-differential operator,*

$$\text{index}(P) = \int_{\Sigma(T^*M)} Todd(m) \wedge ch(\Pi_+(\Sigma p)).$$

PROOF: By Lemma 3.9.8 it suffices to prove this identity if P is the operator of the twisted signature complex. By Theorem 3.1.5,

$$\text{index}(P_V^{\text{signature}}) = \int_M \sum_{2t+4s=m} ch_t(V)2^t \wedge L_s$$

$$= \int_M ch(V) \wedge I(ch(\Pi_+(\Sigma p))) \wedge Todd(m)$$

$$= \int_{\Sigma(T^*M)} ch(V) \wedge ch(\Pi_+(\Sigma p)) \wedge Todd(m)$$

$$= \int_{\Sigma(T^*M)} Todd(m) \wedge ch(\Pi_+(\Sigma p_V)).$$

It is clear $Todd(m)$ is uniquely determined by Lemma 3.9.9. Both the index of an elliptic operator and the formula of Lemma 3.9.9 are multiplicative with respect to products by Lemma 3.9.3 so $Todd(m)$ is a multiplicative characteristic form. We may therefore drop the dependence upon the dimension m and simply refer to $Todd$. We complete the proof of the Atiyah-Singer index theorem by identifying this characteristic form with the real Todd polynomial of $T(M)$.

We work with the Dolbeault complex instead of with the signature complex since the representations involved are simpler. Let m be even, then $(-1)^m = 1$. Let M be a holomorphic manifold with the natural orientation. We orient the fibers of T^*M using the natural orientation which arises from the complex structure on the fibers. If ξ are the fiber coordinates, this gives the orientation:

$$dx_1 \wedge dy_1 \wedge \cdots \wedge dx_n \wedge dy_n \wedge d\xi_1 \wedge \cdots \wedge d\xi_m \qquad \text{where } m = 2n.$$

This gives the total space T^*M an orientation which is $(-1)^n$ times the simplectic orientation. Let S^m denote a fiber sphere of $\Sigma(T^*M)$ with this orientation and let q be the symbol of the Dolbeault complex. Then:

$$\text{index}(\bar{\partial}) = \int_{\Sigma(T^*M)} Todd \wedge ch(\Pi_+\Sigma q) = (-1)^n \int_M Todd \wedge I(ch(\Pi_+\Sigma q)).$$

We define the complex characteristic form

$$S = (-1)^{m/2} I(ch(\Pi_+(\Sigma q)))$$

then the Riemann-Roch formula implies that:

$$S \wedge Todd(m) = Todd(T_cM).$$

It is convenient to extend the definition of the characteristic form S to arbitrary complex vector bundles V. Let $W = \Lambda^*(V)$ and let $q(v): V \to \text{END}(W)$ be defined by Lemma 3.5.2(a) to be the symbol of $\bar{\partial} + \delta''$ if $V = T_c(M)$. We let ext$: V \to \text{HOM}(W, W)$ be exterior multiplication. This is complex linear and we let int be the dual of ext; $\text{int}(\lambda v) = \bar{\lambda}\,\text{int}(v)$ for $\lambda \in \mathbf{C}$. ext is invariantly defined while int requires the choice of a fiber matric. We let $q = \text{ext}(v) - \text{int}(v)$ (where we have deleted the factor of $i/2$ which appears in Lemma 3.5.2 for the sake of simplicity).

We regard q as a section to the bundle $\text{HOM}(V, \text{HOM}(W, W))$. Fix a Riemannian connection on V and covariantly differentiate q to compute $\nabla q \in C^\infty(T^*M \otimes \text{HOM}(V, \text{HOM}(W, W)))$. Since the connection is Riemannian, $\nabla q = 0$; this is not true in general for non-Riemannian connections.

If V is trivial with flat connection, the bundles $\Pi_\pm(\Sigma q)$ have curvature $\pi_\pm d\pi_\pm d\pi_\pm$ as computed in Lemma 2.1.5. If V is not flat, the curvature of

V enters into this expression. The connection and fiber metric on V define a natural metric on T^*V. We use the splitting defined by the connection to decompose T^*V into horizontal and vertical components. These components are orthogonal with respect to the natural metric on V. Over V, q becomes a section to the bundle $\text{HOM}(V, V)$. We let ∇^V denote covariant differentiation over V, then $\nabla^V q$ has only vertical components in T^*V and has no horizontal components.

The calculation performed in Lemma 2.1.5 shows that in this more general setting that the curvatures of $\nabla_\pm$ are given by:

$$\Omega_\pm = \pi_+(\nabla^V \pi_+ \wedge \nabla^V \pi_+ + \rho^* \Omega_W).$$

If we choose a frame for V and W which is covariant constant at a point x_0, then $\nabla^V \pi_+ = d\pi_+$ has only vertical components while $\rho^* \Omega_W$ has only horizontal components. If Ω_V is the curvature of V, then $\Omega_W = \Lambda(\Omega_V)$.

Instead of computing S on the form level, we work with the corresponding invariant polynomial.

LEMMA 3.9.10. *Let A be an $n \times n$ complex skew-adjoint matrix. Let $B = \Lambda(A)$ acting on $\Lambda(\mathbf{C}^n) = \mathbf{C}^{2^n}$. Define:*

$$S(A) = \sum_\nu (-1)^n \left(\frac{i}{2\pi}\right)^\nu \frac{1}{\nu!} \int_{S^{2n}} \text{Tr}\{(\pi_+ d\pi_+ d\pi_+ + B)^\nu\}.$$

If $x_j = i\lambda_j/2\pi$ are the normalized eigenvectors of A, then

$$S(A) = \prod_j \frac{e^{x_j} - 1}{x_j}.$$

PROOF: If $V = V_1 \oplus V_2$ and if $A = A_1 \oplus A_2$, then the Dolbeault complex decomposes as a tensor product by Lemma 3.9.3. The calculations of Lemma 3.9.3 using the decomposition of the bundles $\Pi_\pm$ shows $S(A)$ is a multiplicative characteristic class. To compute the generating function, it suffices to consider the case $n = 1$.

If $n = 1$, $A = \lambda$ so that $B = \begin{pmatrix} 0 & 0 \\ 0 & \lambda \end{pmatrix}$, if we decompose $W = \Lambda^{0,0} \oplus \Lambda^{0,1} = 1 \oplus V$. If $x + iy$ give the usual coordinates on $V = \mathbf{C}$, then:

$$q(x, y) = x \begin{pmatrix} 0 & 1 \\ 1 & 0 \end{pmatrix} + y \begin{pmatrix} 0 & -i \\ i & 0 \end{pmatrix}$$

by Lemma 3.5.2 which gives the symbol of the Dolbeault complex. Therefore:

$$q(x, y, u) = x \begin{pmatrix} 0 & 1 \\ 1 & 0 \end{pmatrix} + y \begin{pmatrix} 0 & -i \\ i & 0 \end{pmatrix} + u \begin{pmatrix} 1 & 0 \\ 0 & -1 \end{pmatrix} = xe_0 + ye_1 + ue_2.$$

We compute:

$$e_0 e_1 e_2 = i \begin{pmatrix} 1 & 0 \\ 0 & 1 \end{pmatrix} \qquad \text{and} \qquad \pi_+ \, d\pi_+ \, d\pi_+ = \frac{i}{2} \pi_+ \, \text{dvol}.$$

Since $n = 1$, $(-1)^n = -1$ and:

$$S(A) = \sum_{j>0} -\frac{i}{2} \left(\frac{i}{2\pi} \right)^j \frac{1}{j!} \int \text{Tr}((\pi_+ \, \text{dvol} + \pi_+ B)^j)$$

$$= \sum_{j>0} -\frac{i}{2} \left(\frac{i}{2\pi} \right)^j \frac{1}{(j-1)!} \int \text{Tr}(\pi_+ B)^{j-1} \, \text{dvol}.$$

We calculate that:

$$\pi_+ B = \frac{1}{2} \begin{pmatrix} 1+u & x-iy \\ x+iy & 1-u \end{pmatrix} \begin{pmatrix} 0 & 0 \\ 0 & \lambda \end{pmatrix} = \frac{1}{2} \begin{pmatrix} 0 & * \\ 0 & (1-u)\lambda \end{pmatrix}$$

$$(\pi_+ B)^{j-1} = 2^{1-j} \lambda^{j-1} (1-u)^{j-1} \begin{pmatrix} 0 & * \\ 0 & 1 \end{pmatrix}$$

where "$*$" indicates a term we are not interested in. We use this identity and re-index the sum to express:

$$S(A) = \sum_{j \geq 0} \frac{1}{4\pi} \left(\frac{i\lambda}{2\pi} \right)^j \frac{1}{j!} 2^{-j} \int_{S^2} (1-u)^j \, \text{dvol}.$$

We introduce the integrating factor of e^{-r^2} to compute:

$$\int_{R^3} u^{2k} e^{-r^2} \, dx \, dy \, du = \pi \int_R u^{2k} e^{-u^2} \, du$$

$$= \int_0^\infty r^{2k+2} e^{-r^2} \, dr \cdot \int_{S^2} u^{2k} \, \text{dvol}$$

$$= (2k+1)/2 \int_0^\infty r^{2k} e^{-r^2} \, dr \cdot \int_{S^2} u^{2k} \, \text{dvol}$$

so that:

$$\int_{S^2} u^{2k} \, \text{dvol} = 4\pi/(2k+1).$$

The terms of odd order integrate to zero so:

$$\int_{S^2} (1-u)^j \, \text{dvol} = 4\pi \sum \binom{j}{2k} \cdot \frac{1}{2k+1} = 4\pi \int_0^1 \sum \binom{j}{2k} t^{2k} \, dt$$

$$= 2\pi \int_0^1 (1+t)^j + (1-t)^j \, dt$$

$$= 2\pi \frac{(1+t)^{j+1} - (1-t)^{j+1}}{j+1} \Big|_0^1$$

$$= 4\pi \cdot \frac{2^j}{j+1}.$$

We substitute this to conclude:

$$S(A) = \sum_{j \geq 0} \left(\frac{i\lambda}{2\pi}\right)^j \frac{1}{(j+1)!}.$$

If we introduce $x = i\lambda/2\pi$ then

$$S(x) = \sum_{j \geq 0} \frac{x^j}{(j+1)!} = \frac{e^x - 1}{x}$$

which gives the generating function for S. This completes the proof of the lemma.

We can now compute $Todd$. The generating function of $Todd(T_c)$ is $x/(1 - e^{-x})$ so that $Todd = S^{-1} \cdot Todd(T_c)$ will have generating function:

$$\frac{x}{1 - e^{-x}} \cdot \frac{x}{e^x - 1} = \frac{x}{1 - e^{-x}} \cdot \frac{-x}{1 - e^x}$$

which is, of course, the generating function for the real $Todd$ class. This completes the proof. We have gone into some detail to illustrate that it is not particularly difficult to evaluate the integrals which arise in applying the index theorem. If we had dealt with the signature complex instead of the Dolbeault complex, the integrals to be evaluated would have been over S^4 instead of S^2 but the computation would have been similar.

CHAPTER 4

GENERALIZED
INDEX THEOREMS
AND SPECIAL TOPICS

Introduction

This chapter is less detailed than the previous three as several lengthy calculations are omitted in the interests of brevity. In sections 4.1 through 4.6, we sketch the interrelations between the Atiyah-Patodi-Singer twisted index theorem, the Atiyah-Patodi-Singer index theorem for manifolds with boundary, and the Lefschetz fixed point formulas.

In section 4.1, we discuss the absolute and relative boundary conditions for the de Rham complex if the boundary is non-empty. We discuss Poincaré duality and the Hodge decomposition theorem in this context. The spin, signature, and Dolbeault complexes do not admit such local boundary conditions and the index theorem in this context looks quite different. In section 4.2, we prove the Gauss-Bonnet theorem for manifolds with boundary and identify the invariants arising from the heat equation with these integrands.

In section 4.3, we introduce the eta invariant as a measure of spectral asymmetry and establish the regularity at $s = 0$. We discuss without proof the Atiyah-Patodi-Singer index theorem for manifolds with boundary. The eta invariant enters as a new non-local ingredient which was missing in the Gauss-Bonnet theorem. In section 4.4, we review secondary characteristic classes and sketch the proof of the Atiyah-Patodi-Singer twisted index theorem with coefficients in a locally flat bundle. We discuss in some detail explicit examples on a 3-dimensional lens space.

In section 4.5, we turn to the Lefschetz fixed point formulas. We treat the case of isolated fixed points in complete detail in regard to the four classical elliptic complexes. We also return to the 3-dimensional examples discussed in section 4.4 to relate the Lefschetz fixed point formulas to the twisted index theorem using results of Donnelly. We discuss in some detail the Lefschetz fixed point formulas for the de Rham complex if the fixed point set is higher dimensional. There are similar results for both the spin and signature complexes which we have omitted for reasons of space. In section 4.6 we use these formulas for the eta invariant to compute the K-theory of spherical space forms.

In section 4.7, we turn to a completely new topic. In a lecture at M.I.T., Singer posed the question:

Suppose $P(G)$ is a scalar valued invariant of the metric so that $P(M) = \int_M P(G)\,\mathrm{dvol}$ is independent of the metric. Then is there a universal constant c so $P(M) = c\chi(M)$?

The answer to this question (and to related questions involving form valued invariants) is yes. This leads immediately to information regarding the higher order terms in the expansion of the heat equation. In section 4.8, we use the functorial properties of the invariants to compute $a_n(x, P)$ for an arbitrary second order elliptic parital differential operator with leading symbol given by the metric tensor for $n = 0, 2, 4$. We list (without proof) the corresponding formula if $n = 6$. This leads to Patodi's formula for $a_n(x, \Delta_p^m)$ discussed in Theorem 4.8.18. In section 4.9 we discuss some results of Ikeda to give examples of spherical space forms which are isospectral but not diffeomorphic. We use the eta invariant to show these examples are not even equivariant cobordant.

The historical development of these ideas is quite complicated and in particular the material on Lefschetz fixed point formulas is due to a number of authors. We have included a brief historical survey at the beginning of section 4.6 to discuss this material.

4.1. The de Rham Complex for Manifolds with Boundary.

In section 1.9, we derived a formula for the index of a strongly elliptic boundary value problem. The de Rham complex admits suitable boundary conditions leading to the relative and absolute cohomolgy groups. It turns out that the other 3 classical elliptic complexes do *not* admit even a weaker condition of ellipticity.

Let $(d + \delta): C^\infty(\Lambda(M)) \rightarrow C^\infty(\Lambda(M))$ be the de Rham complex. In the third chapter we assumed the dimension m of M to be even, but we place no such restriction on the parity here. M is assumed to be compact with smooth boundary dM. Near the boundary, we introduce coordinates (y, r) where $y = (y_1, \ldots, y_{m-1})$ give local coordinates for dM and such that $M = \{ x : r(x) \geq 0 \}$. We further normalize the coordinates by assuming the curves $x(r) = (y_0, r)$ are unit speed geodesics perpendicular to dM for $r \in [0, \delta)$.

Near dM, we decompose any differential form $\theta \in \Lambda(M)$ as

$$\theta = \theta_1 + \theta_2 \wedge dr \qquad \text{where } \theta_i \in \Lambda(dM)$$

are tangential differential forms. We use this decomposition to define:

$$\alpha(\theta) = \theta_1 - \theta_2 \wedge dr.$$

α is self-adjoint and $\alpha^2 = 1$. We define the absolute and relative boundary conditions

$$B_a(\theta) = \theta_2 \qquad \text{and} \qquad B_r(\theta) = \theta_1.$$

We let B denote either B_a or B_r, then $B: \Lambda(M)|_{dM} \rightarrow \Lambda(dM)$. We note that B_a can be identified with orthogonal projection on the -1 eigenspace of α while B_r can be identified with orthogonal projection on the $+1$ eigenspace of α. There is a natural inclusion map $i: dM \rightarrow M$ and $B_r(\theta) = i^*(\theta)$ is just the pull-back of θ. The boundary condition B_r does not depend on the Riemannian metric chosen, while the boundary condition B_a does depend on the metric.

LEMMA 4.1.1. *Let $B = B_a$ or B_r, then $(d + \delta, B)$ is self-adjoint and elliptic with respect to the cone $\mathbf{C} - \mathbf{R}_+ - \mathbf{R}_-$.*

PROOF: We choose a local orthonormal frame $\{e_0, \ldots, e_{m-1}\}$ for T^*M near dM so that $e_0 = dr$. Let

$$p_j = ie_j$$

act on $\Lambda(M)$ by Clifford multiplication. The p_j are self-adjoint and satisfy the commutation relation:

$$p_j p_k + p_k p_j = 2\delta_{jk}.$$

The symbol of $(d + \delta)$ is given by:

$$p(x, \xi) = zp_0 + \sum_{j=1}^{m-1} \varsigma_j p_j.$$

As in Lemma 1.9.5, we define:

$$\tau(y, \varsigma, \lambda) = ip_0 \left(\sum_{j=1}^{m-1} \varsigma_j p_j - \lambda \right), \quad (\varsigma, \lambda) \neq (0, 0) \in T^*(dM) \times \{\mathbf{C} - \mathbf{R}_+ - \mathbf{R}_-\}.$$

We define the new matrices:

$$q_0 = p_0 \quad \text{and} \quad q_j = ip_0 p_j \quad \text{for } 1 \leq j \leq m - 1$$

so that:

$$\tau(y, \varsigma, \lambda) = -i\lambda q_0 + \sum_{j=1}^{m-1} \varsigma_j q_j.$$

The $\{q_j\}$ are self-adjoint and satisfy the commutation relations $q_j q_k + q_k q_j = 2\delta_{jk}$. Consequently:

$$(y, \varsigma, \lambda)^2 = (|\varsigma|^2 - \lambda^2)I.$$

We have $(|\varsigma|^2 - \lambda^2) \in \mathbf{C} - \mathbf{R}_- - 0$ so we can choose $\mu^2 = (|\varsigma|^2 - \lambda^2)$ with $\mathrm{Re}(\mu) > 0$. Then τ is diagonalizable with eigenvalues $\pm\mu$ and $V_\pm(\tau)$ is the span of the eigenvectors of τ corresponding to the eigenvalue $\pm\mu$. (We set $V = \Lambda(M) = \Lambda(T^*M)$ to agree with the notation of section 9.1).

We defined $\alpha(\theta_1 + \theta_2 \wedge dr) = \theta_1 - \theta_2 \wedge dr$. Since Clifford multiplication by e_j for $1 \leq j \leq m - 1$ preserves $\Lambda(dM)$, the corresponding p_j commute with α. Since Clifford multiplication by dr interchanges the factors of $\Lambda(dM) \oplus \Lambda(dM) \wedge dr$, α anti-commutes with p_0. This implies α anti-commutes with all the q_j and consequently anti-commutes with τ. Thus the only common eigenvectors must belong to the eigenvalue 0. Since 0 is not an eigenvalue,

$$V_\pm(\alpha) \cap V_\pm(\tau) = \{0\}.$$

Since B_a and B_r are just orthogonal projection on the ∓ 1 eigenspaces of α, $\mathrm{N}(B_a)$ and $\mathrm{N}(B_r)$ are just the ± 1 eigenspaces of α. Thus

$$B: V_\pm(\tau) \rightarrow \Lambda(dM)$$

is injective. Since $\dim(V_\pm(\tau)) = \dim \Lambda(dM) = 2^{m-1}$, this must be an isomorphism which proves the ellipticity. Since p_0 anti-commutes with α,

$$p_0: V_\pm(\alpha) \rightarrow V_\mp(\alpha)$$

so $(d + \delta)$ is self-adjoint with respect to either B_a or B_r by Lemma 1.9.5.

By Lemma 1.9.1, there is a spectral resolution for the operator $(d+\delta)_B$ in the form $\{\lambda_\nu, \phi_\nu\}_{\nu=1}^\infty$ where $(d + \delta)\phi_\nu = \lambda_\nu\phi_\nu$ and $B\phi_\nu = 0$. The $\phi_\nu \in C^\infty(\Lambda(M))$ and $|\lambda_\nu| \to \infty$. We set $\Delta = (d+\delta)^2 = d\delta + \delta d$ and define:

$$N((d + \delta)_B) = \{\, \phi \in C^\infty(\Lambda(M)) : B\phi = (d + \delta)\phi = 0\,\}$$
$$N((d + \delta)_B)_j = \{\, \phi \in C^\infty(\Lambda^j(M)) : B\phi = (d + \delta)\phi = 0\,\}$$
$$N(\Delta_B) = \{\, \phi \in C^\infty(\Lambda(M)) : B\phi = B(d + \delta)\phi = \Delta\phi = 0\,\}$$
$$N(\Delta_B)_j = \{\, \phi \in C^\infty(\Lambda^j(M)) : B\phi = B(d + \delta)\phi = \Delta\phi = 0\,\}.$$

LEMMA 4.1.2. *Let B denote either the relative or the absolute boundary conditions. Then*
(a) $N((d + \delta)_B) = N(\Delta_B)$.
(b) $N((d + \delta)_B)_j = N(\Delta_B)_j$.
(c) $N((d + \delta)_B) = \bigoplus_j N((d + \delta)_B)_j$.

PROOF: Let $B\phi = B(d + \delta)\phi = \Delta\phi = 0$. Since $(d + \delta)_B$ is self-adjoint with respect to the boundary condition B, we can compute that $\Delta\phi \cdot \phi = (d + \delta)\phi \cdot (d + \delta)\phi = 0$ so $(d + \delta)\phi = 0$. This shows $N(\Delta_B) \subset N((d + \delta)_B)$ and $N(\Delta_B)_j \subset N((d + \delta)_B)_j$. The reverse inclusions are immediate which proves (a) and (b). It is also clear that $N((d+\delta)_B)_j \subset N((d+\delta)_B)$ for each j. Conversely, let $\theta \in N((d + \delta)_B)$. We decompose $\theta = \theta_0 + \cdots + \theta_m$ into homogeneous pieces. Then $\Delta\theta = \sum_j \Delta\theta_j = 0$ implies $\Delta\theta_j = 0$ for each j. Therefore we must check $B\theta_j = B(d + \delta)\theta_j = 0$ since then $\theta_j \in N(\Delta_B)_j = N((d + \delta)_B)_j$ which will complete the proof.

Since B preserves the grading, $B\theta_j = 0$. Suppose $B = B_r$ is the relative boundary conditions so $B(\alpha_1 + \alpha_2 \wedge dr) = \alpha_1|_{dM}$. Then $Bd = dB$ so $B(d + \delta)\theta = dB(\theta) + B\delta(\theta) = B\delta(\theta) = 0$. Since B preserves the homogeneity, this implies $B\delta\theta_j = 0$ for each j. We observed $Bd\theta = dB\theta = 0$ so $Bd\theta_j = 0$ for each j as well. This completes the proof in this case. If $B = B_a$ is the absolute boundary condition, use a similar argument based on the identity $B\delta = \delta B$.

We illustrate this for $m = 1$ by considering $M = [0, 1]$. We decompose $\theta = f_0 + f_1\, dx$ to decompose $C^\infty(\Lambda(M)) = C^\infty(M) \oplus C^\infty(M)\, dx$. It is immediate that:

$$(d + \delta)(f_0, f_1) = (-f_1', f_0')$$

so $(d+\delta)\theta = 0$ implies θ is constant. B_a corresponds to Dirichlet boundary conditions on f_1 while B_r corresponds to Dirichlet boundary conditions on f_0. Therefore:

$$H_a^0([0, 1]; \mathbf{C}) = \mathbf{C} \qquad H_a^1([0, 1]; \mathbf{C}) = 0$$
$$H_r^0([0, 1]; \mathbf{C}) = 0 \qquad H_r^1([0, 1]; \mathbf{C}) = \mathbf{C}.$$

A priori, the dimensions of the vector spaces $H_a^j(M;\mathbf{C})$ and $H_r^j(M;\mathbf{C})$ depend on the metric. It is possible, however, to get a more invariant definition which shows in fact they are independent of the metric. Lemma 1.9.1 shows these spaces are finite dimensional.

Let $d: C^\infty(\Lambda^j) \to C^\infty(\Lambda^{j+1})$ be the de Rham complex. The relative boundary conditions are independent of the metric and are d-invariant. Let $C_r^\infty(\Lambda^p) = \{\theta \in C^\infty \Lambda^p : B_r\theta = 0\}$. There is a chain complex $d: C_r^\infty(\Lambda^p) \to C_r^\infty(\Lambda^{p+1}) \to \cdots$. We define $H_r^p(M;\mathbf{C}) = (\ker d/\text{image } d)_p$ on $C_r^\infty(\Lambda^p)$ to be the cohomology of this chain complex. The de Rham theorem for manifolds without boundary generalizes to identify these groups with the relative simplicial cohomology $H^p(M, dM; \mathbf{C})$. If $\theta \in \mathrm{N}((d+\delta)_{B_r})_p$ then $d\theta = B_r\theta = 0$ so $\theta_j \in H^p(M, dM; \mathbf{C})$. The Hodge decomposition theorem discussed in section 1.5 for manifolds without boundary generalizes to identify $H_r^p(M;\mathbf{C}) = \mathrm{N}(\Delta_{B_r})_p$. If we use absolute boundary conditions and the operator δ, then we can define $H_a^p(M;\mathbf{C}) = \mathrm{N}(\Delta_{B_a})_p = H^p(M;\mathbf{C})$. We summarize these results as follows:

LEMMA 4.1.3. (HODGE DECOMPOSITION THEOREM). *There are natural isomorphisms between the harmonic cohomology with absolute and relative boundary conditions and the simplicial cohomology groups of* M:

$$H_a^p(M;\mathbf{C}) = \mathrm{N}((d+\delta)_{B_a})_j \simeq H^p(M;\mathbf{C})$$

and

$$H_r^p(M;\mathbf{C}) = \mathrm{N}((d+\delta)_{B_r})_j \simeq H^p(M, dM;\mathbf{C}).$$

If M is oriented, we let $*$ be the Hodge operator $*: \Lambda^p \to \Lambda^{m-p}$. Since $*$ interchanges the decomposition $\Lambda(T^*dM) \oplus \Lambda(T^*dM) \wedge dr$, it anti-commutes with α and therefore $B_a(\theta) = 0$ if and only if $B_r(*\theta) = 0$. Since $d\theta = 0$ if and only if $\delta * \theta = 0$ and similarly $\delta\theta = 0$ if and only if $d * \theta = 0$, we conclude:

LEMMA 4.1.4. *Let* M *be oriented and let* $*$ *be the Hodge operator. Then* $*$ *induces a map, called Poincaré duality,*

$$*: H^p(M;\mathbf{C}) \simeq H_a^p(M;\mathbf{C}) \xrightarrow{\simeq} H_r^{m-p}(M,\mathbf{C}) \simeq H^{m-p}(M, dM;\mathbf{C}).$$

We define the Euler-Poincaré characteristics by:

$$\chi(M) = \sum(-1)^p \dim H^p(M;\mathbf{C})$$

$$\chi(dM) = \sum(-1)^p \dim H^p(dM;\mathbf{C})$$

$$\chi(M, dM) = \sum(-1)^p \dim H^p(M, dM;\mathbf{C}).$$

The long exact sequence in cohomology:

$$\cdots H^p(dM;\mathbf{C}) \leftarrow H^p(M;\mathbf{C}) \leftarrow H^p(M, dM;\mathbf{C}) \leftarrow H^{p-1}(dM;\mathbf{C}) \cdots$$

shows that:
$$\chi(M) = \chi(dM) + \chi(M, dM).$$

If m is even, then $\chi(dM) = 0$ as dM is an odd dimensional manifold without boundary so $\chi(M) = \chi(M, dM)$. If m is odd and if M is orientable, then $\chi(M) = -\chi(M, dM)$ by Poincaré duality. $\chi(M)$ is the index of the de Rham complex with absolute boundary conditions; $\chi(M, dM)$ is the index of the de Rham complex with relative boundary conditions. By Lemma 1.9.3, there is a local formula for the Euler-Poincaré characteristic. Since $\chi(M) = -\chi(M, dM)$ if m is odd and M is orientable, by passing to the double cover if necessary we see $\chi(M) = -\chi(M, dM)$ in general if m is odd. This proves:

LEMMA 4.1.5.
(a) If m is even, $\chi(M) = \chi(M, dM)$ and $\chi(dM) = 0$.
(b) If m is odd, $\chi(M) = -\chi(M, dM) = \frac{1}{2}\chi(dM)$.

In contrast to the situation of manifolds without boundary, if we pass to the category of manifolds with boundary, there exist non-zero index problems in all dimensions m.

In the next subsection, we will discuss the Gauss-Bonnet formula for manifolds with boundary. We conclude this subsection with a brief discussion of the more general ellipticity conditions considered by Atiyah and Bott. Let $Q: C^\infty(V_1) \to C^\infty(V_2)$ be an elliptic differential operator of order $d > 0$ on the interior—i.e., if $q(x, \xi)$ is the leading symbol of Q, then $q(x, \xi): V_1 \to V_2$ is an isomorphism for $\xi \neq 0$. Let $W_1 = V_1 \otimes 1^d\big|_{dM}$ be the bundle of Cauchy data. We assume $\dim W_1$ is even and let W_1' be a bundle over dM of dimension $\frac{1}{2}(\dim W_1)$. Let $B: C^\infty(W_1) \to C^\infty(W_1')$ be a tangential pseudo-differential operator. We consider the ODE

$$q(y, 0, \varsigma, D_r)f = 0, \qquad \lim_{r \to \infty} f(r) = 0$$

and let $V_+(\tau)(\varsigma)$ be the bundle of Cauchy data of solutions to this equation. We say that (Q, B) is elliptic with respect to the cone $\{0\}$ if for all $\varsigma \neq 0$, the map:

$$\sigma_g(B)(y, \varsigma): V_+(\tau)(\varsigma) \to W_1'$$

is an isomorphism (i.e., we can find a unique solution to the ODE such that $\sigma_g(B)(y, \varsigma)\underline{\gamma}f = f'$ is pre-assigned in W_1'). $V_+(\tau)$ is a sub-bundle of W_1 and is the span of the generalized eigenvectors corresponding to eigenvalues with positive real parts for a suitable endomorphism $\tau(\varsigma)$ just as in the first order case. σ_g is the graded leading symbol as discussed in section 1.9.

This is a much weaker condition than the one we have been considering since the only complex value involved is $\lambda = 0$. We study the pair

$$(Q, B): C^\infty(V_1) \to C^\infty(V_2) \oplus C^\infty(W_1').$$

Under the assumption of elliptic with respect to the cone $\{0\}$, this operator is Fredholm in a suitable sense with closed range, finite dimensional null-space and cokernel. We let $\text{index}(Q, B)$ be the index of this problem. The Atiyah-Bott theorem gives a formula for the index of this problem.

There exist elliptic complexes which do not admit boundary conditions satisfying even this weaker notion of ellipticity. Let $q(x, \xi)$ be a first order symbol and expand

$$q(y, 0, \xi, z) = q_0 z + \sum_{j=1}^{m-1} q_j \varsigma_j.$$

As in Lemma 1.9.5 we define

$$\tau = i q_0^{-1} \sum_{j=1}^{m-1} q_j \varsigma_j;$$

the ellipticity condition on the interior shows τ has no purely imaginary eigenvalues for $\varsigma \neq 0$. We let $V_\pm(\tau)(\varsigma)$ be the sub-bundle of V corresponding to the span of the generalized eigenvectors of τ corresponding to eigenvalues with positive/negative real part. Then (Q, B) is elliptic if and only if

$$\sigma_g(B)(\varsigma) \colon V_+(\tau)(\varsigma) \to W_1'$$

is an isomorphism for all $\varsigma \neq 0$.

Let $S(T^*(dM)) = \{ \varsigma \in T^*(dM) : |\varsigma|^2 = 1 \}$ be the unit sphere bundle over dM. $V_\pm(\tau)$ define sub-bundles of V over $S(T^*(dM))$. The existence of an elliptic boundary condition implies these sub-bundles are trivial over the fiber spheres. We study the case in which $q^* q = |\varsigma|^2 I$. In this case, $q_0^{-1} = q_0^*$. If we set $p_j = i q_0^{-1} q_j$, then these are self-adjoint and satisfy $p_j p_k + p_k p_j = 2\delta_{jk}$. If m is even, then the fiber spheres have dimension $m - 2$ which will be even. The bundles $V_\pm(\tau)$ were discussed in Lemma 2.1.5 and in particular are non-trivial if

$$\text{Tr}(p_1, \ldots, p_{m-1}) \neq 0.$$

For the spin, signature, and Dolbeault complexes, the symbol is given by Clifford multiplication and $p_1, \ldots, p_{m-1}$ is multiplication by the orientation form (modulo some normalizing factor of i). Since the bundles involved were defined by the action of the orientation form being ± 1, this proves:

LEMMA 4.1.6. Let $Q \colon C^\infty(V_1) \to C^\infty(V_2)$ denote either the signature, the spin, or the Dolbeault complex. Then there does not exist a boundary condition B so that (Q, B) is elliptic with respect to the cone $\{0\}$.

The difficulty comes, of course, in not permitting the target bundle W' to depend upon the variable ς. In the first order case, there is a natural pseudo-differential operator $B(\varsigma)$ with leading symbol given by projection on $V_+(\tau)(\varsigma)$. This operator corresponds to global (as opposed to local) boundary conditions and leads to a well posed boundry value problem for the other three classical elliptic complexes. Because the boundary value problem is non-local, there is an additional non-local term which arises in the index theorem for these complexes. This is the eta invariant we will discuss later.

4.2. The Gauss-Bonnet Theorem
For Manifolds with Boundary.

Let B denote either the absolute or relative boundary conditions for the operator $(d + \delta)$ discussed previously. We let $\chi(M)_B$ be either $\chi(M)$ or $\chi(M, dM)$ be the index of the de Rham complex with these boundary conditions. Let Δ_B^{even} and Δ_B^{odd} be the Laplacian on even/odd forms with the boundary conditions $B\theta = B(d + \delta)\theta = 0$. Let $a_n(x, d + \delta) = a_n(x, \Delta^{\text{even}}) - a_n(x, \Delta^{\text{odd}})$ be the invariants of the heat equation defined in the interior of M which were discussed in Lemma 1.7.4. On the boundary dM, let $a_n(y, d + \delta, B) = a_n(y, \Delta_B^{\text{even}}) - a_n(y, \Delta_B^{\text{odd}})$ be the invariants of the heat equation defined in Lemma 1.9.2. Then Lemma 1.9.3 implies:

$$\chi(M)_B = \text{Tr}\{\exp(-t\Delta_B^{\text{even}})\} - \text{Tr}\exp\{(-t\Delta_B^{\text{odd}})\}$$

$$\sim \sum_{n=0}^{\infty} t^{(n-m)/2} \int_M a_n(x, d + \delta) \, \text{dvol}(x)$$

$$+ \sum_{n=0}^{\infty} t^{(n-m+1)/2} \int_{dM} a_n(y, d + \delta, B) \, \text{dvol}(y).$$

The interior invariants $a_n(x, d+\delta)$ do not depend on the boundary condition so we can apply Lemma 2.4.8 to conclude:

$$a_n(x, d + \delta) = 0 \qquad \text{if } n < m \text{ or if } m \text{ is odd}$$

$$a_m(x, d + \delta) = E_m \qquad \text{is the Euler intergrand if } m \text{ is even.}$$

In this subsection we will prove the Gauss-Bonnet theorem for manifolds with boundary and identify the boundary integrands $a_n(y, d + \delta, B)$ for $n \leq m - 1$.

We let $\mathcal{P}$ be the algebra generated by the $\{g_{ij/\alpha}\}$ variables for $|\alpha| \neq 0$. We always normalize the coordinate system so $g_{ij}(X, G)(x_0) = \delta_{ij}$. We normalize the coordinate system $x = (y, r)$ near the boundary as discussed in section 4.1; this introduces some additional relations on the $g_{ij/\alpha}$ variables we shall discuss shortly. We let $P(Y, G)(y_0)$ be the evaluation of $P \in \mathcal{P}$ on a metric G and relative to the given coordinate system Y on dM. We say that P is invariant if $P(Y, G)(y_0) = P(\bar{Y}, G)(y_0)$ for any two such coordinate systems Y and $\bar{Y}$. We introduce the same notion of homogeneity as that discussed in the second chapter and let $\mathcal{P}_{m,n}^b$ be the finite dimensional vector space of invariant polynomials which are homogeneous of order n on a manifold M of dimension m. The "b" stands for boundary and emphasizes that these are invariants only defined on dM; there is a natural inclusion $\mathcal{P}_{m,n} \rightarrow \mathcal{P}_{m,n}^b$; by restricting the admissible coordinate transformations we increase the space of invariants.

LEMMA 4.2.1. *If B denotes either absolute or relative boundary conditions, then $a_n(y, d + \delta, B)$ defines an element of $P^b_{m,n}$.*

PROOF: By Lemma 1.9.2, $a_n(y, d + \delta, B)$ is given by a local formula in the jets of the metric which is invariant. Either by examining the analytic proof of Lemma 1.9.2 in a way similar to that used to prove Lemma 1.7.5 and 2.4.1 or by using dimensional analysis as was done in the proof of Lemma 2.4.4, we can show that a_n must be homogeneous of order n and polynomial in the jets of the metric.

Our normalizations impose some additional relations on the $g_{ij/\alpha}$ variables. By hypothesis, the curves (y_0, r) are unit speed geodesics perpendicular to dM at $r = 0$. This is equivalent to assuming:

$$\nabla_N N = 0 \qquad \text{and} \qquad g_{jm}(y, 0) = \delta_{jm}$$

where $N = \partial/\partial r$ is the inward unit normal. The computation of the Christoffel symbols of section 2.3 shows this is equivalent to assuming:

$$\Gamma_{mmj} = \tfrac{1}{2}(g_{mj/m} + g_{mj/m} - g_{mm/j}) = 0.$$

If we take $j = m$, this implies $g_{mm/m} = 0$. Since $g_{mm}(y, 0) \equiv 1$, we conclude $g_{mm} \equiv 1$ so $g_{mm/\alpha} \equiv 0$. Thus $\Gamma_{mmj} = g_{mj/m} = 0$. As $g_{mj}(y, 0) = \delta_{mj}$ we conclude $g_{mj} \equiv \delta_{mj}$ and therefore $g_{jm/\alpha} = 0$, $1 \le j \le m$. We can further normalize the coordinate system Y on dM by assuming $g_{jk/l}(Y, G)(y_0) = 0$ for $1 \le j, k, l \le m - 1$. We eliminate all these variables from the algebra defining P; the remaining variables are algebraically independent.

The only 1-jets of the metric which are left are the $\{g_{jk/m}\}$ variables for $1 \le j, k \le m - 1$. The first step in Chapter 2 was to choose a coordinate system in which all the 1-jets of the metric vanish; this proved to be the critical obstruction to studying non-Kaehler holomorphic manifolds. It turns out that the $\{g_{jk/m}\}$ variables cannot be normalized to zero. They are tensorial and give essentially the components of the second fundamental form or shape operator.

Let $\{e_1, e_2\}$ be vector fields on M which are tangent to dM along dM. We define the shape operator:

$$S(e_1, e_2) = (\nabla_{e_1} e_2, N)$$

along dM. It is clear this expression is tensorial in e_1. We compute:

$$(\nabla_{e_1} e_2, N) - (\nabla_{e_2} e_1, N) = ([e_1, e_2], N).$$

Since e_1 and e_2 are tangent to dM along dM, $[e_1, e_2]$ is tangent to dM along dM and thus $([e_1, e_2], N) = 0$ along dM. This implies $S(e_1, e_2) =$

$S(e_2, e_1)$ is tensorial in e_2. The shape operator defines a bilinear map from $T(dM) \times T(dM) \to \mathbf{R}$. We compute

$$(\nabla_{\partial/\partial y_j} \partial/\partial y_k, N) = \Gamma_{jkm} = \tfrac{1}{2}(g_{jm/k} + g_{km/j} - g_{jk/m}) = -\tfrac{1}{2}g_{jk/m}.$$

We can construct a number of invariants as follows: let $\{e_j\}$ be a local orthonormal frame for $T(M)$ such that $e_m = N = \partial/\partial r$. Define:

$$\nabla e_j = \sum_{1 \le k \le m} \omega_{jk} e_k \qquad \text{for } \omega_{jk} \in T^*M \text{ and } \omega_{jk} + \omega_{kj} = 0$$

and

$$\Omega_{jk} = d\omega_{jk} - \sum_{1 \le \nu \le m} \omega_{j\nu} \wedge \omega_{\nu k}.$$

The ω_{jm} variables are tensorial as $\omega_{jm} = \sum_{k=1}^{m-1} S(e_j, e_k) \cdot e^k$. We define:

$$Q_{k,m} = c_{k,m} \sum \varepsilon(i_1, \ldots, i_{m-1}) \Omega_{i_1 i_2} \wedge \cdots \wedge \Omega_{i_{2k-1}, i_{2k}}$$
$$\wedge \omega_{i_{2k+1}, m} \wedge \cdots \wedge \omega_{i_{m-1}, m} \in \Lambda^{m-1}$$

for

$$c_{k,m} = \frac{(-1)^k}{\pi^p k! 2^{k+p} \cdot 1 \cdot 3 \cdots (2p - 2k - 1)} \qquad \text{where } p = \left[\frac{m}{2}\right].$$

The sum defining Q_k is taken over all permutations of $m - 1$ indices and defines an $m - 1$ form over M. If m is even, we define:

$$E_m = \frac{(-1)^p}{\{2^m \pi^p p!\}} \sum \varepsilon(i_1, \ldots, i_m) \Omega_{i_1 i_2} \wedge \cdots \wedge \Omega_{i_{m-1} i_m}$$

as the Euler form discussed in Chapter 2.

$Q_{k,m}$ and E_m are the SO-invariant forms on M. E_m is defined on all of M while $Q_{k,m}$ is only defined near the boundary. Chern noted that if m is even,

$$E_m = -d\left(\sum_k Q_{k,m}\right).$$

This can also be interpreted in terms of the transgression of Chapter 2. Let ∇_1 and ∇_2 be two Riemannian connections on TM. We defined an $m - 1$ form $TE_m(\nabla_1, \nabla_2)$ so that

$$dTE_m(\nabla_1, \nabla_2) = E_m(\nabla_1) - E_m(\nabla_2).$$

Near dM, we split $T(M) = T(dM) \oplus 1$ as the orthogonal complement of the unit normal. We project the Levi-Civita connection on this decomposition, and let ∇_2 be the projected connection. ∇_2 is just the sum of the Levi-Civita connection of $T(dM)$ and the trivial connection on 1 and is flat in the normal direction. As ∇_2 is a direct sum connection, $E_m(\nabla_2) = 0$. $\nabla_1 - \nabla_2$ is essentially just the shape operator. $TE_m = -\sum Q_{k,m}$ and $dTE_m = E_m(\nabla_1) = E_m$. It is an easy exercise to work out the $Q_{k,m}$ using the methods of section 2 and thereby compute the normalizing constants given by Chern.

The Chern-Gauss-Bonnet theorem for manifolds with boundary in the oriented category becomes:

$$\chi(M) = \int_M E_m + \int_{dM} \sum_k Q_{k,m}.$$

In the unoriented category, we regard $E_m \, dvol(x)$ as a measure on M and $\int Q_{k,m} \, dvol(y)$ as a measure on dM. If m is odd, of course, $\chi(M) = \frac{1}{2}\chi(dM) = \frac{1}{2}\int_{dM} E_{m-1}$ so there is no difficulty with the Chern-Gauss-Bonnet theorem in this case.

We derive the Chern-Gauss-Bonnet theorem for manifolds with boundary from the theorem for manifolds without boundary. Suppose m is even and that the metric is product near the boundary. Let $\overline{M}$ be the double of M then $\chi(\overline{M}) = \int_{\overline{M}} E_m = 2\int_M E_m = 2\chi(M) - \chi(d\overline{M}) = 2\chi(M)$ so $\chi(M) = \int_M E_m$. If the metric is not product near the boundary, let $M' = dM \times [-1,0] \cup M$ be the manifold M with a collar sewed on. Let G_0 be the restriction of the metric on M to the boundary and let G_0' be the product metric on the collar $dM \times [-1,0]$. Using a partition of unity, extend the original metric on M to a new metric which agrees with G_0' near $dM \times \{-1\}$ which is the boundary of M'. Then:

$$\chi(M) = \chi(M') = \int_{M_1} E_m = \int_M E_m - \int_{dM \times [-1,0]} d\left(\sum_k Q_{k,m}\right)$$

$$= \int_M E_m + \int_{dM} \sum_k Q_{k,m}$$

by Stoke's theorem; since the Q_k vanish identically near $dM \times \{-1\}$ there is no contribution from this component of the boundary of the collar (we change the sign since the orientation of dM as the boundary of M and as the boundary of $dM \times [-1,0]$ are opposite).

We now study the invariants of the heat equation. We impose no restrictions on the dimension m. If $M = S^1 \times M_1$ and if θ is the usual periodic parameter on S^1, then there is a natural involution on $\Lambda(T^*M)$ given by interchanging ψ with $d\theta \wedge \psi$ for $\psi \in \Lambda(M_1)$. This involution preserves the boundary conditions and the associated Laplacians, but changes the parity of the factors involved. This shows $a_n(y, d + \delta, B) = 0$ for such a product metric. We define:

$$r: P^b_{m,n} \to P^b_{m-1,n}$$

to be the dual of the map $M_1 \to S^1 \times M_1$. Then algebraically:

$$r(g_{ij/\alpha}) = \begin{cases} 0 & \text{if } \deg_1(g_{ij/\alpha}) \neq 0 \\ * & \text{if } \deg_1(g_{ij/\alpha}) = 0 \end{cases}$$

where "$*$" is simply a renumbering to shift all the indices down one. (At this stage, it is inconvenient to have used the last index to indicate the normal direction so that the first index must be used to denote the flat index; denoting the normal direction by the last index is sufficiently cannonical that we have not attempted to adopt a different convention despite the conflict with the notation of Chapter 2). This proves:

LEMMA 4.2.2. *Let B denote either the relative or absolute boundary conditions. Then $a_n(y, d + \delta, B) \in P^b_{m,n}$. Furthermore, $r(a_n) = 0$ where $r: P^b_{m,n} \to P^b_{m-1,n}$ is the restriction map.*

We can now begin to identify $a_n(y, d + \delta, B)$ using the same techniques of invariance theory applied in the second chapter.

LEMMA 4.2.3.. *Let $P \in P^b_{m,n}$. Suppose that $r(P) = 0$. Then:*
(a) *$P = 0$ if $n < m - 1$.*
(b) *If $n = m - 1$, then P is a polynomial in the variables $\{g_{ij/m}, g_{ij/kl}\}$ for $1 \leq i, j, k, l \leq m - 1$. Furthermore, $\deg_j(A) = 2$ for any monomial A of P and for $1 \leq j \leq m - 1$.*

PROOF: As in the proof of Theorem 2.4.7, we shall count indices. Let $P \neq 0$ and let A be a monomial of P. Decompose A in the form:

$$A = g_{u_1 v_1/\alpha_1} \cdots g_{u_k v_k/\alpha_k} g_{i_1 j_1/m} \cdots g_{i_r j_r/m} \qquad \text{for } |\alpha_\nu| \geq 2.$$

(We have chosen our coordinate systems so the only non-zero 1-jets are the $g_{ij/m}$ variables.) Since $r(P) = 0$, $\deg_1(A) \neq 0$. Since P is invariant, $\deg_j(A) > 0$ is even for $1 \leq j \leq m - 1$. This yields the inequalities:

$$2m - 2 \leq \sum_{j \leq m-1} \deg_j(A) \qquad \text{and} \qquad r \leq \deg_m(A).$$

From this it follows that:

$$2m - 2 + r \leq \sum_j \deg_j(A) = 2r + 2k + \sum_\nu |\alpha_\nu| + r = 2r + 2k + n.$$

Since $|\alpha_\nu| \geq 2$ we conclude

$$2k \leq \sum_\nu |\alpha_\nu| = n - r.$$

We combine these inequalities to conclude $2m - 2 + r \leq 2n + r$ so $n \geq m - 1$. This shows $P = 0$ if $n < m - 1$ which proves (a). If $n = m - 1$, all these inequalities must be equalities. $|\alpha_\nu| = 2$, $\deg_m(A) = r$, and $\deg_j(A) = 2$ for $1 \leq j \leq m - 1$. Since the index m only appears in A in the $g_{ij/m}$ variables where $1 \leq i, j \leq m - 1$, this completes the proof of (b).

Lemma 2.5.1 only used invariance under the group SO(2). Since P is invariant under the action of SO$(m - 1)$, we apply the argument used in the proof of Theorem 2.4.7 to choose a monomial A of P of the form:

$$A = g_{11/22} \cdots g_{2k-1,2k-1/kk} g_{k+1,k+1/m} \cdots g_{m-1,m-1/m}$$

where if $k = 0$ the terms of the first kind do not appear and if $k = m - 1$, the terms of the second kind do not appear. Since $m - 1 = 2k + r$, it is clear that $r \equiv m - 1 \mod 2$. We denote such a monomial by A_k. Since $P \neq 0$ implies $c(A_k, P) \neq 0$ for some k, we conclude the dimension of the space of such P is at most the cardinality of $\{A_k\} = \left[\frac{m+1}{2}\right]$. This proves:

LEMMA 4.2.4. Let $r: P^b_{m,m-1} \to P^b_{m-1,m-1}$ be the restriction map defined earlier. Then $\dim N(r) \leq \left[\frac{m+1}{2}\right]$.

We can now show:

LEMMA 4.2.5. Let $P \in P^b_{m,m-1}$ with $r(P) = 0$. Let $i: dM \to M$ be the inclusion map and $i^*: \Lambda^{m-1}(T^*M) \to \Lambda^{m-1}(T^*(dM))$ be the dual map. Let $*_{m-1}$ be the Hodge operator on the boundary so $*_{m-1}: \Lambda^{m-1}(T^*(dM)) \to \Lambda^0(T^*(dM))$. Let $\bar{Q}_{k,m} = *_{m-1}(i^*Q_{k,m})$. Then the $\{\bar{Q}_{k,m}\}$ form a basis for $N(r)$ so we can express P as a linear combination of the $\bar{Q}_{k,m}$.

PROOF: It is clear $r(\bar{Q}_{k,m}) = 0$ and that these elements are linearly independent. We have $[(m+1)/2]$ such elements so by Lemma 4.2.4 they must be a basis for the kernel of r. (If we reverse the orientation we change both the sign of $*$ and Q so $\bar{Q}$ is a scalar invariant.)

We note that if m is odd, then $Q_{m-1,m} = c \cdot \sum \varepsilon(i_1, \ldots, i_{m-1}) \Omega_{i_1 i_2} \wedge \cdots \wedge \Omega_{i_{m-1} i_{m-1}}$ is not the Euler form on the boundary since we are using the Levi-Civita connection on M and not the Levi-Civita connection on dM. However, E_{m-1} can be expressed in terms of the $\bar{Q}_{k,m}$ in this situation.

Before proceeding to discuss the heat equation, we need a uniqueness theorem:

LEMMA 4.2.6. *Let* $P = \sum_k a_k Q_{k,m}$ *be a linear combination of the* $\{Q_{k,m}\}$. *Suppose that* $P \neq 0$. *Then there exists a manifold M and a metric G so* $\int_{dM} P(G)(y) \neq 0$.

PROOF: By assumption not all the $a_k = 0$. Choose k maximal so $a_k \neq 0$. Let $n = m - 2k$ and let $M = S^{2k} \times D^{m-2k}$ with the standard metric. (If $m - 2k = 1$, we let $M = D^m$ and choose a metric which is product near S^{m-1}.) We let indices $1 \leq i \leq 2k$ index a frame for $T^*(S^{2k})$ and indices $2k+1 \leq u \leq m$ index a frame for $T^*(D^{m-2k})$. Since the metric is product, $\Omega_{iu} = \omega_{iu} = 0$ in this situation. Therefore $Q_{j,m} = 0$ if $j < k$. Since $a_j = 0$ for $j > k$ by assumption, we conclude $\int_{dM} P(G) = a_k \int_{dM} Q_{k,m}(G)$, so it suffices to show this integral is non-zero. This is immediate if $n = 1$ as $Q_{m-1,m} = E_{m-1}$ and $m - 1$ is even.

Let $n \geq 2$. We have $Q_{k,m}(G) = E_{2k}(G_1) \cdot Q_{0,m-2k}(G_2)$, since the metric is product. Since

$$\int_{S^{2k}} E_{2k} = \chi(S^{2k}) = 2 \neq 0$$

we must only show $\int_{dD^n} Q_{0,n}$ is non-zero for all $n > 1$. Let θ be a system of local coordinates on the unit sphere S^{n-1} and let r be the usual radial parameter. If ds_e^2 is the Euclidean metric and ds_θ^2 is the spherical metric, then

$$ds_e^2 = r^2 \, ds_\theta^2 + dr \cdot dr.$$

From the description of the shape operator given previously we conclude that $S = -ds_\theta^2$. Let $\{e_1, \ldots, e_{n-1}\}$ be a local oriented orthonormal frame for $T(S^n)$, then $\omega_{in} = -e^i$ and therefore $Q_{0,n} = c \cdot \text{dvol}_{n-1}$ where c is a non-zero constant. This completes the proof.

We combine these results in the following Theorem:

THEOREM 4.2.7. (GAUSS-BONNET FORMULA FOR MANIFOLDS WITH BOUNDARY).
(a) *Let the dimension m be even and let B denote either the relative or the absolute boundary conditions. Let*

$$Q_{k,m} = c_{k,m} \sum \varepsilon(i_1, \ldots, i_{m-1})\Omega_{i_1,i_2} \wedge \cdots \wedge \Omega_{i_{2k-1},i_{2k}}$$
$$\wedge \omega_{i_{2k+1},m} \wedge \cdots \wedge \omega_{i_{m-1},m}$$

for $c_{k,m} = (-1)^k/(\pi^{m/2} \cdot k! \cdot 2^{k+m/2} \cdot 1 \cdot 3 \cdots (m - 2k - 1))$. *Let* $\bar{Q}_{k,m} = *(Q_{k,m}|dM) \in P^b_{m,m-1}$. *Then:*
(i) $a_n(x, d + \delta) = 0$ *for* $n < m$ *and* $a_n(y, d + \delta, B) = 0$ *for* $n < m - 1$,
(ii) $a_m(x, d + \delta) = E_m$ *is the Euler integrand*,
(iii) $a_{m-1}(y, d + \delta, B) = \sum_k \bar{Q}_{k,m}$,
(iv) $\chi(M) = \chi(M, dM) = \int_M E_m \, \text{dvol}(x) + \sum_k \int_{dM} \bar{Q}_{k,m} \, \text{dvol}(y)$.

(b) Let the dimension m be odd and let B_r be the relative and B_a the absolute boundary conditions. Then:

(i) $a_n(x, d + \delta) = 0$ for all n and $a_n(y, d + \delta, B_r) = a_n(y, d + \delta, B_a) = 0$
    for $n < m - 1$,

(ii) $a_{m-1}(y, d + \delta, B_a) = \frac{1}{2} E_{m-1}$ and $a_{m-1}(y, d + \delta, B_r) = -\frac{1}{2} E_{m-1}$,

(iii) $\chi(M) = -\chi(M, dM) = \frac{1}{2} \int_{dM} E_{m-1} \, dvol(y) = \frac{1}{2} \chi(dM)$.

This follows immediately from our previous computations, and Lemmas 4.2.5 and 4.2.6.

The Atayah-Bott theorem gives a generalization of the Atiyah-Singer index theorem for index problems on manifolds with boundary. This theorem includes the Gauss-Bonnet theorem as a special case, but does not include the Atiyah-Patodi-Singer index theorem since the signature, spin, and Dolbeault complexes do not admit local boundary conditions of the form we have been discussing. We will discuss this in more detail in subsection 4.5.

4.3. The Regularity at $s = 0$ of the Eta Invariant.

In this section, we consider the eta invariant defined in section 1.10. This section will be devoted to proving eta is regular at $s = 0$. In the next section we will use this result to discuss the twisted index theorem using coefficients in a locally flat bundle. This invariant appears as a boundary correction term in the index theorem for manifolds with boundary.

We shall assume $P: C^\infty(V) \to C^\infty(V)$ is a self-adjoint elliptic pseudo-differential operator of order $d > 0$. We define

$$\eta(s, P) = \sum_{\lambda_i > 0} (\lambda_i)^{-s} - \sum_{\lambda_i < 0} (-\lambda_i)^{-s} \qquad \text{for } \operatorname{Re}(s) \gg 0$$

and use Theorem 1.10.3 to extend η meromorphically to the complex plane with isolated simple poles on the real axis. We define

$$R(P) = d \cdot \operatorname{Res}_{s=0} \eta(s, P).$$

We will show $R(P) = 0$ so η is regular at $s = 0$. The first step is to show:

LEMMA 4.3.1. Let P and Q be self-adjoint elliptic pseudo-differential operators of order $d > 0$.
(a) $P \cdot (P^2)^v$ is a self-adjoint elliptic pseudo-differential operator for any v and if $2v + 1 > 0$, $R(P) = R(P \cdot (P^2)^v)$.
(b) $R(P \oplus Q) = R(P) + R(Q)$.
(c) There is a local formula $a(x, P)$ in the jets of the symbol of P up to order d so that $R(P) = \int_M a(x, P) \, |\operatorname{dvol}(x)|$.
(d) If P_t is a smooth 1-parameter family of such operators, then $R(P_t)$ is independent of the parameter t.
(e) If P is positive definite, then $R(P) = 0$.
(f) $R(-P) = -R(P)$.

PROOF: We have the formal identity: $\eta(s, P \cdot (P^2)^v) = \eta((2v + 1)s, P)$. Since we normalized the residue by multiplying by the order of the operator, (a) holds. The fact that $P \cdot (P^2)^v$ is again a pseudo-differential operator follows from the work of Seeley. (b) is an immediate consequence of the definition. (c) and (d) were proved in Lemma 1.10.2 for differential operators. The extension to pseudo-differential operators again follows Seeley's work. If P is positive definite, then the zeta function and the eta function coincide. Since zeta is regular at the origin, (e) follows by Lemma 1.10.1. (f) is immediate from the definition.

We note this lemma continues to be true if we assume the weaker condition $\det(p(x, \xi) - it) \neq 0$ for $(\xi, t) \neq (0, 0) \in T^*M \times \mathbf{R}$.

We use Lemma 4.3.1 to interpret $R(P)$ as a map in K-theory. Let $S(T^*M)$ be the unit sphere bundle in T^*M. Let V be a smooth vector bundle over M equipped with a fiber inner product. Let $p: S(T^*M) \to$

END(V) be self-adjoint and elliptic; we assume $p(x, \xi) = p^*(x, \xi)$ and $\det p(x, \xi) \neq 0$ for $(x, \xi) \in S(T^*M)$. We fix the order of homogeneity $d > 0$ and let $p_d(x, \xi)$ be the extension of p to T^*M which is homogeneous of degree d. (In general, we must smooth out the extension near $\xi = 0$ to obtain a C^∞ extension, but we suppress such details in the interests of notational clarity.)

LEMMA 4.3.2. *Let $p: S(T^*M) \to$ END(V) be self-adjoint and elliptic. Let $d > 0$ and let $P: C^\infty(V) \to C^\infty(V)$ have leading symbol p_d. Then $R(P)$ depends only on p and not on the order d nor the particular operator P.*

PROOF: Let P' have order d with the same leading symbol p_d. We form the elliptic family $P_t = tP' + (1 - t)P$. By Lemma 4.3.1, $R(P_t)$ is independent of t so $R(P') = R(P)$. Given two different orders, let $(1 + 2v)d = d'$. Let $Q = P(P^2)^v$ then $R(Q) = R(P)$ by Lemma 4.3.1(a). The leading symbol of Q is $p(p^2)^v$. p^2 is positive definite and elliptic. We construct the homotopy of symbols $q_t(x, \xi) = p(x, \xi)(tp^2(x, \xi) + (1 - t)|\xi|^2)$. This shows the symbol of Q restricted to $S(T^*M)$ is homotopic to the symbol of P restricted to $S(T^*M)$ where the homotopy remains within the class of self-adjoint elliptic symbols. Lemma 4.3.1 completes the proof.

We let $r(p) = R(P)$ for such an operator P. Lemma 4.3.1(d) shows $r(p)$ is a homotopy invariant of p. Let $\Pi_\pm(p)$ be the subspaces of V spanned by the eigenvectors of $p(x, \xi)$ corresponding to positive/negative eigenvalues. These bundles have constant rank and define smooth vector bundles over $S(T^*M)$ so $\Pi_+ \oplus \Pi_- = V$. In section 3.9, an essential step in proving the Atayah-Singer index theorem was to interpret the index as a map in K-theory. To show $R(P) = 0$, we must first interpret it as a map in K-theory. The natural space in which to work is $K(S(T^*M); \mathbf{Q})$ and not $K(\Sigma(T^*M); \mathbf{Q})$.

LEMMA 4.3.3. *Let G be an abelian group and let $R(P) \in G$ be defined for any self-adjoint elliptic pseudodifferential operator. Assume R satisfies properties (a), (b), (d), (e) and (f) [but not necessarily (c)] of Lemma 4.3.1. Then there exists a $\mathbf{Z}$-linear map $r: K(S(T^*M)) \to G$ so that:*

(a) $R(P) = r(\Pi_+(p))$,
(b) *If $\tau: S(T^*M) \to M$ is the natural projection, then $r(\tau^*V) = 0$ for all $V \in K(M)$ so that*

$$r: K(S(T^*M))/K(M) \to G.$$

Remark: We shall apply this lemma in several contexts later, so state it in somewhat greater generality than is needed here. If $G = \mathbf{R}$, we can extend r to a $\mathbf{Q}$ linear map
$$r: K(S(T^*M); \mathbf{Q}) \to \mathbf{R}.$$

PROOF: Let $p: S(T^*M) \to \text{END}(V)$ be self-adjoint and elliptic. We define the bundles $\Pi_{\pm}(p)$ and let $\pi_{\pm}(x, \xi)$ denote orthogonal projection on $\Pi_{\pm}(p)(x, \xi)$. We let $p_0 = \pi_+ - \pi_-$ and define $p_t = tp + (1 - t)p_0$ as a homotopy joining p and p_0. It is clear that the p_t are self-adjoint. Fix (x, ξ) and let $\{\lambda_i, v_i\}$ be a spectral resolution of the matrix $p(x, \xi)$. Then:

$$p(x, \xi)\left(\sum c_i v_i\right) = \sum \lambda_i c_i v_i$$

$$p_0(x, \xi)\left(\sum c_i v_i\right) = \sum \text{sign}(\lambda_i) c_i v_i.$$

Consequently

$$p_t\left(\sum c_i v_i\right) = \sum (t\lambda_i + (1 - t)\,\text{sign}(\lambda_i)) c_i v_i.$$

Since $t\lambda_i + (1 - t)\,\text{sign}(\lambda_i) \neq 0$ for $t \in [0, 1]$, the family p_t is elliptic. We therefore assume henceforth that $p(x, \xi)^2 = I$ on $S(T^*M)$ and $\pi_{\pm} = \frac{1}{2}(1 \pm p)$.

We let k be large and choose $V \in \text{Vect}_k(S(T^*M))$. Choose $W \in \text{Vect}_k(S(T^*M))$ so $V \oplus W \simeq 1^{2k}$. We may choose the metric on 1^{2k} so this direct sum is orthogonal and we define $\pi_{\pm}$ to be orthogonal projection on V and W. We let $p = \pi_+ - \pi_-$ so $\Pi_+ = V$ and $\Pi_- = W$. We define $r(V) = r(p)$. We must show this is well defined on Vect_k. W is unique up to isomorphism but the isomorphism $V \oplus W = 1^{2k}$ is non-canonical. Let $u: V \to \overline{V}$ and $v: W \to \overline{W}$ be isomorphisms where we regard $\overline{V}$ and $\overline{W}$ as orthogonal complements of 1^{2k} (perhaps with another trivialization). We must show $r(p) = r(\bar{p})$. For $t \in [0, 1]$ we let

$$V(t) = \text{span}\big(t \cdot v \oplus (1 - t) \cdot u(v)\big)_{v \in V}$$

$$\subseteq V \oplus W \oplus \overline{V} \oplus \overline{W} = 1^{2k} \oplus 1^{2k} = 1^{4k}.$$

This is a smooth 1-parameter family of bundles connecting $V \oplus 0$ to $0 \oplus \overline{V}$ in 1^{4k}. This gives a smooth 1-parameter family of symbols $p(t)$ connecting $p \oplus (-1_{2k})$ to $(-1_{2k}) \oplus \bar{p}$. Thus $r(p) = r(p \oplus -1_{2k}) = r(p(t)) = r(-1_{2k} \oplus \bar{p}) = r(\bar{p})$ so this is in fact a well defined map $r: \text{Vect}_k(S(T^*M)) \to G$. If V is the trivial bundle, then W is the trivial bundle so p decomposes as the direct sum of two self-adjoint matrices. The first is positive definite and the second negative definite so $r(p) = 0$ by Lemma 4.3.1(f). It is clear that $r(V_1 \oplus V_2) = r(V_1) + r(V_2)$ by Lemma 4.3.1(b) and consequently since $r(1) = 0$ we conclude r extends to an additive map from $\widetilde{K}(S(T^*M)) \to G$. We extend r to be zero on trivial bundles and thus $r: K(S(T^*M)) \to G$.

Suppose $V = \tau^* V_0$ for $V_0 \in \text{Vect}_k(M)$. We choose $W_0 \in \text{Vect}_k(M)$ so $V_0 \oplus W_0 \simeq 1^{2k}$. Then $p = p_+ \oplus p_-$ for $p_+: S(T^*M) \to \text{END}(V_0, V_0)$

and $p_-: S(T^*M) \to \text{END}(W_0, W_0)$. By Lemma 4.3.1, we conclude $r(p) = r(p_+) + r(p_-)$. Since p_+ is positive definite, $r(p_+) = 0$. Since p_- is negative definite, $r(p_-) = 0$. Thus $r(p) = 0$ and $r(\tau^* V_0) = 0$.

This establishes the existence of the map $r: K(S(T^*M)) \to G$ with the desired properties. We must now check $r(p) = r(\Pi_+)$ for general p. This follows by definition if V is a trivial bundle. For more general V, we first choose W so $V \oplus W = 1$ over M. We let $q = p \oplus 1$ on $V \oplus W$, then $r(q) = r(p) + r(1) = r(p)$. However, q acts on a trivial bundle so $r(q) = r(\Pi_+(q)) = r(\Pi_+(p) \oplus \tau^* W) = r(\Pi_+(p)) + r(\tau^* W) = r(\Pi_+(p))$ which completes the proof.

Of course, the bundles $\Pi_\pm(p)$ just measure the infinitesimal spectral asymmetry of P so it is not surprising that the bundles they represent in K-theory are related to the eta invariant. This construction is completely analogous to the construction given in section 3.9 which interpreted the index as a map in K-theory. We will return to this construction again in discussing the twisted index with coefficients in a locally flat bundle.

Such operators arise naturally from considering boundary value problems. Let M be a compact oriented Riemannian manifold of dimension $m = 2k - 1$ and let $N = M \times [0, 1)$; we let $n \in [0, 1)$ denote the normal parameter. Let

$$(d + \delta)_+: C^\infty(\Lambda^+(T^*N)) \to C^\infty(\Lambda^-(T^*N))$$

be the operator of the signature complex. The leading symbol is given by Clifford multiplication. We can use $c(dn)$, where c is Clifford multiplication, to identify these two bundles over N. We express:

$$(d + \delta)_+ = c(dn)(\partial/\partial n + A).$$

The operator A is a tangential differential operator on $C^\infty(\Lambda^+(T^*N))$; it is called the tangential operator of the signature complex. Since we have $c(dn) * c(dn) = -1$, the symbol of A is $-ic(dn)c(\xi)$ for $\xi \in T^*(M)$. It is immediate that the leading symbol is self-adjoint and elliptic; since A is natural this implies A is a self-adjoint elliptic partial differential operator on M.

Let $\{e_1, \ldots, e_m\}$ be a local oriented orthonormal frame for T^*M. Define:

$$\omega_m = i^k e_1 * \cdots * e_m \quad \text{and} \quad \omega_{m+1} = i^k(-dn) * e_1 * \cdots * e_m = -dn * \alpha_m$$

as the local orientations of M and N. ω_m is a central element of $\text{CLIF}(M)$; $\omega_m^2 = \omega_{m+1}^2 = 1$. If $\phi \in \Lambda(M)$ define $\tau_\pm(\phi) = \phi \pm c(\omega_{m+1})\phi$. This gives an isomorphism $\tau_{\pm} : \Lambda(M) \to \Lambda^{\pm}(N)$. We compute:

$$
(\tau_+)^{-1} \{ -c(dn)c(\xi) \} \tau_+ \phi
$$
$$
= (\tau_+)^{-1} \{ -c(dn)c(\xi)\phi - c(dn)c(\xi)c(-dn)c(\omega_m)\phi \}
$$
$$
= (\tau_+)^{-1} \{ -c(dn)c(\omega_m)c(\omega_m)c(\xi) + c(\omega_m)c(\xi)\phi \}
$$
$$
= (\tau_+)^{-1} \{ (c(\omega_{m+1}) + 1)c(\omega_m)c(\xi)\phi \}
$$
$$
= c(\omega_m)c(\xi)\phi.
$$

If we use τ_+ to regard A as an operator on $C^{\infty}(\Lambda(M))$ then this shows that A is given by the diagram:

$$
C^{\infty}(\Lambda(M)) \overset{\nabla}{\to} C^{\infty}(T^*(M \otimes \Lambda(M)) \overset{c}{\to} C^{\infty}(\Lambda(M)) \overset{\omega}{\to} C^{\infty}(\Lambda(M))
$$

where $\omega = \omega_m$. This commutes with the operator $(d + \delta)$; $A = \omega(d + \delta)$.

Both ω and $(d+\delta)$ reverse the parity so we could decompose $A = A^{\text{even}} \oplus A^{\text{odd}}$ acting on smooth forms of even and odd degree. We let $p = \sigma_L(A) = ic(\omega * \xi) = \sum_j \xi_j f_j$. The $\{f_j\}$ are self-adjoint matrices satisfying the commutation relation

$$
f_j f_k + f_k f_j = 2\delta_{jk}.
$$

We calculate:

$$
f_1 \cdots f_m = i^m c(\omega^m * e_1 * \cdots * e_m) = i^{m-k} c(\omega^m * \omega) = i^{m-k} c(1)
$$

so that if we integrate over a fiber sphere with the natural (not simplectic) orientation,

$$
\int_{S^{m-1}} ch(\Pi_+(p)) = i^{k-1} 2^{1-k} \operatorname{Tr}(f_1 \cdots f_m)
$$
$$
= i^{k-1} 2^{1-k} i^{m-k} 2^m = i^{k-1} 2^{1-k} i^{k-1} 2^{2k-1}
$$
$$
= (-1)^{k-1} \cdot 2^k.
$$

In particular, this is non-zero, so this cohomology class provides a cohomology extension to the fiber.

As a $H^*(M; \mathbf{Q})$ module, we can decompose $H^*(S(M); \mathbf{Q}) = H^*(M; \mathbf{Q}) \oplus xH^*(M; \mathbf{Q})$ where $x = ch(\Pi_+(p))$. If we twist the operator A by taking coefficients in an auxilary bundle V, then we generate $xH^*(M; \mathbf{Q})$. The same argument as that given in the proof of Lemma 3.9.8 permits us to interpret this in K-theory:

LEMMA 4.3.4. *Let M be a compact oriented Riemannian manifold of dimension $m = 2k - 1$. Let A be the tangential operator of the signature complex on $M \times [0, 1)$. If $\{e_j\}$ is an oriented local orthonormal basis for $T(M)$, let $\omega = i^k e_1 * \cdots * e_m$ be the orientation form acting by Clifford multiplication on the exterior algebra. $A = \omega(d + \delta)$ on $C^\infty(\Lambda(M))$. If the symbol of A is p,*

$$\int_{S^{m-1}} ch(\Pi_+(p)) = 2^k(-1)^{k-1}.$$

The natural map $K(M; \mathbf{Q}) \to K(S(T^*(M); \mathbf{Q})$ is injective and the group $K(S(T^*M); \mathbf{Q})/K(M; \mathbf{Q})$ is generated by the bundles $\{\Pi_+(A_V)\}$ as V runs over $K(M)$.

Remark: This operator can also be represented in terms of the Hodge operator. On $C^\infty(\Lambda^{2p})$ for example it is given by $i^k(-1)^{p+1}(*d - d*)$. We are using the entire tangential operator (and not just the part acting on even or odd forms). This will produce certain factors of 2 differing from the formulas of Atiyah-Patodi-Singer. If M admits a SPIN$_c$ structure, one can replace A by the tangential operator of the SPIN$_c$ complex; in this case the corresponding integrand just becomes $(-1)^{k-1}$.

We can use this representation to prove:

LEMMA 4.3.5. *Let $\dim M$ be odd and let $P: C^\infty(V) \to C^\infty(V)$ be a self-adjoint elliptic pseudo-differential operator of order $d > 0$. Then $R(P) = 0$—i.e., $\eta(s, P)$ is regular at $s = 0$.*

PROOF: We first suppose M is orientable. By Lemma 4.3.4, it suffices to prove $R(A_V) = 0$ since r is defined in K-theory and would then vanish on the generators. However, by Lemma 4.3.1(c), the residue is given by a local formula. The same analysis as that done for the heat equation shows this formula must be homogeneous of order m in the jets of the metric and of the connection on V. Therefore, it must be expressible in terms of Pontrjagin forms of TM and Chern forms of V by Theorem 2.6.1. As m is odd, this local formula vanishes and $R(A_V) = 0$. If M is not orientable, we pass to the oriented double cover. If $P: C^\infty(V) \to C^\infty(V)$ over M, we let $P': C^\infty(V') \to C^\infty(V')$ be the lift to the oriented double cover. Then $R(P) = \frac{1}{2}R(P')$. But $R(P') = 0$ since the double cover is oriented and thus $R(P) = 0$. This completes the proof.

This result is due to Atayah, Patodi, and Singer. The trick used in section 3.9 to change the parity of the dimension by taking products with a problem over the circle does not go through without change as we shall see. Before considering the even dimensional case, we must first prove a product formula.

LEMMA 4.3.6. *Let M_1 and M_2 be smooth manifolds. Let $P: C^\infty(V_1) \to C^\infty(V_2)$ be an elliptic complex over M_1. Let $Q: C^\infty(V) \to C^\infty(V)$ be a self-adjoint elliptic operator over M_2. We assume P and Q are differential operators of the same order and form:*

$$R = \begin{pmatrix} Q & P^* \\ P & -Q \end{pmatrix} \qquad \text{on } C^\infty(V_1 \otimes V \oplus V_2 \otimes V),$$

then $\eta(s, R) = \text{index}(P) \cdot \eta(s, Q)$.

PROOF: This lemma gives the relationship between the index and the eta invariant which we will exploit henceforth. We perform a formal computation. Let $\{\lambda_\nu, \phi_\nu\}_{\nu=1}^\infty$ be a spectral resolution of the operator Q on $C^\infty(V)$. We let $\Delta = P^* P$ and decompose $C^\infty(V_1) = N(\Delta) + R(\Delta)$. We let $\{\mu_j, \theta_j\}$ be a spectral resolution of Δ restricted to $N(\Delta)^\perp = R(\Delta)$. The μ_j are positive real numbers; $\{\mu_j, P\theta_j / \sqrt{\mu_j}\}$ form a spectral resolution of $\Delta' = PP^*$ on $N(\Delta')^\perp = R(\Delta') = R(P)$.

We decompose $L^2(V_1 \otimes V) = N(\Delta) \otimes L^2(V) \oplus R(\Delta) \otimes L^2(V)$ and $L^2(V_2 \otimes V) = N(\Delta') \otimes L^2(V) \oplus R(\Delta') \otimes L^2(V)$. In $R(\Delta) \otimes L^2(V) \oplus R(\Delta') \otimes L^2(V)$ we study the two-dimensional subspace that is spanned by the elements: $\{\theta_j \otimes \phi_\nu, P\theta_j / \sqrt{\mu_j} \otimes \phi_\nu\}$. The direct sum of these subspaces as j, ν vary is $R(\Delta) \otimes L^2(V) \oplus R(\Delta') \otimes L^2(V)$. Each subspace is invariant under the operator R. If we decompose R relative to this basis, it is represented by the 2×2 matrix:

$$\begin{pmatrix} \lambda_\nu & \sqrt{\mu_j} \\ \sqrt{\mu_j} & -\lambda_\nu \end{pmatrix}.$$

This matrix has two eigenvalues with opposite signs: $\pm\sqrt{\lambda_\nu^2 + \mu_j}$. Since $\lambda_\nu > 0$ these eigenvalues are distinct and cancel in the sum defining eta. Therefore the only contribution to eta comes from $N(\Delta) \otimes L^2(V)$ and $N(\Delta') \otimes L^2(V)$. On the first subspace, R is $1 \otimes Q$. Each eigenvalue of Q is repeated $\dim N(\Delta)$ times so the contribution to eta is $\dim N(\Delta) \eta(s, Q)$. On the second subspace, R is $1 \otimes -Q$ and the contribution to eta is $-\dim N(\Delta) \eta(s, Q)$. When we sum all these contributions, we conclude that:

$$\eta(s, R) = \dim N(\Delta) \eta(s, Q) - \dim N(\Delta') \eta(s, Q) = \text{index}(P) \eta(s, Q).$$

Although this formal cancellation makes sense even if P and Q are not differential operators, R will not be a pseudo-differential operator if P and Q are pseudo-differential operators in general. This did not matter when we studied the index since the index was constant under approximations. The eta invariant is a more delicate invariant, however, so we cannot use the same trick. Since the index of any differential operator on the circle is zero, we cannot use Lemmas 4.3.5 and 4.3.6 directly to conclude $R(P) = 0$ if m is even.

We recall the construction of the operator on $C^\infty(S^1)$ having index 1. We fix $a \in \mathbf{R}$ as a real constant and fix a positive order $d \in \mathbf{Z}$. We define:

$$Q_0 = -i\partial/\partial\theta, \qquad Q_1(a) = (Q_0^{2d} + a^2)^{1/2d}$$
$$Q_2(a) = \tfrac{1}{2}(e^{-i\theta}(Q_0 + Q_1) + Q_0 - Q_1)$$
$$Q(a) = Q_2(a) \cdot (Q_1(a))^{d-1}.$$

The same argument as that given in the proof of Lemma 3.9.4 shows these are pseudo-differential operators on $C^\infty(S^1)$. $Q_2(a)$ and $Q(a)$ are elliptic families. If $a = 0$, then $Q_2(0)$ agrees with the operator of Lemma 3.9.4 so index $Q_2(0) = 1$. Since the index is continuous under perturbation, index $Q_2(a) = 1$ for all values of a. $Q_1(a)$ is self-adjoint so its index is zero. Consequently index $Q(a) = $ index $Q_2(a) + $ index $Q_1(a) = 1$ for all a.

We let $P: C^\infty(V) \to C^\infty(V)$ be an elliptic self-adjoint partial differential operator of order $d > 0$ over a manifold M. On $M \times S^1$ we define the operators:

$$Q_0 = -i\partial/\partial\theta, \qquad Q_1 = (Q_0^{2d} + P^2)^{1/2d}$$
$$Q_2 = \tfrac{1}{2}(e^{-i\theta}(Q_0 + Q_1) + Q_0 - Q_1), \qquad Q = Q_2 \cdot Q_1^{d-1}$$

on $C^\infty(V)$. These are pseudo-differential operators over $M \times S^1$ since $Q_0^{2d} + P^2$ has positive definite leading order symbol. We define R by:

$$R = \begin{pmatrix} P & Q^* \\ Q & -P \end{pmatrix}: \quad C^\infty(V \oplus V) \to C^\infty(V \oplus V).$$

This is a pseudo-differential operator of order d which is self-adjoint. We compute:

$$R^2 = \begin{pmatrix} P^2 + Q^*Q & 0 \\ 0 & P^2 + QQ^* \end{pmatrix}$$

so

$$\sigma_L(R^2)(\xi, z) = \begin{pmatrix} p(\xi)^2 + q^*q(\xi, z) & 0 \\ 0 & p(\xi)^2 + qq^*(\xi, z) \end{pmatrix}$$

for $\xi \in T^*M$ and $z \in T^*S^1$. Suppose R is not elliptic so $\sigma_L(R)(\xi, z)v = 0$ for some vector $v \in V \oplus V$. Then $\{\sigma_L(R)(\xi, z)\}^2 v = 0$. We decompose $v = v_1 \oplus v_2$. We conclude:

$$(p(\xi)^2 + q^*q(\xi, z))v_1 = (p(\xi)^2 + qq^*(\xi, z))v_2 = 0.$$

Using the condition of self-adjointness this implies:

$$p(\xi)v_1 = q(\xi, z)v_1 = p(\xi)v_2 = q^*(\xi, z)v_2 = 0.$$

We suppose $v \neq 0$ so not both v_1 and v_2 are zero. Since P is elliptic, this implies $\xi = 0$. However, for $\xi = 0$, the operators Q_1 and Q_2 agree with the operators of Lemma 3.9.4 and are elliptic. Therefore $q(\xi, z)v_1 = q^*(\xi, z)v_2 = 0$ implies $z = 0$. Therefore the operator R is elliptic.

Lemma 4.3.6 generalizes in this situation to become:

LEMMA 4.3.7. *Let P, Q, R be defined as above, then $\eta(s, R) = \eta(s, P)$.*

PROOF: We let $\{\lambda_\nu, \phi_\nu\}$ be a spectral resolution of P on $C^\infty(V)$ over M. This gives an orthogonal direct sum decomposition:

$$L^2(V) \text{ over } M \times S^1 = \bigoplus_\nu L^2(S^1) \otimes \phi_\nu.$$

Each of these spaces is invariant under both P and R. If R_ν denotes the restriction of R to this subspace, then

$$\eta(s, R) = \sum_\nu \eta(s, R_\nu).$$

On $L^2(\phi_\nu)$, P is just multiplication by the real eigenvalue λ_ν. If we replace P by λ_ν we replace Q by $Q(\lambda_\nu)$, so R_ν becomes:

$$R_\nu = \begin{pmatrix} \lambda_\nu & Q^*(\lambda_\nu) \\ Q(\lambda_\nu) & -\lambda_\nu \end{pmatrix}.$$

We now apply the argument given to prove Lemma 4.3.6 to conclude:

$$\eta(s, R_\nu) = \text{sign}(\lambda_\nu)|\lambda_\nu|^{-s} \text{index}(Q(\lambda_\nu)).$$

Since index $Q(\lambda_\nu) = 1$, this shows $\eta(s, R) = \text{sign}(\lambda_\nu)|\lambda_\nu|^{-s}$ and completes the proof.

We can now generalize Lemma 4.3.5 to all dimensions:

THEOREM 4.3.8. *Let $P: C^\infty(V) \to C^\infty(V)$ be a self-adjoint elliptic pseudo-differential operator of order $d > 0$. Then $R(P) = 0$—i.e., $\eta(s, P)$ is regular at $s = 0$.*

PROOF: This result follows from Lemma 4.3.5 if $\dim M = m$ is odd. If $\dim M = m$ is even and if P is a differential operator, then we form the pseudo-differential operator P over $M \times S^1$ with $\eta(s, R) = \eta(s, P)$. Then $\eta(s, R)$ is regular at $s = 0$ implies $\eta(s, P)$ is regular at $s = 0$. This proves

Theorem 4.3.8 for differential operators. Of course, if P is only pseudo-differential, then R need not be pseudo-differential so this construction does not work. We complete the proof of Theorem 4.3.8 by showing the partial differential operators of even order generate $K(S(T^*M); \mathbf{Q})/K(M; \mathbf{Q})$ if m is even. (We already know the operators of odd order generate if m is odd and if M is oriented).

Consider the involution $\xi \mapsto -\xi$ of the tangent space. This gives a natural $\mathbf{Z}_2$ action on $S(T^*M)$. Let $\pi: S(T^*M) \to S(T^*M)/\mathbf{Z}_2 = \mathbf{R}P(T^*M)$ be the natural projection on the quotient projective bundle. Since m is even, $m-1$ is odd and π^* defines an isomorphism between the cohomology of two fibers

$$\pi^*: H^*(\mathbf{R}P^{m-1}; \mathbf{Q}) = \mathbf{Q} \oplus \mathbf{Q} \to H^*(S^{m-1}; \mathbf{Q}) = \mathbf{Q} \oplus \mathbf{Q}.$$

The Kunneth formula and an appropriate Meyer-Vietoris sequence imply that

$$\pi^*: H^*(\mathbf{R}P(T^*M); \mathbf{Q}) \to H^*(S(T^*M); \mathbf{Q})$$

is an isomorphism in cohomology for the total spaces. We now use the Chern isomorphism between cohomology and K-theory to conclude there is an isomorphism in K-theory

$$\pi^*: K(\mathbf{R}P(T^*M); \mathbf{Q}) \simeq K(S(T^*M); \mathbf{Q}).$$

Let $S_0(k)$ be the set of all $k \times k$ self-adjoint matrices A such that $A^2 = 1$ and $\mathrm{Tr}(A) = 0$. We noted in section 3.8 that if k is large, $\widetilde{K}(X) = [X, S_0(2k)]$. Thus $\widetilde{K}(S(T^*M); \mathbf{Q}) = \widetilde{K}(\mathbf{R}P(T^*M); \mathbf{Q})$ is generated by maps $p: S(T^*M) \to S_0(2k)$ such that $p(x, \xi) = p(x, -\xi)$. We can approximate any even map by an even polynomial using the Stone-Weierstrass theorem. Thus we may suppose p has the form:

$$p(x, \xi) = \sum_{\substack{|\alpha| \le n \\ |\alpha| \text{ even}}} p_\alpha(x) \xi^\alpha \Big|_{S(T^*M)},$$

where the $p_\alpha: M \to S_0(2k)$ and where n is large. As α is even, we can replace ξ^α by $\xi^\alpha |\xi|^{\{n-|\alpha|\}/2}$ and still have a polynomial with the same values on $S(T^*M)$. We may therefore assume that p is a homogeneous even polynomial; this is the symbol of a partial differential operator which completes the proof.

If m is odd and if M is orientable, we constructed specific examples of operators generating $K(S(T^*M); \mathbf{Q})/K(M; \mathbf{Q})$ using the tangential operator of the signature complex with coefficients in an arbitrary coefficient bundle. If m is even, it is possible to construct explicit second order oprators generating this K-theory group. One can then prove directly that eta is regular at $s = 0$ for these operators as they are all "natural" in a certain suitable sense. This approach gives more information by explcitly exhibiting the generators; as the consturction is quite long and technical we have chosen to give an alternate argument based on K-theory and refer to (Gilkey, The residue of the global eta function at the origin) for details.

We have given a global proof of Theorem 4.3.8. In fact such a treatment is necessary since in general the local formulas giving the residue at $s = 0$ are non-zero. Let $P: C^\infty(V) \rightarrow C^\infty(V)$ be self-adjoint and elliptic. If $\{\lambda_\nu, \phi_\nu\}$ is a spectral resolution of P, we define:

$$\eta(s, P, x) = \sum_\nu \text{sign}(\lambda_\nu)|\lambda_\nu|^{-s}(\phi_\nu, \phi_\nu)(x)$$

so that:

$$\eta(s, P) = \int_M \eta(s, P, x)\, \text{dvol}(x).$$

Thus $\text{Res}_{s=0}\, \eta(s, P, x) = a(P, x)$ is given by a local formula.

We present the following example (Gilkey, The residue of the local eta function at the origin) to show this local formula need not vanish identically in general.

Example 4.3.9. Let

$$e_1 = \begin{pmatrix} 1 & 0 \\ 0 & -1 \end{pmatrix}, \qquad e_2 = \begin{pmatrix} 0 & 1 \\ 1 & 0 \end{pmatrix}, \qquad e_3 = \begin{pmatrix} 0 & i \\ -i & 0 \end{pmatrix}$$

be Clifford matrices acting on $\mathbf{C}^2$. Let T^m be the m-dimensional torus with periodic parameters $0 \le x_j \le 2$ for $1 \le j \le m$.

(a) Let $m = 3$ and let $b(x)$ be a real scalar. Let $P = i\sum_j e_j \partial/\partial x_j + b(x)I$. Then P is self-adjoint and elliptic and $a(x, P) = c\Delta b$ where $c \ne 0$ is some universal constant.

(b) Let $m = 2$ and let b_1 and b_2 be imaginary scalar functions. Let $P = e_1 \partial^2/\partial x_1^2 + e_2 \partial^2/\partial x_2^2 + 2e_3 \partial^2/\partial x_1 \partial x_2 + b_1 e_1 + b_2 e_2$. Then P is self-adjoint and elliptic and $a(x, P) = c'(\partial b_1/\partial x_2 - \partial b_2/\partial x_1)$ where $c' \ne 0$ is some universal constant.

(c) By twisting this example with a non-trivial index problem and using Lemma 4.3.6, we can construct examples on T_m so that $a(x, P)$ does not vanish identically for any dimension $m \ge 2$.

The value of eta at the origin plays a central role in the Atiyah-Patodi-Singer index theorem for manifolds with boundary. In section 2, we discussed the transgression briefly. Let ∇_i be two connections on $T(N)$. We defined $TL_k(\nabla_1, \nabla_2)$ so that $dTL_k(\nabla_1, \nabla_2) = L_k(\nabla_1) - L_k(\nabla_2)$; this is a secondary characteristic class. Let ∇_1 be the Levi-Civita connection of N and near $M = dN$ let ∇_2 be the product connection arising from the product metric. As $L_k(\nabla_2) = 0$ we see $dTL_k(\nabla_1, \nabla_2) = 0$. This is the analogous term which appeared in the Gauss-Bonnet theorem; it can be computed in terms of the first and second fundamental forms. For example, if dim $N = 4$

$$L_1(R) = \frac{-1}{24 \cdot \pi^2} \operatorname{Tr}(R \wedge R) \qquad TL_1(R, \omega) = \frac{-1}{24 \cdot 8 \cdot \pi^2} \operatorname{Tr}(R \wedge \omega)$$

where R is the curvature 2-form and ω is the second fundamental form.

THEOREM 4.3.10 (ATIYAH-PATODI-SINGER INDEX THEOREM FOR MANIFOLDS WITH BOUNDARY). *Let N be a $4k$ dimensional oriented compact Riemannian manifold with boundary M. Then:*

$$\operatorname{signature}(N) = \int_N L_k - \int_M TL_k - \tfrac{1}{2}\eta(0, A)$$

where A is the tangential operator of the signature complex discussed in Lemma 4.3.4.

In fact, the eta invariant more generally is the boundary correction term in the index theorem for manifolds with boundary. Let N be a compact Riemannian manifold with boundary M and let $P: C^\infty(V_1) \to C^\infty(V_2)$ be an elliptic first order differential complex over N. We take a metric which is product near the boundary and identify a neighborhood of M in N with $M \times [0, 1)$. We suppose P decomposes in the form $P = \sigma(dn)(\partial/\partial n + A)$ on this collared neighborhood where A is a self-adjoint elliptic first order operator over M. This is in fact the case for the signature, Dolbeault, or spin compexes. Let B be the spectral projection on the non-negative eigenvalues of A. Then:

THEOREM 4.3.11. (THE ATIYAH-PATODI-SINGER INDEX THEOREM). *Adopt the notation above. P with boundary condition B is an elliptic problem and*

$$\operatorname{index}(P, B) = \int_N \{a_n(x, P^*P) - a_n(x, PP^*)\} - \tfrac{1}{2}\{\eta(0, A) + \dim \operatorname{N}(A)\}.$$

In this expression, $n = \dim N$ and the invariants a_n are the invariants of the heat equation discussed previously.

Remark: If M is empty then this is nothing but the formula for index(P) discussed previously. If the symbol does not decompose in this product structure near the boundary of N, there are corresponding local boundary correction terms similar to the ones discussed previously. If one takes P to be the operator of the signature complex, then Theorem 4.3.10 can be derived from this more general result by suitably interpreting index$(P, B) = $ signature$(N) - \tfrac{1}{2} \dim \operatorname{N}(A)$.

4.4. The Eta Invariant with Coefficients
In a Locally Flat Bundle.

The eta invariant plays a crucial role in the index theorem for manifolds with boundary. It is also possible to study the eta invariant with coefficients in a locally flat bundle to get a generalization of the Atiyah-Singer theorem.

Let $\rho: \pi_1(M) \to U(k)$ be a unitary representation of the fundamental group. Let $\overline{M}$ be the universal cover of M and let $\overline{m} \to g\overline{m}$ for $\overline{m} \in \overline{M}$ and $g \in \pi_1(M)$ be the acion of the deck group. We define the bundle V_ρ over M by the identification:

$$V_\rho = \overline{M} \times \mathbf{C}^k \bmod (\overline{m}, z) = (g\overline{m}, \rho(g)z).$$

The transition functions of V_ρ are locally constant and V_ρ inherits a natural unitary structure and connection ∇_ρ with zero curvature. The holonomy of ∇_ρ is just the representation ρ. Conversely, given a unitary bundle with locally constant transition functions, we can construct the connection ∇ to be unitary with zero curvature and recover ρ as the holonomy of the connection. We assume ρ is unitary to work with self-adjoint operators, but all the constructions can be generalized to arbitrary representations in $GL(k, \mathbf{C})$.

Let $P: C^\infty(V) \to C^\infty(V)$. Since the transition functions of V_ρ are locally constant, we can define P_ρ on $C^\infty(V \otimes V_\rho)$ uniquely using a partition of unity if P is a differential operator. If P is only pseudo-differential, P_ρ is well defined modulo infinitely smoothing terms. We define:

$$\tilde{\eta}(P) = \tfrac{1}{2}\{\eta(P) + \dim N(P)\} \quad \bmod \mathbf{Z}$$
$$\mathrm{ind}(\rho, P) = \tilde{\eta}(P_\rho) - k\tilde{\eta}(P) \quad \bmod \mathbf{Z}.$$

LEMMA 4.4.1. $\mathrm{ind}(\rho, P)$ is a homotopy invariant of P. If we fix ρ, we can interpret this as a map

$$\mathrm{ind}(\rho, *): K(S(T^*M))/K(M) \to \mathbf{R} \bmod \mathbf{Z}$$

such that $\mathrm{ind}(\rho, P) = \mathrm{ind}(\rho, \Pi_+(\sigma_L(P)))$.

PROOF: We noted previously that $\tilde{\eta}$ was well defined in $\mathbf{R} \bmod \mathbf{Z}$. Let $P(t)$ be a smooth 1-parameter family of such operators and let $P'(t) = \dfrac{d}{dt} P(t)$. In Theorem 1.10.2, we proved:

$$\frac{d}{dt}\tilde{\eta}(P_t) = \int_M a(x, P(t), P'(t))\, \mathrm{dvol}(x)$$

was given by a local formula. Let $P_k = P \otimes 1^k$ acting on $V \otimes 1^k$. This corresponds to the trivial representation of $\pi_1(M)$ in $U(k)$. The operators

P_ρ and P_k are locally isomorphic modulo ∞ smoothing terms which don't affect the local invariant. Thus $a(x, P_k(t), P_k'(t)) - a(x, P_\rho(t), P_\rho'(t)) = 0$. This implies $\dfrac{d}{dt}\,\mathrm{ind}(\rho, P_t) = 0$ and completes the proof of homotopy invariance. If the leading symbol of P is definite, then the value $\eta(0, P)$ is given by a local formula so $\mathrm{ind}(\rho, *) = 0$ in this case, as the two local formulas cancel. This verifies properties (d) and (e) of Lemma 4.3.1; properties (a), (b) and (f) are immediate. We therefore apply Lemma 4.3.3 to regard

$$\mathrm{ind}(\rho, *)\colon K(S(T^*M))/K(M) \to \mathbf{R} \bmod \mathbf{Z}$$

which completes the proof.

In sections 4.5 and 4.6 we will adopt a slightly different notation for this invariant. Let G be a group and let $R(G)$ be the group representation ring generated by the unitary representations of G. Let $R_0(G)$ be the ideal of representations of virtual dimension 0. We extend $\tilde\eta\colon R(G) \to \mathbf{R} \bmod \mathbf{Z}$ to be a $\mathbf{Z}$-linear map. We let $\mathrm{ind}(\rho, P)$ denote the restriction to $R_0(G)$. If ρ is a representation of dimension j, then $\mathrm{ind}(\rho, P) = \mathrm{ind}(\rho - j \cdot 1, P) = \eta(P_\rho) - j\eta(P)$. It is convenient to use both notations and the context determines whether we are thinking of virtual representations of dimension 0 or the projection of an actual representation to $R_0(G)$. $\mathrm{ind}(\rho, P)$ is not topological in ρ, as the following example shows. We will discuss K-theory invariants arising from the ρ dependence in Lemma 4.6.5.

Example 4.4.2: Let $M = S^1$ be the circle with periodic parameter $0 \le \theta \le 2\pi$. Let $g(\theta) = e^{i\theta}$ be the generator of $\pi_1(M) \simeq \mathbf{Z}$. We let ε belong to $\mathbf{R}$ and define:

$$\rho_\varepsilon(g) = e^{2\pi i \varepsilon}$$

as a unitary representation of $\pi_1(M)$. The locally flat bundle V_ρ is topologically trivial since any complex bundle over S^1 is trivial. If we define a locally flat section to $S^1 \times \mathbf{C}$ by:

$$\vec{s}(\theta) = e^{i\varepsilon\theta}$$

then the holonomy defined by $\vec{s}$ gives the representation ρ_ε since

$$\vec{s}(2\pi) = e^{2\pi i \varepsilon}\,\vec{s}(0).$$

We let $P = -i\partial/\partial\theta$ on $C^\infty(S^1)$. Then

$$P_\varepsilon = P_{\rho_\varepsilon} = e^{+i\varepsilon\theta} P e^{-i\varepsilon\theta} = P - \varepsilon.$$

The spectrum of P is $\{n\}_{n\in\mathbf{Z}}$ so the spectrum of P_ε is $\{n - \varepsilon\}_{n\in\mathbf{Z}}$. Therefore:

$$\eta(s, P_\varepsilon) = \sum_{n-\varepsilon\neq 0} \mathrm{sign}(n - \varepsilon)|n - \varepsilon|^{-s}.$$

We differentiate with respect to ε to get:

$$\frac{d}{d\varepsilon}(s, P_\varepsilon) = s \sum_{n-\varepsilon \neq 0} |n - \varepsilon|^{-s-1}.$$

We evaluate at $s = 0$. Since the sum defining this shifted zeta function ranges over all integers, the pole at $s = 0$ has residue 2 so we conclude:

$$\frac{d}{d\varepsilon}\tilde{\eta}(P_\varepsilon) = 2 \cdot \frac{1}{2} = 1 \qquad \text{and} \qquad \text{ind}(\rho_\varepsilon, P) = \int_0^\varepsilon 1\, d\varepsilon = \varepsilon.$$

We note that if we replace ε by $\varepsilon + j$ for $j \in \mathbf{Z}$, then the representation is unchanged and the spectrum of the operator P_ε is unchanged. Thus reduction mod $\mathbf{Z}$ is essential in making $\text{ind}(\rho_\varepsilon, P)$ well defined in this context.

If V_ρ is locally flat, then the curvature of ∇_ρ is zero so $ch(V_\rho) = 0$. This implies V_ρ is a torsion element in K-theory so $V_\rho \otimes 1^n \simeq 1^{kn}$ for some integer n. We illustrate this with

Example 4.4.3: Let $M = \mathbf{R}P_3 = S^3/\mathbf{Z}_2 = SO(3)$ be real projective space in dimension 3 so $\pi_1(M) = \mathbf{Z}_2$. Let $\rho: Z_2 \to U(1)$ be the non-trivial representation with $\rho(g) = -1$ where g is the generator. Let $L = S^3 \times \mathbf{C}$ and identify $(x, z) = (-x, -z)$ to define a line bundle L_ρ over $\mathbf{R}P_3$ with holonomy ρ. We show L_ρ is non-trivial. Suppose the contrary, then L_ρ is trivial over $\mathbf{R}P_2$ as well. This shows there is a map $f: S^2 \to S^1$ with $-f(-x) = f(x)$. If we restrict f to the upper hemisphere of S^2, then $f: D_+^2 \to S^1$ satisfies $f(x) = -f(-x)$ on the boundary. Therefore f has odd degree. Since f extends to D_+, f must have zero degree. This contradiction establishes no such f exists and L_ρ is non-trivial.

The bundle $L_\rho \oplus L_\rho$ is $S^3 \times \mathbf{C}$ modulo the relation $(x, z) = (-x, -z)$. We let $g(x): S^3 \to SU(2)$ be the identity map:

$$g(x) = \begin{pmatrix} x_0 + ix_1 & -x_2 + ix_3 \\ x_2 + ix_3 & x_0 - ix_1 \end{pmatrix}$$

then $g(-x) = -g(x)$. Thus g descends to give a global frame on $L_\rho \oplus L_\rho$ so this bundle is topologically trivial and L_ρ represents a $\mathbf{Z}_2$ torsion class in $K(\mathbf{R}P_3)$. Since L_ρ is a line bundle and is not topologically trivial, $L_\rho - 1$ is a non-zero element of $\tilde{K}(\mathbf{R}P_3)$. This construction generalizes to define L_ρ over $\mathbf{R}P_n$. We use the map $g: S^n \to U(2^k)$ defined by Clifford algebras so $g(x) = -g(-x)$ to show $2^k L_\rho - 2^k = 0$ in $\tilde{K}(\mathbf{R}P_n)$ where $k = [n/2]$; we refer to Lemma 3.8.9 for details.

We suppose henceforth in this section that the bundle V_ρ is topologically trivial and let $\vec{s}$ be a global frame for V_ρ. (In section 4.9 we will study the more general case). We can take $P_k = P \otimes 1$ relative to the frame $\vec{s}$ so that both P_k and P_ρ are defined on the same bundle with the same leading symbol. We form the 1-parameter family $tP_k + (1-t)P_\rho = P(t, \rho, \vec{s})$ and define:

$$\mathrm{ind}(\rho, P, \vec{s}) = \int_0^1 \frac{d}{dt} \tilde{\eta}(P(t, \rho, \vec{s})) \, dt = \int_M a(x, \rho, P, \vec{s}) \, \mathrm{dvol}(x)$$

where

$$a(x, \rho, P, \vec{s}) = \int_0^1 a(x, P(t, \rho, \vec{s}), P'(t, \rho, \vec{s})) \, dt.$$

The choice of a global frame permits us to lift ind from $\mathbf{R}$ mod $\mathbf{Z}$ to $\mathbf{R}$. If we take the operator and representation of example 4.4.2, then $\mathrm{ind}(\rho_\varepsilon, P, \vec{s}) = \varepsilon$ and thus in particular the lift depends on the global frame chosen (or equivalently on the particular presentation of the representation ρ on a trivial bundle). This also permits us to construct non-trivial real valued invariants even on simply connected manifolds by choosing suitable inequivalent global trivializations of V_ρ.

LEMMA 4.4.4.
(a) $\mathrm{ind}(\rho, P, \vec{s}) = \int_M a(x, \rho, P, \vec{s}) \, \mathrm{dvol}(x)$ *is given by a local formula which depends on the connection 1-form of ∇_ρ relative to the global frame $\vec{s}$ and on the symbol of the operator P.*
(b) *If $\vec{s}_t$ is a smooth 1-parameter family of global sections, then* $\mathrm{ind}(\rho, P, \vec{s}_t)$ *is independent of the parameter t.*

PROOF: The first assertion follows from the definition of $a(x, \rho, P, \vec{s})$ given above and from the results of the first chapter. This shows $\mathrm{ind}(\rho, P, \vec{s}_t)$ varies continuously with t. Since its mod $\mathbf{Z}$ reduction is $\mathrm{ind}(\rho, P)$, this mod $\mathbf{Z}$ reduction is constant. This implies $\mathrm{ind}(\rho, P, \vec{s}_t)$ itself is constant.

We can use $\mathrm{ind}(\rho, P, \vec{s})$ to detect inequivalent trivializations of a bundle and thereby study the homotopy $[M, \mathrm{U}(k)]$ even if M is simply connected. This is related to spectral flow.

The secondary characteristic classes are cohomological invariants of the representation ρ. They are normally $\mathbf{R}$ mod $\mathbf{Z}$ classes, but can be lifted to $\mathbf{R}$ and expressed in terms of local invariants if the bundle V_ρ is given a fixed trivialization. We first recall the definition of the Chern character. Let W be a smooth vector bundle with connection ∇. Relative to some local frame, we let ω be the connection 1-form and $\Omega = d\omega - \omega \wedge \omega$ be the curvature. The Chern character is given by $ch_k(\nabla) = \left(\frac{i}{2\pi}\right)^k \frac{1}{k!} \mathrm{Tr}(\Omega^k)$. This is a closed $2k$ form independent of the frame $\vec{s}$ chosen. If ∇_i are two connections for $i = 0, 1$, we form $\nabla_t = t\nabla_1 + (1-t)\nabla_0$. If $\theta = \omega_1 - \omega_0$, then θ transforms like a tensor. If Ω_t is the curvature of the connection ∇_t, then:

$$ch_k(\nabla_1) - ch_k(\nabla_0) = \int_0^1 \frac{d}{dt} ch_k(\nabla_t) \, dt = d(Tch_k(\nabla_1, \nabla_0))$$

where the transgression Tch_k is defined by:

$$Tch_k(\nabla_1, \nabla_0) = \left(\frac{i}{2\pi}\right)^k \frac{1}{(k-1)!} \operatorname{Tr}\left\{\int_0^1 \theta\Omega_t^{k-1}\, dt\right\}.$$

We refer to the second chapter for further details on this construction.

We apply this construction to the case in which both ∇_1 and ∇_0 have zero curvature. We choose a local frame so $\omega_0 = 0$. Then $\omega_1 = \theta$ and $\Omega_1 = d\theta - \theta \wedge \theta = 0$. Consequently:

$$\omega_t = t\theta \qquad \text{and} \qquad \Omega_t = td\theta - t^2\theta \wedge \theta = (t - t^2)\theta \wedge \theta$$

so that:

$$Tch_k(\nabla_1, \nabla_0) = \left(\frac{i}{2\pi}\right)^k \frac{1}{(k-1)!} \int_0^1 (t - t^2)^{k-1}\, dt \cdot \operatorname{Tr}(\theta^{2k-1}).$$

We integrate by parts to evaluate this coefficient:

$$\int_0^1 (t - t^2)^{k-1}\, dt = \int_0^1 t^{k-1}(1-t)^{k-1}\, dt = \frac{k-1}{k}\int_0^1 t^k(1-t)^{k-2}\, dt$$

$$= \frac{(k-1)!}{k \cdot (k+1)\cdots(2k-2)}\int_0^1 t^{2k-2}\, dt$$

$$= \frac{(k-1)!\,(k-1)!}{(2k-1)!}.$$

Therefore

$$Tch_k(\nabla_1, \nabla_0) = \left(\frac{i}{2\pi}\right)^k \frac{(k-1)!}{(2k-1)!} \cdot \operatorname{Tr}(\theta^{2k-1}).$$

We illustrate the use of secondary characteristic classes by giving another version of the Atiyah-Singer index theorem. Let $Q: C^\infty(1^k) \to C^\infty(1^k)$ be an elliptic complex. Let $\vec{s}_\pm$ be global frames on $\Pi_+(\Sigma q)$ over $D_+(T^*M)$ so that $\vec{s}_- = q^t(x, \xi)\vec{s}_+$ on $D_+(T^*M)\cap D_-(T^*M) = S(T^*M)$. (The clutching function is q; to agree with the notation adopted in the third section we express $\vec{s}_- = q^t\vec{s}_+$ as we think of q being a matrix acting on column vectors of $\mathbf{C}^k$. The action on the frame is therefore the transpose action). We choose connections $\nabla_\pm$ on $\Pi_+(\Sigma q)$ so $\nabla_\pm(\vec{s}_\pm) = 0$ on $D_\pm(T^*M)$. Then:

$$\operatorname{index}(Q) = (-1)^m \int_{\Sigma(T^*M)} Todd(M) \wedge ch(\nabla_-)$$

$$= (-1)^m \int_{D_+(M)} Todd(M) \wedge ch(\nabla_-).$$

However, on D_+ we have $\Omega_+ = 0$ so we can replace $ch(\nabla_-)$ by $ch(\nabla_-) - ch(\nabla_+)$ without changing the value of the integral. $ch(\nabla_-) - ch(\nabla_+) = d\,Tch(\nabla_-, \nabla_+)$ so an application of Stokes theorem (together with a careful consideration of the orientations involved) yields:

$$\text{index}(Q) = (-1)^m \int_{S(T^*M)} Todd(M) \wedge Tch(\nabla_-, \nabla_+).$$

Both connections have zero curvature near the equator $S(T^*M)$. The fibers over D_+ are glued to the fibers over D_- using the clutching function q. With the notational conventions we have established, if f_+ is a smooth section relative to the frame $\vec{s}_+$ then the corresponding representation is qf_+ relative to the frame $\vec{s}_-$. Therefore $\nabla_-(f_+) = q^{-1}dq \cdot f_+ + df_+$ and consequently

$$\nabla_- - \nabla_+ = \theta = q^{-1}dq.$$

(In obtaining the Maurer-Cartan form one must be careful which convention one uses—right versus left—and we confess to having used both conventions in the course of this book.) We use this to compute:

$$Tch_k(\nabla_-, \nabla_+) = \sum c_k \,\text{Tr}((q^{-1}dq)^{2k-1})$$

for

$$c_k = \left(\frac{i}{2\pi}\right)^k \frac{(k-1)!}{(2k-1)!}.$$

Let $\theta = g^{-1}dg$ be the Maurer-Cartan form and let $Tch = \sum_k c_k \,\text{Tr}(\theta^{2k-1})$.

This defines an element of the odd cohomology of $GL(\cdot, \mathbf{C})$ such that $Tch(\nabla_-, \nabla_+) = q^*(Tch)$. We summarize these computations as follows:

LEMMA 4.4.5. *Let $\theta = g^{-1}dg$ be the Maurer-Cartan form on the general linear group. Define:*

$$Tch = \sum_k \left(\frac{i}{2\pi}\right)^k \frac{(k-1)!}{(2k-1)!} \cdot \text{Tr}(\theta^{2k-1})$$

as an element of the odd cohomology. If $Q: C^\infty(1\cdot) \to C(1\cdot)$ is an elliptic complex defined on the trivial bundle, let q be the symbol so $q: S(T^*M) \to GL(\cdot, \mathbf{C})$. Then $\text{index}(Q) = (-1)^m \int_{S(T^*M)} Todd(M) \wedge q^*(Tch)$.

We compute explicitly the first few terms in the expansion:

$$Tch = \frac{i}{2\pi}\,\text{Tr}(\theta) + \frac{-1}{24\pi^2}\,\text{Tr}(\theta^3) + \frac{-i}{960\pi^3}\,\text{Tr}(\theta^5) + \cdots.$$

In this version, the Atiyah-Singer theorem generalizes to the case of $dM \neq \emptyset$ as the Atiyah-Bott theorem. We shall discuss this in section 4.5.

We can now state the Atiyah-Patodi-Singer twisted index theorem:

Theorem 4.4.6. *Let $P: C^\infty(V) \to C^\infty(V)$ be an elliptic self-adjoint pseudo-differential operator of order $d > 0$. Let $\rho: \pi_1(M) \to U(k)$ be a unitary representation of the fundamental group and assume the associated bundle V_ρ is toplogically trivial. Let $\vec{s}$ be a global frame for V_ρ and let $\nabla_0(\vec{s}) \equiv 0$ define the connection ∇_0. Let ∇_ρ be the connection defined by the representation ρ and let $\theta = \nabla_\rho(\vec{s})$ so that:*

$$Tch(\nabla_\rho, \nabla_0) = \sum_k \left(\frac{i}{2\pi}\right)^k \frac{(k-1)!}{(2k-1)!} \cdot \mathrm{Tr}(\theta^{2k-1}).$$

Then:

$$\mathrm{ind}(\rho, P, \vec{s}) = (-1)^m \int_{S(T^*M)} Todd(M) \wedge ch(\Pi_+ p) \wedge Tch(\nabla_\rho, \nabla_0).$$

$S(T^*M)$ is given the orientation induced by the simplectic orientation $dx_1 \wedge d\xi_1 \wedge \cdots \wedge dx_m \wedge d\xi_m$ on T^*M where we use the outward pointing normal so $N \wedge \omega_{2m-1} = \omega_{2m}$.

We postpone the proof of Theorem 4.4.6 for the moment to return to the examples considered previously. In example 4.4.2, we let $\rho = e^{2\pi i \varepsilon}$ on the generator of $\pi_1 S^1 \simeq \mathbf{Z}$. We let "1" denote the usual trivialization of the bundle $S^1 \times \mathbf{C}$ so that:

$$\nabla_\rho(1) = -i\varepsilon d\theta \qquad \text{and} \qquad Tch(\nabla_\rho, \nabla_0) = \varepsilon d\theta/2\pi.$$

The unit sphere bundle decomposes $S(T^*M) = S^1 \times \{1\} \cup S^1 \times \{-1\}$. The symbol of the operator P is multiplication by the dual variable ξ so $ch(\Pi_+ p) = 1$ on $S^1 \times \{1\}$ and $ch(\Pi_+ p) = 0$ on $S^1 \times \{-1\}$. The induced orientation on $S^1 \times \{1\}$ is $-d\theta$. Since $(-1)^m = -1$, we compute:

$$-\int_{S(T^*M)} Todd(M) \wedge ch(\Pi_+ p) \wedge Tch(\nabla_\rho, \nabla_0) = \int_0^{2\pi} \varepsilon \, d\theta/2\pi = \varepsilon.$$

This also gives an example in which Tch_1 is non-trivial.

In example 4.4.3, we took the non-trivial representation ρ of $\pi_1(\mathbf{R}P_3) = \mathbf{Z}_2$ to define a non-trivial complex line bundle V_ρ such that $V_\rho \oplus V_\rho = 1^2$. The appropriate generalization of this example is to 3-dimensional lens spaces and provides another application of Theorem 4.4.6:

Example 4.4.7: Let $m = 3$ and let n and q be relatively prime positive integers. Let $\lambda = e^{2\pi i/n}$ be a primitive n^{th} root of unity and let $\gamma = \mathrm{diag}(\lambda, \lambda^q)$ generate a cyclic subgroup Γ of $U(2)$ of order n. If $\rho_s(\gamma) = \lambda^s$, then $\{\rho_s\}_{0 \le s < n}$ parametrize the irreducible representations of Γ. Γ acts without fixed points on the unit sphere S^3. Let $L(n, q) = S^3/\Gamma$ be the quotient manifold. As $\pi_1(S^3) = 0$, we conclude $\pi_1(L(n,q)) = \Gamma$. Let V_s be the line bundle corresponding to the representation ρ_s. It is defined from $S^3 \times \mathbf{C}$ by the equivalence relation $(z_1, z_2, w) = (\lambda z_1, \lambda^q z_2, \lambda^s w)$.

Exactly the same arguments (working mod n rather than mod 2) used to show V_1 is non-trivial over $\mathbf{R}P_3$ show V_s is non-trivial for $0 < s < n$. In fact $\tilde{K}(L(n,q)) = \mathbf{Z}_n$ as we shall see later in Corollary 4.6.10 and the bundle $(V_1 - 1)$ generates the reduced K-theory group. A bundle $V_{s_1} \oplus \cdots \oplus V_{s_t}$ is topologically trivial if and only if $s_1 + \cdots + s_t \equiv 0 \ (n)$.

The bundle $V_s \oplus V_{-s}$ is topologically trivial. We define:

$$g(z_1, z_2) = \begin{pmatrix} z_1^s & z_2^{sq'} \\ \bar{z}_2^{sq'} & -\bar{z}_1^s \end{pmatrix} \qquad \text{where } qq' \equiv 1 \ (n).$$

It is immediate that:

$$g(\lambda z_1, \lambda^q z_2) = \begin{pmatrix} \lambda^s z_1^s & \lambda^s z_2^{sq'} \\ \lambda^{-s} \bar{z}_2^{sq'} & -\lambda^{-s} \bar{z}_1^s \end{pmatrix} = \gamma g(z_1, z_2)$$

so we can regard g as an equivariant frame to $V_s \otimes V_{-s}$. If $\{\nabla_s \oplus \nabla_{-s}\}$ denotes the connection induced by the locally flat structure and if $\nabla_0 \oplus \nabla_0$ denotes the connection defined by the new frame, then:

$$\theta = \{\nabla_s \oplus \nabla_{-s}\} - \{\nabla_0 \oplus \nabla_0\} = dg \cdot g^{-1}.$$

Suppose first $s = q = q' = 1$ so $g: S^3 \to \mathrm{SU}(2)$ is the identity map. $\mathrm{Tr}(\theta^3)$ is a right invariant 3-form and is therefore a constant multiple of the volume element of $\mathrm{SU}(2)$. We calculate at $z = (1, 0)$ in $\mathbf{C}^2$. Let

$$e_0 = \begin{pmatrix} 1 & 0 \\ 0 & -1 \end{pmatrix}, \quad e_1 = \begin{pmatrix} i & 0 \\ 0 & i \end{pmatrix}, \quad e_2 = \begin{pmatrix} 0 & 1 \\ 1 & 0 \end{pmatrix}, \quad e_3 = \begin{pmatrix} 0 & i \\ -i & 0 \end{pmatrix}$$

so that $g(z) = g(x) = x_0 e_0 + x_1 e_1 + x_2 e_2 + x_3 e_3$. At $(1, 0)$ we have:

$$(dg \cdot g^{-1})^3 = \{(e_1 \, dx_1 + e_2 \, dx_2 + e_3 \, dx_3)e_0\}^3.$$

e_0 commutes with e_1 and anti-commutes with e_2 and e_3 so that, as $e_0^2 = 1$,

$$(dg \cdot g^{-1})^3 = (e_1 \, dx_1 + e_2 \, dx_2 + e_3 \, dx_3)(e_1 \, dx_1 - e_2 \, dx_2 - e_3 \, dx_3) \times$$
$$(e_1 \, dx_1 + e_2 \, dx_2 + de_3 \, dx_3)e_0$$
$$= (-e_1 e_2 e_3 + e_1 e_3 e_2 - e_2 e_1 e_3 - e_2 e_3 e_1 + e_3 e_1 e_2 + e_3 e_2 e_1) \times$$
$$e_0 \, dx_1 \wedge dx_2 \wedge dx_3$$
$$= -6 e_1 e_2 e_3 e_0 \cdot \mathrm{dvol} = -6 \begin{pmatrix} 1 & 0 \\ 0 & 1 \end{pmatrix} \mathrm{dvol}$$

so that $\text{Tr}(\theta^3) = -12 \cdot \text{dvol}$. Therefore:

$$\int_{S^3} Tch_2(\nabla_1 \oplus \nabla_{-1}, \nabla_0 \oplus \nabla_0) = \frac{12}{24\pi^2} \cdot \text{volume}(S^3) = 1.$$

We can study other values of (s, q) by composing with the map $(z_1, z_2) \mapsto (z_1^s, z_2^{sq'})/|(z_1^s, z_2^{sq'}|$. This gives a map homotopic to the original map and doesn't change the integral. This is an $s^2 q'$-to-one holomorophic map so the corresponding integral becomes $s^2 q'$. If instead of integrating over S^3 we integrate over $L(n, q)$, we must divide the integral by n so that:

$$\int_{L(n,q)} Tch_2(\nabla_s \oplus \nabla_{-s}, \nabla_0 \oplus \nabla_0) = \frac{s^2 q'}{n}.$$

Let P be the tangential operator of the signature complex. By Lemma 4.3.2, we have:

$$\int_{S^2} ch(\Pi_+(P)) = -4.$$

S^3 is parallelizable. The orientation of $S(T^*S^3)$ is $dx_1 \wedge dx_2 \wedge d\xi_2 \wedge dx_3 \wedge d\xi_3 = -dx_1 \wedge dx_2 \wedge dx_3 \wedge d\xi_2 \wedge d\xi_3$ so $S(T^*S^3) = -S^3 \times S^2$ given the usual orientation. Thus

$$(-1)^3 \int_{S(T^*S^3)} Tch_2(\nabla_s \oplus \nabla_{-s}, \nabla_0 \oplus \nabla_0) \wedge ch(\Pi_+ P)$$
$$= (-1)(-1)(-1) \cdot 4 \cdot s^2 q'/n = -4s^2 q'/n.$$

Consequently by Theorem 4.4.6, we conclude, since $Todd(S^3) = 1$,

$$\text{ind}(\rho_s + \rho_{-s}, \rho_0 + \rho_0, P) = -4s^2 q'/n$$

using the given framing.

The operator P splits into an operator on even and odd forms with equal eta invariants and a corresponding calculation shows that $\text{ind}(\rho_s + \rho_{-s}, \rho_0 + \rho_0, P_{\text{even}}) = -2s^2 q'/n$. There is an orientation preserving isometry $T: L(p, q) \to L(p, q)$ defined by $T(z_1, z_2) = (\bar{z}_1, \bar{z}_2)$. It is clear T interchanges the roles of ρ_s and ρ_{-s} so that as $\mathbf{R} \mod \mathbf{Z}$-valued invariants, $\text{ind}(\rho_s, P_{\text{even}}) = \text{ind}(\rho_{-s}, P_{\text{even}})$ so that:

$$\text{ind}(\rho_s, P) = 2\,\text{ind}(\rho_s, P_{\text{even}}) = \text{ind}(\rho_s, P_{\text{even}}) + \text{ind}(\rho_{-s}, P_{\text{even}})$$
$$= -2s^2 q'/n.$$

This gives a formula in $\mathbf{R} \mod \mathbf{Z}$ for the index of a representation which need not be topologically trivial. We will return to this formula to discuss generalizations in Lemma 4.6.3 when discussing $\text{ind}(*, *)$ for general spherical space forms in section 4.6.

We sketch the proof of Theorem 4.4.6. We shall omit many of the details in the interests of brevity. We refer to the papers of Atiyah, Patodi, and Singer for complete details. An elementary proof is contained in (Gilkey, The eta invariant and the secondary characteristic classes of locally flat bundles). Define:

$$\mathrm{ind}_1(\rho, P, s) = (-1)^m \int_{S(T^*M)} Todd(M) \wedge ch(\Pi_+p) \wedge Tch(\nabla_\rho, \nabla_0)$$

then Lemmas 4.3.6 and 4.3.7 generalize immediately to:

LEMMA 4.4.8.

(a) Let M_1 and M_2 be smooth manifolds. Let $P: C^\infty(V_1) \to C^\infty(V_2)$ be an elliptic complex over M_1. Let $Q: C^\infty(V) \to C^\infty(V)$ be a self-adjoint elliptic operator over M_2. We assume P and Q are differential operators of the same order and form $R = \begin{pmatrix} Q & P^* \\ P & -Q \end{pmatrix}$ over $M = M_1 \times M_2$. Let ρ be a representation of $\pi_1(M_2)$. Decompose $\pi_1(M) = \pi_1(M_1) \oplus \pi_1(M_2)$ and extend ρ to act trivially on $\pi_1(M_1)$. Then we can identify V_ρ over M with the pull-back of V_ρ over M_2. Let s be a global trivialization of V_ρ then:

$$\mathrm{ind}(\rho, R, \vec{s}) = \mathrm{index}(P)\,\mathrm{ind}(\rho, Q, \vec{s})$$
$$\mathrm{ind}_1(\rho, R, \vec{s}) = \mathrm{index}(P)\,\mathrm{ind}_1(\rho, Q, \vec{s}).$$

(b) Let $M_1 = S^1$ be the circle and let (R, Q) be as defined in Lemma 4.3.7, then:

$$\mathrm{ind}(\rho, R, \vec{s}) = \mathrm{ind}(\rho, P, \vec{s})$$
$$\mathrm{ind}_1(\rho, R, \vec{s}) = \mathrm{ind}_1(\rho, P, \vec{s}).$$

PROOF: The assertions about ind follow directly from Lemmas 4.3.6 and 4.3.7. The assertions about ind_1 follow from Lemma 3.9.3(d) and from the Atiyah-Singer index theorem.

Lemma 4.4.8(c) lets us reduce the proof of Theorem 4.4.6 to the case $\dim M$ odd. Since both ind and ind_1 are given by local formulas, we may assume without loss of generality that M is also orientable. Using the same arguments as those given in subsection 4.3, we can interpret both ind and ind_1 as maps in K-theory once the representation ρ and the global frame $\vec{s}$ are fixed. Consequently, the same arguments as those given for the proof of Theorem 4.3.8 permit us to reduce the proof of Theorem 4.4.6 to the case in which $P = A_V$ is the operator discussed in Lemma 4.3.4.

$\mathrm{ind}(\rho, P, \vec{s})$ is given by a local formula. If we express everything with respect to the global frame $\vec{s}$, then $P_k = P \otimes 1$ and P_ρ is functorially expressible in terms of P and in terms of the connection 1-form $\omega = \nabla_\rho \vec{s}$;

$$\mathrm{ind}(\rho, P, \vec{s}) = \int_M a(x, G, \omega)\,\mathrm{dvol}(x).$$

The same arguments as those given in discussing the signature complex show $a(x, G, \omega)$ is homogeneous of order n in the jets of the metric and of the connection 1-form. The local invariant changes sign if the orientation is reversed and thus $a(x, G, \omega) \, \mathrm{dvol}(x)$ should be regarded as an m-form not as a measure. The additivity of η with respect to direct sums shows $a(x, G, \omega_1 \oplus \omega_2) = a(x, G, \omega_1) + a(x, G, \omega_2)$ when we take the direct sum of representations. Arguments similar to those given in the second chapter (and which are worked out elsewhere) prove:

LEMMA 4.4.9. Let $a(x, G, \omega)$ be an m-form valued invariant which is defined on Riemannian metrics and on 1-form valued tensors ω so that $d\omega - \omega \wedge \omega = 0$. Suppose a is homogeneous of order m in the jets of the metric and the tensor ω and suppose that $a(x, G, \omega_1 \oplus \omega_2) = a(x, G, \omega_1) + a(x, G, \omega_2)$. Then we can decompose:

$$a(x, G, \omega) = \sum_\nu f_\nu(G) \wedge Tch_\nu(\omega) = \sum_\nu f_\nu(G) \wedge c_\nu \, \mathrm{Tr}(\omega^{2\nu - 1})$$

where $f_\nu(G)$ are real characteristic forms of $T(M)$ of order $m + 1 - 2\nu$.

We use the same argument as that given in the proof of the Atiyah-Singer index theorem to show there must exist a local formula for $\mathrm{ind}(\rho, P, s)$ which has the form:

$$\mathrm{ind}(\rho, P, s) =$$

$$\int_{S(T^*M)} \sum_{4i+2j+2k=m+1} Td'_{i,m}(M) \wedge ch_j(\Pi_+ p) \wedge Tch_k(\nabla_\rho, \nabla_0).$$

When the existence of such a formula is coupled with the product formula given in Lemma 4.4.8(a) and with the Atiyah-Singer index theorem, we deduce that the formula must actually have the form:

$$\mathrm{ind}(\rho, P, s) =$$

$$\int_{S(T^*M)} \sum_{j+2k=m+1} \left(Todd(M) \wedge ch(\Pi_+ p) \right)_j \wedge c(k) \, Tch_k(\nabla_\rho, \nabla_0)$$

where $c(k)$ is some universal constant which remains to be determined.

If we take an Abelian representation, all the Tch_k vanish for $k > 1$. We already verified that the constant $c(1) = 1$ by checking the operator of example 4.4.2 on the circle. The fact that the other normalizing constants are also 1 follows from a detailed consideration of the asymptotics of the heat equation which arise; we refer to (Gilkey, The eta invariant and the secondary characteristic classes of locally flat bundles) for further details regarding this verification. Alternatively, the Atiyah-Patodi-Singer index theorem for manifolds with boundary can be used to check these normal-izing constants; we refer to (Atiyah, Patodi, Singer: Spectral asymmetry and Riemannian geometry I–III) for details.

The Atiyah-Singer index theorem is a formula on $K(\Sigma(T^*M))$. The Atiyah-Patodi-Singer twisted index theorem can be regarded as a formula on $K(S(T^*M)) = K^1(\Sigma(T^*M))$. These two formulas are to be regarded as suspensions of each other and are linked by Bott periodicity in a purely formal sense which we shall not make explicit.

If V_ρ is not topologically trivial, there is no local formula for $\text{ind}(\rho, P)$ in general. It is possible to calculate this using the Lefschetz fixed point formulas as we will discuss later.

We conclude this section by discussing the generalization of Theorem 4.4.6 to the case of manifolds with boundary. It is a fairly straightforward computation using the methods of section 3.9 to show:

LEMMA 4.4.9. *We adopt the notation of Theorem 4.4.6. Then for any $k \geq 0$,*

$$\text{ind}(\rho, P, \vec{s}) = (-1)^m \int_{\Sigma^{2k}(T^*M)} Todd(M) \wedge ch(\Pi_+(\Sigma^{2k}(p))) \wedge Tch(\nabla_\rho, \nabla_0).$$

Remark: This shows that we can stabilize by suspending as often as we please. The index can be computed as an integral over $\Sigma^{2k+1}(T^*M)$ while the twisted index is an integral over $\Sigma^{2k}(T^*M)$. These two formulas are at least formally speaking the suspensions of each other.

It turns out that the formula of Theorem 4.4.6 does *not* generalize directly to the case of manifolds with boundary, while the formula of Lemma 4.4.9 with $k = 1$ does generalize. Let M be a compact manifold with boundary dM. We choose a Riemannian metric on M which is prod-uct near dM. Let $P: C^\infty(V) \to C^\infty(V)$ be a first order partial differential operator with leading symbol p which is formally self-adjoint. We suppose $p(x, \xi)^2 = |\xi|^2 \cdot 1_V$ so that p is defined by Clifford matrices. If we decom-pose $p(x, \xi) = \sum_j p_j(x)\xi_j$ relative to a local orthonormal frame for $T^*(M)$, then the p_j are self-adjoint and satisfy the relations $p_j p_k + p_k p_j = 2\delta_{jk}$.

Near the boundary we decompose $T^*(M) = T^*(dM) \oplus 1$ into tangential and normal directions. We let $\xi = (\varsigma, z)$ for $\varsigma \in T^*(dM)$ reflect this decomposition. Decompose $p(x, \xi) = \sum_{1 \leq j \leq m-1} p_j(x)\xi_j + p_m z$. Let t be a real parameter and define:

$$\tau(x, \varsigma, t) = i \cdot p_m \left\{ \sum_{1 \leq j \leq m-1} p_j(x)\varsigma_j - it \right\}$$

as the endomorphism defined in Chapter 1. We suppose given a self-adjoint endomorphism q of V which anti-commutes with τ. Let B denote the orthogonal projection $\frac{1}{2}(1 + q)$ on the $+1$ eigenspace of q. (P, B) is a self-adjoint elliptic boundary value problem. The results proved for manifolds without boundary extend to this case to become:

THEOREM 4.4.10. *Let (P, B) be an elliptic first order boundary value problem. We assume P is self-adjoint and $\sigma_L(p)^2 = |\xi|^2 \cdot I_V$. We assume $B = \frac{1}{2}(1+q)$ where q anti-commutes with $\tau = i \cdot p_m \left(\sum_{1 \leq j \leq m-1} p_j \varsigma_j - it \right)$. Let $\{\lambda_\nu\}_{\nu=1}^\infty$ denote the spectrum of the operator P_B. Define:*

$$\eta(s, P, B) = \sum_\nu \text{sign}(\lambda_\nu)|\lambda_\nu|^{-s}.$$

(a) $\eta(s, P, B)$ is well defined and holomorphic for $\text{Re}(s) \gg 0$.
(b) $\eta(s, P, B)$ admits a meromorphic extension to $\mathbf{C}$ with isolated simple poles at $s = (n - m)/2$ for $n = 0, 1, 2, \ldots$. The residue of η at such a simple pole is given by integrating a local formula $a_n(x, P)$ over M and a local formula $a_{n-1}(x, P, B)$ over dM.
(c) The value $s = 0$ is a regular value.
(d) If (P_u, B_u) is a smooth 1-parameter family of such operators, then the derivative $\dfrac{d}{du}\eta(0, P_u, B_u)$ is given by a local formula.
Remark: This theorem holds in much greater generality than we are stating it; we restrict to operators and boundary conditions given by Clifford matrices to simplify the discussion. The reader should consult (Gilkey-Smith) for details on the general case.

Such boundary conditions always exist if M is orientable and m is even; they may not exist if m is odd. For m even and M orientable, we can take $q = p_1 \ldots p_{m-1}$. If m is odd, the obstruction is $\text{Tr}(p_1 \ldots p_m)$ as discussed earlier. This theorem permits us to define $\text{ind}(\rho, P, B)$ if ρ is a unitary representation of $\pi_1(M)$ just as in the case where dM is empty. In $\mathbf{R} \bmod \mathbf{Z}$, it is a homotopy invariant of (P, B). If the bundle V_ρ is topologically trivial, we can define

$$\text{ind}(\rho, P, B, \vec{s})$$

as a real-valued invariant which is given by a local formula integrated over M and dM.

The boundary condition can be used to define a homotopy of Σp to an elliptic symbol which doesn't depend upon the tangential fiber coordinates.

Homotopy 4.4.11: Let (P, B) be as in Theorem 4.4.10 with symbols given by Clifford matrices. Let u be an auxilary parameter and define:

$$\tau(x, \varsigma, t, u) = \cos\left(\tfrac{\pi}{2} \cdot u\right) \tau(x, \varsigma, t) + \sin\left(\tfrac{\pi}{2} \cdot u\right) q \qquad \text{for } u \in [-1, 0].$$

It is immediate that $\tau(x, \varsigma, t, 0) = \tau(x, \varsigma, t)$ and $\tau(x, \varsigma, t, -1) = -q$. As τ and q are self-adjoint and anti-commute, this is a homotopy through self-adjoint matrices with eigenvalues $\pm \left\{\cos^2\left(\tfrac{\pi}{2} \cdot u\right)(|\varsigma|^2 + t^2) + \sin^2\left(\tfrac{\pi}{2} \cdot u\right)\right\}^{1/2}$.

Therefore $\tau(x, \varsigma, t, u) - iz$ is a non-singular elliptic symbol for $(\varsigma, t, z) \neq (0, 0, 0)$. We have

$$\Sigma p(x, \varsigma, z, t) = p(x, \varsigma, z) - it = ip_m \cdot \{\tau(x, \varsigma, t) - iz\}$$

so we define the homotopy

$$\Sigma p(x, \varsigma, z, t, u) = ip_m \cdot \{\tau(x, \varsigma, t, u) - iz\} \qquad \text{for } u \in [-1, 0].$$

We identifiy a neighborhood of dM in M with $dM \times [0, \delta)$. We sew on a collared neighborhood $dM \times [-1, 0]$ to define $\widetilde{M}$. We use the homotopy just defined to define an elliptic symbol $(\Sigma p)_B$ on $\Sigma(T^* \widetilde{M})$ which does not depend upon the tangential fiber coordinates ς on the boundary $dM \times \{-1\}$. We use the collaring to construct a diffeomorphism of M and $\widetilde{M}$ to regard $(\Sigma p)_B$ on $\Sigma(T^* M)$ where this elliptic symbol is independent of the tangential fiber coordinates on the boundary. We call this construction Homotopy 4.4.11.

We can now state the generalization of Theorem 4.4.6 to manifolds with boundary:

THEOREM 4.4.12. *We adopt the notation of Theorem 4.4.10 and let (P, B) be an elliptic self-adjoint first order boundary value problem with the symbols given by Clifford matrices. Let ρ be a representation of $\pi_1(M)$. Suppose V_ρ is topologically trivial and let $\vec{s}$ be a global frame. Then*

$$\text{ind}(\rho, P, B, \vec{s}) =$$

$$(-1)^m \int_{\Sigma^{2k}(T^* M)} Todd(M) \wedge ch(\Pi_+(\Sigma^{2k-1}((\Sigma p)_B))) \wedge Tch(\nabla_\rho, \nabla_0),$$

for any $k \geq 1$. Here $(\Sigma p)_B$ is the symbol on $\Sigma(T^* M)$ defined by Homotopy 4.4.11 so that it is an elliptic symbol independent of the tangential fiber variables ς on the boundary.

Remark: This theorem is in fact true in much greater generality. It is true under the much weaker assumption that P is a first order formally self-adjoint elliptic differential operator and that (P, B) is self-adjoint and strongly elliptic in the sense discussed in Chapter 1. The relevant homotopy is more complicated to discuss and we refer to (Gilkey-Smith) for both details on this generalization and also for the proof of this theorem.

We conclude by stating the Atiyah-Bott formula in this framework. Let $P: C^\infty(V_1) \to C^\infty(V_2)$ be a first order elliptic operator and let B be an elliptic boundary value problem. The boundary value problem gives a homotopy of Σp through *self-adjoint* elliptic symbols to a symbol independent of the tangential fiber variables. We call the new symbol $\{\Sigma p\}_B$. The Atiyah-Bott formula is

$$\text{index}(P, B) = (-1)^m \int_{\Sigma(T^* M)} Todd(M) \wedge ch(\Pi_+((\Sigma p)_B)).$$

4.5. Lefschetz Fixed Point Formulas.

In section 1.8, we discussed the Lefschetz fixed point formulas using heat equation methods. In this section, we shall derive the classical Lefschetz fixed point formulas for a non-degenerate smooth map with isolated fixed points for the four classical elliptic complexes. We shall also discuss the case of higher dimensional fixed point sets for the de Rham complex. The corresponding analysis for the signature and spin complexes is much more difficult and is beyond the scope of this book, and we refer to (Gilkey, Lefschetz fixed point formulas and the heat equation) for further details. We will conclude by discussing the theorem of Donnelly relating Lefschetz fixed point formulas to the eta invariant.

A number of authors have worked on proving these formulas. Kotake in 1969 discussed the case of isolated fixed points. In 1975, Lee extended the results of Seeley to yield the results of section 1.8 giving a heat equation approach to the Lefschetz fixed point formulas in general. We also derived these results independently not being aware of Lee's work. Donnelly in 1976 derived some of the results concerning the existence of the asymptotic expansion if the map concerned was an isometry. In 1978 he extended his results to manifolds with boundary.

During the period 1970 to 1976, Patodi had been working on generalizing his results concerning the index theorem to Lefschetz fixed point formulas, but his illness and untimely death in December 1976 prevented him from publishing the details of his work on the G-signature theorem. Donnelly completed Patodi's work and joint papers by Patodi and Donnelly contain these results. In 1976, Kawasaki gave a proof of the G-signature theorem in his thesis on V-manifolds. We also derived all the results of this section independently at the same time. In a sense, the Lefschetz fixed point formulas should have been derived by heat equation methods at the same time as the index theorem was proved by heat equation methods in 1972 and it remains a historical accident that this was not done. The problem was long over-due for solution and it is not surprising that it was solved simultaneously by a number of people.

We first assume $T: M \to M$ is an isometry.

LEMMA 4.5.1. *If T is an isometry, then the fixed point set of T consists of the disjoint union of a finite number of totally geodesic submanifolds $N_1, \ldots$. If N is one component of the fixed point set, the normal bundle ν is the orthogonal complement of $T(N)$ in $T(M)|_N$. ν is invariant under dT and $\det(I - dT_\nu) > 0$ so T is non-degenerate.*

PROOF: Let $a > 0$ be the injectivity radius of M so that if $\text{dist}(x, y) < a$, then there exists a unique shortest geodesic γ joining x to y in M. If $T(x) = x$ and $T(y) = y$, then $T\gamma$ is another shortest geodesic joining x to y so $T\gamma = \gamma$ is fixed pointwise and γ is contained in the fixed point set. This shows the fixed point set is totally geodesic. Fix $T(x) = x$ and decompose $T(M)_x = V_1 \oplus \nu$ where $V_1 = \{ v \in T(M)_x : dT(x)v = v \}$. Since $dT(x)$ is orthogonal, both V_1 and ν are invariant subspaces and dT_ν is an orthogonal matrix with no eigenvalue 1. Therefore $\det(I - dT_\nu) > 0$. Let γ be a geodesic starting at x so $\gamma'(0) \in V_1$. Then $T\gamma$ is a geodesic starting at x with $T\gamma'(0) = dT(x)\gamma'(0) = \gamma'(0)$ so $T\gamma = \gamma$ pointwise and $\exp_x(V_1)$ parametrizes the fixed point set near x. This completes the proof.

We let M be oriented and of even dimension $m = 2n$. Let $(d+\delta) \colon C^\infty(\Lambda^+)$ $\to C^\infty(\Lambda^-)$ be the operator of the signature complex. We let $H^\pm(M; \mathbf{C}) = N(\Delta_\pm)$ on $C^\infty(\Lambda^\pm)$ so that signature$(M) = \dim H^+ - \dim H^-$. If T is an orientation preserving isometry, then $T^*d = dT^*$ and $T^* * = *T^*$ where "$*$" denotes the Hodge operator. Therefore T^* induces maps $T^\pm$ on $H^\pm(M; \mathbf{C})$. We define:

$$L(T)_{\text{signature}} = \mathrm{Tr}(T^+) - \mathrm{Tr}(T^-).$$

Let $A = dT \in SO(m)$ at an isolated fixed point. Define

$$\mathrm{defect}(A, \text{signature}) = \{\mathrm{Tr}(\Lambda^+(A)) - \mathrm{Tr}(\Lambda^-(A))\}/\det(I - A)$$

as the contribution from Lemma 1.8.3. We wish to calculate this characteristic polynomial. If $m = 2$, let $\{e_1, e_2\}$ be an oriented orthonormal basis for $T(M) = T^*(M)$ such that

$$Ae_1 = (\cos\theta)e_1 + (\sin\theta)e_2 \qquad \text{and} \qquad Ae_2 = (\cos\theta)e_2 - (\sin\theta)e_1.$$

The representation spaces are defined by:

$$\Lambda^+ = \mathrm{span}\{1 + ie_1e_2, e_1 + ie_2\} \qquad \text{and} \qquad \Lambda^- = \mathrm{span}\{1 - ie_1e_2, e_1 - ie_2\}$$

so that:

$$\mathrm{Tr}(\Lambda^+(A)) = 1 + e^{-i\theta}, \quad \mathrm{Tr}(\Lambda^-(A)) = 1 - e^{-i\theta}, \quad \det(I - A) = 2 - 2\cos\theta$$

and consequently:

$$\mathrm{defect}(A, \text{signature}) = (-2i\cos\theta)/(2 - 2\cos\theta) = -i\cot(\theta/2).$$

More generally let $m = 2n$ and decompose $A = dT$ into a product of mutually orthogonal and commuting rotations through angles θ_j corresponding to complex eigenvalues $\lambda_j = e^{i\theta_j}$ for $1 \leq j \leq n$. The multiplicative nature of the signature complex then yields the defect formula:

$$\mathrm{defect}(A, \text{signature}) = \prod_{j=1}^{n}\{-i\cot(\theta_j/2)\} = \prod_{j=1}^{n}\frac{\lambda_j + 1}{\lambda_j - 1}.$$

This is well defined since the condition that the fixed point be isolated is just $0 < \theta_j < 2\pi$ or equivalently $\lambda_j \neq 1$.

There are similar characteristic polynomials for the other classical elliptic complexes. For the de Rham complex, we noted in Chapter 1 the corresponding contribution to be $\pm 1 = \operatorname{sign}(\det(I - A))$. If M is spin, let $A: C^\infty(\Delta^+) \to C^\infty(\Delta^-)$ be the spin complex. If T is an isometry which can be lifted to a spin isometry, we can define $L(T)_{\text{spin}}$ to be the Lefschetz number of T relative to the spin complex. This lifts A from $\mathrm{SO}(m)$ to $\mathrm{SPIN}(m)$. We define:

$$\operatorname{defect}(A, \text{spin}) = \{\operatorname{Tr}(\Delta^+(A)) - \operatorname{Tr}(\Delta^-(A))\}/\det(I - A)$$

and use Lemma 3.2.5 to calculate if $m = 2$ that:

$$\operatorname{defect}(A, \text{spin}) = (e^{-i\theta/2} - e^{i\theta/2})/(2 - 2\cos\theta)$$

$$= -i\sin(\theta/2)/(1 - \cos\theta) = -\frac{i}{2}\operatorname{cosec}(\theta/2)$$

$$= \left\{\sqrt{\bar{\lambda}} - \sqrt{\lambda}\right\}/(2 - \lambda - \bar{\lambda}) = \sqrt{\lambda}/(\lambda - 1).$$

Using the multiplicative nature we get a similar product formula in general.

Finally, let M be a holomorphic manifold and let $T: M \to M$ be a holomorphic map. Then T and $\bar\partial$ commute. We let $L(T)_{\text{Dolbeault}} = \sum_q (-1)^q \operatorname{Tr}(T^*$ on $H^{0,q})$ be the Lefschetz number of the Dolbeault complex. Just because T has isolated fixed points does not imply that it is non-degenerate; the map $z \mapsto z + 1$ defines a map on the Riemann sphere S^2 which has a single isolated degenerate fixed point at ∞. We suppose $A = L(T) \in \mathrm{U}(\frac{m}{2})$ is in fact non-degenerate and define

$$\operatorname{defect}(A, \text{Dolbeault}) = \{\operatorname{Tr}(\Lambda^{0,\text{even}}(A)) - \operatorname{Tr}(\Lambda^{0,\text{odd}}(A))\}\det(I - A_{\text{real}}).$$

If $m = 2$, it is easy to calculate

$$\operatorname{defect}(A, \text{Dolbeault}) = (1 - e^{i\theta})/(2 - 2\cos\theta) = (1 - \lambda)/(2 - \lambda - \bar\lambda)$$

$$= \lambda/(\lambda - 1)$$

with a similar multiplicative formula for $m > 2$. We combine these result with Lemma 1.8.3 to derive the classical Lefschetz fixed point formulas:

THEOREM 4.5.2. *Let $T: M \to M$ be a non-degenerate smooth map with isolated fixed points at $F(T) = \{x_1, \ldots, x_r\}$. Then*
(a) $L(T)_{\text{de Rham}} = \sum_j \operatorname{sign}(\det(I - dT))(x_j)$.
(b) *Suppose T is an orientation preserving isometry. Define:*

$$\operatorname{defect}(A, \text{signature}) = \prod_j \{-i \cdot \cot\theta_j\} = \prod_j \frac{\lambda_j + 1}{\lambda_j - 1},$$

then $L(T)_{\text{signature}} = \sum_j \text{defect}(dT(x_j), \text{signature})$.

(c) Suppose T is an isometry preserving a spin structure. Define:

$$\text{defect}(A, \text{spin}) = \prod_j \left\{ -\frac{i}{2} \text{cosec}(\theta_j/2) \right\} = \prod_j \frac{\sqrt{\lambda_j}}{\lambda_j - 1},$$

then $L(T)_{\text{spin}} = \sum_j \text{defect}(dT(x_j), \text{spin})$.

(d) Suppose T is holomorphic. Define:

$$\text{defect}(A, \text{Dolbeault}) = \prod_j \frac{\lambda_j}{\lambda_j - 1},$$

then $L(T)_{\text{Dolbeault}} = \sum_j \text{defect}(dT(x_j), \text{Dolbeault})$.

We proved (a) in Chapter 1. (b)–(d) follow from the calculations we have just given together with Lemma 1.8.3. In defining the defect, the rotation angles of A are $\{\theta_j\}$ so $Ae_1 = (\cos\theta)e_1 + (\sin\theta)e_2$ and $Ae_2 = (-\sin\theta)e_1 + (\cos\theta)e_2$. The corresponding complex eigenvalues are $\{\lambda_j = e^{i\theta_j}\}$ where $1 \leq j \leq m/2$. The formula in (a) does not depend on the orientation. The formula in (b) depends on the orientation, but not on a unitary structure. The formula in (c) depends on the particular lift of A to spin. The formula in (d) depends on the unitary structure. The formulas (b)–(d) are all for even dimensional manifolds while (a) does not depend on the parity of the dimension. There is an elliptic complex called the PIN_c complex defined over non-orientable odd dimensional manifolds which also has an interesting Lefschetz number relative to an orientation reversing isometry.

Using Lemmas 1.8.1 and 1.8.2 it is possible to get a local formula for the Lefschetz number concentrated on the fixed pont set even if T has higher dimensional fixed point sets. A careful analysis of this situation leads to the G-signature theorem in full generality. We refer to the appropriate papers of (Gilkey, Kawasaki, and Donnelly) for details regarding the signature and spin complexes. We shall discuss the case of the de Rham complex in some detail.

The interesting thing about the Dolbeault complex is that Theorem 4.5.2 does *not* generalize to yield a corresponding formula in the case of higher dimensional fixed point sets in terms of characteristic classes. So far, it has not proven possible to identify the invariants of the heat equation with generalized cohomology classes in this case. The Atiyah-Singer index theorem in its full generality also does not yield such a formula. If one assumes that T is an isometry of a Kaehler metric, then the desired result follows by passing first to the SPIN_c complex as was done in Chapter 3. However, in the general case, no heat equation proof of a suitable generalization is yet known. We remark that Toledo and Tong do have a formula for $L(T)_{\text{Dolbeault}}$ in this case in terms of characteristic classes, but their method of proof is quite different.

Before proceeding to discuss the de Rham complex in some detail, we pause to give another example:

Example 4.5.3: Let T_2 be the 2-dimensional torus $S^1 \times S^1$ with usual periodic parameters $0 \le x, y \le 1$. Let $T(x, y) = (-y, x)$ be a rotation through $90°$. Then
(a) $L(T)_{\text{de Rham}} = 2$.
(b) $L(T)_{\text{signature}} = 2i$.
(c) $L(T)_{\text{spin}} = 2i/\sqrt{2}$.
(d) $L(T)_{\text{Dolbeault}} = 1 - i$.

PROOF: We use Theorem 4.5.2 and notice that there are two fixed points at $(0,0)$ and $(1/2, 1/2)$. We could also compute directly using the indicated action on the cohomology groups. This shows that although the signature of a manifold is always zero if $m \equiv 2$ (4), there do exist non-trivial $L(T)_{\text{signature}}$ in these dimensions. Of course, there are many other examples.

We now consider the Lefschetz fixed point formula for the de Rham complex if the fixed point set is higher dimensional. Let $T: M \to M$ be smooth and non-degenerate. We assume for the sake of simplicity for the moment that the fixed point set of T has only a single component N of dimension n. The general case will be derived by summing over the components of the fixed point set. Let ν be the sub-bundle of $T(M)|_N$ spanned by the generalized eigenvectors of dT corresponding to eigenvalues other than 1. We choose the metric on M so the decomposition $T(M)|_N = T(N) \oplus \nu$ is orthogonal. We further normalize the choice of metric by assuming that N is a totally goedesic submanifold.

Let $a_k(x, T, \Delta_p)$ denote the invariants of the heat equation discussed in Lemma 1.8.1 so that

$$\text{Tr}(T^* e^{-t\Delta_p}) \sim \sum_{k=0}^{\infty} t^{(k-n)/2} \int_N a_k(x, T, \Delta_p) \, \text{dvol}(x).$$

Let $a_k(x, T)_{\text{de Rham}} = \sum (-1)^p a_k(x, T, \Delta_p)$, then Lemma 1.8.2 implies:

$$L(T)_{\text{de Rham}} \sim \sum_{k=0}^{\infty} t^{(k-n)/2} \int_N a_k(x, T)_{\text{de Rham}} \, \text{dvol}(x).$$

Thus:

$$\int_N a_k(x, T)_{\text{de Rham}} \, \text{dvol}(x) = \begin{cases} 0 & \text{if } k \ne \dim N \\ L(T)_{\text{de Rham}} & \text{if } k = \dim N. \end{cases}$$

We choose coordinates $x = (x_1, \ldots, x_n)$ on N. We extend these co-ordinates to a system of coordinates $z = (x, y)$ for M near N where $y = (y_1, \ldots, y_{m-n})$. We adopt the notational convention:

indices $1 \leq i, j, k \leq m$ index a frame for $T(M)|_N$

$1 \leq a, b, c \leq n$ index a frame for $T(N)$

$n < u, v, w \leq m$ index a frame for ν.

If α is a multi-index, we deompose $\alpha = (\alpha_N, \alpha_\nu)$ into tangential and normal components.

We let $g_{ij/\alpha}$ denote the jets of the metric tensor. Let $T = (T_1, \ldots, T_m)$ denote the components of the map T relative to the coordinate system Z. Let $T_{i/\beta}$ denote the jets of the map T; the $T_{i/j}$ variables are tensorial and are just the components of the Jacobian. Define:

$$\mathrm{ord}(g_{ij/\alpha}) = |\alpha|, \qquad \mathrm{ord}(T_{i/\beta}) = |\beta| - 1, \qquad \deg_v(T_{i/\beta}) = \delta_{i,v} + \beta(v).$$

Let $\mathcal{T}$ be the polynomial algebra in the formal variables $\{g_{ij/\alpha}, T_{i/\beta}\}$ for $|\alpha| > 0$, $|\beta| > 1$ with coefficients $c(g_{ij}, T_{i/j})|\det(I - dT_\nu)|^{-1}$ where the $c(g_{ij}, T_{i/j})$ are smooth in the g_{ij} and $T_{i/j}$ variables. The results of section 1.8 imply $a_k(x, T)_{\text{de Rham}} \in \mathcal{T}$. If Z is a coordinate system, if G is a metric, and if T is the germ of a non-degenerate smooth map we can evaluate $p(Z, T, G)(x)$ for $p \in \mathcal{T}$ and $x \in N$. p is said to be invariant if $p(Z, T, G)(x) = p(Z', T, G)(x)$ for any two coordinate systems Z, Z' of this form. We let $\mathcal{T}_{m,n}$ be the sub-algebra of $\mathcal{T}$ of all invariant polynomials and we let $\mathcal{T}_{m,n,k}$ be the sub-space of all invariant polynomials which are homogeneous of degree k using the grading defined above. It is not difficult to show there is a direct sum decomposition $\mathcal{T}_{m,n} = \bigoplus_k \mathcal{T}_{m,n,k}$ as a graded algebra.

LEMMA 4.5.4. $a_k(x, T)_{\text{de Rham}} \in \mathcal{T}_{m,n,k}$.

PROOF: The polynomial dependence upon the jets involved together with the form of the coefficients follows from Lemma 1.8.1. Since $a_k(x, T)_{\text{de Rham}}$ does not depend upon the coordinate system chosen, it is invariant. We check the homogeneity using dimensional analysis as per usual. If we re-place the metric by a new metric $c^2 G$, then a_k becomes $c^{-k} a_k$. We replace the coordinate system Z by $Z' = cZ$ to replace $g_{ij/\alpha}$ by $c^{-|\alpha|} g_{ij/\alpha}$ and $T_{i/\beta}$ by $c^{-|\beta|+1} T_{i/\beta}$. This completes the proof. The only feature different from the analysis of section 2.4 is the transformation rule for the variables $T_{i/\beta}$.

The invariants a_k are multiplicative.

LEMMA 4.5.5. Let $T': M' \to M'$ be non-degenerate with fixed submanifold N'. Define $M = S^1 \times M'$, $N = S^1 \times N'$, and $T = I \times T'$. Then

$T: M \to M$ is non-degenerate with fixed submanifold N. We give M the product metric. Then $a_k(x, T)_{\text{de Rham}} = 0$.

PROOF: We may decompose

$$\Lambda^p(M) = \Lambda^0(S^1) \otimes \Lambda^p(M') \oplus \Lambda^1(S^1) \otimes \Lambda^{p-1}(M') \simeq \Lambda^p(M') \oplus \Lambda^{p-1}(M').$$

This decomposition is preserved by the map T. Under this decomposition, the Laplacian Δ_p^M splits into $\Delta_p' \oplus \Delta_p''$. The natural bundle isomorphism identifies $\Delta_p'' = \Delta_{p-1}'$. Since $a_k(x, T, \Delta_p^M) = a_k(x, T, \Delta_p') + a_k(x, T, \Delta_p'') = a_k(x, T, \Delta_p') + a_k(x, T, \Delta_{p-1}')$, the alternating sum defining $a_k(x, T)_{\text{de Rham}}$ yields zero in this example which completes the proof.

We normalize the coordinate system Z by requiring that $g_{ij}(x_0) = \delta_{ij}$. There are further normalizations we shall discuss shortly. There is a natural restriction map $r: \mathcal{T}_{m,n,k} \to \mathcal{T}_{m-1,n-1,k}$ defined algebraically by:

$$r(g_{ij/\alpha}) = \begin{cases} 0 & \text{if } \deg_1(g_{ij/\alpha}) \neq 0 \\ g_{ij/\alpha}' & \text{if } \deg_1(g_{ij/\alpha}) = 0 \end{cases}$$

$$r(T_{i/\beta}) = \begin{cases} 0 & \text{if } \deg_1(T_{i/\beta}) \neq 0 \\ T_{i/\beta}' & \text{if } \deg_1(T_{i/\beta}) = 0. \end{cases}$$

In this expression, we let g' and T' denote the renumbered indices to refer to a manifold of one lower dimension. (It is inconvenient to have used the last $m - n$ indices for the normal bundle at this point, but again this is the cannonical convention which we have chosen not to change). In particular we note that Lemma 4.5.5 implies $r(a_k(x, T)_{\text{de Rham}}) = 0$. Theorem 2.4.7 generalizes to this setting as:

LEMMA 4.5.6. Let $p \in \mathcal{T}_{m,n,k}$ be such that $r(p) = 0$.
(a) If k is odd or if $k < n$ then $p = 0$.
(b) If $k = n$ is even we let E_n be the Euler form of the metric on N. Then $a_n(x, T)_{\text{de Rham}} = |\det(I - dT_\nu)|^{-1} f(dT_\nu) E_n$ for some $GL(m - n)$ invariant smooth function $f(*)$.

We postpone the proof of this lemma for the moment to complete our discussion of the Lefschetz fixed point formula for the de Rham complex.

THEOREM 4.5.7. Let $T: M \to M$ be non-degenerate with fixed point set consisting of the disjoint union of the submanifolds $N_1, N_2, \ldots, N_r$. Let N denote one component of the fixed point set of dimension n and let $a_k(x, T)_{\text{de Rham}}$ be the invariant of the heat equation. Then:
(a) $a_k(x, T)_{\text{de Rham}} = 0$ for $k < n$ or if k is odd.
(b) If $k = n$ is even, then $a_n(x, T)_{\text{de Rham}} = \text{sign}(\det(I - dT_\nu)) E_n$.
(c) $L(T)_{\text{de Rham}} = \sum_\nu \text{sign}(\det(I - dT)) \chi(N_\nu)$ (Classical Lefschetz fixed point formula).

PROOF: (a) follows directly from Lemmas 4.5.4–4.5.7. We also conclude that if $k = n$ is even, then $a_n(x, T)_{\text{de Rham}} = h(dT_\nu) E_n$ for some invariant function h. We know $h(dT_\nu) = \text{sign}(\det(I - dT))$ if $n = m$ by Theorem 1.8.4. The multiplicative nature of the de Rham complex with respect to products establishes this formula in general which proves (b). Since

$$L(T)_{\text{de Rham}} = \sum_\mu \text{sign}(\det(I - dT_\nu)) \int_{N_\mu} E_{n_\mu} \, \text{dvol}(x_\mu)$$

by Theorem 1.8.2, (c) follows from the Gauss-Bonnet theorem already established. This completes the proof of Theorem 4.5.7.

Before beginning the proof of Lemma 4.5.6, we must further normalize the coordinates being considered. We have assumed that the metric chosen makes N a totally geodesic submanifold. (In fact, this assumption is inessential, and the theorem is true in greater generality. The proof, however, is more complicated in this case). This implies that we can normalize the coordinate system chosen so that $g_{ij/k}(x_0) = 0$ for $1 \leq i, j, k \leq m$. (If the submanifold is not totally geodesic, then the second fundamental form enters in exactly the same manner as it did for the Gauss-Bonnet theorem for manifolds with boundary).

By hypothesis, we decomposed $T(M)|_N = T(N) \oplus \nu$ and decomposed $dT = I \oplus dT_\nu$. This implies that the Jacobian matrix $T_{i/j}$ satisfies:

$$T_{a/b} = \delta_{a/b}, \qquad T_{a/u} = T_{u/a} = 0 \text{ along } N.$$

Consequently:

$$T_{a/\beta_N \beta_\nu} = 0 \text{ for } |\beta_\nu| < 2 \qquad \text{and} \qquad T_{u/\beta_N \beta_\nu} = 0 \text{ for } |\beta_\nu| = 0.$$

Consequently, for the non-zero $T_{i/\beta}$ variables:

$$\sum_{a \leq n} \deg_a(T_{i/\beta}) \leq \text{ord}(T_{i/\beta}) = |\beta| - 1.$$

Let p satisfy the hypothesis of Lemma 4.5.6. Let $p \neq 0$ and let A be a monomial of p. We decompose:

$$A = f(dT_\nu) \cdot A_1 \cdot A_2 \quad \text{where} \quad \begin{cases} A_1 = g_{i_1 j_1/\alpha_1} \cdots g_{i_r j_r/\alpha_r} & \text{for } |\alpha_\nu| \geq 2 \\ A_2 = A_{k_1/\beta_1} \cdots A_{k_s/\beta_s} & \text{for } |\beta_\nu| \geq 2. \end{cases}$$

Since $r(p) = 0$, $\deg_1 A \neq 0$. Since p is invariant under orientation reversing changes of coordinates on N, $\deg_1 A$ must be even. Since p is invariant under coordinate permutations on N, $\deg_a A \geq 2$ for $1 \leq a \leq n$. We now count indices:

$$2n \leq \sum_{1 \leq a \leq n} \deg_a(A) = \sum_{1 \leq a \leq n} \deg_a(A_1) + \sum_{1 \leq a \leq n} \deg_a(A_2)$$

$$\leq 2r + \text{ord}(A_1) + \text{ord}(A_2) \leq 2\,\text{ord}(A_1) + 2\,\text{ord}(A_2) = 2\,\text{ord}(A).$$

This shows $p = 0$ if $\text{ord}(A) < n$.

We now study the limiting case $\text{ord}(A) = n$. All these inequalities must have been equalities. This implies in particular that $\text{ord}(A_2) = 2\,\text{ord}(A_2)$ so $\text{ord}(A_2) = 0$ and A_2 does not appear. Furthermore, $4r = \text{ord}(A_1)$ implies A_1 is a polynomial in the 2-jets of the metric. Finally, $\sum_{1 \leq a \leq n} \deg_a(A_1) = 4r$ implies that A_1 only depends upon the $\{g_{ab/cd}\}$ variables. Since $p \neq 0$ implies $n = 2r$, we conclude $p = 0$ if $k = n$ is odd. This completes the proof of (a).

To prove (b), we know that p is a polynomial in the $\{g_{ab/cd}\}$ variables with coefficients which depend on the $\{T_{u/v}\} = dT_\nu$ variables. Exactly the same arguments as used in section 2.5 to prove Theorem 2.4.7 now show that p has the desired form. This completes the proof; the normal and tangential indices decouple completely.

The Lefschetz fixed point formulas have been generalized by (Donnelly, The eta invariant of G-spaces) to the case of manifolds with boundary. We briefly summarize his work. Let N be a compact Riemannian manifold with boundary M. We assume the metric is product near the boundary so a neighborhood of M in N has the form $M \times [0, \varepsilon)$. Let n be the normal parameter. Let G be a finite group acting on N by isometries; G must preserve M as a set. We suppose G has no fixed points on M and only isolated fixed points in the interior of N for $g \neq I$. Let $F(g)$ denote the set of fixed points for $g \neq I$. Let $\overline{M} = M/G$ be the resulting quotient manifold. It bounds a V-manifold in the sense of Kawasaki, but does not necessarily bound a smooth manifold as N/G need not be a manifold.

Let $P: C^\infty(V_1) \to C^\infty(V_2)$ be an elliptic first order differential complex over N. Near the boundary, we assume P has the form $P = p(dn)(\partial/\partial n + A)$ where A is a self-adjoint elliptic first order differential operator over M whose coefficients are independent of the normal parameter. We also assume the G action on N extends to an action on this elliptic complex. Then $gA = Ag$ for all $g \in G$. Decompose $L^2(V_1|_M) = \bigoplus_\lambda E(\lambda)$ into the finite dimensional eigenspaces of A. Then g induces a representation on each $E(\lambda)$ and we define:

$$\eta(s, A, g) = \sum_\lambda \text{sign}(\lambda) \cdot |\lambda|^{-s} \text{Tr}(g \text{ on } E(\lambda))$$

as the equivariant version of the eta invariant. This series converges absolutely for $\text{Re}(s) \gg 0$ and has a meromorphic extension to $\mathbf{C}$. It is easy to see using the methods previously developed that this extension is regular for all values of s since g has no fixed points on M for $g \neq I$. If $g = I$, this is just the eta invariant previously defined.

Let B be orthonormal projection on the non-negative spectrum of A. This defines a non-local elliptic boundary value problem for the operator

P. Since *g* commutes with the operator *A*, it commutes with the boundary conditions. Let $L(P, B, g)$ be the Lefschetz number of this problem. Theorems 4.3.11 and 1.8.3 generalize to this setting to become:

THEOREM 4.5.8 (DONNELLY). *Let N be a compact Riemannian manifold with boundary M. Let the metric on N be product near M. Let $P: C^\infty(V_1) \to C^\infty(V_2)$ be a first order elliptic differential complex over N. Assume near M that P has the form $P = p(dn)(\partial/\partial n + A)$ where A is a self-adjoint elliptic tangential operator on $C^\infty(V_1)$ over the boundary M. Let $L^2(V_1|_M) = \bigoplus_\lambda E(\lambda)$ be a spectral resolution of A and let B be orthonormal projection on the non-negative spectrum of A. (P, B) is an elliptic boundary value problem. Assume given an isometry $g: N \to N$ with isolated fixed points $x_1, \ldots, x_r$ in the interior of N. Assume given an action g on V_i so $gP = Pg$. Then $gA = Ag$ as well. Define:*

$$\eta(s, A, g) = \sum_\lambda \operatorname{sign}(\lambda)|\lambda|^{-s} \cdot \operatorname{Tr}(g \text{ on } E(\lambda)).$$

This converges for $\operatorname{Re}(s) \gg 0$ and extends to an entire function of s. Let $L(P, B, g)$ denote the Lefschetz number of g on this elliptic complex and let

$$\operatorname{defect}(P, g)(x_i) = \{(\operatorname{Tr}(g \text{ on } V_1) - \operatorname{Tr}(g \text{ on } V_2))/\det(I - dT)\}(x_i)$$

then:

$$L(P, B, g) = \left\{ \sum_i \operatorname{defect}(P, G)(x_i) \right\} - \frac{1}{2}\{\eta(0, A, g) + \operatorname{Tr}(g \text{ on } N(A))\}.$$

Remark: Donnelly's theorem holds in greater generality as one does not need to assume the fixed points are isolated and we refer to (Donnelly, The eta invariant of *G*-spaces) for details.

We use this theorem to compute the eta invariant on the quotient manifold $\overline{M} = M/G$. Equivariant eigensections for *A* over *M* correspond to the eigensections of $\bar{A}$ over $\overline{M}$. Then:

$$\tilde{\eta}(\bar{A}) = \frac{1}{2}\{\eta(0, \bar{A}) + \dim N(\bar{A})\}$$

$$= \frac{1}{|G|} \cdot \frac{1}{2} \cdot \sum_{g \in G} \{\eta(0, A, g) + \operatorname{Tr}(g \text{ on } N(A))\}$$

$$= \frac{1}{|G|} \sum_{g \in G} \{-L(P, B, g)\} + \frac{1}{|G|} \int_N (a_n(x, P^*P) - a_n(x, PP^*)) \, dx$$

$$+ \frac{1}{|G|} \sum_{\substack{g \in G \\ g \neq I}} \sum_{x \in F(g)} \operatorname{defect}(P, g)(x).$$

The first sum over the group gives the equivariant index of P with the given boundary condition. This is an integer and vanishes in $\mathbf{R}$ mod $\mathbf{Z}$. The second contribution arises from Theorem 4.3.11 for $g = I$. The final contribution arises from Theorem 4.5.8 for $g \neq I$ and we sum over the fixed points of g (which may be different for different group elements).

If we suppose $\dim M$ is even, then $\dim N$ is odd so $a_n(x, P^*P) - a_n(x, PP^*) = 0$. This gives a formula for $\eta(\bar{A})$ over $\overline{M}$ in terms of the fixed point data on N. By replacing V_i by $V_i \otimes 1^k$ and letting G act by a representation ρ of G in $\mathrm{U}(k)$, we obtain a formula for $\eta(\bar{A}_\rho)$.

If $\dim M$ is odd and $\dim N$ is even, the local interior formula for $\mathrm{index}(P, B)$ given by the heat equation need not vanish identically. If we twist by a virtual representation ρ, we alter the defect formulas by multiplying by $\mathrm{Tr}(\rho(g))$. The contribution from Theorem 4.3.11 is multiplied by $\mathrm{Tr}(\rho(1)) = \dim \rho$. Consequently this term disappears if ρ is a representation of virtual dimension 0. This proves:

THEOREM 4.5.9. *Let N be a compact Riemannian manifold with boundary M. Let the metric on N be product near M. Let $P: C^\infty(V_1) \to C^\infty(V_2)$ be a first order elliptic differential complex over N. Assume near M that P has the form $P = p(dn)(\partial/\partial + A)$ where A is a self-adjoint elliptic tangential operator on $C^\infty(V_1)$ over M. Let G be a finite group acting by isometries on N. Assume for $g \neq I$ that g has only isolated fixed points in the interior of N, and let $F(g)$ denote the fixed point set. Assume given an action on V_i so $gP = Pg$ and $gA = Ag$. Let $\overline{M} = M/G$ be the quotient manifold and $\bar{A}$ the induced self-adjoint elliptic operator on $C^\infty(\bar{V}_1 = V_1/G)$ over $\overline{M}$. Let $\rho \in R(G)$ be a virtual representation. If $\dim M$ is odd, we assume $\dim \rho = 0$. Then:*

$$\tilde{\eta}(\bar{A}_\rho) = \frac{1}{|G|} \sum_{\substack{g \in G \\ g \neq I}} \sum_{s \in F(g)} \mathrm{Tr}(\rho(g)) \, \mathrm{defect}(P, g)(x) \quad \mathrm{mod} \ \mathbf{Z}$$

for

$$\mathrm{defect}(P, g)(x) = \{(\mathrm{Tr}(g \text{ on } V_1) - \mathrm{Tr}(g \text{ on } V_2))/ \det(I - dg)\}(x).$$

We shall use this result in the next section and also in section 4.9 to discuss the eta invariant for sherical space forms.

4.6. The Eta Invariant and the K-Theory of Spherical Space Forms.

So far, we have used K-theory as a tool to prove theorems in analysis. K-theory and the Chern isomorphism have played an important role in our discussion of both the index and the twisted index theorems as well as in the regularity of eta at the origin. We now reverse the process and will use analysis as a tool to compute K-theory. We shall discuss the K-theory of spherical space forms using the eta invariant to detect the relevant torsion.

In section 4.5, Corollary 4.5.9, we discussed the equivariant computation of the eta invariant. We apply this to spherical space forms as follows. Let G be a finite group and let $\tau\colon G \to U(l)$ be a fixed point free representation. We suppose $\det(I - \tau(g)) \neq 0$ for $g \neq I$. Such a representation is necessarily faithful; the existence of such a representation places severe restrictions on the group G. In particular, all the Sylow subgroups for odd primes must be cyclic and the Sylow subgroup for the prime 2 is either cyclic or generalized quaternionic. These groups have all been classified by (Wolf, Spaces of Constant Curvature) and we refer to this work for further details on the subject.

$\tau(G)$ acts without fixed points on the unit sphere S^{2l-1} in $\mathbf{C}^l$. Let $M = M(\tau) = S^{2l-1}/\tau(G)$. We suppose $l > 1$ so, since S^{2l-1} is simply connected, τ induces an isomorphism between G and $\pi_1(M)$. M inherits a natural orientation and Riemannian metric. It also inherits a natural Cauchy-Riemann and SPIN_c structure. (M is not necessarily a spin manifold). The metric has constant positive sectional curvature. Such a manifold is called a spherical space form; all odd dimensional compact manifolds without boundary admitting metrics of constant positive sectional curvature arise in this way. The only even dimensional spherical space forms are the sphere S^{2l} and the projective space $\mathbf{R}P^{2l}$. We concentrate for the moment on the odd dimensional case; we will return to consider $\mathbf{R}P^{2l}$ later in this section.

We have the geometrical argument:

$$T(S^{2l-1}) \oplus 1 = T(\mathbf{R}^{2l})|_{S^{2l-1}} = S^{2l-1} \times \mathbf{R}^{2l} = S^{2l-1} \times \mathbf{C}^l$$

is the trivial complex bundle of dimension l. The defining representation τ is unitary and acts naturally on this bundle. If V_τ is the locally flat complex bundle over $M(\tau)$ defined by the representation of $\pi_1(M(\tau)) = G$, then this argument shows

$$(V_\tau)_{\mathrm{real}} = T(M(\tau)) \oplus 1;$$

this is, of course, the Cauchy-Riemann structure refered to previously. In particular V_τ admits a nowhere vanishing section so we can split $V_\tau = V_1 \oplus 1$ where V_1 is an orthogonal complement of the trivial bundle corresponding to the invariant normal section of $T(\mathbf{R}^{2l})|_{S^{2l-1}}$. Therefore

$$\sum_\nu (-1)^\nu \Lambda^\nu(V_\tau) = \sum_\nu (-1)^\nu \Lambda^\nu(V_1 \oplus 1) = \sum_\nu (-1)^\nu \{\Lambda^\nu(V_1) \oplus \Lambda^{\nu-1}(V_1)\} = 0$$

in $K(M)$. This bundle corresponds to the virtual representation $\alpha = \sum_\nu (-1)^\nu \Lambda^\nu(\tau) \in R_0(G)$. This proves:

LEMMA 4.6.1. *Let τ be a fixed point free representation of a finite group G in $U(l)$ and let $M(\tau) = S^{2l-1}/\tau(G)$. Let $\alpha = \sum_\nu (-1)^\nu \Lambda^\nu(\tau) \in R_0(G)$. Then:*

$$T(M(\tau)) \oplus 1 = (V_\tau)_{\text{real}} \qquad \text{and} \qquad V_\alpha = 0 \text{ in } \widetilde{K}(M(\tau)).$$

The sphere bounds a disk D^{2l} in C^l. The metric is not product near the boundary of course. By making a radial change of metric, we can put a metric on D^{2l} agreeing with the standard metric at the origin and with a product metric near the boundary so that the action of $O(2l)$ continues to be an action by isometries. The transition functions of V_τ are unitary so $T(M(\tau))$ inherits a natural SPIN_c structure. Let $* = $ signature or Dolbeault and let A_* be the tangential operator of the appropriate elliptic complex over the disk. $\tau(G)$ acts on D^{2l} and the action extends to an action on both the signature and Dolbeault complexes. There is a single fixed point at the origin of the disk. Let $\text{defect}(\tau(g), *)$ denote the appropriate term from the Lefschetz fixed point formulas. Let $\{\lambda_\nu\}$ denote the complex eigenvalues of $\tau(g)$, and let $\tau(g)_r$ denote the corresponding element of $SO(2l)$. It follows from section 4.5 that:

$$\text{defect}(\tau(g), \text{signature}) = \prod_\nu \frac{\lambda_\nu + 1}{\lambda_\nu - 1},$$

$$\text{defect}(\tau(g), \text{Dolbeault}) = \frac{\det(\tau(g))}{\det(\tau(g) - I)} = \prod_\nu \frac{\lambda_\nu}{\lambda_\nu - 1}.$$

We apply Corollary 4.5.9 to this situation to compute:

4.6.2. *Let $\tau: G \to U(l)$ be a fixed point free representation of a finite group. Let $M(\tau) = S^{2l-1}/\tau(G)$ be a spherical space form. Let $* = $ signature or Dolbeault and let A_* be the tangential operator of the appropriate elliptic complex. Let $\rho \in R_0(G)$ be a virtual representation of dimension 0. Then:*

$$\tilde{\eta}((A_*)_\rho) = \frac{1}{|G|} \sum_{\substack{g \in G \\ g \neq I}} \text{Tr}(\rho(g)) \, \text{defect}(\tau(g), *) \qquad \text{in } \mathbf{R} \text{ mod } \mathbf{Z}.$$

Remark: A priori, this identity is in $\mathbf{R}$ mod $\mathbf{Z}$. It is not difficult to show that this generalized Dedekind sum is always $\mathbf{Q}$ mod $\mathbf{Z}$ valued and that $|G|^l \tilde{\eta} \in \mathbf{Z}$ so one has good control on the denominators involved.

The perhaps somewhat surprising fact is that this invariant is polynomial. Suppose $G = \mathbf{Z}_n$ is cyclic. Let $x = (x_1, \ldots, x_l)$ be a collection of indeterminates. Let Td and L be the Todd and Hirzebruch polynomials discussed previously. We define:

$$Td_0(\vec{x}) = 1 \qquad\qquad L_0(\vec{x}) = 1$$

$$Td_1(\vec{x}) = \frac{1}{2}\sum x_j \qquad\qquad L_1(\vec{x}) = 0$$

$$Td_2(\vec{x}) = \frac{1}{12}\left\{\sum_{j<k} x_j x_k + \left(\sum_j x_j\right)^2\right\} \qquad L_2(\vec{x}) = \frac{1}{3}\sum x_i^2$$

where we have renumbered the L_k polynomials to be homogeneous of degree k. Let s be another parameter which represents the first Chern class of a line bundle. The integrands of the index formula are given by:

$$P_l(s;\vec{x};\text{signature}) = \sum_{j+k=l} s^j L_k(\vec{x}) 2^j / j!$$

$$P_l(s;\vec{x};\text{Dolbeault}) = \sum_{j+k=l} s^j \, Td_k(\vec{x})/j! \,.$$

Let $\mu(l)$ denote the least common denominator of these rational polynomials.

We identify $\mathbf{Z}_n$ with the group of nth roots of unity in $\mathbf{C}$. Let $\rho_s(\lambda) = \lambda^s$ for $0 \leq s < n$ parameterize the irreducible representations. If $\vec{q} = (q_1, \ldots, q_l)$ is a collection of integers coprime to n, let $\tau = \rho_{q_1} \oplus \cdots \oplus \rho_{q_l}$ so $\tau(\lambda) = \text{diag}(\lambda^{q_1}, \ldots, \lambda^{q_l})$. This is a fixed pont free representation; up to unitary equivalence any fixed point free representation has this form. Let $L(n;\vec{q}) = M(\tau) = S^{2l-1}/\tau(\mathbf{Z}_n)$ be the corresponding spherical space form; this is called a lens space.

LEMMA 4.6.3. Let $M = L(n;\vec{q})$ be a lens space of dimension $2l-1$. Let $e \in \mathbf{Z}$ satisfy $eq_1 \ldots q_l \equiv 1 \bmod n \cdot \mu(l)$. Let $* = $ signature or Dolbeault and let A_* be the tangential operator of the corresponding elliptic complex. Let $P_l(s;x;*)$ denote the corresponding rational polynomial as defined above. Then

$$\text{ind}(\rho_s - \rho_0, A_*) = -\frac{e}{n}\{P_l(s;n,\vec{q};*) - P_l(0;n,\vec{q};*)\} \bmod \mathbf{Z}.$$

Remark: If M admits a spin structure, there is a corresponding formula for the tangential operator of the spin complex. This illustrates the close relationship between the Lefschetz fixed point formulas, the Atiyah-Singer index theorem, and the eta invariant, as it ties together all these elements. If $l = 2$ so $M = L(n;1,q)$ then

$$\text{ind}(\rho_s - \rho_0, A_{\text{signature}}) \equiv -\frac{q'}{n}\cdot(2s)^2\cdot\frac{1}{2} \equiv \frac{-q'\cdot 2s^2}{n}$$

which is the formula obatined previously in section 4.4.

PROOF: We refer to (Gilkey, The eta invariant and the K-theory of odd dimensional spherical space forms) for the proof; it is a simple residue calculation using the results of Hirzebruch-Zagier.

This is a very computable invariant and can be calculated using a computer from either Lemma 4.6.2 or Lemma 4.6.3. Although Lemma 4.6.3 at first sight only applies to cyclic groups, it is not difficult to use the Brauer induction formula and some elementary results concerning these groups to obtain similar formulas for arbitrary finite groups admitting fixed point free representations.

Let M be a compact manifold without boundary with fundamental group G. If ρ is a unitary representation of G and if P is a self-adjoint elliptic differential operator, we have defined the invariant $\tilde{\eta}(P_\rho)$ as an $\mathbf{R}$ mod $\mathbf{Z}$ valued invariant. (In fact this invariant can be defined for arbitrary representations and for elliptic pseudo-differential operators with leading symbol having no purely imaginary eigenvalues on $S(T^*M)$ and most of what we will say will go over to this more general case. As we are only interested in finite groups it suffices to work in this more restricted category).

Let $R(G)$ be the group representation ring of unitary virtual representations of G and let $R_0(G)$ be the ideal of virtual representations of virtual dimension 0. The map $\rho \mapsto V_\rho$ defines a ring homomorphism from $R(G)$ to $K(M)$ and $R_0(G)$ to $\widetilde{K}(M)$. We shall denote the images by $K_{\text{flat}}(M)$ and $\widetilde{K}_{\text{flat}}(M)$; these are the rings generated by virtual bundles admitting locally flat structures, or equivalently by virtual bundles with constant transition functions. Let P be a self-adjoint and elliptic differential oprator. The map $\rho \mapsto \tilde{\eta}(A_\rho)$ is additive with respect to direct sums and extends to a map $R(G) \to \mathbf{R}$ mod $\mathbf{Z}$ as already noted. We let $\text{ind}(\rho, P)$ be the map form $R_0(G)$ to $\mathbf{R}$ mod $\mathbf{Z}$. This involves a slight change of notation from section 4.4; if ρ is a representation of G, then

$$\text{ind}(\rho - \dim(\rho) \cdot 1, P)$$

denotes the invariant previously defined by $\text{ind}(\rho, P)$. This invariant is constant under deformations of P within this class; the Atiyah-Patodi-Singer index theorem for manifolds with boundary (Theorem 4.3.11) implies it is also an equivariant cobordism invariant. We summarize its relevant properties:

LEMMA 4.6.4. *Let M be a compact smooth manifold without boundary. Let P be a self-adjoint elliptic differential operator over M and let $\rho \in R_0(\pi_1(M))$.*
(a) Let $P(a)$ be a smooth 1-parameter family of such operators, then $\text{ind}(\rho, P(a))$ is independent of a in $\mathbf{R}$ mod $\mathbf{Z}$. If G is finite, this is $\mathbf{Q}$ mod $\mathbf{Z}$ valued.

(b) Let P be a first order operator. Suppose there exists a compact manifold N with $dN = M$. Suppose there is an elliptic complex $Q: C^\infty(V_1) \to C^\infty(V_2)$ over N so P is the tangential part of Q. Suppose the virtual bundle V_ρ can be extended as a locally flat bundle over N. Then $\mathrm{ind}(\rho, P) = 0$.

PROOF: The first assertion of (a) follows from Lemma 4.4.1. The locally flat bundle V_ρ is rationally trivial so by multipying by a suitable integer we can actually assume V_ρ corresponds to a flat structure on the difference of trivial bundles. The index is therefore given by a local formula. If we lift to the universal cover, we multiply this local formula by $|G|$. On the universal cover, the index vanishes identically as $\pi_1 = 0$. Thus an integer multiple of the index is 0 in $\mathbf{R}/\mathbf{Z}$ so the index is in $\mathbf{Q}/\mathbf{Z}$ which proves (a). To prove (b) we take the operator Q with coefficients in V_ρ. The local formula of the heat equation is just multiplied by the scaling constant $\dim(\rho) = 0$ since V_ρ is locally flat over N. Therefore Theorem 4.3.11 yields the identity $\mathrm{index}(Q, B, \mathrm{coeff\ in\ } V_\rho) = 0 - \mathrm{ind}(\rho, P)$. As the index is always an integer, this proves (b). We will use (b) in section 4.9 to discuss isospectral manifolds which are not diffeomorphic.

Examle 4.4.2 shows the index is *not* an invariant in K-theory. We get K-theory invariants as follows:

LEMMA 4.6.5. *Let M be a compact manifold without boundary. Let P be an elliptic self-adjoint pseudo-differential operator. Let $\rho_\nu \in R_0(\pi_1(M))$ and define the associative symmetric bilinear form on $R_0 \otimes R_0$ by:*

$$\mathrm{ind}(\rho_1, \rho_2, P) = \mathrm{ind}(\rho_1 \otimes \rho_2, P).$$

This takes values in $\mathbf{Q}$ mod $\mathbf{Z}$ and extends to an associative symmetric bilinear form $\mathrm{ind}(*, *, P): \widetilde{K}_{\mathrm{flat}}(M) \otimes \widetilde{K}_{\mathrm{flat}}(M) \to \mathbf{Q}$ mod $\mathbf{Z}$. If we consider the dependence upon P, then we get a trilinear map

$$\mathrm{ind}: \widetilde{K}_{\mathrm{flat}}(M) \otimes \widetilde{K}_{\mathrm{flat}}(M) \otimes K(S(T^*M))/K(M) \to \mathbf{Q} \ \mathrm{mod}\ \mathbf{Z}.$$

PROOF: The interpretation of the dependence in P as a map in K-theory follows from 4.3.3 and is therefore omitted. Any virtual bundle admitting a locally flat structure has vanishing rational Chern character and must be a torsion class. Once we have proved the map extends to K-theory, it will follow it must be $\mathbf{Q}$ mod $\mathbf{Z}$ valued. We suppose given representations ρ_1, $\hat{\rho}_1$, ρ_2 and a bundle isomorphism $V_{\rho_1} = V_{\hat{\rho}_1}$. Let $j = \dim(\rho_1)$ and $k = \dim(\rho_2)$. If we can show

$$\mathrm{ind}((\rho_1 - j) \otimes (\rho_2 - k), P) = \mathrm{ind}((\hat{\rho}_1 - j) \otimes (\rho_2 - k), P)$$

then the form will extend to $\widetilde{K}_{\mathrm{flat}}(M) \otimes \widetilde{K}_{\mathrm{flat}}(M)$ and the lemma will be proved.

We calculate that:

$$\text{ind}((\rho_1 - j) \otimes (\rho_2 - k), P)$$
$$= \tilde{\eta}(P_{\rho_1 \otimes \rho_2}) + j \cdot k \cdot \tilde{\eta}(P) - j \cdot \tilde{\eta}(P_{\rho_2}) - k \cdot \tilde{\eta}(P_{\rho_1})$$
$$= \tilde{\eta}\{(P_{\rho_1})_{\rho_2}\} - k \cdot \tilde{\eta}(P_{\rho_1}) - j \cdot \{\tilde{\eta}(P_{\rho_2}) - k \cdot \tilde{\eta}(P)\}$$
$$= \text{ind}(\rho_2, P_{\rho_1}) - j \cdot \text{ind}(\rho_2, P).$$

By hypothesis the bundles defined by ρ_1 and $\hat{\rho}_1$ are isomorphic. Thus the two operators P_{ρ_1} and $P_{\hat{\rho}_1}$ are homotopic since they have the same leading symbol. Therefore $\text{ind}(\rho_2, P_{\rho_1}) = \text{ind}(\rho_2, P_{\hat{\rho}_1})$ which completes the proof. We remark that this bilinear form is also associative with respect to multiplication by $R(G)$ and $K_{\text{flat}}(M)$.

We will use this lemma to study the K-theory of spherical space forms.

LEMMA 4.6.6. *Let τ be a fixed point free representation of a finite group G. Define*

$$\text{ind}_\tau(\rho_1, \rho_2) = \frac{1}{|G|} \sum_{\substack{g \in G \\ g \neq I}} \text{Tr}(\rho_1 \otimes \rho_2)(g) \frac{\det(\tau(g))}{\det(\tau(g) - I)}.$$

Let $\alpha = \sum_\nu (-1)^\nu \Lambda^\nu(\tau) \in R_0(G)$. Let $\rho_1 \in R_0(G)$ and suppose $\text{ind}_\tau(\rho_1, \rho_2) = 0$ in $\mathbf{Q}$ mod $\mathbf{Z}$ for all $\rho_2 \in R_0(G)$. Then $\rho_2 \in \alpha R(G)$.

PROOF: The virtual representation α is given by the defining relation that $\text{Tr}(\alpha(g)) = \det(I - \tau(g))$. $\det(\tau)$ defines a 1-dimensional representation of G; as this is an invertible element of $R(G)$ we see that the hypothesis implies

$$\frac{1}{|G|} \sum_{\substack{g \in G \\ g \neq I}} \text{Tr}(\rho_1 \otimes \rho_2) / \det(I - \tau(g)) \in \mathbf{Z} \qquad \text{for all } \rho_2 \in R_0(G).$$

If f and $\tilde{f}$ are any two class functions on G, we define the symmetric inner product $(f, \tilde{f}) = \frac{1}{|G|} \sum_{g \in G} f(g) \tilde{f}(g)$. The orthogonality relations show that f is a virtual character if and only if $(f, \text{Tr}(\rho)) \in \mathbf{Z}$ for all $\rho \in R(\mathbf{Z})$. We define:

$$f(g) = \text{Tr}(\rho_1(g)) / \det(I - \tau(g)) \qquad \text{for } g \neq I$$
$$f(g) = - \sum_{\substack{h \in G \\ h \neq I}} \text{Tr}(\rho_1(h)) / \det(I - \tau(h)) \qquad \text{if } g = I.$$

Then $(f, 1) = 0$ by definition. As $(f, \rho_2) \in \mathbf{Z}$ by hypothesis for $\rho_2 \in R_0(G)$ we see $(f, \rho_2) \in \mathbf{Z}$ for all $\rho_2 \in R(G)$ so f is a virtual character. We let $\text{Tr}(\rho)(g) = f(g)$. The defining equation implies:

$$\text{Tr}(\rho \otimes \alpha)(g) = \text{Tr}(\rho_1(g)) \qquad \text{for all } g \in G.$$

This implies $\rho_1 = \rho \otimes \alpha$ and completes the proof.

We can now compute the K-theory of odd dimensional spherical space forms.

THEOREM 4.6.7. *Let* $\tau: G \to \mathrm{U}(l)$ *be a fixed point free representation of a finite group* G. *Let* $M(\tau) = S^{2l-1}/\tau(G)$ *be a spherical space form. Suppose* $l > 1$. *Let* $\alpha = \sum_\nu (-1)^\nu \Lambda^\nu(\tau) \in R_0(G)$. *Then* $\widetilde{K}_{\mathrm{flat}}(M) = R_0(G)/\alpha R(G)$ *and*

$$\mathrm{ind}(*, *, A_{\mathrm{Dolbeault}}): \widetilde{K}_{\mathrm{flat}}(M) \otimes \widetilde{K}_{\mathrm{flat}}(M) \to \mathbf{Q} \bmod \mathbf{Z}$$

is a non-singular bilinear form.

Remark: It is a well known topological fact that for such spaces $\widetilde{K} = \widetilde{K}_{\mathrm{flat}}$ so we are actually computing the reduced K-theory of odd dimensional spherical space forms (which was first computed by Atiyah). This particular proof gives much more information than just the isomorphism and we will draw some corollaries of the proof.

PROOF: We have a surjective map $R_0(G) \to \widetilde{K}_{\mathrm{flat}}(M)$. By Lemma 4.6.1 we have $\alpha \mapsto 0$ so $\alpha R(G)$ is in the kernel of the natural map. Conversely, suppose $V_\rho = 0$ in $\widetilde{K}$. By Lemma 4.6.5 we have $\mathrm{ind}(\rho, \rho_1, A_{\mathrm{Dolbeault}}) = 0$ for all $\rho_1 \in R_0(G)$. Lemma 4.6.2 lets us identify this invariant with $\mathrm{ind}_\tau(\rho, \rho_1)$. Lemma 4.6.6 lets us conclude $\rho \in \alpha R(G)$. This shows the kernel of this map is precisely $\alpha R(G)$ which gives the desired isomorphism. Furthermore, $\rho \in \ker(\mathrm{ind}_\tau(\rho, *))$ if and only if $\rho \in \alpha R(G)$ if and only if $V_\rho = 0$ so the bilinear form is non-singular on $\widetilde{K}$.

It is possible to prove a number of other results about the K-theory ring using purely group theoretic methods; the existence of such a non-singular associative symmetric $\mathbf{Q} \bmod \mathbf{Z}$ form is an essential ingredient.

COROLLARY 4.6.8. *Adopt the notation of Theorem 4.6.7.*
(a) $\widetilde{K}(M)$ *only depends on* (G, l) *as a ring and not upon the particular* τ *chosen.*
(b) The index of nilpotency for this ring is at most l—*i.e., if* $\rho_\nu \in R_0(G)$ *then* $\prod_{1 \le \nu \le l} \rho_\nu \in \alpha R_0(G)$ *so the product of* l *virtual bundles of* $\widetilde{K}(M)$ *always gives* 0 *in* $\widetilde{K}(M)$.
(c) Let $V \in \widetilde{K}(M)$. *Then* $V = 0$ *if and only if* $\pi^*(V) = 0$ *for all possible covering projections* $\pi: L(n; \vec{q}) \to M$ *by lens spaces.*

There is, of course, a great deal known concerning these rings and we refer to (N. Mahammed, K-theorie des formes spheriques) for further details.

If $G = \mathbf{Z}_2$, then the resulting space is $\mathbf{R}P^{2l-1}$ which is projective space. There are two inequivalent unitary irreducible representations ρ_0, ρ_1 of G. Let $x = \rho_1 - \rho_0$ generate $R_0(\mathbf{Z}_2) = \mathbf{Z}$; the ring structure is given by

$x^2 = -2x$. Let A be the tangential operator of the Dolbeault complex:

$$\text{ind}(x, A) = \frac{1}{2} \text{Tr}(x(-1)) \cdot \det(-I_l)/\det((-I)_l - I_l)$$
$$= -2^{-l}$$

using Lemma 4.6.2. Therefore $\text{ind}(x, x, A) = 2^{-l+1}$ which implies that $\tilde{K}(\mathbf{R}P^{2l-1}) = \mathbf{Z}/2^{l-1}\mathbf{Z}$. Let $L = V_{\rho_1}$ be the tautological bundle over $\mathbf{R}P^{2l-1}$. It is $S^{2l-1} \times \mathbf{C}$ modulo the relation $(x, z) = (-x, -z)$. Using Clifford matrices we can construct a map $e: S^{2l-1} \to U(2^{l-1})$ such that $e(-x) = -e(x)$. This gives an equivariant trivialization of $S^{2l-1} \times \mathbf{C}^{2^{l-1}}$ which descends to give a trivialization of $2^{l-1} \cdot L$. This shows explicitly that $2^{l-1}(L - 1) = 0$ in $K(\mathbf{R}P^{2l-1})$; the eta invariant is used to show no lower power suffices. This proves:

COROLLARY 4.6.9. *Let* $M = \mathbf{R}P^{2l-1}$ *then* $\tilde{K}(M) \simeq \mathbf{Z}/2^{l-1}\mathbf{Z}$ *where the ring structure is* $x^2 = -2x$.

Let $l = 2$ and let $G = \mathbf{Z}_n$ be cyclic. Lemma 4.6.3 shows

$$\text{ind}(\rho_s - \rho_0, A_{\text{Dolbeault}}) = \frac{-q'}{n} \cdot \frac{s^2}{2}.$$

Let $x = \rho_1 - \rho_0$ so $x^2 = \rho_2 - \rho_0 - 2(\rho_1 - \rho_0)$ and

$$\text{ind}(x, x, A_{\text{Dolbeault}}) = \frac{-q'}{n} \cdot \frac{4 - 2}{2}$$

is a generator of $\mathbf{Z}[\frac{1}{n}]$ mod $\mathbf{Z}$. As $x^2 = 0$ in $\tilde{K}$ and as $x \cdot R(G) = R_0(G)$ we see:

COROLLARY 4.6.10. *Let* $M = L(n; 1, q)$ *be a lens space of dimension 3.* $\tilde{K}(M) = \mathbf{Z}_n$ *with trivial ring structure.*

We have computed the K-theory for the odd dimensional spherical space forms. $\tilde{K}(S^{2l}) = \mathbf{Z}$ and we gave a generator in terms of Clifford algebras in Chapter 3. To complete the discussion, it suffices to consider even dimensional real projective space $M = \mathbf{R}P^{2l}$. As $\tilde{H}^{\text{even}}(M; \mathbf{Q}) = 0$, $\tilde{K}$ is pure torsion by the Chern isomorphism. Again, it is known that the flat and regular K-theory coincide. Let $x = L - 1 = V_{\rho_1} - V_{\rho_0}$. This is the restriction of an element of $\tilde{K}(\mathbf{R}P^{2l+1})$ so $2^l x = 0$ by Corollary 4.6.9. It is immediate that $x^2 = -2x$. We show $\tilde{K}(\mathbf{R}P^{2l+1}) = \mathbf{Z}/2^l\mathbf{Z}$ by giving a surjective map to a group of order 2^l.

We construct an elliptic complex Q over the disk D^{2l+1}. Let $\{e_0, \ldots, e_{2l}\}$ be a collection of $2^l \times 2^l$ skew-adjoint matrices so $e_j e_k + e_k e_j = -2\delta_{jk}$. Up to unitary equivalence, the only invariant of such a collection is $\text{Tr}(e_0 \ldots e_{2l}) =$

$\pm (2i)^l$. There are two inequivalent collections; the other is obtained by taking $\{-e_0, \ldots, -e_{2l}\}$. Let Q be the operator $\sum_j e_j \partial/\partial x_j$ acting to map $Q: C^\infty(V_1) \to C^\infty(V_2)$ where V_i are trivial bundles of dimension 2^l over the disk. Let $g(x) = -x$ be the antipodal map. Let g act by $+1$ on V_1 and by -1 on V_2, then $gQ = Qg$ so this is an equivariant action. Let A be the tangential operator of this complex on S^{2l} and $\bar{A}$ the corresponding operator on $S^{2l}/\mathbf{Z}_2 = M$. $\bar{A}$ is a self-adjoint elliptic first order operator on $C^\infty(1^{2^l})$. If we replace Q by $-Q$, the tangential operator is unchanged so $\bar{A}$ is invariantly defined independent of the choice of the $\{e_j\}$. (In fact, $\bar{A}$ is the tangential operator of the PIN_c complex.)

The dimension is even so we can apply Corollary 4.5.9 to conclude

$$\tilde{\eta}(\bar{A}) = \frac{1}{2} \cdot \det(I_{2l+1} - (-I_{2l+1}))^{-1}\{2^l - (-2^l)\} = 2^{-l-1}.$$

If we interchange the roles of V_1 and V_2, we change the sign of the eta invariant. This is equivalent to taking coefficients in the bundle $L = V_{\rho_1}$. Let $x = L - 1$ then $\mathrm{ind}(\rho_1 - \rho_0, \bar{A}) = -2^{-l-1} - 2^{-l-1} = -2^{-l}$.

The even dimensional rational cohomology of $S(T^*M)$ is generated by H^0 and thus $K(S(T^*M))/K(M) \otimes \mathbf{Q} = 0$. Suppose $2^{l-1} \cdot x = 0$ in $\tilde{K}(M)$. Then there would exist a local formula for $2^{l-1}\mathrm{ind}(\rho_1 - \rho_0, *)$ so we could lift this invariant from $\mathbf{Q}$ mod $\mathbf{Z}$ to $\mathbf{Q}$. As this invariant is defined on the torsion group $K(S(T^*M))/K(M)$ it would have to vanish. As $2^{l-1}\mathrm{ind}(\rho_1 - \rho_0, \bar{A}) = -\frac{1}{2}$ does not vanish, we conclude $2^{l-1} \cdot x$ is non-zero in K-theory as desired. This proves:

COROLLARY 4.6.11. $\tilde{K}(\mathbf{R}P^{2l}) \simeq \mathbf{Z}/2^l\mathbf{Z}$. If $x = L - 1$ is the generator, then $x^2 = -2x$.

We can squeeze a bit more out of this construction to compute the K-theory of the unit sphere bundle $K(S(T^*M))$ where M is a spherical space form. First suppose $\dim M = 2l-1$ is odd so $M = S^{2l-1}/\tau(G)$ where τ is a unitary representation. Then M has a Cauchy-Riemann structure and we can decompose $T^*(M) = 1 \oplus V$ where V admits a complex structure; $T(M) \oplus 1 = (V_\tau)_{\mathrm{real}}$. Thus $S(T^*M)$ has a non-vanishing section and the exact sequence $0 \to K(M) \to K(S(T^*M)) \to K(S(T^*M))/K(M) \to 0$ splits. The usual clutching function construction permits us to identify $K(S(T^*M))/K(M)$ with $K(V)$. As V admits a complex structure, the Thom isomorphism identifies $K(V) = x \cdot K(M)$ where x is the Thom class. This gives the structure $K(S(T^*M)) = K(M) \oplus xK(M)$. The bundle x over $S(T^*M)$ can be taken to be $\Pi_+(p)$ where p is the symbol of the tangential operator of the Dolbeault complex. The index form can be regarded as a pairing $\tilde{K}(M) \otimes x \cdot \tilde{K}(M) \to \mathbf{Q}$ mod $\mathbf{Z}$ which is non-degenerate.

It is more difficult to analyse the even dimensional case. We wish to compute $K(S(T^*\mathbf{R}P^{2l}))$. $2^l \mathrm{ind}(\rho_1 - \rho_0, *)$ is given by a local formula since $2^l x = 0$. Thus this invariant must be zero as $K(S(T^*M))/K(M)$ is torsion. We have constructed an operator $\bar{A}$ so $\text{ind}(\rho_1 - \rho_0, \bar{A}) = -2^{-l}$ and thus $\text{ind}(\rho_1 - \rho_0, *): K(S(T^*M))/K(M) \to \mathbf{Z}[2^{-l}]/\mathbf{Z}$ is surjective with the given range. This is a cyclic group of order 2^l so equivalently

$$\text{ind}(\rho_1 - \rho_0, *): K(S(T^*M))/K(M) \to \mathbf{Z}/2^l\mathbf{Z} \to 0.$$

Victor Snaith (private communication) has shown us that the existence of such a sequence together with the Hirzebruch spectral sequence in K-theory shows

$$0 \to K(M) \to K(S(T^*M)) \text{ is exact and } |K(S(T^*M))/K(M)| = 2^l$$

so that $\text{index}(\rho_1 - \rho_0, *)$ becomes part of a short exact sequence:

$$0 \to K(M) \to K(S(T^*M)) \to \mathbf{Z}/2^l\mathbf{Z} \to 0.$$

To compute the structure of $K(S(T^*M)) = \mathbf{Z} \oplus \tilde{K}(S(T^*M))$, we must determine the group extension.

Let $x = L - 1 = V_{\rho_1} - V_{\rho_0}$ generate $\tilde{K}(M)$. Let $y = \Pi_+(\sigma_L\bar{A})) - 2^{l-1} \cdot 1$, then this exact sequence together with the computation $\text{ind}(\rho_0 - \rho_1, \bar{A}) = 2^{-l}$ shows $\tilde{K}(S(T^*M))$ is generated by x and y. We know $2^l x = 0$ and that $|\tilde{K}(S(T^*M))| = 4^l$; to determine the additive structure of the group, we must find the order of y. Consider the de Rham complex $(d+\delta): C^\infty(\Lambda^{\text{even}}(D)) \to C^\infty(\Lambda^{\text{odd}}(D))$ over the disk. The antipodal map acts by $+1$ on Λ^e and by -1 on Λ°. Let $\bar{A}_1$ be the tangential operator of this complex. We decompose $(d + \delta)$ into 2^l operators each of which is isomorphic to $\pm Q$. This indeterminacy does not affect the tangential operator and thus $\bar{A}_1 = 2^l\bar{A}$.

The symbol of $\bar{A}_1$ on $\Lambda^{\text{even}}(D)$ is $-ic(dn)c(\varsigma)$ for $\varsigma \in T^*(M)$. Let $\tau\{\theta_{\text{even}} + \theta_{\text{odd}}\} = \theta_{\text{even}} + c(dn)\theta_{\text{odd}}$ provide an isomorphism between $\Lambda(M)$ and $\Lambda^{\text{even}}(D|_M)$. We may regard $\bar{A}_1$ as an operator on $C^\infty(\Lambda(M))$ with symbol $\bar{a}_1$ given by:

$$\begin{aligned}
\bar{a}_1(x,\varsigma)(\theta_{\text{even}} + \theta_{\text{odd}}) &= \{\tau^{-1} \cdot -ic(dn)c(\varsigma) \cdot \tau\}\{\theta_{\text{even}} + \theta_{\text{odd}}\} \\
&= \{\tau^{-1} \cdot -ic(dn)c(\varsigma)\}\{\theta_{\text{even}} + c(dn)\theta_{\text{odd}}\} \\
&= \tau^{-1}\{-ic(dn)c(\varsigma)\theta_{\text{even}} - ic(\varsigma)\theta_{\text{odd}}\} \\
&= -ic(\varsigma)\{\theta_{\text{even}} + \theta_{\text{odd}}\}
\end{aligned}$$

so that $\bar{A}_1 = -(d + \delta)$ on $C^\infty(\Lambda(M))$. Let $\varepsilon(\theta_{\text{odd}}) = \theta_{\text{odd}} - ic(\varsigma)\theta_{\text{odd}}$ provide an isomorphism between $\Lambda^{\text{odd}}(M)$ and $\Pi_+(\bar{a}_1)$. This shows:

$$2^l \cdot y = \Lambda^{\text{odd}}(M) - 2^{2l-1} \cdot 1.$$

Let $\gamma(W) = W - \dim(W) \cdot 1$ be the natural projection of $K(M)$ on $\widetilde{K}(M)$. We wish to compute $\gamma(\Lambda^{\mathrm{odd}}(M))$. As complex vector bundles we have $T^*(M) \oplus 1 = (2l+1) \cdot L$ so that we have the relation $\Lambda^j(T^*M) \oplus \Lambda^{j-1}(T^*M) = \binom{2l+1}{j} \cdot L^j$. This yields the identities:

$$\gamma(\Lambda^j(T^*M)) + \gamma(\Lambda^{j-1}(T^*M)) = 0 \qquad \text{if } j \text{ is even}$$

$$\gamma(\Lambda^j(T^*M)) + \gamma(\Lambda^{j-1}(T^*M)) = \binom{2l+1}{j} \cdot \gamma(L) \qquad \text{if } j \text{ is odd.}$$

Thus $\gamma(\Lambda^j(T^*M)) = \gamma(\Lambda^{j-2}(T^*M)) + \binom{2l+1}{j}\gamma(L)$ if j is odd. This leads to the identity:

$$\gamma(\Lambda^{2j+1}(T^*M)) = \left\{ \binom{2l+1}{2j+1} + \binom{2l+1}{2j-1} + \cdots + \binom{2l+1}{1} \right\} \cdot \gamma(L)$$

$$\gamma(\Lambda^{\mathrm{odd}}(T^*M)) = \left\{ l\binom{2l+1}{1} + (l-1)\binom{2l+1}{3} + \cdots + \binom{2l+1}{2l-1} \right\} \cdot \gamma(L).$$

We complete this calculation by evaluating this coefficient. Let

$$f(t) = \frac{1}{2}\left((t+1)^{2l+1} - (t-1)^{2l+1} \right)$$

$$= t^{2l}\binom{2l+1}{1} + t^{2l-2}\binom{2l+1}{3} + \cdots + t^2\binom{2l+1}{2l+1} + 1,$$

$$f'(t) = \frac{1}{2}(2l+1)\left((t+1)^{2l} - (t-1)^{2l} \right)$$

$$= 2 \cdot \left\{ t^{2l-1} \cdot l \cdot \binom{2l+1}{1} + t^{2l-3} \cdot (l-1) \cdot \binom{2l+1}{3} + \cdots + t \cdot 1 \cdot \binom{2l+1}{2l-1} \right\}.$$

We evaluate at $t = 1$ to conclude:

$$l\binom{2l+1}{1} + (l-1)\binom{2l+1}{3} + \cdots + \binom{2l+1}{2l-1}$$
$$= \tfrac{1}{2}f'(1) = \tfrac{1}{4}(2l+1)2^{2l} = (2l+1)4^{l-1}.$$

Therefore:

$$2^l \cdot y = (2l+1) \cdot 4^{l-1}(L-1).$$

If $l = 1$, this gives the relation $2y = 3x = x$ so $\widetilde{K}(S(T^*M)) = \mathbf{Z}_4$. In fact, $S(T^*M) = S^3/\mathbf{Z}_4$ is a lens space so this calculation agrees with Corollary 4.6.10. If $l > 1$, then $2^l \mid 4^{l-1}$ so $2^l \cdot y = 0$. From this it follows $\widetilde{K}(S(T^*M)) = \mathbf{Z}/2^l\mathbf{Z} \oplus \mathbf{Z}/2^l\mathbf{Z}$ and the short exact sequence actually splits in this case. This proves:

THEOREM 4.6.12 (V. SNAITH). Let $X = S(T^*(\mathbf{R}P^{2l}))$ be the unit tangent bundle over even dimenional real projective space. If $l = 1$ then $K(X) = \mathbf{Z} \oplus \mathbf{Z}_4$. Otherwise $K(X) = \mathbf{Z} \oplus \mathbf{Z}/2^l\mathbf{Z} \oplus \mathbf{Z}/2^l\mathbf{Z}$. The map

ind$(*, *)$ *gives a perfect pairing* $K(\mathbf{R}P^{2l}) \otimes K(X)/K(\mathbf{R}P^{2l}) \to \mathbf{Q}$ mod $\mathbf{Z}$.
The generators of $K(X)$ *are* $\{1, x, y\}$ *for* $x = L - 1$ *and* $y = \gamma\Pi_+(\bar{a})$.

Remark: This gives the additive structure. We have $x^2 = -2x$. We can calculate $x \cdot y$ geometrically. Let $p(u) = i \sum v_j \cdot e_j$. We regard $p \colon 1^v \to L^v$ for $v = 2^l$ over $\mathbf{R}P_2l$. Let $\bar{a}_L = a \otimes I$ on L^v. Then $pa_Lp = \bar{a}_L$ so $\Pi_\pm(\bar{a}) \otimes L = \Pi_\mp(\bar{a})$. Therefore:

$$
\begin{aligned}
(L - 1) \otimes \left(\Pi_+(\bar{a}) - 1^{2^{l-1}}\right) &= -2^{l-1}(L - 1) + (L \otimes \Pi_+(\bar{a}) - \Pi_+(\bar{a})) \\
&= 2^{l-1}(L - 1) + \Pi_-(\bar{a}) - \Pi_+(\bar{a}) \\
&= 2^{l-1}(L - 1) + \Pi_+(\bar{a}) + \Pi_-(\bar{a}) - 2\Pi_+(\bar{a}) \\
&= 2^{l-1} \cdot x - 2y
\end{aligned}
$$

so that $x \cdot y = 2^{l-1} \cdot x - 2y$. This gives at least part of the ring structure; we do not know a similar simple geometric argument to compute $y \cdot y$.

4.7. Singer's Conjecture for the Euler Form.

In this section, we will study a partial converse to the Gauss-Bonnet theorem as well as other index theorems. This will lead to information regarding the higher order terms which appear in the heat equation. In a lecture at M.I.T., I. M. Singer proposed the following question:

Suppose that $P(G)$ is a scalar valued invariant of the metric such that $P(M) = \int_M P(G)\,\mathrm{dvol}$ is independent of the metric. Then is there some universal constant c so that $P(M) = c\chi(M)$?

Put another way, the Gauss-Bonnet theorem gives a local formula for a topological invariant (the Euler characteristic). Is this the only theorem of its kind? The answer to this question is yes and the result is due to E. Miller who settled the conjecture using topological means. We also settled the question in the affirmative using local geometry independently and we would like to present at least some the the ideas involved. If the invariant is allowed to depend upon the orientation of the manifold, then the characteristic numbers also enter as we shall see later.

We let $P(g_{ij/\alpha})$ be a polynomial invariant of the metric; real analytic or smooth invariants can be handled similarly. We suppose $P(M) = \int_M P(G)\,\mathrm{dvol}$ is independent of the particular metric chosen on M.

LEMMA 4.7.1. *Let P be a polynomial invariant of the metric tensor with coefficients which depend smoothly on the g_{ij} variables. Suppose $P(M)$ is independent of the metric chosen. We decompose $P = \sum P_n$ for $P_n \in \mathcal{P}_{m,n}$ homogeneous of order n in the metric. Then $P_n(M)$ is independent of the metric chosen separately for each n; $P_n(M) = 0$ for $n \neq m$.*

PROOF: This lets us reduce the questions involved to the homogeneous case. If we replace the metric G by $c^2 G$ then $P_n(c^2 G) = c^{-n} P_n(G)$ by Lemma 2.4.4. Therefore $\int_M P(c^2 G)\,\mathrm{dvol}(c^2 G) = \sum_n c^{m-n} \int_M P_n(G)$. Since this is independent of the constant c, $P_n(M) = 0$ for $n \leq m$ and $P_m(M) = P(M)$ which completes the proof.

If Q is 1-form valued, we let $P = \operatorname{div} Q$ be scalar valued. It is clear that $\int_M \operatorname{div} Q(G)\,\mathrm{dvol}(G) = 0$ so $P(M) = 0$ in this case. The following gives a partial converse:

LEMMA 4.7.2. *Let $P \in \mathcal{P}_{m,n}$ for $n \neq m$ satisfy $P(M) = \int_M P(G)\,\mathrm{dvol}$ is independent of the metric G. Then there exists $Q \in \mathcal{P}_{m,n-1,1}$ so that $P = \operatorname{div} Q$.*

PROOF: Since $n \neq m$, $P(M) = 0$. Let $f(x)$ be a real valued function on M and let G_t be the metric $e^{tf(x)} G$. If $n = 0$, then P is constant so $P = 0$ and the lemma is immediate. We assume $n > 0$ henceforth. We let

$P(t) = P(e^{tf(x)}G)$ and compute:

$$\frac{d}{dt}\left(P(e^{tf(x)}G)\operatorname{dvol}(e^{tf(x)}G)\right) = \frac{d}{dt}\left(P(e^{tf(x)}G)e^{tmf(x)/2}\right)\operatorname{dvol}(G)$$
$$= Q(G, f)\operatorname{dvol}(G).$$

$Q(f, G)$ is a certain expression which is linear in the derivatives of the scaling function f. We let $f_{;i_1...i_p}$ denote the multiple covariant derivatives of f and let $f_{:i_1...i_p}$ denote the symmetrized covariant derivatives of f. We can express

$$Q(f, G) = \sum Q_{i_1...i_p} f_{:i_1...i_p}$$

where the sum ranges over symmetric tensors of length less than n. We formally integrate by parts to express:

$$Q(f, G) = \operatorname{div} R(f, G) + \sum(-1)^p Q_{i_1...i_p:i_1...i_p} f.$$

By integrating over the structure group $O(m)$ we can ensure that this process is invariantly defined. If $S(G) = \sum(-1)^p Q_{i_1...i_p:i_1...i_p}$ then the identity:

$$0 = \int_M Q(f, G)\operatorname{dvol}(G) = \int_M S(G)f\operatorname{dvol}(G)$$

is true for every real valued function f. This implies $S(G) = 0$ so $Q(f, G) = \operatorname{div} R(f, G)$. We set $f = 1$. Since

$$e^{tm/2}P(e^tG) = e^{(m-n)t/2}P(G),$$

we conclude $Q(1, G) = \frac{m-n}{2}P(G)$ so $P(G) = \frac{2}{m-n}\operatorname{div} R(1, G)$ which completes the proof.

There is a corresponding lemma for form valued invariants. The proof is somewhat more complicated and we refer to (Gilkey, Smooth invariants of a Riemannian manifold) for further details:

LEMMA 4.7.3.
(a) Let $P \in \mathcal{P}_{m,n,p}$. We assume $n \neq p$ and $dP = 0$. If $p = m$, we assume $\int_M P(G)$ is independent of G for every G on M. Then there exists $Q \in \mathcal{P}_{m,n-1,p-1}$ so that $dQ = P$.
(b) Let $P \in \mathcal{P}_{m,n,p}$. We assume $n \neq m - p$ and $\delta P = 0$. If $p = 0$ we assume $\int_M P(G)\operatorname{dvol}$ is independent of G for every G in M. Then there exists $Q \in \mathcal{P}_{m,n-1,p+1}$ so that $\delta Q = P$.

Remark: (a) and (b) are in a sense dual if one works with $SO(m)$ invariance and not just $O(m)$ invariance. We can use this Lemma together with the results of Chapter 2 to answer the generalized Singer's conjecture:

THEOREM 4.7.4.

(a) Let P be a scalar valued invariant so that $P(M) = \int_M P(G)\,\mathrm{dvol}$ is independent of the particular metric chosen. Then we can decompose $P = c \cdot E_m + \mathrm{div}\,Q$ where E_m is the Euler form and where Q is a 1-form valued invariant. This implies $P(M) = c\chi(M)$.

(b) Let P be a p-form valued invariant so that $dP = 0$. If $p = m$, we assume $P(M) = \int_M P(G)$ is independent of the particular metric chosen. Then we can decompose $P = R + dQ$. Q is $p - 1$ form valued and R is a Pontrjagin form.

PROOF: We decompose $P = \sum P_j$ into terms which are homogeneous of order j. Then each P_j satisfies the hypothesis of Theorem 4.7.4 separately so we may assume without loss of generality that P is homogeneous of order n. Let P be as in (a). If $n \neq m$, then $P = \mathrm{div}\,Q$ be Lemma 4.7.2. If $n = m$, we let $P_1 = r(P) \in \mathcal{P}_{m-1,m}$. It is immediate $\int_{M_1} P_1(G_1)\,\mathrm{dvol}(G_1) = \frac{1}{2\pi}\int_{S^1 \times M_1} P(1 \times G_1)$ is independent of the metric G_1 so P_1 satisfies the hypothesis of (a) as well. Since $m - 1 \neq m$ we conclude $P_1 = \mathrm{div}\,Q_1$ for $Q_1 \in \mathcal{P}_{m-1,m-1,1}$. Since r is surjective, we can choose $Q \in \mathcal{P}_{m,m-1,1}$ so $r(Q) = Q_1$. Therefore $r(P - \mathrm{div}\,Q) = P_1 - \mathrm{div}\,Q_1 = 0$ so by Theorem 2.4.7, $P - \mathrm{div}\,Q = cE_m$ for some constant c which completes the proof. (The fact that $r : \mathcal{P}_{m,n,p} \to \mathcal{P}_{m-1,n,p} \to 0$ is, of course, a consequence of H. Weyl's theorem so we are using the full force of this theorem at this point for the first time).

If P is p-form valued, the proof is even easier. We decompose P into homogeneous terms and observe each term satisfies the hypothesis separately. If P is homogeneous of degree $n \neq p$ then $P = dQ$ be Lemma 4.7.3. If P is homogeneous of degree $n = p$, then P is a Pontrjagin form by Lemma 2.5.6 which completes the proof in this case.

The situation in the complex catagory is not as satisfactory.

THEOREM 4.7.5. Let M be a holomorphic manifold and let P be a scalar valued invariant of the metric. Assume $P(M) = \int_M P(G)\,\mathrm{dvol}$ is independent of the metric G. Then we can express $P = R + \mathrm{div}\,Q + \mathcal{E}$. Q is a 1-form valued invariant and $P = *R'$ where R' is a Chern form. The additional error term $\mathcal{E}$ satisfies the conditions: $r(\mathcal{E}) = 0$ and $\mathcal{E}$ vanishes for Kaehler metric. Therefore $P(M)$ is a characteristic number if M admits a Kaehler metric.

The additional error term arises because the axiomatic characterization of the Chern forms given in Chapter 3 were only valid for Kaehler metrics. $\mathcal{E}$ in general involves the torsion of the holomorphic connection and to show $\mathrm{div}\,Q' = \mathcal{E}$ for some Q' is an open problem. Using the work of E. Miller, it does follow that $\int \mathcal{E}\,\mathrm{dvol} = 0$ but the situation is not yet completely resolved.

We can use these results to obtain further information regarding the higher terms in the heat expansion:

THEOREM 4.7.6.
(a) Let $a_n(x, d + \delta)$ denote the invariants of the heat equation for the de Rham complex. Then

(i) $a_n(x, d + \delta) = 0$ if m or n is odd or if $n < m$,
(ii) $a_m(x, d + \delta) = E_m$ is the Euler form,
(iii) If m is even and if $n > m$, then $a_n(x, d + \delta) \not\equiv 0$ in general. However, there does exist a 1-form valued invariant $q_{m,n}$ so $a_n = \operatorname{div} q_{m,n}$ and $r(q_{m,n}) = 0$.

(b) Let $a_n^{\text{sign}}(x, V)$ be the invariants of the heat equation for the signature complex with coefficients in a bundle V. Then

(i) $a_n^{\text{sign}}(x, V) = 0$ for $n < m$ or n odd,
(ii) $a_m^{\text{sign}}(x, V)$ is the integrand of the Atiyah-Singer index theorem,
(iii) $a_n^{\text{sign}}(x, V) = 0$ for n even and $n < m$. However, there exists an $m - 1$ form valued invariant $q_{m,n}^{\text{sign}}(x, V)$ so that $a_n^{\text{sign}} = d(q_{m,n}^{\text{sign}})$.

A similar result holds for the invariants of the spin complex.

(c) Let $a_n^{\text{Dolbeault}}(x, V)$ denote the invariants of the heat equation for the Dolbeault complex with coefficients in V. We do not assume that the metric in question is Kaehler. Then:

(i) $a_n^{\text{Dolbeault}}(x, V) = \operatorname{div} Q_{m,n}$ where $Q_{m,n}$ is a 1-form valued invariant for $n \neq m$,
(ii) $a_m^{\text{Dolbeault}}(x, V) = \operatorname{div} Q_{m,n}+$ the integrand of the Riemann-Roch theorem.

The proof of all these results relies heavily on Lemma 4.7.3 and 4.7.2 and will be omitted. The results for the Dolbeault complex are somewhat more difficult to obtain and involve a use of the SPIN$_c$ complex.

We return to the study of the de Rham complex. These arguments are due to L. Willis. Let $m = 2n$ and let $a_{m+2}(x, d+\delta)$ be the next term above the Euler form. Then $a_{m+2} = \operatorname{div} Q_{m+1}$ where $Q_{m+1} \in P_{m,m+1,1}$ satisfies $r(Q_{m+1}) = 0$. We wish to compute a_{m+2}. The first step is:

LEMMA 4.7.7. Let m be even and let $Q \in P_{m,m+1,1}$ be 1-form valued. Suppose $r(Q) = 0$. Then Q is a linear combination of dE_m and Φ_m defined by:

$$\Phi_m = \sum_{k,\rho,\tau} \operatorname{sign}(\rho) \operatorname{sign}(\tau) \cdot \left\{ (-8\pi)^{m/2} \left(\frac{m}{2} - 1 \right)! \right\}^{-1}$$

$$\times R_{\rho(1)\rho(2)\tau(1)k;k} R_{\rho(3)\rho(4)\tau(3)\tau(4)} \cdots R_{\rho(m-1)\rho(m)\tau(m-1)\tau(m)} e^{\tau(2)} \in \Lambda^1$$

where $\{e^1, \ldots, e^m\}$ are a local orthonormal frame for $T^* M$.

PROOF: We let Ae^k be a monomial of P. We express A in the form:

$$A = g_{i_1 j_1 / \alpha_1} \cdots g_{i_r j_r / \alpha_r}.$$

By Lemma 2.5.5, we could choose a monomial B of P with $\deg_k(B) = 0$ for $k > 2\ell(A) = 2r$. Since $r(Q) = 0$, we conclude $2r \geq m$. On the other hand, $2r \leq \sum |\alpha_\nu| = m + 1$ so we see

$$m \leq 2r \leq m + 1.$$

Since m is even, we conclude $2r = m$. This implies one of the $|\alpha_\nu| = 3$ while all the other $|\alpha_\nu| = 2$. We choose the notation so $|\alpha_1| = 3$ and $|\alpha_\nu| = 2$ for $\nu > 1$. By Lemma 2.5.1, we may choose A in the form:

$$A = g_{ij/111} g_{i_2 j_2 / k_2 l_2} \cdots g_{i_r j_r / k_r l_r}.$$

We first suppose $\deg_1(A) = 3$. This implies A appears in the expression Ae^1 in P so $\deg_j(A) \geq 2$ is even for $j > 1$. We estimate $2m + 1 = 4r + 1 = \sum_j \deg_j(A) = 3 + \sum_{j>1} \deg_j(A) \geq 3 + 2(m - 1) = 2m + 1$ to show $\deg_j(A) = 2$ for $j > 1$. If $m = 2$, then $A = g_{22/111}$ which shows $\dim N(r) = 1$ and Q is a multiple of dE_2. We therefore assume $m > 2$. Since $\deg_j(A) = 2$ for $j > 1$, we can apply the arguments used to prove Theorem 2.4.7 to the indices $j > 1$ to show that

$$g_{22/111} g_{33/44} \cdots g_{m-1/mm} e^1$$

is a monomial of P.

Next we suppose $\deg_1(A) > 3$. If $\deg_1(A)$ is odd, then $\deg_j(A) \geq 2$ is even for $j \geq 2$ implies $2m+1 = 4r+1 = \sum_j \deg_j(A) \geq 5+2(m-1) = 2m+3$ which is false. Therefore $\deg_1(A)$ is even. We choose the notation in this case so Ae^2 appears in P and therefore $\deg_2(A)$ is odd and $\deg_j(A) \geq 2$ for $j > 2$. This implies $2m + 1 = 4r + 1 = \deg_j(A) \geq 4 + 1 + \sum_{j>2} \deg_j(A) \geq 5 + 2(m - 2) = 2m + 1$. Since all the inequalities must be equalities we conclude

$$\deg_1(A) = 4, \quad \deg_2(A) = 1, \quad \deg_3(A) = 2 \text{ for } j > 2.$$

We apply the arguments of the second chapter to choose A of this form so that every index $j > 2$ which does not touch either the index 1 or the index 2 touches itself. We choose A so the number of indices which touch themselves in A is maximal. Suppose the index 2 touches some other index than the index 1. If the index 2 touches the index 3, then the index 3 cannot touch itself in A. An argument using the fact $\deg_2(A) = 1$ and using the arguments of the second chapter shows this would contradict the maximality of A and thus the index 1 must touch the index 2 in A. We use non-linear changes of coordinates to show $g_{12/111}$ cannot divide A. Using non-linear changes of coordinates to raplace $g_{44/12}$ by $g_{12/44}$ if necessary, we conclude A has the form

$$A = g_{33/111}g_{12/44}g_{55/66}\cdots g_{m-1,m-1/mm}.$$

This shows that if $m \geq 4$, then $\dim N(r: P_{m,m+1,1} \to P_{m-1,m+1,1}) \leq 2$. Since these two invariants are linearly independent for $m \geq 4$ this completes the proof of the lemma.

We apply the lemma and decompose

$$a_{m+2}(x, d+\delta) = c_1(m)(E_m)_{;kk} + c_2(m)\operatorname{div}\Phi_m.$$

The $c_1(m)$ and $c_2(m)$ are universal constants defending only on the dimension m. We consider a product manifold of the form $M = S^2 \times M^{m-2}$ to define a map:

$$r_{(2)}: P_{m,n,0} \to \bigoplus_{q \leq n} P_{m-2,q,0}.$$

This restriction map does not preserve the grading since by throwing derivatives to the metric over S^2, we can lower the order of the invariant involved.

Let $x_1 \in S^2$ and $x_2 \in M^{m-2}$. The multiplicative nature of the de Rham complex implies:

$$a_n(x, d+\delta) = \sum_{j+k=m} a_j(x_1, d+\delta)_{S^2}\, a_k(x_2, d+\delta)_{M^{m-2}}.$$

However, $a_j(x_1, d+\delta)$ is a constant since S^2 is a homogeneous space. The relations $\int_{S^2} a_j(x_1, d+\delta) = 2\delta_{j,2}$ implies therefore

$$a_n(x, d+\delta) = \frac{1}{2\pi}a_{n-2}(x_2, d+\delta)$$

so that

$$r_{(2)}a_{m+2}^m = \frac{1}{2\pi}a_m^{m-2}.$$

Since $r_{(2)}(E_m)_{;kk} = \frac{1}{2\pi}(E_{m-2})_{;kk}$ and $r_{(2)}\operatorname{div}\Phi_m = \frac{1}{2\pi}\operatorname{div}\Phi_{m-2}$, we conclude that in fact the universal constants c_1 and c_2 do not depend upon m. (If $m = 2$, these two invariants are not linearly independent so we adjust $c_1(2) = c_1$ and $c_2(2) = c_2$). It is not difficult to use Theorem 4.8.16 which will be discussed in the next section to compute that if $m = 4$, then:

$$a_6(x, d+\delta) = \frac{1}{12}(E_m)_{;kk} - \frac{1}{6}\operatorname{div}\Phi_m.$$

We omit the details of the verification. This proves:

THEOREM 4.7.8. *Let m be even and let $a_{m+2}(x, d + \delta)$ denote the invariant of the Heat equation for the de Rham complex. Then:*

$$a_{m+2} = \frac{1}{12}(E_m)_{;kk} - \frac{1}{6}\operatorname{div}\Phi_m.$$

We gave a different proof of this result in (Gilkey, Curvature and the heat equation for the de Rham complex). This proof, due to Willis, is somewhat simpler in that it uses Singer's conjecture to simplify the invariance theory involved.

There are many other results concerning the invariants which appear in the heat equation for the de Rham complex. Gunther and Schimming have given various shuffle formulas which generalize the alternating sum defined previously. The combinatorial complexities are somewhat involved so we shall simply give an example of the formulas which can be derived.

THEOREM 4.7.9. *Let m be odd. Then*

$$\sum(-1)^p(m - p)a_n(x, \Delta_p^m) = \begin{cases} 0 & \text{if } n < m - 1 \\ E_{m-1} & \text{if } n = m - 1. \end{cases}$$

PROOF: We remark that there are similar formulas giving the various Killing curvatures E_k for $k < m$ in all dimensions. Since $r: P_{m,n} \to P_{m-1,n}$ for $n < m$ is an isomorphism, it suffices to prove this formula under restriction. Since $r(a_n(x, \Delta_p^m)) = a_n(x, \Delta_p^{m-1}) + a_n(x, \Delta_{p-1}^{m-1})$ we must study

$$\sum(-1)^p(m - p)(a_n(x, \Delta_p^{m-1}) + a_n(x, \Delta_{p-1}^{m-1})) = (-1)^p a_n(x, \Delta_p^{m-1})$$

and apply Theorem 2.4.8.

We remark that all the shuffle formulas of Gunther and Schimming (including Theorem 4.7.9) have natural analogues for the Dolbeault complex for a Kaehler metric and the proofs are essentially the same and rely on Theorem 3.6.9 and 3.4.10.

4.8. Local Formulas for the Invariants
Of the Heat Equation.

In this subsection, we will compute $a_n(x, \Delta_p^m)$ for $n = 0, 1, 2$. In principle, the combinatorial formulas from the first chapter could be used in this calculation. In practice, however, these formulas rapidly become much too complicated for practical use so we shall use instead some of the functorial properties of the invariants involved.

Let $P: C^\infty(V) \to C^\infty(V)$ be a second order elliptic operator with leading symbol given by the metric tensor. This means we can express:

$$P = -\left(\sum_{i,j} g^{ij} \partial^2/\partial x_i \partial x_j + \sum_k A_k \partial/\partial x_k + B\right)$$

where the A_k, B are endomorphisms of the bundle V and where the leading term is scalar. This category of differential operators includes all the Laplacians we have been considering previously. Our first task is to get a more invariant formulation for such an operator.

Let ∇ be a connection on V and let $E \in C^\infty(\text{END}(V))$ be an endomorphism of V. We use the Levi-Civita connection on M and the connection ∇ on V to extend ∇ to tensors of all orders. We define the differential operator P_∇ by the diagram:

$$P_\nabla: C^\infty(V) \xrightarrow{\nabla} C^\infty(T^*M \otimes V) \xrightarrow{\nabla} C^\infty(T^*M \otimes T^*M \otimes V) \xrightarrow{-g \otimes 1} C^\infty(V).$$

Relative to a local orthonormal frame for T^*M, we can express

$$P_\nabla(f) = -f_{;ii}$$

so this is the trace of second covariant differentiation. We define

$$P(\nabla, E) = P_\nabla - E$$

and our first result is:

LEMMA 4.8.1. *Let* $P: C^\infty(V) \to C^\infty(V)$ *be a second order operator with leading symbol given by the metric tensor. There exists a unique connection on V and a unique endomorphism so $P = P(\nabla, E)$.*

PROOF: We let indices i, j, k index a coordinate frame for $T(M)$. We shall not introduce indices to index a frame for V but shall always use matrix notation. The Christoffel symbols $\Gamma_{ij}{}^k = -\Gamma_i{}^k{}_j$ of the Levi-Civita connection are given by:

$$\Gamma_{ij}{}^k = \tfrac{1}{2} g^{kl}(g_{il/j} + g_{jl/i} - g_{ij/l})$$
$$\nabla_{\partial/\partial x_i}(\partial/\partial x_j) = \Gamma_{ij}{}^k(\partial/\partial x_k)$$
$$\nabla_{\partial/\partial x_i}(dx_j) = \Gamma_i{}^j{}_k(dx_k)$$

where we sum over repeated indices. We let $\vec{s}$ be a local frame for V and let ω_i be the components of the connection 1-form ω so that:

$$\nabla(f \cdot \vec{s}) = dx^i \otimes (\partial f / \partial x_i + \omega_i(f)) \cdot \vec{s}.$$

With this notational convention, the curvature is given by:

$$\Omega_{ij} = \omega_{j/i} - \omega_{i/j} + \omega_i \omega_j - \omega_j \omega_i \qquad (\Omega = d\omega + \omega \wedge \omega).$$

We now compute:

$$\nabla^2(f \cdot \vec{s}) = dx^j \otimes dx^i \otimes (\partial^2 f / \partial x_i \partial x_j + \omega_j \partial f / \partial x_i + \Gamma_j{}^k{}_i \partial f / \partial x_k$$
$$+ \omega_i \partial f / \partial x_j + \omega_{i/j} f + \omega_j \omega_i f + \Gamma_j{}^k{}_i \omega_k f) \cdot \vec{s}$$

from which it follows that:

$$P_\nabla(f \cdot \vec{s}) = - \{ g^{ij} \partial f / \partial x_i \partial x_j + (2g^{ij}\omega_i - g^{jk}\Gamma_{jk}{}^i) \partial f / \partial x_i$$
$$+ (g^{ij}\omega_{i/j} + g^{ij}\omega_i\omega_j - g^{ij}\Gamma_{ij}{}^k\omega_k) f \} \cdot \vec{s}$$

We use this identity to compute:

$$(P - P_\nabla)(f \cdot \vec{s}) = -\{(A_i - 2g^{ij}\omega_j + g^{jk}\Gamma_{jk}{}^i)\partial f / \partial x_i + (*)f\} \cdot \vec{s}$$

where we have omitted the 0th order terms. Therefore $(P - P_\nabla)$ is a 0th order operator if and only if

$$A_i - 2g^{ij}\omega_j + g^{jk}\Gamma_{jk}{}^i = 0 \quad \text{or equivalently} \quad \omega_i = \tfrac{1}{2}(g_{ij}A_j + g_{ij}g^{kl}\Gamma_{kl}{}^j).$$

This shows that the $\{\omega_i\}$ are uniquely determined by the condition that $(P - P_\nabla)$ is a 0th order operator and specificies the connection ∇ uniquely.

We define:

$$E = P_\nabla - P \quad \text{so} \quad E = B - g^{ij}\omega_{i/j} - g^{ij}\omega_i\omega_j + g^{ij}\omega_k\Gamma_{ij}{}^k.$$

We fix this connection ∇ and endomorphism E determined by P. We summarize these formulas as follows:

COROLLARY 4.8.2. *If (∇, E) are determined by the second order operator $P = -(g^{ij}\partial^2 f / \partial x_i \partial x_j + A_i \partial j / \partial x_i + Bf) \cdot \vec{s}$, then*

$$\omega_i = \tfrac{1}{2}(g_{ij}A_j + g_{ij}g^{kl}\Gamma_{kl}{}^j)$$
$$E = B - g^{ij}\omega_{i/j} - g^{ij}\omega_i\omega_j + g^{ij}\omega_k\Gamma_{ij}{}^k.$$

We digress briefly to express the Laplacian Δ_p in this form. If ∇ is the Levi-Civita connection acting on p-forms, it is clear that $\Delta_p - P_\nabla$ is a first order operator whose leading symbol is linear in the 1-jets of the metric. Since it is invariant, the leading symbol vanishes so $\Delta_p - P_\nabla$ is a 0th order and the Levi-Civita connection is in fact the connection determined by the operator Δ_p. We must now compute the curvature term. The operator $(d + \delta)$ is defined by Clifford multiplication:

$$(d + \delta) \colon C^\infty(\Lambda(T^*M)) \to C^\infty(T^*M \otimes \Lambda(T^*M))$$

$$\xrightarrow{\text{Clifford multiplication}} C^\infty(\Lambda(T^*M)).$$

If we expand $\theta = f_I \, dx_I$ and $\nabla\theta = dx_i \otimes (\partial f_I / \partial x_i + \Gamma_{iI}{}^J f_I) \, dx_J$ then it is immediate that if "$*$" denotes Clifford multiplication,

$$(d + \delta)(f_I \, dx_I) = (f_{I/i} \, dx_i * dx_I) + (f_I \Gamma_{iIJ} \, dx_i * dx_J)$$
$$\Delta(f_I dx_I) = (f_{I/ij} \, dx_j * dx_i * dx_I) + (f_I \Gamma_{iIJ/j} \, dx_j * dx_i * dx_J) + \cdots$$

where we have omitted terms involving the 1-jets of the metric. Similary, we compute:

$$P_\nabla(f_I \, dx_I) = -g^{ij} f_{I/ij} \, dx_I - g^{ij} \Gamma_{iIJ/j} f_I \, dx_J + \cdots.$$

We now fix a point x_0 of M and let X be a system of geodesic polar coordinates centered at x_0. Then $\Gamma_{iIJ/i} = 0$ and $\Gamma_{iIJ/j} = \frac{1}{2} R_{jiIJ}$ gives the curvature tensor at x_0. Using the identities $dx_i * dx_j + dx_j * dx_i = -2\delta_{ij}$ we see $f_{I/ij} \, dx_j * dx_i = -f_{I/ii}$ and consequently:

$$(P_\nabla - \Delta)(f_I \, dx_I) = -\tfrac{1}{2} R_{ijIJ} f_I \, dx_i * dx_j * dx_J = \sum_{i<j} R_{ijIJ} f_I \, dx_j * dx_i * dx_J.$$

This identity holds true at x_0 in geodesic polar coordinates. Since both sides are tensorial, it holds in general which proves:

LEMMA 4.8.3. *Let Δ_p be the Laplacian acting on p-forms and let R_{ijIJ} be the curvature tensor of the Levi-Civita connection. Then*

$$(P_\nabla - \Delta_p)(f_I \, dx_I) = \sum_{i<j} R_{ijIJ} f_I \, dx_j * dx_i * dx_J.$$

We now return to the problem of computing the invariants $a_n(x, P)$.

LEMMA 4.8.4. *Let P be an operator as in Lemma 4.8.1. Then $a_0(x, P) = (4\pi)^{-m/2} \dim V$.*

PROOF: We first consider the operator $P = -\partial^2 / \partial\theta^2$ on the unit circle $[0, 2\pi]$. The eigenvalues of P are $\{n^2\}_{n \in \mathbf{Z}}$. Since a_0 is homogeneous of order 0 in the jets of the symbol, $a_0(\theta, P) = a_0(P)/\operatorname{vol}(M)$ is constant. We compute:

$$\sum_n e^{-tn^2} \sim t^{-1/2} \int_M a_0(\theta, P)\, d\theta = t^{-1/2}(2\pi)a_0(\theta, P).$$

However, the Riemann sums approximating the integral show that

$$\sqrt{\pi} = \int_{-\infty}^{\infty} e^{-tx^2}\, dx = \lim_{t \to 0} \sum_n e^{-(\sqrt{t}\, n)^2} \sqrt{t}$$

from which it follows immediately that $a_0(\theta, P) = \sqrt{\pi}/(2\pi) = (4\pi)^{-1/2}$. More generally, by taking the direct sum of these operators acting on $C^\infty(S^1 \times \mathbf{C}^k)$ we conclude $a_0(\theta, P) = (4\pi)^{-1/2} \dim V$ which completes the proof if $m = 1$.

More generally, $a_0(x, P)$ is homogeneous of order 0 in the jets of the symbol so $a_0(x, P)$ is a constant which only depends on the dimension of the manifold and the dimension of the vector bundle. Using the additivity of Lemma 1.7.5, we conclude $a_0(x, P)$ must have the form:

$$a_0(x, P) = c(m) \cdot \dim V.$$

We now let $M = S^1 \times \cdots \times S^1$, the flat m-torus, and let $\Delta = -\sum_\nu \partial^2/\partial\theta_\nu^2$. The product formula of Lemma 1.7.5(b) implies that:

$$a_0(x, \Delta) = \prod_\nu a_0(\theta_\nu, -\partial^2/\partial\theta_\nu^2) = (4\pi)^{-m/2}$$

which completes the proof of the lemma.

The functorial properties of Lemma 1.7.5 were essential to the proof of Lemma 4.8.4. We will continue to exploit this functoriality in computing a_2 and a_4. (In principal one could also compute a_6 in this way, but the calculations become of formidable difficulty and will be omitted).

It is convenient to work with more tensorial objects than with the jets of the symbol of P. We let $\dim V = k$ and $\dim M = m$. We introduce formal variables $\{R_{i_1 i_2 i_3 i_4; \ldots}, \Omega_{i_1 i_2; \ldots}, E_{; \ldots}\}$ for the covariant derivatives of the curvature tensor of the Levi-Civita connection, of the curvature tensor of ∇, and of the covariant derivatives of the endomorphism E. We let S be the non-commutative algebra generated by these variables. Since there are relations among these variables, S isn't free. If $S \in \mathcal{S}$ and if e is a local orthonormal frame for $T^*(M)$, we define $S(P)(e)(x) = S(G, \nabla, E)(e)(x)(\operatorname{END}(V))$ by evaluation. We say S is invariant if $S(P) = S(P)(a)$ is independent of the orthonormal frame e chosen for $T(M)$. We define:

$$\operatorname{ord}(R_{i_1 i_2 i_4 i_4; i_5 \ldots i_k}) = k + 2$$
$$\operatorname{ord}(\Omega_{i_1 i_2; i_3 \ldots i_k}) = k$$
$$\operatorname{ord}(E_{; i_1 \ldots i_k}) = k + 2$$

to be the degree of homogeneity in the jets of the symbol of P, and let $S_{m,n,k}$ be the finite dimensional subspace of S consisting of the invariant polynomials which are homogeneous of order n.

If we apply H. Weyl's theorem to this situation and apply the symmetries involved, it is not difficult to show:

LEMMA 4.8.5.

(a) $S_{m,2,k}$ is spanned by the two polynomials $R_{ijij}I$, E.

(b) $S_{m,4,k}$ is spanned by the eight polynomials

$$R_{ijij;kk}I, \quad R_{ijij}R_{klkl}I, \quad R_{ijik}R_{ljlk}I, \quad R_{ijkl}R_{ijkl}I,$$
$$E^2, \qquad E_{;ii}, \qquad R_{ijij}E, \qquad \Omega_{ij}\Omega_{ij}.$$

We omit the proof in the interests of brevity; the corresponding spanning set for $S_{m,6,k}$ involves 46 polynomials.

The spaces $S_{m,n,k}$ are related to the invariants $a_n(x, P)$ of the heat equation as follows:

LEMMA 4.8.6. Let (m, n, k) be given and let P satisfy the hypothesis of Lemma 4.8.1. Then there exists $S_{m,n,k} \in S_{m,n,k}$ so that $a_n(x, P) = \mathrm{Tr}(S_{m,n,k})$.

PROOF: We fix a point $x_0 \in M$. We choose geodesic polar coordinates centred at x_0. In such a coordinate system, all the jets of the metric at x_0 can be computed in terms of the $R_{i_1i_2i_3i_4;\ldots}$ variables. We fix a frame s_0 for the fiber V_0 over x_0 and extend s_0 by parallel translation along all the geodesic rays from x_0 to get a frame near x_0. Then all the derivatives of the connection 1-form at x_0 can be expressed in terms of the $R_{\ldots}$ and $\Omega_{\ldots}$ variables at x_0. We can solve the relations of Corollary 4.8.2 to express the jets of the symbol of P in terms of the jets of the metric, the jets of the connection 1-form, and the jets of the endomorphism E. These jets can all be expressed in terms of the variables in S at x_0 so any invariant endomorphism which is homogeneous of order n belongs to $S_{m,n,k}$. In Lemma 1.7.5, we showed that $a_n(x, P) = \mathrm{Tr}(e_n(x, P))$ was the trace of an invariant endomorphism and this completes the proof; we set $S_{m,n,k} = e_n(x, P)$.

We use Lemmas 4.8.5 and 4.8.6 to expand $a_n(x, P)$. We regard scalar invariants of the metric as acting on V by scalar multiplication; alternatively, such an invariant R_{ijij} could be replaced by $R_{ijij}I_V$.

LEMMA 4.8.7. Let $m = \dim M$ and let $k = \dim V$. Then there exist universal constants $c_i(m, k)$ so that if P is as in Lemma 4.8.1,

(a) $a_2(x, P) = (4\pi)^{-m/2} \mathrm{Tr}(c_1(m, k)R_{ijij} + c_2(m, k)E)$,

(b)

$$a_4(x, P) = (4\pi)^{-m/2} \operatorname{Tr}(c_3(m, k) R_{ijij;kk} + c_4(m, k) R_{ijij} R_{klkl}$$
$$+ c_5(m, k) R_{ijik} R_{ljlk} + c_6(m, k) R_{ijkl} R_{ijkl}$$
$$+ c_7(m, k) E_{;ii} + c_8(m, k) E^2$$
$$+ c_9(m, k) E R_{ijij} + c_{10}(m, k) \Omega_{ij} \Omega_{ij}).$$

This is an important simplification because it shows in particular that terms such as $\operatorname{Tr}(E)^2$ do not appear in a_4.

The first observation we shall need is the following:

LEMMA 4.8.8. *The constants* $c_i(m, k)$ *of Lemma 4.8.7 can be chosen to be independent of the dimension* m *and the fiber dimension* k.

PROOF: The leading symbol of P is scalar. The analysis of Chapter 1 in this case immediately leads to a combinatorial formula for the coefficients in terms of certain trignometric integrals $\int \xi^\alpha e^{-|\xi|^2} d\xi$ and the fiber dimension does not enter. Alternatively, we could use the additivity of $e_n(x, P_1 \oplus P_2) = e_n(x, P_1) \oplus e_n(x, P_2)$ of Lemma 1.7.5 to conclude the formulas involved must be independent of the dimension k. We may therefore write $c_i(m, k) = c_i(m)$.

There is a natural restriction map $r: S_{m,n,k} \to S_{m-1,n,k}$ defined by restricting to operators of the form $P = P_1 \otimes 1 + I_V \otimes (-\partial^2/\partial\theta^2)$ over $M = M_1 \times S^1$. Algebraically, we simply set to zero any variables involving the last index. The multiplicative property of Lemma 1.7.5 implies

$$r(S_{m,n,k}) = \sum_{p+q=n} S_{m-1,p,k}(P_1) \otimes S_{1,q,1}(-\partial^2/\partial\theta^2).$$

Since all the jets of the symbol of $-\partial^2/\partial\theta^2$ vanish for $q > 0$, $S_{1,q,1} = 0$ for $q > 0$ and $a_0 = (4\pi)^{-1/2}$ by Lemma 4.8.4. Therefore

$$r(S_{m,n,k}) = (4\pi)^{-1/2} S_{m-1,n,k}.$$

Since we have included the normalizing constant $(4\pi)^{-m/2}$ in our definition, the constants are independent of the dimension m for $m \geq 4$. If $m = 1, 2, 3$, then the invariants of Lemma 4.8.6 are not linearly independent so we choose the constants to agree with $c_i(m, k)$ in these cases.

We remark that if P is a higher order operator with leading symbol given by a power of the metric tensor, then there a similar theory expressing a_n in terms of invariant tensorial expressions. However, in this case, the coefficients depend upon the dimension m in a much more fundamental way than simply $(4\pi)^{-m/2}$ and we refer to (Gilkey, the spectral geometry of the higher order Laplacian) for further details.

Since the coefficients do not depend on (m, k), we drop the somewhat cumbersome notation $S_{m,n,k}$ and return to the notation $e_n(x, P)$ discussed in the first chapter so $\operatorname{Tr}(e_n(x, P)) = a_n(x, P)$. We use the properties of the exponential function to compute:

LEMMA 4.8.9. *Using the notation of Lemma 4.8.7,*

$$c_2 = 1, \quad c_8 = \frac{1}{2}, \quad c_9 = c_1.$$

PROOF: Let P be as in Lemma 4.8.1 and let a be a real constant. We construct the operator $P_a = P - a$. The metric and connection are unchanged; we must replace E by $E + a$. Since $e^{-t(P-a)} = e^{-tP}e^{ta}$ we conclude:

$$e_n(x, P - a) = \sum_{p+q=n} e_p(x, P)a^q/q!$$

by comparing terms in the asymptotic expansion. We shall ignore factors of $(4\pi)^{-m/2}$ henceforth for notational convenience. Then:

$$e_2(x, P - a) = e_2(x, P) + c_2 a = e_2(x, P) + e_0(x, P)a = e_2(x, P) + a.$$

This implies that $c_2 = 1$ as claimed. Next, we have

$$\begin{aligned}
e_4(x, P - a) &= e_4(x, P) + c_8 a^2 + 2c_8 aE + c_9 aR_{ijij} \\
&= e_4(x, P) + e_2(x, P)a + e_0(x, P)a^2/2 \\
&= e_4(x, P) + (c_1 R_{ijij} + c_2 E)a + a^2/2
\end{aligned}$$

which implies $c_8 = 1/2$ and $c_1 = c_9$ as claimed.

We now use some recursion relations derived in (Gilkey, Recursion relations and the asymptotic behavior of the eigenvalues of the Laplacian). To illustrate these, we first suppose $m = 1$. We consider two operators:

$$A = \partial/\partial x + b, \quad A^* = -\partial/\partial x + b$$

where b is a real scalar function. This gives rise to operators:

$$\begin{aligned}
P_1 &= A^* A = -(\partial^2/\partial x^2 + (b' - b^2)) \\
P_2 &= AA^* = -(\partial^2/\partial x^2 + (-b' - b^2))
\end{aligned}$$

acting on $C^\infty(S^1)$. The metric and connection defined by these operators is flat. $E(P_1) = b' - b^2$ and $E(P_2) = -b' - b^2$.

LEMMA 4.8.10.

$$(n - 1)\big(e_n(x, P_1) - e_n(x, P_2)\big) = \partial/\partial x\{\partial/\partial x + 2b\}e_{n-2}(x, P_1).$$

PROOF: Let $\{\lambda_\nu, \theta_\nu\}$ be a complete spectral resolution of P_1. We ignore any possible zero spectrum since it won't contribute to the series we shall be constructing. Then $\{\lambda_\nu, A\theta_\nu/\sqrt{\lambda_\nu}\}$ is a complete spectral resolution of P_2. We compute:

$$\frac{d}{dt}\{K(t,x,x,P_1) - K(t,x,x,P_2)\}$$

$$= \sum e^{-t\lambda_\nu}\{-\lambda_\nu\theta_\nu\theta_\nu + A\theta_\nu A\theta_\nu\}$$

$$= \sum e^{-t\lambda_\nu}\{-P_1\theta_\nu \cdot \theta_\nu + A\theta_\nu A\theta_\nu\}$$

$$= \sum e^{-t\lambda_\nu}\{\theta_\nu''\theta_\nu + (b' - b^2)\theta_\nu^2 + \theta_\nu'\theta_\nu' + 2b\theta_\nu'\theta_\nu + b^2\theta_\nu^2\}$$

$$= \sum e^{-t\lambda_\nu}\left(\tfrac{1}{2}\partial/\partial x\right)(\partial/\partial x + 2b)\{\theta_\nu^2\}$$

$$= \tfrac{1}{2}\partial/\partial x(\partial/\partial x + 2b)K(t,x,x,P_1).$$

We equate terms in the asymptotic expansions

$$\sum t^{(n-3)/2}\frac{n-1}{2}(e_n(x,P_1) - e_n(x,P_2))$$

$$\sim \frac{1}{2}\sum t^{(n-1)/2}\partial/\partial x(\partial/\partial x + 2b)e_n(x,P_1)$$

to complete the proof of the lemma.

We apply this lemma to compute the coefficient c_7. If $n = 4$, then we conclude:

$$e_4(x,P_1) = c_7(b' - b^2)'' + c_8(b' - b^2)^2 = c_7b''' + \text{ lower order terms}$$
$$e_4(x,P_2) = c_7(-b' - b^2)'' + c_8(-b' - b^2)^2 = -c_7b''' + \text{ lower order terms}$$
$$e_2(x,P_1) = b' - b^2$$

so that:

$$3(e_4(x,P_1) - e_4(x,P_2)) = 6c_7b''' + \text{ lower order terms}$$
$$\partial/\partial x(\partial/\partial x + 2b)(b' - b^2) = b''' + \text{ lower order terms}$$

from which it follows that $c_7 = 1/6$. It is also convenient at this stage to obtain information about e_6. If we let $e_6 = cE'''' + $ lower order terms then we express:

$$e_6(x,P_1) - e_6(x,P_2) = 2cb^{(5)}$$
$$\partial/\partial x(\partial/\partial x + 2b)(e_4) = c_7b^{(5)}$$

from which it follows that the constant c is $(4\pi)^{-1/2} \cdot c_7/10 = (4\pi)^{-1/2}/60$. We summarize these results as follows:

LEMMA 4.8.11. *We can expand a_n in the form:*
(a) $a_2(x, P) = (4\pi)^{-m/2} \operatorname{Tr}(c_1 R_{ijij} + E).$
(b)

$$a_4(x, P) = (4\pi)^{-m/2} \operatorname{Tr}(c_3 R_{ijij;kk} + c_4 R_{ijij} R_{klkl} + c_5 R_{ijik} R_{ljlk}$$
$$+ c_6 R_{ijkl} R_{ijkl} + E_{;kk}/6 + E^2/2$$
$$+ c_1 R_{ijij} E + c_{10} \Omega_{ij} \Omega_{ij}).$$

(c)

$$a_6(x, P) = (4\pi)^{-m/2} \operatorname{Tr}(E_{;kkll}/60$$
$$+ c_{11} R_{ijij;kkll} + \text{lower order terms}).$$

PROOF: (a) and (b) follow immediately from the computations previously performed. To prove (c) we argue as in the proof of 4.8.7 to show $a_6(x, P) = (4\pi)^{-m/2} \operatorname{Tr}(cE_{;kk} + c_{11} R_{ijij;kk} + \text{lower order terms})$ and then use the evaluation of c given above.

We can use a similar recursion relation if $m = 2$ to obtain further information regarding these coefficients. We consider the de Rham complex, then:

LEMMA 4.8.12. *If $m = 2$ and Δ_p is the Laplacian on $C^\infty(\Lambda^p(T^*M))$, then:*

$$\frac{(n-2)}{2} \{a_n(x, \Delta_0) - a_n(x, \Delta_1) + a_n(x, \Delta_2)\} = a_{n-2}(x, \Delta_0)_{;kk}.$$

PROOF: This recursion relationship is due to McKean and Singer. Since the invariants are local, we may assume M is orientable and $a_n(x, \Delta_0) = a_n(x, \Delta_2)$. Let $\{\lambda_\nu, \theta_\nu\}$ be a spectral resolution for the non-zero spectrum of Δ_0 then $\{\lambda_\nu, *\theta_\nu\}$ is a spectral resolution for the non-zero spectrum of Δ_2 and $\{\lambda_\nu, d\theta_\nu/\sqrt{\lambda_\nu}, \delta * \theta_\nu/\sqrt{\lambda_\nu}\}$ is a spectral resolution for the non-zero spectrum of Δ_1. Therefore:

$$\frac{d}{dt} \left(K(t, x, x, \Delta_0) - K(t, x, x, \Delta_1) + K(t, x, x, \Delta_2) \right)$$
$$= \sum e^{-t\lambda_\nu} (-2\lambda_\nu \theta_\nu \theta_\nu + d\theta_\nu \cdot d\theta_\nu + \delta * \theta_\nu \cdot \delta * \theta_\nu)$$
$$= \sum e^{-t\lambda_\nu} (-2\Delta_0 \theta_\nu \cdot \theta_\nu + 2d\theta_\nu \cdot d\theta_\nu) = K(t, x, x, \Delta_0)_{;kk}$$

from which the desired identity follows.

Before using this identity, we must obtain some additional information about Δ_1.

LEMMA 4.8.13. *Let m be arbitrary and let $\rho_{ij} = -R_{ijik}$ be the Ricci tensor. Then $\Delta\theta = -\theta_{;kk} + \rho(\theta)$ for $\theta \in C^\infty(\Lambda^1)$ and where $\rho(\theta)_i = \rho_{ij}\theta_j$. Thus $E(\Delta_2) = -\rho$.*

PROOF: We apply Lemma 4.8.3 to conclude

$$E(\theta) = \frac{1}{2}R_{abij}\theta_i e_b * e_a * e_j$$

from which the desired result follows using the Bianchi identities.

We can now check at least some of these formulas. If $m = 2$, then $E(\Delta_1) = R_{1212}I$ and $E(\Delta_0) = E(\Delta_2) = 0$. Therefore:

$$a_2(x, \Delta_0) - a_2(x, \Delta_1) + a_2(x, \Delta_2) = (4\pi)^{-1}\{(1 - 2 + 1)c_1 R_{ijij} - 2R_{1212}\}$$
$$= -R_{1212}/2\pi$$

which is, in fact, the integrand of the Gauss-Bonnet theorem. Next, we compute, supressing the factor of $(4\pi)^{-1}$:

$$a_4(x, \Delta_0) - a_4(x, \Delta_1) + a_4(x, \Delta_2)$$
$$= (c_3 R_{ijij;kk} + c_4 R_{ijij}R_{klkl} + c_5 R_{ijik}R_{ljlk}$$
$$+ c_6 R_{ijkl}R_{ijkl})(1 - 2 + 1)$$
$$+ \frac{1}{6}(-2)(R_{1212;kk}) - \frac{2}{2}R_{1212}R_{1212}$$
$$- 4c_1 R_{1212}R_{1212} - c_{10}\,\mathrm{Tr}(\Omega_{ij}\Omega_{ij})$$
$$= -\frac{1}{3}(R_{1212;kk}) - (1 + 4c_1 - 4c_{10})(R_{1212})^2$$

$$a_2(x, \Delta_0)_{;kk} = 2c_1 R_{1212;kk}$$

so that Lemma 4.2.11 applied to the case $n = 4$ implies:

$$a_4(x, \Delta_0) - a_4(x, \Delta_1) + a_4(x, \Delta_2) = a_2(x, \Delta_0)_{;kk}$$

from which we derive the identities:

$$c_1 = -\frac{1}{6} \quad \text{and} \quad 1 + 4c_1 - 4c_{10} = 0$$

from which it follows that $c_{10} = 1/12$. We also consider a_6 and Lemma 4.8.11(c)

$$a_6(x, \Delta_0) - a_6(x, \Delta_1) + a_6(x, \Delta_2) = c_{11}R_{ijij;kkll}(1 - 2 + 1) - \frac{2}{60}R_{1212;kkll}$$
$$+ \text{ lower order terms}$$
$$a_4(x, \Delta_0)_{;kk} = 2c_3 R_{1212;jjkk} + \text{ lower order terms}$$

so Lemma 4.2.12 implies that $2(-2/60) = 2c_3$ so that $c_3 = -1/30$.

This leaves only the constants c_4, c_5, and c_6 undetermined. We let $M = M_1 \times M_2$ be the product manifold with $\Delta_0 = \Delta_0^1 + \Delta_0^2$. Then the product formulas of Lemma 1.7.5 imply:

$$a_4(\Delta_0, x) = a_4(\Delta_0^1, x_1) + a_4(\Delta_0^2, x_2) + a_2(\Delta_0^1, x_1)a_2(\Delta_0^2, x_2).$$

The only term in the expression for a_4 giving rise to cross terms involving derivatives of both metrics is $c_4 R_{ijij} R_{klkl}$. Consequently, $2c_4 = c_1^2 = 1/36$ so $c_4 = 1/72$. We summarize these computations as follows:

LEMMA 4.8.14. We can expand α_n in the form:
(a) $a_2(x, P) = (4\pi)^{-m/2} \operatorname{Tr}(-R_{kjkj} + 6E)/6$.
(b)
$$a_4(x, P) = \frac{(4\pi)^{-m/2}}{360} \operatorname{Tr}(-12R_{ijij;kk} + 5R_{ijij}R_{klkl} + c_5 R_{ijik}R_{ljlk}$$
$$+ c_6 R_{ijkl}R_{ijkl} - 60ER_{ijij} + 180E^2 + 60E_{;kk} + 30\Omega_{ij}\Omega_{ij}).$$

We have changed the notation slightly to introduce the common denominator 360.

We must compute the universal constants c_5 and c_6 to complete our determination of the formula a_4. We generalize the recursion relations of Lemmas 4.8.10 and 4.8.12 to arbitrary dimensions as follows. Let $M = T_m$ be the m-dimensional torus with usual periodic parameters $0 \le x_i \le 2$ for $i = 1, \ldots, m$. We let $\{e_i\}$ be a collection of Clifford matrices so $e_i e_j + e_j e_i = 2\delta_{ij}$. Let $h(x)$ be a real-valued function on T_m and let the metric be given by:

$$ds^2 = e^{-h}(dx_1^2 + \cdots + dx_m^2), \quad \text{dvol} = e^{-hm/2}dx_1 \ldots dx_m.$$

We let the operator A and A^* be defined by:

$$A = e^{mh/4} \sum e_j \frac{\partial}{\partial x_j} e^{(2-m)h/4}$$

$$A^* = e^{(2+m)h/4} \sum e_j \frac{\partial}{\partial x_j} e^{-mh/4}$$

and define:

$$P_1 = A^*A = -e^{(2+m)h/4} \sum_j \frac{\partial^2}{\partial x_j^2} e^{(2-m)h/4}$$

$$= -e^h \left\{ \sum \frac{\partial^2}{\partial x_j^2} + \frac{1}{2}(2-m)h_{/j}\frac{\partial}{\partial x_j} \right.$$

$$\left. + \frac{1}{16}(4(2-m)h_{/jj} + (2-m)^2 h_{/j}h_{/j}) \right\}$$

$$P_2 = AA^* = e^{mh/4} \sum_j e_j \frac{\partial}{\partial x_j} e^h \sum_k e_k \frac{\partial}{\partial x_k} e^{-mh/4}.$$

LEMMA 4.8.15. *With the notation given above,*

$$(n - m)\{a_n(x, P_1) - a_n(x, P_2)\} = e^{hm/2}\frac{\partial^2}{\partial x_k^2}e^{h(2-m)/2}a_{n-2}(x, P).$$

PROOF: We let P_0 be the scalar operator

$$-e^{(2+m)h/4}\sum_j \frac{\partial^2}{\partial x_j^2}e^{(2-m)h/4}.$$

If the representation space on which the e_j act has dimension u, then $P_1 = P_0 \otimes I_u$. We let v_s be a basis for this representation space. Let $\{\lambda_\nu, \theta_\nu\}$ be a spectral resolution for the operator P_0 then $\{\lambda_\nu, \theta_\nu \otimes v_s\}$ is a spectral resolution of P_1. We compute:

$$\frac{d}{dt}\left(\operatorname{Tr}K(t, x, x, P_1) - \operatorname{Tr}(t, x, x, P_2)\right)$$
$$= \sum_{\nu,s} e^{-t\lambda_\nu}\{-(P_1\theta_\nu \otimes u_s, \theta_\nu \otimes u_s) + (A\theta_\nu \otimes u_s, A\theta_\nu \otimes u_s)\}$$

where $(\ ,\ )$ denotes the natural inner product $(u_s, u_{s'}) = \delta_{s,s'}$. The e_j are self-adjoint matrices. We use the identity $(e_j u_s, e_k u_s) = \delta_{jk}$ to compute:

$$= \sum_\nu v \cdot e^{-t\lambda_\nu}e^h\left\{\theta_{\nu/kk}\theta_\nu + \frac{1}{2}(2-m)h_{/k}\theta_{\nu/k}\theta_\nu + \frac{1}{4}(2-m)h_{/kk}\theta_\nu\theta_\nu\right.$$
$$+ \frac{1}{16}(2-m)^2 h_{/k}h_{/k}\theta_\nu\theta_\nu + \theta_{\nu/k}\theta_{\nu/k} + \frac{1}{2}(2-m)h_{/k}\theta_{\nu/k}\theta_\nu$$
$$\left. + \frac{1}{16}(2-m)^2 h_{/k}h_{/k}\theta_\nu\theta_\nu\right\}$$
$$= \sum_\nu v \cdot e^{-t\lambda_\nu}e^{hm/2}\frac{1}{2}\sum_j \frac{\partial^2}{\partial x_j^2}e^{h(2-m)/2}(\theta_\nu, \theta_\nu)$$
$$= \frac{1}{2}e^{hm/2}\sum_j \frac{\partial^2}{\partial x_j^2}e^{h(2-m)/2}\operatorname{Tr}K(t, x, x, P_1).$$

We compare coefficients of t in the two asymptotic expansions to complete the proof of the lemma.

We apply Lemma 4.8.15 to the special case $n = m = 6$. This implies that

$$\sum_j \frac{\partial^2}{\partial x_j^2}e^{-2h}a_4(x, P_1) = 0.$$

Since $a_x(x, P_1)$ is a formal polynomial in the jets of h with coefficients which are smooth functions of h, and since a_4 is homogeneous of order 4, this identity for all h implies $a_4(x, P_1) = 0$. This implies $a_4(x, P_0) = \frac{1}{v}a_4(x, P_1) = 0$.

Using the formulas of lemmas 4.8.1 and 4.8.2 it is an easy exercise to calculate $R_{ijkl} = 0$ if all 4 indices are distinct. The non-zero curvatures are given by:

$$R_{iji}{}^k = -\tfrac{1}{4}(2h_{/jk} + h_{/j}h_{/k}) \text{ if } j \neq k \qquad \text{(don't sum over } i)$$

$$R_{iji}{}^j = -\tfrac{1}{4}\Big(2h_{/ii} + 2h_{/jj} - \sum_{k\neq i, k\neq j} h_{/k}h_{/k}\Big) \qquad \text{(don't sum over } i, j)$$

$$\Omega_{ij} = 0$$

$$E(P_0) = e^h \sum_k (-h_{/kk} + h_{/k}h_{/k}).$$

When we contract the curvature tensor to form scalar invariants, we must include the metric tensor since it is not diagonal. This implies:

$$\tau = \sum_{i,j} g^{ii} R_{iji}{}^j = e^h \sum_k (-5h_{/kk} + 5h_{/k}h_{/k}) = 5E(P_1)$$

which implies the helpful identities:

$$-12\tau_{;kk} + 60E_{;kk} = 0 \qquad \text{and} \qquad 5R_{ijij}R_{klkl} - 60R_{ijij}E + 180E^2 = 5E^2$$

so that we conclude:

$$5E^2 + c_5 R_{ijik}R_{ljlk} + c_6 R_{ijkl}R_{ijkl} = 0.$$

We expand:

$$R_{ijik}R_{ljlk} = e^{2h}\left(\tfrac{15}{2}h_{/11}h_{/11} + 8h_{/12}h_{/12} + \text{ other terms}\right)$$

$$R_{ijkl}R_{ijkl} = e^{2h}\left(5h_{/11}h_{/11} + 8h_{/12}h_{/12} + \text{ other terms}\right)$$

$$E^2 = e^{2h}\left(h_{/11}h_{/11} + 0 \cdot h_{/12}h_{/12} + \text{ other terms}\right)$$

so we conclude finally:

$$15c_5 + 10c_6 + 10 = 0 \qquad \text{and} \qquad c_5 + c_6 = 0.$$

We solve these equations to conclude $c_5 = -2$ and $c_6 = 2$ which proves finally:

THEOREM 4.8.16. *Let P be a second order differential operator with leading symbol given by the metric tensor. Let $P = P_\nabla - E$ be decomposed as in 4.8.1. Let $a_n(x, P)$ be the invariants of the heat equation discussed in Chapter 1.*

(a) $a_0(x, P) = (4\pi)^{-m/2} \operatorname{Tr}(I)$.

(b) $a_2(x, P) = (4\pi)^{-m/2} \operatorname{Tr}(-R_{ijij} + 6E)/6$.

(c)

$$a_4(x, P) = \frac{(4\pi)^{-m/2}}{360} \times$$
$$\operatorname{Tr}(-12R_{ijij;kk} + 5R_{ijij}R_{klkl} - 2R_{ijik}R_{ljlk} + 2R_{ijkl}R_{ijkl}$$
$$- 60R_{ijij}E + 180E^2 + 60E_{;kk} + 30\Omega_{ij}\Omega_{ij}).$$

(d)

$$a_6(x, P) = (4\pi)^{-m/2} \times$$

$$\operatorname{Tr}\Bigg\{ \frac{1}{7!}\Big(-18R_{ijij;kkll} + 17R_{ijij;k}R_{ulul;k} - 2R_{ijik;l}R_{ujuk;l}$$

$$- 4R_{ijik;l}R_{ujul;k} + 9R_{ijku;l}R_{ijku;l} + 28R_{ijij}R_{kuku;ll}$$

$$- 8R_{ijik}R_{ujuk;ll} + 24R_{ijik}R_{ujul;kl} + 12R_{ijkl}R_{ijkl;uu}\Big)$$

$$+ \frac{1}{9 \cdot 7!}\Big(-35R_{ijij}R_{klkl}R_{pqpq} + 42R_{ijij}R_{klkp}R_{qlqp}$$

$$- 42R_{ijij}R_{klpq}R_{klpq} + 208R_{ijik}R_{julu}R_{kplp}$$

$$- 192R_{ijik}R_{uplp}R_{jukl} + 48R_{ijik}R_{julp}R_{kulp}$$

$$- 44R_{ijku}R_{ijlp}R_{kulp} - 80R_{ijku}R_{ilkp}R_{jlup}\Big)$$

$$+ \frac{1}{360}\Big(8\Omega_{ij;k}\Omega_{ij;k} + 2\Omega_{ij;j}\Omega_{ik;k} + 12\Omega_{ij}\Omega_{ij;kk} - 12\Omega_{ij}\Omega_{jk}\Omega_{ki}$$

$$- 6R_{ijkl}\Omega_{ij}\Omega_{kl} + 4R_{ijik}\Omega_{jl}\Omega_{kl} - 5R_{ijij}\Omega_{kl}\Omega_{kl}\Big)$$

$$+ \frac{1}{360}\Big(6E_{;iijj} + 60EE_{;ii} + 30E_{;i}E_{;i} + 60E^3 + 30E\Omega_{ij}\Omega_{ij}$$

$$- 10R_{ijij}E_{;kk} - 4R_{ijik}E_{;jk} - 12R_{ijij;k}E_{;k} - 30R_{ijij}E^2$$

$$- 12R_{ijij;kk}E + 5R_{ijij}R_{klkl}E$$

$$- 2R_{ijik}R_{ijkl}E + 2R_{ijkl}R_{ijkl}E\Big).\Bigg\}$$

PROOF: We have derived (a)–(c) explicitly. We refer to (Gilkey, The spectral geometry of a Riemannian manifold) for the proof of (d) as it is quite long and complicated. We remark that our sign convention is that $R_{1212} = -1$ on the sphere of radius 1 in R^3.

We now begin our computation of $a_n(x, \Delta_p^m)$ for $n = 0, 2, 4$.

LEMMA 4.8.17. *Let $m = 4$, and let Δ_p be the Laplacian on p-forms. Decompose*

$$a_2(x, \Delta_p) = (4\pi)^{-2} \cdot c_0(p) \cdot R_{ijij}/6$$
$$a_4(x, \Delta_p) = (4\pi)^{-2}\{c_1(p)R_{ijij;kk} + c_2(p)R_{ijij}R_{klkl}$$
$$c_3(p)R_{ijik}R_{ljlk} + c_4(p)R_{ijkl}R_{ijkl}\}/360.$$

Then the c_i are given by the following table:

	c_0	c_1	c_2	c_3	c_4
$p = 0$	-1	-12	5	-2	2
$p = 1$	2	12	-40	172	-22
$p = 2$	6	48	90	-372	132
$p = 3$	2	12	-40	172	-22
$p = 4$	-1	-12	5	-2	2
$\sum(-1)^p$	0	0	180	-720	180

PROOF: By Poincare duality, $a_n(x, \Delta_p) = a_n(x, \Delta_{4-p})$ so we need only check the first three rows as the corresponding formulas for Δ_3 and Δ_4 will follow. In dimension 4 the formula for the Euler form is $(4\pi)^{-2}(R_{ijij}R_{klkl} - 4R_{ijik}R_{ljlk} + R_{ijkl}R_{ijkl})/2$ so that the last line follows from Theorem 2.4.8. If $p = 0$, then $E = \Omega = 0$ so the first line follows from Theorem 4.8.15. If $p = 1$, then $E = -\rho_{ij} = R_{ikij}$ is the Ricci tensor by Lemma 4.8.13. Therefore:

$$\mathrm{Tr}(-12R_{ijij;kk} + 60E_{;kk}) = -48R_{ijij;kk} + 60R_{ijij;kk}$$
$$= 12R_{ijij;kk}$$
$$\mathrm{Tr}(5R_{ijij}R_{klkl} - 60R_{ijij}E) = 20R_{ijij}R_{klkl} - 60R_{ijij}R_{klkl}$$
$$= -40R_{ijij}R_{klkl}$$
$$\mathrm{Tr}(-2R_{ijik}R_{lklk} + 180E^2) = -8R_{ijik}R_{ljlk} + 180R_{ijik}R_{ljlk}$$
$$= 172R_{ijik}R_{ljlk}$$
$$\mathrm{Tr}(2R_{ijkl}R_{ijkl} + 30\Omega_{ij}\Omega_{ij}) = 8R_{ijkl}R_{ijkl} - 30R_{ijkl}R_{ijkl}$$
$$= -22R_{ijkl}R_{ijkl}$$

which completes the proof of the second line. Thus the only unknown is $c_k(2)$. This is computed from the alternating sum and completes the proof.

More generally, we let $m > 4$. Let $M = M_4 \times T_{m-4}$ be a product manifold, then this defines a restriction map $r_{m-4}: \mathcal{P}_{m,n} \to \mathcal{P}_{4,n}$ which is an isomorphism for $n \leq 4$. Using the multiplication properties given in Lemma 1.7.5, it follows:

$$a_n(x, \Delta_p^m) = \sum_{\substack{i+j=n \\ p_1+p_2=p}} a_i(x, \Delta_{p_1}^4) a_j(x, \Delta_{p_2}^{m-4}).$$

On the flat torus, all the invariants vanish for $j > 0$. By Lemma 4.8.4 $a_0(x, \Delta_{p_2}^{m-4}) = (4\pi)^{-m/2} \binom{m-4}{p_2}$ so that:

$$a_n(x, \Delta_u^4) = \sum_{u+v=p} (4\pi)^{(m-4)/2} \binom{m-4}{v} a_n(x, \Delta_u^4) \qquad \text{for } n \leq 4$$

where $a_n(x, \Delta_u^4)$ is given by Lemma 4.8.16. If we expand this as a polynomial in m for small values of u we conclude:

$$a_2(x, \Delta_1^m) = \frac{(4\pi)^{-m/2}}{6}(6 - m)R_{ijij}$$

$$a_4(x, \Delta_1^m) = \frac{(4\pi)^{-m/2}}{360}\{(60 - 12m)R_{ijij;kk} + (5m - 60)R_{ijij}R_{klkl}$$

$$+ (180 - 2m)R_{ijik}R_{ljlk} + (2m - 30)R_{ijkl}R_{ijkl}\}$$

and similarly for $a_2(x, \Delta_2^m)$ and $a_4(x, \Delta_2^m)$. In this form, the formulas also hold true for $m = 2, 3$. We summarize our conclusions in the following theorem:

THEOREM 4.8.18. *Let Δ_p^m denote the Laplacian acting on the space of smooth p-forms on an m-dimensional manifold. We let R_{ijkl} denote the curvature tensor with the sign convention that $R_{1212} = -1$ on the sphere of radius 1 in $\mathbf{R}^3$. Then:*
(a) $a_0(x, \Delta_p^m) = (4\pi)^{-m/2}\binom{m}{p}$.
(b) $a_2(x, \Delta_p^m) = \dfrac{(4\pi)^{-m/2}}{6}\left\{\binom{m-2}{p-2} + \binom{m-2}{p} - 4\binom{m-2}{p-1}\right\}(-R_{ijij})$.
(c) Let

$$a_4(x, \Delta_p^m) = \frac{(4\pi)^{-m/2}}{360}\big(c_1(m,p)R_{ijij;kk} + c_2(m,p)R_{ijij}R_{klkl}$$

$$+ c_3(m,p)R_{ijik}R_{ljlk} + c_4(m,p)R_{ijkl}R_{ijkl}\big).$$

Then for $m \geq 4$ the coefficients are:

$$c_1(m,p) = -12\left[\binom{m-4}{p} + \binom{m-4}{p-4}\right] + 12\left[\binom{m-4}{p-1} + \binom{m-4}{p-3}\right] + 48\binom{m-4}{p-2}$$

$$c_2(m,p) = 5\left[\binom{m-4}{p} + \binom{m-4}{p-4}\right] - 40\left[\binom{m-4}{p-1} + \binom{m-4}{p-3}\right] + 90\binom{m-4}{p-2}$$

$$c_3(m,p) = -2\left[\binom{m-4}{p} + \binom{m-4}{p-4}\right] + 172\left[\binom{m-4}{p-1} + \binom{m-4}{p-3}\right] - 372\binom{m-4}{p-2}$$

$$c_4(m,p) = 2\left[\binom{m-4}{p} + \binom{m-4}{p-4}\right] - 22\left[\binom{m-4}{p-1} + \binom{m-4}{p-3}\right] + 132\binom{m-4}{p-2}.$$

(d)

$$a_0(x, \Delta_0^m) = (4\pi)^{-m/2}$$

$$a_2(x, \Delta_0^m) = \frac{(4\pi)^{-m/2}}{6}(-R_{ijij})$$

$$a_4(x, \Delta_0^m) = \frac{(4\pi)^{-m/2}}{360}(-12R_{ijij;kk} + 5R_{ijij}R_{klkl} - 2R_{ijik}R_{ljlk}$$
$$+ 2R_{ijkl}R_{ijkl})$$

(e)

$$a_0(x, \Delta_1^m) = (4\pi)^{-m/2}$$

$$a_2(x, \Delta_1^m) = \frac{(4\pi)^{-m/2}}{6}(6 - m)R_{ijij}$$

$$a_4(x, \Delta_1^m) = \frac{(4\pi)^{-m/2}}{360}\{(60 - 12m)R_{ijij;kk} + (5m - 60)R_{ijij}R_{klkl}$$
$$+ (180 - 2m)R_{ijik}R_{ljlk}$$
$$+ (2m - 30)R_{ijkl}R_{ijkl})\}$$

(f)

$$a_0(x, \Delta_2^m) = \frac{(4\pi)^{-m/2}}{2}m(m - 1)$$

$$a_2(x, \Delta_2^m) = \frac{(4\pi)^{-m/2}}{12}(-m^2 + 13m - 24)R_{ijij}$$

$$a_4(x, \Delta_2^m) = \frac{(4\pi)^{-m/2}}{720}\{(-12m^2 + 108m - 144)R_{ijijk;kk}$$
$$+ (5m^2 - 115m + 560)R_{ijij}R_{klkl}$$
$$+ (-2m^2 + 358m - 2144)R_{ijik}R_{ljlk}$$
$$+ (2m^2 - 58m + 464)R_{ijkl}R_{ijkl}\}$$

These results are, of course, not new. They were first derived by Patodi. We could apply similar calculations to determine a_0, a_2, and a_4 for any operator which is natural in the sense of Epstein and Stredder. In particular, the Dirac operator can be handled in this way.

In principal, we could also use these formulas to compute a_6, but the lower order terms become extremely complicated. It is not too terribly difficult, however, to use these formulas to compute the terms in a_6 which involve the 6 jets of the metric and which are bilinear in the 4 and 2 jets of the metric. This would complete the proof of the result concerning a_6 if $m = 4$ discussed in section 4.7.

4.9. Spectral Geometry.

Let M be a compact Riemannian manifold without boundary and let $\text{spec}(M, \Delta)$ denote the spectrum of the scalar Laplacian where each eigenvalue is repeated according to its multiplicity. Two manifolds M and $\overline{M}$ are said to be *isospectral* if $\text{spec}(M, \Delta) = \text{spec}(\overline{M}, \Delta)$. The leading term in the asymptotic expansion of the heat equation is $(4\pi)^{-m/2} \cdot \text{vol}(M) \cdot t^{-m/2}$ so that if M and $\overline{M}$ are isospectral,

$$\dim M = \dim \overline{M} \qquad \text{and} \qquad \text{volume}(M) = \text{volume}(\overline{M})$$

so these two quantitites are spectral invariants. If $P \in \mathcal{P}_{m,n,0}$ is an invariant polynomial, we define $P(M) = \int_M P(G) |\,\text{dvol}\,|$. (This depends on the metric in general.) Theorem 4.8.18 then implies that $R_{ijij}(M)$ is a spectral invasriant since this appears with a non-zero coefficient in the asymptotic expansion of the heat equation.

The scalar Laplacian is not the only natural differential operator to study. (We use the word "natural" in the technical sense of Epstein and Stredder in this context.) Two Riemannian manifolds M and $\overline{M}$ are said to be *strongly isospectral* if $\text{spec}(M, P) = \text{spec}(\overline{M}, P)$ for all natural operators P. Many of the global geometry properties of the manifold are reflected by their spectral geometry. Patodi, for example, proved:

THEOREM 4.9.1 (PATODI). *Let* $\text{spec}(M, \Delta_p) = \text{spec}(\overline{M}, \Delta_p)$ *for* $p = 0, 1, 2$. *Then:*
(a)

$$\dim M = \dim \overline{M}, \qquad\qquad \text{volume}(M) = \text{volume}(\overline{M}),$$
$$R_{ijij}(M) = R_{ijij}(\overline{M}), \qquad\qquad R_{ijij} R_{klkl}(M) = R_{ijij} R_{klkl}(\overline{M}),$$
$$R_{ijik} R_{ljlk}(M) = R_{ijik} R_{ljlk}(\overline{M}), \qquad R_{ijkl} R_{ijkl}(M) = R_{ijkl} R_{ijkl}(\overline{M}).$$

(b) If M has constant scalar curvature c, then so does $\overline{M}$.
(c) If M is Einstein, then so is $\overline{M}$.
(d) If M has constant sectional curvature c, then so does $\overline{M}$.

PROOF: The first three identities of (a) have already been derived. If $m \geq 4$, the remaining 3 integral invariants are independent. We know:

$$a_4(\Delta_p^m) = c_2(m, p) R_{ijij} R_{klkl} + c_3(m, p) R_{ijik} R_{ljlk}(M)$$
$$+ c_4(m, p) R_{ijkl} R_{ijkl}(M).$$

As $p = 0, 1, 2$ the coefficients form a 3×3 matrix. If we can show the matrix has rank 3, we can solve for the integral invariants in terms of the spectral invariants to prove (a). Let $\nu = m - 4$ and $c = (4\pi)^{-\nu/2}$. then our computations in section 4.8 show

$$c_j(m, 0) = c \cdot c_j(4, 0)$$
$$c_j(m, 1) = c \cdot \nu \cdot c_j(4, 0) + c \cdot c_j(4, 1)$$
$$c_j(m, 2) = c \cdot \nu \cdot (\nu - 1)/2 \cdot c_j(4, 0) + c \cdot \nu \cdot c_j(4, 1) + c \cdot c_j(4, 2)$$

so that the matrix for $m > 4$ is obtained from the matrix for $m = 4$ by elementary row operations. Thus it suffices to consider the case $m = 4$. By 4.8.17, the matrix there (modulo a non-zero normalizing constant) is:

$$\begin{pmatrix} 5 & -2 & 2 \\ -40 & 172 & -22 \\ 90 & -372 & 132 \end{pmatrix}.$$

The determinant of this matrix is non-zero from which (a) follows. The case $m = 2$ and $m = 3$ can be checked directly using 4.8.18.

To prove (b), we note that M has constant scalar curvature c if and only if $(2c + R_{ijij})^2(M) = 0$ which by (a) is a spectral invariant. (c) and (d) are similar.

From (d) follows immediately the corollary:

COROLLARY 4.9.2. Let M and $\overline{M}$ be strongly isospectral. If M is isometric to the standard sphere of radius r, then so is $\overline{M}$.

PROOF: If M is a compact manifold with sectional curvature $1/r$, then the universal cover of M is the sphere of radius r. If $\text{vol}(M)$ and $\text{vol}(S(r))$ agree, then M and the sphere are isometric.

There are a number of results which link the spectral and the global geometry of a manifold. We list two of these results below:

THEOREM 4.9.3. Let M and $\overline{M}$ be strongly isospectral manifolds. Then:
(a) If M is a local symmetric space (i.e., $\nabla R = 0$), then so is $\overline{M}$.
(b) If the Ricci tensor of M is parallel (i.e., $\nabla \rho = 0$), then the Ricci tensor of $\overline{M}$ is parallel. In this instance, the eigenvalues of ρ do not depend upon the particular point of the manifold and they are the same for M and $\overline{M}$.

Although we have chosen to work in the real category, there are also isospectral results available in the holomorphic category:

THEOREM 4.9.4. Let M and $\overline{M}$ be holomorphic manifolds and suppose $\text{spec}(M, \Delta_{p,q}) = \text{spec}(\overline{M}, \Delta_{p,q})$ for all (p, q).
(a) If M is Kaehler, then so is $\overline{M}$.
(b) If M is $\mathbf{CP}_n$, then so is $\overline{M}$.

At this stage, the natural question to ask is whether or not the spectral geometry completely determines M. This question was phrased by Kac in the form: *Can you hear the shape of a drum?* It is clear that if there exists an isometry between two manifolds, then they are strongly isospectral. That the converse need not hold was shown by Milnor, who gave examples of isospectral tori which were not isometric. In 1978 Vigneras gave examples of isospectral manifolds of constant negative curvature which are not isometric. (One doesn't know yet if they are strongly isospectral.) If the dimension is at least 3, then the manifolds have different fundamental groups, so are not homotopic. The fundamental groups in question are all infinite and the calculations involve some fairly deep results in quaternion algebras.

In 1983 Ikeda constructed examples of spherical space forms which were strongly isospectral but not isometric. As de Rham had shown that diffeomorphic spherical space forms are isometric, these examples are not diffeomorphic. Unlike Vigneras' examples, Ikeda's examples involve finite fundamental groups and are rather easily studied. In the remainder of this section, we will present an example in dimension 9 due to Ikeda illustrating this phenomenon. These examples occur much more generally, but this example is particularly easy to study. We refer to (Gilkey, On spherical space forms with meta-cyclic fundamental group which are isospectral but not equivariant cobordant) for more details.

Let G be the group of order 275 generated by two elements A, B subject to the relations:

$$A^{11} = B^{25} = 1 \quad \text{and} \quad BAB^{-1} = A^3.$$

We note that $3^5 \equiv 1 \mod 11$. This group is the semi-direct product $\mathbf{Z}_{11} \propto \mathbf{Z}_{25}$. The center of G is generated by B^5 and the subgroup generated by A is a normal subgroup. We have short exact sequences:

$$0 \to \mathbf{Z}_{11} \to G \to \mathbf{Z}_{25} \to 0 \quad \text{and} \quad 0 \to \mathbf{Z}_{55} \to G \to \mathbf{Z}_5 \to 0.$$

We can obtain an explicit realization of G as a subgroup of U(5) as follows. Let $\alpha = e^{2\pi i/11}$ and $\beta = e^{2\pi i/25}$ be primitive roots of unity. Let $\{e_j\}$ be the standard basis for $\mathbf{C}^5$ and let $\sigma \in$ U(5) be the permutation matrix $\sigma(e_j) = e_{j-1}$ where the index j is regarded as defined mod 5. Define a representation:

$$\pi_k(A) = \text{diag}(\alpha, \alpha^3, \alpha^9, \alpha^5, \alpha^4) \quad \text{and} \quad \pi_k(B) = \beta^k \cdot \sigma.$$

It is immediate that $\pi_k(A)^{11} = \pi_k(B)^{25} = 1$ and it is an easy computation that $\pi_k(B)\pi_k(A)\pi_k(B)^{-1} = \pi_k(A)^3$ so this extends to a representation of G for $k = 1, 2, 3, 4$. (If H is the subgroup generated by $\{A, B^5\}$, we let $\rho(A) = \alpha$ and $\rho(B^5) = \beta^{5k}$ be a unitary representation of H. The representation $\pi_k = \rho^G$ is the induced representation.)

In fact these representations are fixed point free:

LEMMA 4.9.5. *Let G be the group of order 275 generated by $\{A, B\}$ with the relations $A^{11} = B^{25} = 1$ and $BAB^{-1} = A^3$. The center of G is generated by B^5; the subgroup generated by A is normal. Let $\alpha = e^{2\pi i/11}$ and $\beta = e^{2\pi i/25}$ be primitive roots of unity. Define representations of G in $U(5)$ for $1 \leq k \leq 4$ by: $\pi_k(A) = \mathrm{diag}(\alpha, \alpha^3, \alpha^9, \alpha^5, \alpha^4)$ and $\pi_k(B) = \beta^k \cdot \sigma$ where σ is the permutation matrix defined by $\sigma(e_i) = e_{i-1}$.*
(a) Enumerate the elements of G in the form $A^a B^b$ for $0 \leq a < 11$ and $0 \leq b < 25$. If $(5, b) = 1$, then $A^a B^b$ is conjugate to B^b.
(b) The representations π_k are fixed point free.
(c) The eigenvalues of $\pi_k(A^a B^b)$ and $\pi_1(A^a B^{kb})$ are the same so these two matrices are conjugate in $U(5)$.

PROOF: $A^{-j} B^b A^j = A^{j(3^b-1)} B^b$. If $(5, b) = 1$, then $3^b - 1$ is coprime to 11 so we can solve the congruence $j(3^b - 1) \equiv a \mod 11$ to prove (a). Suppose $(5, b) = 1$ so the eigenvalues of $\pi_k(A^a B^b)$ and $\pi_k(B^b)$ coincide. Let $\{\varepsilon_1, \dots, \varepsilon_5\}$ be the 5th roots of unity; these are the eigenvalues of σ and of σ^b. Thus the eigenvalues of $\pi_k(B^b)$ are $\{\beta^{kb}\varepsilon_1, \dots, \beta^{kb}\varepsilon_5\}$ and are primitive 25th roots of unity. Thus $\det(\pi_k(B^b) - I) \neq 0$ and $\pi_k(B^b)$ has the same eigenvalues as $\pi_1(B^{kb})$. To complete the proof, we conside an element $A^a B^{5b}$. π_k is diagonal with eigenvalues $\beta^{5kb}\{\alpha^a, \alpha^{3a}, \alpha^{9a}, \alpha^{5a}, \alpha^{4a}\}$. If $(a, 11) = 1$ and $(b, 5) = 1$ these are all primitive 55th roots of unity; if $(a, 11) = 1$ and $(b, 5) = 5$ these are all primitive 11th roots of unity; if $(a, 11) = 11$ and $(b, 5) = 1$ these are all primitive 5th roots of unity. This shows $\pi_k(A^a B^{5b})$ is fixed point free and has the same eigenvalues as $\pi_1(A^a B^{5kb})$ which completes the proof.

We form the manifolds $M_k = S^9/\pi_k(G)$ with fundamental group G. These are all spherical space forms which inherit a natural orientation and metric from S^9 as discussed previously.

LEMMA 4.9.6. *Adopt the notation of Lemma 4.9.5. Let $M = S^9/\pi_k(G)$ be spherical space forms. Then M_1, M_2, M_3 and M_4 are all strongly isospectral.*

PROOF: Let P be a self-adjoint elliptic differential operator which is natural in the category of oriented R̃iemannian manifolds. Let P_0 denote this operator on S^9, and let P_k denote the corresponding operator on M_k. (For example, we could take P to be the Laplacian on p-forms or to be the tangential operator of the signature complex). Let $\lambda \in \mathbf{R}$ and let $E_0(\lambda)$ and $E_k(\lambda)$ denote the eigenspaces of P_0 and P_k. We must show $\dim E_k(\lambda)$ is independent of k for $1 \leq k \leq 4$. The unitary group acts on S^9 by orientation preserving isometries. The assumption of naturality lets us extend this to an action we shall denote by $e_0(\lambda)$ on $E_0(\lambda)$. Again, the assumption of naturality implies the eigenspace $E_k(\lambda)$ is just the subspace of $E_0(\lambda)$ invariant under the action of $e_0(\lambda)(\pi_k(G))$. We can calculate the dimension of this invariant subspace by:

$$\dim e_k(\lambda) = \frac{1}{|G|} \sum_{g \in G} \text{Tr}\{e_0(\lambda)(\pi_k(g)\}$$

$$= \frac{1}{275} \sum_{a,b} \text{Tr}\{e_0(\lambda)(\pi_k(A^a B^b)\}.$$

We apply Lemma 4.9.5 to conclude $\pi_k(A^a B^b)$ is conjugate to $\pi_1(A^a B^{kb})$ in $U(5)$ so the two traces are the same and

$$\dim e_k(\lambda) = \frac{1}{275} \sum_{a,b} \text{Tr}\{e_0(\lambda)\pi_1(A^a B^{kb})\}$$

$$= \frac{1}{275} \sum_{a,b} \text{Tr}\{e_0(\lambda)\pi_1(A^a B^b)\} = \dim e_1(\lambda)$$

since we are just reparameterizing the group. This completes the proof.

Let $\rho \in R_0(\mathbf{Z}_5)$. We regard $\rho \in R_0(G)$ by defining $\rho(A^a B^b) = \rho(b)$. This is nothing but the pull-back of ρ using the natural map $0 \to \mathbf{Z}_{55} \to G \to \mathbf{Z}_5 \to 0$.

LEMMA 4.9.7. Let $\rho \in R_0(\mathbf{Z}_5)$ and let G be as in Lemma 4.9.5. Let $\psi: G \to G$ be a group automorphism. Then $\rho \cdot \psi = \rho$ so this representation of G is independent of the marking chosen.

PROOF: The Sylow 11-subgroup is normal and hence unique. Thus $\psi(A) = A^a$ for some $(a, 11) = 1$ as A generates the Sylow 11-subgroup. Let $\psi(B) = A^c B^d$ and compute:

$$\psi(A^3) = A^{3a} = \psi(B)\psi(A)\psi(B)^{-1} = A^c B^d A^a B^{-d} A^{-c} = A^{a \cdot 3^d}.$$

Since $(a, 11) = 1$, $3^d \equiv 1$ (11). This implies $d \equiv 1$ (5). Therefore $\psi(A^u B^v) = A^* B^{dv}$ so that $\rho\psi(A^u B^v) = \rho(dv) = \rho(v)$ as $\rho \in R_0(\mathbf{Z}_5)$ which completes the proof.

These representations are canonical; they do not depend on the marking of the fundamental group. This defines a virtual locally flat bundle V_ρ over each of the M_k. Let P be the tangential operator of the signature complex; $\text{ind}(\rho, \text{signature}, M_k)$ is an oriented diffeomorphism of M_k. In fact more is true. There is a canonical $\mathbf{Z}_5$ bundle over M_k corresponding to the sequence $G \to \mathbf{Z}_5 \to 0$ which by Lemma 4.9.7 is independent of the particular isomorphism of $\pi_1(M_k)$ with G chosen. Lemma 4.6.4(b) shows this is a $\mathbf{Z}_5$-cobordism invariant. We apply Lemma 4.6.3 and calculate for $\rho \in R_0(\mathbf{Z}_5)$ that:

$$\text{ind}(\rho, \text{signature}, M_k) = \frac{1}{275} \cdot \sideset{}{'}\sum_{a,b} \text{Tr}(\rho(b)) \cdot \text{defect}(\pi_k(A^a B^b), \text{signature}).$$

$\sum'$ denotes the sum over $0 \le a < 11$, $0 \le b < 25$, $(a, b) \ne (0, 0)$. Since this is an element of $R_0(\mathbf{Z}_5)$, we may suppose $(5, b) = 1$. As $A^a B^b$ is conjugate to B^b by 4.9.4, we can group the 11 equal terms together to see

$$\text{ind}(\rho, \text{signature}, M_k) = \frac{1}{25} \sum_b{}' \text{Tr}(\rho(b)) \cdot \text{defect}(\pi_k(B^b), \text{signature})$$

where we sum over $0 \le b < 25$ and $(5, b) = 1$.

If $(5, b) = 1$, then the eigenvalues of $\pi_k(B^b)$ are $\{\beta^{bk}\varepsilon_1, \dots, \beta^{bk}\varepsilon_5\}$. Thus

$$\text{defect}(\pi_k(B^b), \text{signature}) = \prod_\nu \frac{\beta^{bk}\varepsilon_\nu + 1}{\beta^{bk}\varepsilon_\nu - 1}$$

$$= \frac{\beta^{5bk} + 1}{\beta^{5bk} - 1}$$

since the product ranges over the primitive 5th roots of unity. Let $\gamma = \beta^5 = e^{2\pi i/5}$, then we conclude

$$\text{ind}(\rho, \text{signature}, M_k) = \frac{1}{25} \cdot \sum_b{}' \text{Tr}(\rho(b)) \cdot \frac{\gamma^{kb} + 1}{\gamma^{kb} - 1}$$

$$= \frac{1}{5} \cdot \sum_{0 < b < 5} \text{Tr}(\rho(b)) \cdot \frac{\gamma^{kb} + 1}{\gamma^{kb} - 1},$$

if we group equal terms together.

We now calculate for the specific example $\rho = \rho_1 - \rho_0$:

$$\text{ind}(\rho_1 - \rho_0, \text{signature}, M_k) = \frac{1}{5} \sum_{b=1}^{4} (\gamma^b - 1) \cdot \frac{\gamma^{kb} + 1}{\gamma^{kb} - 1}$$

$$= \frac{1}{5} \sum_{b=1}^{4} (\gamma^{b\bar{k}} - 1) \cdot \frac{\gamma^b + 1}{\gamma^b - 1}$$

if we let $k\bar{k} \equiv 1$ (5). We perform the indicated division; $(x^{\bar{k}} - 1)/(x - 1) = x^{\bar{k}-1} + \cdots + 1$ so we obtain

$$= \frac{1}{5} \sum_{b=1}^{4} (\gamma^{b\bar{k}-b} + \gamma^{b\bar{k}-2b} + \cdots + 1)(\gamma^b + 1).$$

This expression is well defined even if $b = 0$. If we sum over the entire group, we get an integer by the orthogonality relations. The value at 0 is $+2\bar{k}/5$ and therefore $\text{ind}(\rho_1 - \rho_0, \text{signature}, M_k) = -2\bar{k}/5 \bmod \mathbf{Z}$.

We choose the orientation arising from the given orientation on S^9. If we reverse the orientation, we change the sign of the tangential operator of the signature complex which changes the sign of this invariant.

THEOREM 4.9.8. *Let $M_k = S^9/\pi_k(G)$ with the notation of Lemma 4.9.5. These four manifolds are all strongly isospectral for $k = 1, 2, 3, 4$. There is a natural $\mathbf{Z}_5$ bundle over each M_k. $\mathrm{ind}(\rho_1 - \rho_0, \mathrm{signature}, M_k) = -2\bar{k}/5$ mod $\mathbf{Z}$ where $k\bar{k} \equiv 1$ (5). As these 5 values are all different, these 4 manifolds are* <u>not</u> *$\mathbf{Z}_5$ oriented equivariant cobordant. Thus in particular there is no orientation preserving diffeomorphism between any two of these manifolds. $M_1 = -M_4$ and $M_2 = -M_3$. There is no diffeomorphism between M_1 and M_2 so these are different topological types.*

PROOF: If we replace β by $\bar{\beta}$ we replace π_k by π_{5-k} up to unitary equivalence. The map $z \mapsto \bar{z}$ reverses the orientation of $\mathbf{C}^5$ and thus $M_k = -M_{5-k}$ as oriented Riemannian manifolds. This change just alters the sign of $\mathrm{ind}(*, \mathrm{signature}, *)$. Thus the given calculation shows $M_1 \neq \pm M_2$. The statement about oriented equivariant cobordism follows from section 4.6. (In particular, these manifolds are not oriented G-cobordant either.) This gives an example of strongly isospectral manifolds which are of different topological types.

Remark: These examples, of course, generalize; we have chosen to work with a particular example in dimension 9 to simplify the calculations involved.

BIBLIOGRAPHY

S. Agmon, "Lectures on Elliptic Boundary Value Problems", Math. Studies No. 2, Van Nostrand, 1965.

M. F. Atiyah, "K Theory", W. A. Benjamin, Inc., 1967.

—————, "Elliptic Operators and Compact Groups", Springer Verlag, 1974.

—————, *Bott Periodicity and the Index of Elliptic Operators*, Quart. J. Math. **19** (1968), 113–140.

—————, *Algebraic Topology and Elliptic Operators*, Comm. Pure Appl. Math. **20** (1967), 237–249.

M. F. Atiyah and R. Bott, *A Lefschetz Fixed Point Formula for Elliptic Differential Operators*, Bull. Amer. Math. Soc. **72** (1966), 245–250.

—————, *A Lefschetz Fixed Point Formula for Elliptic Complexes I*, Ann. of Math. **86** (1967), 374–407; *II. Applications*, **88** (1968), 451–491.

—————, *The Index Theorem for Manifolds with Boundary*, in "Differential Analysis (Bombay Colloquium)", Oxford, 1964.

M. F. Atiyah, R. Bott and V. K. Patodi, *On the Heat Equation and Index Theorem*, Invent. Math. **19** (1973), 279–330.

M. F. Atiyah, R. Bott and A. Shapiro, *Clifford Modules*, Topology **3** Suppl. 1 (1964), 3–38.

M. F. Atiyah, H. Donnelly, and I. M. Singer, *Geometry and Analysis of Shimizu L-Functions*, Proc. Nat. Acad. Sci. USA **79** (1982), p. 5751.

—————, *Eta Invariants, Signature Defects of Cusps and Values of L-Functions*, Ann. of Math. **118** (1983), 131–171.

M. F. Atiyah and F. Hirzebruch, *Riemann-Roch Theorems for Differentiable Manifolds*, Bull. Amer. Math. Soc. **69** (1963), 422–433.

M. F. Atiyah, V. K. Patodi, and I. M. Singer, *Spectral Asymmetry and Riemannian Geometry I*, Math. Proc. Camb. Phil. Soc. **77** (1975), 43–69; *II*, **78** (1975), 405–432; *III*, **79** (1976), 71–99.

—————, *Spectral Asymmetry and Riemannian Geometry*, Bull. London Math. Soc. **5** (1973), 229–234.

M. F. Atiyah and G. B. Segal, *The Index of Elliptic Operators II*, Ann. of Math. **87** (1968), 531–545.

M. F. Atiyah and I. M. Singer, *The Index of Elliptic Operators I*, Ann. of Math. **87** (1968), 484–530; *III*, **87** (1968), 546–604; *IV*, **93** (1971), 119–138; *V*, **93** (1971), 139–149.

—————, *The Index of Elliptic Operators on Compact Manifolds*, Bull. Amer. Math. Soc. **69** (1963), 422–433.

—————, *Index Theory for Skew-Adjoint Fredholm Operators*, Inst. Hautes Études Sci. Publ. Math. **37** (1969).

P. Baum and J. Cheeger, *Infinitesimal Isometries and Pontrjagin Numbers*, Topology **8** (1969), 173–193.

P. Berard and M. Berger, *Le Spectre d'Une Variété Riemannienne en 1981*, (preprint).

M. Berger, *Sur le Spectre d'Une Variété Riemannienne*, C. R. Acad. Sci. Paris Sér. I Math. **163** (1963), 13–16.

—————, *Eigenvalues of the Laplacian*, Proc. Symp. Pure Math. **16** (1970), 121–125.

M. Berger, P. Gauduchon and E. Mazet, "Le Spectre d'Une Variété Riemannienne", Springer Verlag, 1971.

B. Booss, "Topologie und Analysis", Springer-Verlag, 1977.

B. Booss and S. Rempel, *Cutting and Pasting of Elliptic Operators*, C. R. Acad. Sci. Paris Sér. I Math. **292** (1981), 711–714.

B. Booss and B. Schulze, *Index Theory and Elliptic Boundary Value Problems, Remarks and Open Problems*, (Polish Acad. Sci. 1978 preprint).

R. Bott, *Vector Fields and Characteristic Numbers*, Michigan Math. J. **14** (1967), 231–244.

————, *A Residue Formula for Holomorphic Vector Fields*, J. Diff. Geo. **1** (1967), 311–330.

————, "Lectures on K(X)", Benjamin, 1969.

R. Bott and R. Seeley, *Some Remarks on the Paper of Callias*, Comm. Math. Physics **62** (1978), 235–245.

L. Boutet de Monvel, *Boundary Problems for Pseudo-Differential Operators*, Acta Math. **126** (1971), 11–51.

A. Brenner and M. Shubin, *Atiyah-Bott-Lefschetz Theorem for Manifolds with Boundary*, J. Funct. Anal. (1982), 286–287.

C. Callias, *Axial Anomalies and Index Theorems on Open Spaces*, Comm. Math. Physics **62** (1978), 213–234.

J. Cheeger, *A Lower Bound for the Smallest Eigenvalue of the Laplacian*, in "Problems in Analysis", Princeton Univ. Press, 1970, pp. 195–199.

————, *On the Hodge Theory of Riemannian Pseudo-Manifolds*, Proc. Symp. in Pure Math. **36** (1980), 91–146.

————, *Analytic Torsion and the Heat Equation*, Ann. of Math. **109** (1979), 259–322.

————, *Analytic Torsion and Reidemeister Torsion*, Proc. Nat. Acad. Sci. USA **74** (1977), 2651–2654.

————, *Spectral Geometry of Singular Riemannian Spaces*, J. Diff. Geo. **18** (1983), 575–651.

————, *On the Spectral Geometry of Spaces with Cone-like Singularities*, Proc. Nat. Acad. Sci. USA **76** No. 5 (1979), 2103–2106.

J. Cheeger and M. Taylor, *On the Diffraction of Waves by Conical Singularities*, Comm. Pure Appl. Math. **25** (1982), 275–331.

J. Cheeger and J. Simons, *Differential Characters and Geometric Invariants*, (preprint).

S. S. Chern, *A Simple Intrinsic Proof of the Gauss-Bonnet Formula for Closed Riemannian Manifolds*, Ann. of Math. **45** (1944), 741–752.

————, *Pseudo-Riemannian Geometry and the Gauss-Bonnet Formula*, Separata do Vol 35 #1 Dos Anais da Academia Brasileira de Ciencias Rio De Janeiro (1963), 17–26.

————, *Geometry of Characteristic Classes*, in "Proc. 13th Biennial Seminar", Canadian Math. Congress, 1972, pp. 1–40.

————, *On the Curvatura Integra in a Riemannian Manifold*, Ann. of Math. **46** (1945), 674–684.

————, "Complex Manifolds Without Potential Theory", Springer Verlag, 1968.

S. S. Chern and J. Simons, *Characteristic Forms and Geometric Invariants*, Ann. of Math. **99** (1974), 48–69.

————, *Some Cohomology Classes in Principal Fiber Bundles and Their Applications to Riemannian Geometry*, Proc. Nat. Acad. Sci. USA **68** (1971), 791–794.

A. Chou, *The Dirac Operator on Spaces with Conical Singularities*, Ph. D. Thesis, S. U. N. Y., Stonybrook, 1982.

Y. Colin de Verdierre, *Spectre du Laplacian et Longuers des Geodesiques Periodiques II*, Compositio Math. **27** (1973), 159–184.

G. de Rham, "Varietes Differentiables", Hermann, 1954, pp. 127–131.

————, *Complexes a Automorphismes et Homeomorphie Differentiable*, Annales de L'Institut Fourier **2** (1950), 51–67.

J. Dodziuk, *Eigenvalues of the Laplacian and the Heat Equation*, Amer. Math. Monthly **88** (1981), 687–695.

A. Domic, *An A Priori Inequality for the Signature Operator*, Ph. D. Thesis, M. I. T., 1978.

H. Donnelly, *Eta Invariant of a Fibered Manifold*, Topology **15** (1976), 247–252.

————, *Eta Invariants for G-Spaces*, Indiana Math. J. **27** (1978), 889–918.

_____, *Spectral Geometry and Invariants from Differential Topology*, Bull. London Math. Soc. **7** (1975), 147–150.

_____, *Spectrum and the Fixed Point Set of Isometries I*, Math. Ann. **224** (1976), 161–170.

_____, *Symmetric Einstein Spaces and Spectral Geometry*, Indiana Math. J. **24** (1974), 603–606.

_____, *Eta Invariants and the Boundaries of Hermitian Manifolds*, Amer. J. Math. **99** (1977), 879–900.

H. Donnelly and V. K. Patodi, *Spectrum and the Fixed Point Set of Isometries II*, Topology **16** (1977), 1–11.

J. S. Dowker and G. Kennedy, *Finite Temperature and Boundary Effects in Static Space-Times*, J. Phys A. Math. **11** (1978), 895–920.

J. J. Duistermaat, "Fourier Integral Operators", Courant Institute Lecture Notes, 1973.

J. J. Duistermaat and V. W. Guillemin, *The Spectrum of Positive Elliptic Operators and Periodic Bicharacteristics*, Invent. Math. **29** (1975), 39–79.

_____, *The Spectrum of Positive Elliptic Operators and Periodic Geodesics*, Proc. Symp. Pure Math. **27** (1975), 205–209.

D. Eck, "Gauge Natural and Generalized Gauge Theories", Memoirs of the Amer. Math. Soc. No. 33, 1981.

D. Epstein, *Natural Tensors on Riemannian Manifolds*, J. Diff. Geo. **10** (1975), 631–646.

H. D. Fegan, *The Laplacian with a Character as a Potential and the Clebsch-Gordon Numbers*, (preprint).

_____, *The Heat Equation and Modula Forms*, J. Diff. Geo. **13** (1978), 589–602.

_____, *The Spectrum of the Laplacian on Forms over a Lie Group*, Pacific J. Math. **90** (1980), 373–387.

H. D. Fegan and P. Gilkey, *Invariants of the Heat Equation*, Pacific J. Math. (to appear).

S. A. Fulling, *The Local Geometric Asymptotics of Continuum Eigenfunction Expansions I*, Siam J. Math. Anal. **31** (1982), 891–912.

M. Gaffney, *Hilbert Space Methods in the Theory of Harmonic Integrals*, Trans. Amer. Math. Soc. **78** (1955), 426–444.

P. B. Gilkey, *Curvature and the Eigenvalues of the Laplacian for Elliptic Complexes*, Adv. in Math. **10** (1973), 344–381.

_____, *Curvature and the Eigenvalues of the Dolbeault Complex for Kaehler Manifolds*, Adv. in Math. **11** (1973), 311–325.

_____, *The Boundary Integrand in the Formula for the Signature and Euler Characteristic of a Riemannian Manifold with Boundary*, Adv. in Math. **15** (1975), 344–360.

_____, *The Spectral Geometry of a Riemannian Manifold*, J. Diff. Geo. **10** (1975), 601–618.

_____, *Curvature and the Eigenvalues of the Dolbeault Complex for Hermitian Manifolds*, Adv. in Math. **21** (1976), 61–77.

_____, *Lefschetz Fixed Point Formulas and the Heat Equation*, in "Partial Differential Equations and Geometry. Proceedings of the Park City Conference", Marcel Dekker, 1979, pp. 91–147.

_____, *Recursion Relations and the Asymptotic Behavior of the Eigenvalues of the Laplacian*, Compositio Math. **38** (1979), 281–240.

_____, *Curvature and the Heat Equation for the de Rham Complex*, in "Geometry and Analysis", Indian Academy of Sciences, 1980, pp. 47–80.

_____, *The Residue of the Local Eta Function at the Origin*, Math. Ann. **240** (1979), 183–189.

_____, *The Spectral Geometry of the Higher Order Laplacian*, Duke Math. J. **47** (1980), 511–528; correction, **48** (1981), p. 887.

_____, *Local Invariants of a Pseudo-Riemannian Manifold*, Math. Scand. **36** (1975), 109–130.

_____, *The Residue of the Global Eta Function at the Origin*, Adv. in Math. **40** (1981), 290–307.

_____, *The Eta Invariant for Even Dimensional Pin-c Manifolds*, (preprint).

_____, *On Spherical Space Forms with Meta-cyclic Fundamental Group which are Isospectral but not Equivariant Cobordant*, (preprint).

_____, *The Eta Invariant and the K-Theory of Odd Dimensional Spherical Space Forms*, Invent. Math. (to appear).

P. B. Gilkey and L. Smith, *The Eta Invariant for a Class of Elliptic Boundary Value Problems*, Comm. Pure Appl. Math. **36** (1983), 85–132.

_____, *The Twisted Index Theorem for Manifolds with Boundary*, J. Diff. Geo. **18** (1983), 393–344.

C. Gordon and E. Wilson, *Isospectral Deformations of Compact Solvmanifolds*, (preprint).

A. Gray, *Comparison Theorems for the Volumes of Tubes as Generalizations of the Weyl Tube Formula*, Topology **21** (1982), 201–228.

P. Greinger, *An Asymptotic Expansion for the Heat Equation*, Proc. Symp. Pure Math. **16** (1970), 133–135.

_____, *An Asymptotic Expansion for the Heat Equation*, Archiv Rational Mechanics Anal. **41** (1971), 163–218.

M. Gromov and H. Lawson, *Positive Scalar Curvature and the Dirac Operator on Complete Riemannian Manifolds*, (preprint).

_____, *Classification of simply connected manifolds of positive scalar curvature*, Ann. of Math. **111** (1980), 423–434.

P. Gunther and R. Schimming, *Curvature and Spectrum of Compact Riemannian Manifolds*, J. Diff. Geo. **12** (1977), 599–618.

S. Helgason, *Eigenspaces of the Laplacian, Integral Representations and Irreducibility*, J. Funct. Anal. **17** (1974), 328–353.

F. Hirzebruch, "Topological Methods in Algebraic Geometry", Springer Verlag, 1966.

_____, *The Signature Theorem: Reminiscenes and Recreation*, in "Prospects in Mathematics", Ann. of Math. Studies 70, Princeton Univ. Press, 1971, pp. 3–31.

F. Hirzebruch and D. Zagier, "The Atiyah-Singer Theorem and Elementary Number Theory", Publish or Perish, Inc., 1974.

N. Hitchin, *Harmonic Spinors*, Adv. in Math. **14** (1974), 1–55.

L. Hormander, *Fourier Integral Operators I*, Acta Math. **127** (1971), 79–183.

_____, *Pseudo-differential Operators and Non-Elliptic Boundary Problems*, Ann. of Math. **83** (1966), 129–209.

_____, *Pseudo-Differential Operators and Hypoelliptic Equations*, Comm. Pure Appl. Math. **10** (1966), 138–183.

_____, *Pseudo-Differential Operators*, Comm. Pure Appl. Math. **18** (1965), 501–517.

A. Ikeda, *On Spherical Space Forms which are Isospectral but not Isometric*, J. Math. Soc. Japan **35** (1983), 437–444.

_____, *On the Spectrum of a Riemannian Manifold of Positive Constant Curvature*, Osaka J. Math. **17** (1980), 75–93; *II*, Osaka J. Math. **17** (1980), 691–701.

_____, *On Lens Spaces which are Isospectral but not Isometric*, Ec. Norm. (1980), 303–315.

_____, *Isospectral Problem for Spherical Space Forms*, (preprint).

A. Ikeda and Y. Taniguchi, *Spectra and Eigenforms of the Laplacian on Sphere and Projective Space*, Osaka J. Math. **15** (1978), 515–546.

M. Kac, *Can One Hear the Shape of a Drum?*, Amer. Math. Monthly **73** (1966), 1–23.

T. Kambe, *The Structure of K-Lambda-Rings of the Lens Space and Their Application*, J. Math. Soc. Japan **18** (1966), 135–146.

M. Karoubi, *K-theorie Equivariante des Fibres en Spheres,*, Topology **12** (1973), 275–281.

_____, "K-Theory, An Introduction", Springer-Verlag, 1978.

K. Katase, *Eta Function on Three-Sphere*, Proc. Japan. Acad. **57** (1981), 233–237.

T. Kawasaki, *An Analytic Lefschetz Fixed Point Formula and its Application to V-Manifolds, General Defect Formula and Hirzebruch Signature Theorem*, Ph. D. Thesis, Johns Hopkins University, 1976.

_____, *The Riemann-Roch Formula for Complex V-Manifolds*, Topology **17** (1978), 75–83.

G. Kennedy, *Some finite Temperature Quantum Field Theory Calculations in Curved Manifolds with Boundaries*, Ph. D. Thesis, University of Manchester, 1979.

J. J. Kohn and L. Nirenberg, *An Algebra of Pseudo-differential Operators*, Comm. Pure Appl. Math. **18** (1965), 269–305.

M. Komuro, *On Atiyah-Patodi-Singer Eta Invariant for S-1 Bundles over Riemann Surfaces*, J. Fac. Sci. Tokyo **30** (1984), 525–548.

T. Kotake, *The Fixed Point Theorem of Atiyah-Bott via Parabolic Operators*, Comm. Pure Appl. Math. **22** (1969), 789–806.

_____, *An Analytic Proof of the Classical Riemann-Roch theorem*, Proc. Symp. Pure Math. **16** (1970), 137–146.

R. Kulkarni, "Index Theorems of Atiyah-Bott-Patodi and Curvature Invariants", Montreal University, 1975.

H. Kumano-go, "Pseudo-differential Operators", M. I. T. Press, 1974.

S. C. Lee, *A Lefschetz Formula for Higher Dimensional Fixed Point Sets*, Ph. D. Thesis, Brandeis University, 1975.

A. Lichnerowicz, *Spineurs Harmoniques*, C. R. Acad. Sci. Paris Sér. I Math. **257** (1963), 7–9.

N. Mahammed, *K-theorie des Formes Spheriques Tetradriques*, C. R. Acad. Sci. Paris Sér. I Math. **281** (1975), 141–144.

_____, *A Pros e la K-theorie des Espaces Lenticulaires*, C. R. Acad. Sci. Paris Sér. I Math. **271** (1970), 639–642.

_____, *K-Theorie des Formes Spheriques*, Thesis L'Universite des Sciences et Techniques de Lille, 1975.

H. P. McKean and I. M. Singer, *Curvature and the Eigenvalues of the Laplacian*, J. Diff. Geo. **1** (1967), 43–69.

E. Miller, Ph. D. Thesis, M. I. T.

R. Millman, *Manifolds with the Same Spectrum*, Amer. Math. Monthly **90** (1983), 553–555.

J. Millson, *Chern-Simons Invariants of Constant Curvature Manifolds*, Ph. D. Thesis, U. C., Berkeley, 1974.

_____, *Closed Geodesics and the Eta Invariant*, Ann. of Math. **108** (1978), 1–39.

_____, *Examples of Nonvanishing Chern-Simons Invariant*, J. Diff. Geo. **10** (1974), 589–600.

J. Milnor, "Lectures on Characteristic Classes", Princeton Notes, 1967.

_____, *Eigenvalues of the Laplace Operator on Certain Manifolds*, Proc. Nat. Acad. Sci. USA **51** (1964), 775–776.

S. Minakshisundaram and A. Pleijel, *Some Properties of the Eigenfunctions of the Laplace Operator on Riemannian Manifolds*, Canadian J. Math. **1** (1949), 242–256.

A. Nestke, *Cohomology of Elliptic Complexes and the Fixed Point Formula of Atiyah and Bott*, Math. Nachr. **104** (1981), 289–313.

L. Nirenberg, "Lectures on Linear Partial Differential Equations", Regional Conference Series in Math. No. 17, Amer. Math. Soc., 1972.

R. Palais, "Seminar on the Atiyah-Singer Index Theorem", Ann. Math. Study 57, Princeton Univ. Press, 1965.

V. K. Patodi, *Curvature and the Eigenforms of the Laplace Operator*, J. Diff. Geo **5** (1971), 233–249.

————————, *An Analytic Proof of the Riemann-Roch-Hirzebruch Theorem for Kaehler Manifolds*, J. Diff. Geo. **5** (1971), 251–283.

————————, *Holomorphic Lefschetz Fixed Point Formula*, Bull. Amer. Math. Soc. **79** (1973), 825–828.

————————, *Curvature and the Fundamental Solution of the Heat Operator*, J. Indian Math. Soc. **34** (1970), 269–285.

M. Pinsky, *On the Spectrum of Cartan-Hadamard Manifolds*, Pacific J. Math. **94** (1981), 223-230.

D. B. Ray, *Reidemeister Torsion and the Laplacian on Lens Spaces*, Adv. in Math. **4** (1970), 109–126.

D. B. Ray and I. M. Singer, *R Torsion and the Laplacian on Riemannian Manifolds*, Adv. in Math. **7** (1971), 145–210.

————————, *Analytic Torsion for Complex Manifolds*, Ann. of Math. **98** (1973), 154–177.

S. Rempel, *An Analytical Index formula for Elliptic Pseudo-differential Boundary Problems*, Math. Nachr. **94** (1980), 243–275.

S. Rempel and B. Schultze, *Complex Powers for Pseudo-differential Boundary Problems*, Math. Nachr. **111** (1983), 41–109.

T. Sakai, *On Eigenvalues of Laplacian and Curvature of Riemannian Manifold*, Tohoku Math. Journal **23** (1971), 589–603.

————————, *On the Spectrum of Lens Spaces*, Kodai Math. Sem. **27** (1976), 249–257.

R. T. Seeley, *Complex Powers of an Elliptic Operator*, Proc. Symp. Pure Math. **10** (1967), 288–307.

————————, *Analytic Extension of the Trace Associated with Elliptic Boundary Problems*, Amer. J. Math. **91** (1969), 963–983.

————————, *Topics in Pseudo-Differential Operators*, in "Pseudo-Differential Operators", C. I. M. E., 1968, pp. 168–305.

————————, *The Resolvant of an Elliptic Boundary Problem*, Amer. J. Math. **91** (1969), 889–920.

————————, *Singular Integrals and Boundary Problems*, Amer. J. Math. **88** (1966), 781–809.

————————, *A Proof of the Atiyah-Bott-Lefschetz Fixed Point Formula*, An. Acad. Brasil Cience **41** (1969), 493–501.

————————, *Integro-differential Operators on Vector Bundles*, Trans. Amer. Math. Soc. **117**, 167–204.

————————, *Asymptotic Expansions at Cone-like Singularities*, (preprint).

J. Simons, *Characteristic Forms and Transgression II–Characters Associated to a Connection*, (preprint).

I. M. Singer, *Connections Between Geometry, Topology, and Analysis*, in "Colloquium Lectures at 82 Meeting", Amer. Math. Soc., 1976.

————————, *Elliptic Operators on Manifolds*, in "Pseudo-differential Operators", C. I. M. E, 1968, pp. 335–375.

D. E. Smith, *The Atiyah-Singer Invariant, Torsion Invariants, and Group Actions on Spheres*, Trans. Amer. Math. Soc. **177** (1983), 469–488.

L. Smith, *The Asymptotics of the Heat Equation for a Boundary Value Problem*, Invent. Math. **63** (1981), 467–494.

B. Steer, *A Note on Fuchsian Groups, Singularities and the Third Stable Homotopy Group of Spheres*, (preprint).

P. Stredder, *Natural Differential Operators on Riemannian Manifolds and Representations of the Orthogonal and Special Orthogonal Group*, J. Diff. Geo. **10** (1975), 647–660.

M. Tanaka, *Compact Riemannian Manifolds which are Isospectral to Three-dimensional Lens Spaces*, in "Minimal Submanifolds and Geodesics", North-Holland, 1979, pp. 273–283; *II*, Proc. Fac. Sci. Tokai Univ. **14** (1979), 11–34.

D. Toledo, *On the Atiyah-Bott formula for Isolated Fixed Points*, J. Diff. Geo. **8** (1973), 401–436.

D. Toledo and Y. L. Tong, *A Parametrix for the D-Bar and Riemann-Roch in Čech Theory*, Topology **15** (1976), 273–301.

—————————, *The Holomorphic Lefschetz Formula*, Bull. Amer. Math. Soc. **81** (1975), 1133–1135.

Y. L. Tong, *De Rham's Integrals and Lefschetz Fixed Point Formula for D-Bar Cohomology*, Bull. Amer. Math. Soc. **78** (1972), 420–422.

F. Treves, "An Introduction to Pseudo-differential and Fourier Integral Operators", Plenum Press, 1980.

H. Urakawa, *Bounded Domains which are Isospectral but not Isometric*, Ec. Norm. **15** (1982), 441–456.

M. F. Vigneras, *Varietes Riemanniennes Isospectrales et non Isometriques*, Ann. of Math. **112** (1980), 21–32.

H. Weyl, "The Classical Groups", Princeton Univ. Press, 1946.

————, *On the Volume of Tubes*, Amer. J. Math. **61** (1939), 461–472.

H. Widom, *Asymptotic Expansions for Pseudo-differential Operators on Bounded Domains*, (preprint).

—————————, *A Trace Formula for Wiener-Hopf Operators*, J. Operator Th. **8** (1982), 279-298.

L. Willis, *Invariants of Riemannian Geometry and the Asymptotics of the Heat Equation*, Ph. D. Thesis, Univ. of Oregon, 1984.

M. Wodzicki, *Spectral Asymmetry and Zeta Functions*, Invent. Math. **66** (1982), 115–135.

K. Wojciechowski, *Spectral Asymmetry, Elliptic Boundary Value Problem, and Cutting and Pasting of Elliptic Operators*, (preprint).

—————————, *Spectral Flow and the General Linear Conjugation Problem*, (preprint).

T. Yoshida, *The Eta Invariant of Hyperbolic 3-manifolds*, (preprint).

D. Zagier, "Equivariant Pontrjagin Classes and Applications to Orbit Spaces", Springer-Verlag, 1972.

————, *Higher Dimensional Dedekind Sums*, Math. Ann. **202** (1973), 149–172.

INDEX

A-roof ($\hat{A}$) genus, 99

Absolute boundary conditions for the de Rham complex, 243

Algebra of operators, 13

Allard, William, 45

Almost complex structure, 180

Arithmetic genus, 190, 200

Arzela-Ascoli Theorem, 8

Asymptotic expansion of symbol, 51

Asymptotics
of elliptic boundary conditions, 72
of heat equation, 54

Atiyah-Bott Index Theorem, 247

Atiyah-Patodi-Singer Theorem, 268, 276

Atiyah-Singer Index Theorem, 233

Axiomatic characterization
of Chern forms, 202
of Euler forms, 121
of Pontrjagin forms, 130

Bigrading exterior algebra, 107

Bott periodicity, 221

Bundle of Cauchy data, 70

Bundles of positive part of symbol, 76

Canonical bundle, 186

Cauchy data, 70

Čech $\mathbf{Z}_2$ cohomology, 166

Change of coordinates, 25

Characteristic forms, 91

Characteristic ring, 93, 98, 103

Chern
character, 91, 174, 217
class of complex projective space, 111
form, 91
isomorphism in K-theory, 217

Christoffel symbols, 104

Clifford
algebra, 148
matrices, 95
multiplication, 148

Clutching functions, 218

Cohomology
of complex projective space, 109, 200
of elliptic complex, 37
of Kaehler manifolds, 195

Collared manifold, 253

Compact operators, 31

Complex
characteristic ring, 93
projective space, 109
tangent space, 107

Connected sum, 154

Connection 1-form, 89

Continuity of operators, 12

Convolution, 2

Cotangent space, 25

Cup product, 151

Curvature
of spin bundle, 172
2-form, 90

d-bar ($\bar{\partial}$), 107

Defect formulas, 286, 296

de Rham
cohomology, 91, 197
complex, 39, 149, 246

de Rham, signature and spin bundles, 172

Dirac matrices, 95

Dirichlet boundary conditions, 71, 245

Dolbeault complex, 183

Dolbeault and de Rham cohomology, 197

Donnelly Theorem, 293

Double suspension, 220

Dual norm, 30

Elliptic
boundary conditions, 72
complex, 37
operators, 23

Eta
function, 81, 258, 263
invariant, 258, 270

Euler
characteristic of elliptic complex, 37
form, 100, 121, 174, 308

Euler-Poincaré characteristic, 246

Exterior
algebra, 148
multiplication, 39, 148

Flat K-theory, 298

Form valued polynomial, 129

Fourier transform, 3

Fredholm
index, 32
operators, 32

Fubini-Study metric, 111, 195, 200
Functional calculus, 48

Gärding's inequality, 23
Gauss-Bonnet Theorem, 122, 179, 256
Geodesically convex, 166
Graded
  boundary conditions, 70
  symbol, 71
  vector bundle, 37
Greens operator, 44
Gunther and Schimming, 313

Half spin representations, 161
Heat equation, 47
  local formulas, 326
Hirzebruch
  L-polynomial, 99, 115, 175
  Signature Theorem, 154
Hodge
  Decomposition Theorem, 38, 246
  manifolds, 196
  * operator, 246
Holomorphic
  connection, 108
  manifold, 107
Holonomy of representation, 270
Homotopy groups of U(n), 221
Hyperplane bundle, 109
Hypoelliptic, 23

Index
  and heat equation, 47
  form in K-theory, 299
  formula (local), 58
  in K-theory, 225
  of representation, 271
  R/Z valued, 270
Index Theorem
  for de Rham complex, 122
  for Dolbeault complex, 189
  for signature complex, 154
  for spin complex, 179
Instanton bundles, 96
Integration along fibers, 235
Interior multiplication, 40, 148
Invariant
  form valued polynomial, 134
  polynomials, 119
Invariants of heat equation, 326
Isospectral manifolds, 331

K-theory, 216, 225
  of locally flat bundles, 298
  of projective spaces, 302
  of spheres, 221
  of spherical space forms, 301
  sphere bundle, 267
Kaehler
  manifolds, 195
  2-form, 193
Kernel
  of heat equation, 54
  of operator, 19

L-polynomial, 99, 115, 175
Lefschetz
  fixed point formulas, 63, 68, 286
  number of elliptic complex, 62
Lens space, 276, 297
Levi-Cività connection, 104
Local
  formula for derivative of eta, 82
  index formula, 58
Locally flat bundle, 270

Maurer-Cartan form, 275
Mellin transform, 78
Monopole bundles, 96

Neuman series, 29
Neumann boundary conditions, 71, 245
Nirenberg-Neulander Theorem, 182

Operators depending on a complex
    parameter, 50
Orientation form, 150

Parametrix, 29, 52
Peetre's Inequality, 10
Plancherel Formula, 5
Poincaré duality, 41, 246
Pontrjagin form, 98, 174, 308
Projective space
  arithmetic genus, 200
  cohomology, 109
  signature, 152
Pseudo-differential operators, 11
Pseudolocal, 20

R/Z valued index, 270
Real characteristic ring, 99
Regularity of eta at 0, 263
Relative boundary conditions for the
    de Rham complex, 243
Rellich Lemma, 8
Residue of eta at $s = 0$, 84, 258, 266

Resolvant of operator, 52
restriction of invariant, 120, 255
Riemann zeta function, 80
Riemann-Roch Theorem, 189, 202
Ring of invariant polynomials, 119

Schwartz class, 2
Secondary characteristic classes, 273
Serre duality, 199
Shape operator, 251
Shuffle formulas, 313
Signature
    complex, 150
      tangential operator, 261
    of projective space, 153
    spin and de Rham bundles, 172
    Theorem, 154
Simple covers, 166
Singer's conjecture for the Euler form, 308
Snaith's Theorem, 305
Sobolev
    Lemma, 7
    spaces, 6
Spectral
    flow, 273
    geometry, 331
Spectrum
    of compact operator, 42
    of self-adjoint elliptic operator, 43
Sphere
    bundle K-theory, 267
    Chern character, 96
    Euler form, 101
    K-theory, 221
Spherical space form, 295, 333

Spin
    and other elliptic complexes, 172
    complex, 176
    group, 159
    structures, 165
      on projective spaces, 170
$SPIN_c$, 186
Spinors, 159
Splitting principal, 190
Stable isomorphism of vector bundles, 216
Stationary phase, 64
Stieffel-Whitney classes, 167
Suspension
    of topological space, 215
    of symbol, 224
Symbol
    de Rham complex, 40
    Spin complex, 176
    Dolbeault complex, 184
    of operator on a manifold, 25

Tangential operator signature
      complex, 261
Tautological bundle of CP_n, 109, 195
Todd form, 97, 114, 236
Topological charge, 223
Trace heat operator, 47
Transgression, 92, 252

$U(n)$, homotopy groups of, 221

Weyl spanning set, 121
Weyl's theorem on invariants of orthogonal
      group, 132

Zeta function, 80